Tube Lore

A REFERENCE FOR USERS AND COLLECTORS

Ludwell Sibley

First Edition
First printing, September 1996

ISBN 0-9654683-0-5

Flemington, NJ 08822

On the cover:
Top left: classic Western Electric "tennis-ball" repeater tubes, flanking a 211A power triode.
Top right: the WE 316A "doorknob" UHF triode.
Lower left: three Majestic spray-shielded types.
Lower right: from the original release of RCA metal tubes, the abandoned 6D5 and the successful 6F6.

Printed by Chernay Printing, Inc., Coopersburg, PA

CONTENTS

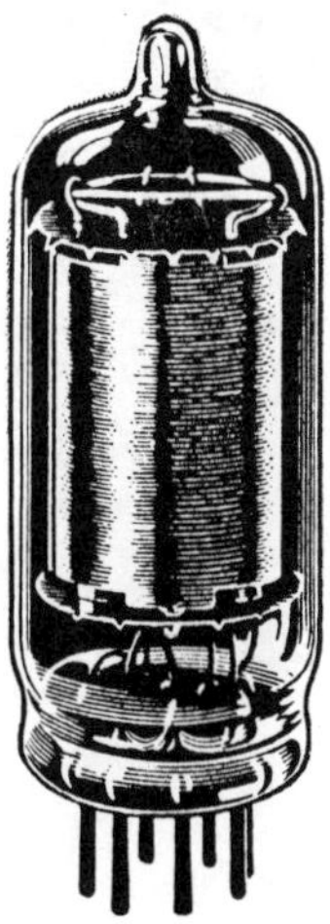

FOREWORD

This book is intended to aid the present-day user and collector of electron tubes by providing historical insight and specific technical data. It supplies information, from a variety of sources, on the charaacteristics and design relationships of tubes. With this insight, the user (equipment restorer, audiophile, etc.) may be aided in finding suitable replacement tubes as the supply grows smaller, and the collector may be able to determine the nature and uses of a fresh acquisition.

This material is naturally concerned with "gridded" tubes of the familiar receiving-transmitting sort. However, tube history includes a rich variety of specialized types, important in specific applications: klystrons and magnetrons in radar, thyratrons and ignitrons in industrial electronics, camera tubes and CRTs in television, and so on. The data tables include these specialized types to add depth to the historical picture.

There is no simple way to organize a guide like this. After outlining some of the major historical trends in design, the data chapters are set up under three major themes: types preceding centralized registration of types by the Radio Manufacturers Association (today's Electronic Industries Association), receiving tubes registered by the RMA/EIA, and special-purpose tubes registered by the RMA/EIA. Supplementary chapters deal with nonregistered special-purpose tubes as produced by multiple makers and as unique to particular manufacturers.

Other topics have their own chapters. A "military" section covers tube-coding systems used historically by the U. S. Army and Navy, and Canadian "REL" types of WW II. A "user's guide" supplies information on identifying, dating, repair-ing, reactivating, testing, and substituting tubes. For equipment restorers, a chapter is dedicated to television sweep tubes that received wide use in audio gear and amateur transmitters.

A final section covers recent auction prices paid for audio, transmitting, and collectible tubes, based on the author's notes on-site. There is a small but active market for collectible types, and a vigorous one for audio tubes. As with all auction "actuals," these are snapshots of single transactions and should be used with caution.

This reference heavily supplements, but cannot replace, conventional tube manuals like the RCA HB-30, nor guides like the H. W. Sams Tube Substitution Handbook. Because it is not tied to the products of a particular manufacturer, it contains information unobtainable elsewhere. (Tubemakers did not publish handbooks covering types available only from their competitors!) It represents the "vector sum" of information from a wide variety of sources - "vector" because errors and ambiguities are often present in old-time tube material, and may send the unwary reader in the wrong direction.

Data sources included the multi-binder tube manuals issued by General Electric, RCA, Tung-Sol and Western Electric; and booklets and data sheets from Amperex, CBS-Hytron, Eimac, Machlett, Mullard, National Union, Raytheon, Standard Telephones & Cables, Sylvania, Taylor, Telefunken, Westinghouse, United Electronics, Victoreen, and the like. For some obscure types, the only available data source was product announcements and advertisements in *Electronics*, *Radio Engineering*, and *Sylvania News*. Tube descriptions in the *Bell Laboratories Record*, *Bell System Technical Journal*, and *RCA Engineer* were quite helpful.

Information on some of the less familiar military tubes of WW II came from the Catalog of Navy Material - Bureau of Ships Section - Part II - Armed Forces Cross Index of Electron Tube Types, Aug. 1950, from TO 08-125-25, Vacuum-Tube Complements of Signal Equipment and Test Data, and from contemporary equipment manuals - with care taken to resolve errors and uncertainties in these hurriedly produced sources. Data on early radar tubes came in part from the MIT Radiation Laboratory Series of radar textbooks, particularly Vol. 6 (Microwave Magnetrons), Vol. 7 (Klystrons and Microwave Triodes), Vol. 14 (Microwave Duplexers), Vol. 17 (Components Handbook), and Vol. 22 (Cathode Ray Tube Displays). Some of the more obscure radar-jammer magnetrons are covered in the Harvard University Radio Research Laboratory countermeasures textbook Very High-Frequency Techniques. Finally, in many cases it was helpful to examine actual tubes.

Great appreciation goes to tube experts who provided research resources for this effort. Contributors, at one time or another, were: Attila Balaton; John Dilks, K2TQN; William Donzelli; Bro. Patrick Dowd, W2GK; Peter Dugal; Alan Klase, N3FRQ; Charles A. Meyer; Lauren Peckham; John Walker; and Jim Whartenby. Bob Dennison, W2HBE, helped out with some crisp tube photos. Particular thanks go to Alan Douglas; John Kelly, N3GVF; and Jerry Vanicek, who all contributed research inputs in specific areas.

Ludwell Sibley

TRENDS IN TUBE DESIGN

The physical design of tubes in the '20s and early '30s relied heavily on techniques adopted from the manufacture of electric lamps: the use of a "press" to support the internal structure, with wires of "dumet" alloy running through the press. There were a huge variety of improvements in detail - a tube was, after all, far more complex than a lamp - but the basic structure was unchanged. However, new trends in design began in the '30s.

THE ACORN TUBE

RCA brought out an important innovation in early 1935: the "acorn" tube. Sold initially under the "RCA - De Forest" brand, the first offering was the 955 triode. This small all-glass tube, with its short leads and low capacitance, was an important advance in tubes for VHF-UHF applications. (The earliest 955s had domed tops, not the flat top that is commonly seen.) Acorn technology spread to foreign manufacturers through the usual licensing arrangements, and was ready - on both sides - when the demands of VHF radar appeared in the late '30s. The full line of North American acorn tubes, as far as can be determined, was as follows:

953	Australian (AWA?) 6-V diode
954	6-V pentode (Br. HA2)
955	6-V, 1.6-W triode (Br. ZA2)
956	6-V remote-cutoff pentode
957	1.25-V triode
958, A	1.25-V xmtg. triode
959	1.25-V remote-cutoff pentode
1650	Low-leakage 955
5731	Premium 955
9004	6-V diode
9005	6-V diode, low-capacitance
6F4	6-V, 2.0-W triode
6L4	6-V, 1.7-W triode
CW10	Canadian 6-V diode

British makers also produced 4-volt variants. Japanese makers (e. g., Hitachi) made 95x-types, designating them "UN-95x."

THE METAL TUBE

RCA and GE announced a new technology in April 1934: the metal tube. Developed in secrecy by GE, the new design used a steel shell, with individual glass-filled eyelets for the lead-in wires, on a new and convenient octal base. This product line gave GE and other receiver makers a complete tube complement for the "1935" line of sets. The "original nine" were:

5Z4 rectifier	6F6 power pentode
6A8 and 6L7 converters	6H6 duodiode
6C5 and 6F5 triodes	6J7 and 6K7 pentodes

A tenth type, the 6D5 power triode, was withdrawn before major production.

The other tube makers were unhappy to have their product lines and tooling obsoleted. Some, like Raytheon, signed up to adopt the new line. Others disparaged RCA's invention - the multiple eyelet seals were sure to leak, they said - and hurried to repackage their existing big-pin products with octal bases. Yet others like Arcturus and Triad quickly devised what they termed the "perfected" metal tube: the metal-glass (MG) design, where a small conventional stem-sealed glass tube was hidden inside a steel shell, on the same octal wafer base as GE-RCA used.

The MG push was short-lived. The RMA issued registration in July, 1935 on the 5Z4MG, 6A8MG, 6C5MG, 6D5MG, 6F5MG, 6F6MG, 6H6MG, 6J7MG, 6K7MG - copies of RCA's line - plus the 25Z5MG rectifier for AC-DC sets. Later registrations through March 1936 covered the 6N6MG, 6N7MG, 6Q7MG, and (today unknown) 6Z6MG rectifier. For those wanting to "collect the set" today, eventual adders were the 6J5MG, 6X5MG, 25A6MG, and 25Z6MG.

A better variant on the MG idea was the Arcturus "Coronet" type, which used a new glass "button" stem construction, with the lead-in wires sealed in a circle. These were used in a line of octal-based replacement types, to be used with an adapter to "modernize" existing radios. The line was: 24, 27, 51, 56, 57, 58, 75, 77, 78, 80, 85, 2A6, and 25Z5. Inherently octal items (no adapter needed) were: 5Y3 and 25A6 (also replacing 43 with an adapter). Later octal additions were versions of the RCA "original nine," 6L6 (!), 6N6, 6N7, 6Q7, 6R7, 6X5, and 25Z6.

Meanwhile RCA overcame great start-up difficulties and got true metal tubes into full production, introducing additional types as well. To overcome the high cost of the eyelet base, a button stem was put into use in 1936. In this design, the glass stem disc with its lead-in wires was sealed into the bottom of the metal envelope. The button-stem idea had great benefits in later tube designs, as we will see. Metal tubes continued to be registered up to the start of WW II, and even a few variant types afterward, but other designs took over.

THE LOKTAL TUBE

Sylvania was a strong competitor to RCA, with design ideas of its own. In May 1938 it released the 1231 (later 7V7) video pentode, the first of a line of advanced "lock-in" tubes. These extended the button-stem idea, with solid pins extending through the glass instead of wires. A stamped metal cup provided an octal-like locating key, accepted a locking spring in the socket, and gave electrical shielding. The bulb was a T9 style (tubular, 9/8" dia.) such as was used on octal tubes by then.

Besides repackaged octal types, the loktal idea allowed construction of some good VHF tubes with short leads - a 7E5, examined side-on, appears empty because the elements are below the locking cup. Loktals had their faults - the small pins were sensitive to corrosion, and anyone who tried to get the tubes out of a car radio without an extractor tool was in for a hard time. They did give radio manufacturers like Philco and Zenith a sales "hook." Registrations of loktals boomed in the late '30s but essentially ended by 1947.

GE and RCA dabbled with versions of the loktal. One was the "LM" loktal-metal. Between mid-1939 and early 1940 RCA registered the 7A7LM, 7B6LM, and 7B8LM, while GE contributed the 12B7LM. RCA's tubes used metal shells on an "octalox" bakelite wafer base with soldered pins. RCA also introduced an "LT" format via the 7B5LT, 7C5LT, 14A7/12B7, 35A5LT and 35Z3LT. These were glass tubes with a metal-banded base and the loktal-like wafer of the "LM." These mutants were abandoned ca. 1943 in favor of true loktals.

THE MINIATURE TUBE

RCA came up with another innovation in November, 1939: the all-glass miniature tube. Here the glass button stem, with molded-in pins, combined with a small bulb to give a welcome reduction in size. The initial offering comprised 1R5 pentagrid converter, 1T4 pentode, 1S5 diode - AF pentode, and 1S4 output pentode. These formed the tube complement in RCA Victor's BP10 personal radio and a host of others. The miniature format was on the scene just in time to go to war, in the BC-611, BC-745, and MAB transceivers. RCA also repackaged its 954, 955, and 956 acorn tubes in miniature form, yielding the 9001, 9002, and 9003.

Miniature tubes required dexterity from the assemblers, but their technical performance led to demands for more types - WE's 6AK5; RCA's 6C4, 6J6; and so on. The initial idea was that the sealing tip might be moved to the base of the tube - RCA's early tube-data sheets cautioned "The center hole in sockets designed for this base provides for the possibility that this tube type may be manufactured with the exhaust-tube tip at the base end." That didn't happen, but such an improvement did appear later in the Compactron.

The list of miniature types by the end of the war was 32, covering a full line of receiver types. It was clear that the seven-pin format limited the introduction of higher-powered or complex new designs, so the nine-pin style followed. The first nine-pin registered was RCA's 12AU7 (October 1946), edging out WE's 2C51 by five weeks.

The "miniature" story wasn't over in the '40s; when the multi-pin Compactron and novar formats appeared, Sylvania's response was the 10-pin miniature of 1960-62. This used an extra pin in the center of the base. It didn't go far - the 6C9-17C9 duotetrode, 6J9 triple-triode, and 19Q9 triode-pentode were the only users. Not content to obsolete tube testers this one way, the tubemongers also devised a 10-pin "decal" miniature base with all pins on the same circle. This appeared on just a few types, mostly of foreign origin, of 1964-66:

5U9 - 6U9 - 8U9/PCF201 triode-pentode
5X9 - 6X9/ECF200 - 8X9/PCF200 triode-pentode
6AB9 - 15AB9 - 17AB9 duotetrode
6AF9 - 11AF9 duopentode
6V9/ECH200 triode-hexode
11Y9/LFL200 - 16Y9/PFL200 - 17Y9 duopentode

THE SUBMINIATURE TUBE

Certain special uses for tubes pressed for extra-small size - the civilian example being the hearing aid, the military example being the proximity fuze. Hytron was probably the first U. S. maker with a significant product in this area, with tubes apparently related to British types from Hivac. These were miniature-sized types with small bakelite bases. Another maker was Microtubes, Inc., which developed a line of true subminiature hearing-aid types before being absorbed by Zenith. WE made T5 quasi-subminiatures (360A-362A) for its hearing aids. But the real position in early subminiature tubes was held by Raytheon, which had an extensive hearing-aid line based on a flattened-tubular construction with flat lead press. Raytheon had a product line for pocket AM radios in late 1945 (the 2E31, 2E35, 2E41, and 2G21), slightly ahead of Sylvania's line (the 1C8, 1Q6, 1V5, and 1W5).

Sylvania got extensive experience with T3 proximity-fuze tubes during WW II, and promoted them heavily afterward. They and Raytheon seem to have had a battle for supremacy of the round T3 format vs. the flattened ("T2X3") style. Raytheon claimed, for example, that its flat press was relatively safe from air leakage. The T3 format more or less won out in the early '60s - in time to be made irrelevant by transistors. The first market for subminiatures, hearing aids, was also the first one lost to solid-state, in 1953-56. The market for later subminiatures was mainly military. It was largely a GE - Raytheon - Sylvania - Tung-Sol show, with RCA barely visible.

THE LIGHTHOUSE TRIODE

Sometimes new technologies come along just when needed. That was certainly true with GE's UHF planar triode, the "lighthouse" tube. Assembled with disc-shaped seal/contact rings fired to stacked sections of glass tubing, this format was well suited to grounded-grid operation in cavity-type circuits at frequencies above 1 GHz. First available in 1940-41, it made it profitable to redesign radar receivers, replacing earlier tubes. For example, the Westinghouse designers developing the Navy's ASB 500-MHz radar started out with acorn tubes - the best available product - working at a frequency perhaps twice what was reasonable in terms of noise and gain. The reworked ASB-7 model had GE's new product in it, far better suited to the task. Likewise with Western Electric's FA-FB-FC-FD radars, which initially used a WE "doorknob" tube - about 10 dB noisier - in the receiver front end. RCA's SC got lighthouse tubes in the SC-2 model. It is likely that the radars got 50% better range from this change.

GE's "best seller" line was the 2B22 diode (ex-559); the 2C40 power/receiving triode (ex-446), the 2C43 small power triode (ex-464), and 3C22 125-watt power triode. These led to the 2C39/A/B planar triode which, while not a "lighthouse" type, was highly successful and long-used. Postwar, RCA's "pencil" planar triodes (5675 et al.), WE's 416-type, and GE's ceramic planar triodes (6299 et al.) were further advances along this line.

THE BUTTON-STEM OCTAL

The button stem found application in another way: in upgrading the heat capacity and ruggedness of glass octal tubes. The first such tubes were probably the WE 373A-374A of 1941 and the Bell Labs - designed 6AR6 of 1945.

Beginning in the early '50s, variants of existing types appeared with button stems. Beneficiaries of this construction included hard-worked types like the 5U4GB, 6L6GB, 6SN7GTB, and 25L6GT. The thick lead-in pins provided a significant new path to draw heat out from the elements - the newly designed 3DJ4 rectifier had dual pins for each plate and each end of the filament. The normal bakelite base was either shortened or, on some types toward the end, contracted to a mere disc.

THE NUVISTOR

RCA disclosed a unique tube design in March 1959: the nuvistor. This was a thimble-sized metal-ceramic format in which the tube elements were jig-assembled on a ceramic wafer, then brazed into a metal shell in a vacuum (no need to pump the tube down). With tiny elements spaced closely, it was a rather good VHF amplifier or oscillator, usable above 1 GHz. Noise level was low for a tube device, typically 6 dB at 470 MHz and 10 dB at 890 MHz. The same small elements gave unusual freedom from microphonic effects. The elements were cantilever-supported from the base, without micas or welds. Nuvistors were mostly triodes, although there were a few tetrodes as well.

The original development plan included a beam power tube which, from the photos, looked like a ceramic-metal cylinder perhaps 1.5" wide and 2" tall, with a metal dome for the plate and novar-sized pins - quite unlike production nuvistors. Reasons for its abandonment are unknown. There was also a project to develop a *half-size* nuvistor.

Nuvistors were featured in the tuners of 1961 "New Vista" TV sets from RCA Victor. They also found their way into VHF surveillance receivers, RF preamplifiers, FM tuners, aircraft radios, radar IF preamplifiers, TV camera preamplifiers, sonobuoys, and other applications where their size and performance were helpful. Production passed the 1,000,000 point by July 1961, even though this was essentially a niche-market device. New types came out up to 1967.

The complete nuvistor line, as far as can be determined, was the following:

2CW4	6CW4 for series-string TV sets.
2DS4	6DS4 for series-string TV sets.
2DV4	6DV4 for series-string TV sets.
6CW4	High-μ triode, for G-G amps. Gave NF of 5.5 dB at Ch. 13 in RCA KRK-90A TV tuner, or about 1 dB better than the 6GK5, 1961's best conventional tube.
6DS4	Vari-μ triode to handle strong signals.
6DV4	Med.-μ UHF osc. triode, gold-plated shell and pins.
13CW4	Triode for antenna-mounted TV preamps like the Winegard "Colortron." Heater took only 60 mA, convenient for remote-powering.
7586	Med.-μ triode, first nuvistor announced, for general industrial/UHF use, 6.3 V.
7587	Sharp-cutoff UHF tetrode w/ plate cap, 6.3 V.
7895	High-μ triode, 6.3 V.
8056	Med.-μ triode optimized for 24-V plate supply, 6.3 V.
8058	High-μ UHF triode, optimized for rcvg. grounded-grid amps., grid on shell, 6.3 V.
8203	Xmtg. triode, 6.3 V.
8227	Xmtg. tetrode, 6.3 V.
8380	Power tetrode for sonobuoys, 6.0-8.5 V as specified by buyer.
8382	Power triode, uses and heater as for 8380.
8393	Med.-μ triode, 13.5 V.
8441	Power triode, high-μ 8382.
8456	Power triode, med.-μ 8382.
8627, A	Power triode, 6.3 V.
8628	Triode, for audio preamps, heater 6.3 V.
8808	High-μ power triode, w/ plate cap, grid on shell, for grounded-grid amps, 6.3 V.

THE COMPACTRON

By the early '60s it was clear that the transistor would eventually invade the mass market for tubes, specifically the television industry. Here GE provided an advance that surely forestalled the shift to solid-state by three or four years: the Compactron. The central element of this design was a 12-pin "duodecar" base. Coupled with a T9 bulb, this gave enough space and pins to build multiple tubes: triple triodes, dual pentodes, duodiode-duotriodes, etc. The sealing tip was at the top on the initial offerings, but "A" versions appeared with the tip moved to the center of the base - as had been hoped for 15+ years earlier with the miniature types.

The new format could also be expanded to T12 size and used for all-glass sweep tubes, either with or without a top cap. (More details on these appear in the sweep-tube chapter.) Combined sweep and damper tubes were feasible also.

The first Compactrons were registered in 1961, appearing in GE TV sets in early 1962. They were largely repackaged conventional types. The first ones and their prototypes were:

6AX3	6AX4GTB	6G11	6DT6 + 6CU5
6B10	6CG7 + duodiode	6GF5	6FW5
6FJ7	6DN7	6J11	2 6EW6s
6K11	12AX7 + ½ 12AU7	6AV11	1-½ 12AU7s
6GE5	6DQ6B	6AR11	2 6GM6s
2AH2	2.5-volt 3A3	6AF11	6CX8 + ½ 12AT7
6AS11	6CX8 + ½ 12AT7	6AL11	6DT6 + 6AQ5

Tung-Sol was an early second designer of Compactrons, starting with its 6Q11 (a 6K11-equivalent).

THE NOVAR

RCA was not idle at this point. It brought out the "novar" format in late 1961. This was Compactron-like, but with nine thick pins on wide spacings. The 7868 audio power tube, registered on June 19, was the first. Others that followed were mainly TV sweep types, both single-ended and capped, and damper diodes. However, there were types for audio (8278), general purposes (6GF7, 6GJ5, 6GT5) and even transmitting (8458, 8579).

The novar and Compactron formats were not truly new: Western Electric had done much the same ten years before with its 418A, 429A, 436A and 437A - with the sealing tip out the base to boot. The main difference was WE's use of thicker (50-mil, vs. 40-mil) pins. GE and RCA get credit for mass-market application.

THE "9-T9"

Sylvania devised the "9-T9" format, in which a 9-pin miniature-style base was used with a tipped T9 bulb. Released in 1960-63 mainly for TV use, types in this style were the 6EW7-10EW7, 6FD7-10FD7-13FD7, 6FR7, 6GC5, 6HC8, 6HZ8, 6JT8-10JT8, and 6KU8. There were also the audio types 6GM5, 7695 and 7754. There wasn't total consistency in the industry: GE's 12HG7 was a 9-T9, while RCA's and Sylvania's 12HG7/12GN7A was in miniature 9-pin format.

HIGH-RELIABILITY TYPES

Throughout the history of the tube, the makers struggled to minimize its shortcomings: high power consumption, susceptibility to microphonic effects from vibration, exposure to mechanical shock, sensitivity to variations in heater voltage, nonuniformity in electrical parameters, etc. Modified tubes optimized to fight these failings formed a huge body of types - there are at least ten "premium" versions of the humble 12AU7. Tube manufacturers occupied an odd dual position: they pushed the use of a few "preferred" types (theirs, of course), but were constantly introducing specials.

With the popularization of public-address systems in the '30s came the possibility of marketing tubes on the basis of improved performance or reliability. National Union offered a "Sound X/TRA" line of premium audio tubes in 1940. Raytheon countered with a "Sound Special" line in 1941. These special variants are quite rare today. The two lines were:

Sound X/TRA	Sound Special		Both
5Z4GT/G	SS6C5G	SS6L7G	5U4G
6J5GT/G	SS6C8G	SS6N6G	5V4G
6N7GT/G	SS6F6G	SS6N7G	5Y3G
6V6G	SS6K6G	SS6X5G	6J7G
6V6GT/G	SS6L6G	SS83	6J7GT
6W5G			
30			

Early reliability programs included RCA's 1946 introduction of the 12SW7-12SX7-12SY7 line for aircraft radio, and of the 1948 "Special Red" line (5690-5693).

A late-'40s attack on tube failures was sponsored by Aeronautical Radio, Inc., on behalf of the airlines. The response from the manufacturers was the "ARINC line" of high-reliability miniature types: 5654, 5670, 5686, 5725, 5726, 5727, 5749, 5750, 5751, 5814A, and 6201.

The military departments made their own attacks on the reliability problem. The Air Force sponsored an '"A" program' leading to "A" versions of existing tubes. The Navy, not necessarily too concerned about the taxpayers, had a separate "WA" program and ordered "USN" variants with tighter testing than the "JAN-1-E" tri-service requirements. The result was tubes with clumsy identifiers like "5654/6AK5W/6096." "WA" and "WB" variants weren't necessarily made differently from a basic "W" ruggedized version, but got more tests and inspections.

Mobile radio, with its varying heater voltages and strenuous vibration/shock environment, was another opportunity for improvement. Each maker had its "mobile-rated" line of specials: RCA's "Command" line, GE's "Five-Star" series, Sylvania's "Gold Brand" types, and Amperex' "10,000-hour" types.

When tube computers appeared in the late '40s, a new failing was found and fixed: a tube that ran for a time with plate current cut off might develop "sleeping sickness": failure due to an interface layer growing on the cathode, like a built-in cathode resistor. The result was special product lines for computer use. GE's line as of 1955 was the 5844, 5915A, 5965, 6211, and 6463. RCA's similar offering as of 1956 was the 5915, 5963, 5964, 5965, 6197, and 6211; later 6350 and 6887 as well.

ELECTRICAL TRENDS

We have covered, so far, trends in physical construction of tubes. Certain trends in electrical design are of interest as well.

Until the late '30s, AC-DC receivers were built with ordinary 6.3-V 300-mA tubes, plus the 43 output tube and 25Z5 rectifier (and variants). This required a ballast resistor or resistor line cord to make up the difference between tube voltage and 120 V. This changed with RCA's introduction in 1939 of a 150-mA complement, familiar today as the "All-American Five": 12SA7 converter, 12SK7 IF, 12SQ7 detector-audio, 50L6GT output, and 35Z5GT rectifier. Sylvania's 1940 loktal answer was the 14Q7-14A7-14B6-50A5-35Y4. Development of RCA's miniature AC-DC lineup was delayed by war, but was complete in 1945: 12BE6-12BA6-12AT6-35W4-50B5. But that was not the end of the AC-DC story: in 1960, Sylvania released a *100-mA* line: 18FX6-18FW6-18FY6-32ET5-36AM3. Other line-ups were planned, but abandoned, for a 4-tube AM radio, FM radio, and a phonograph. The new complement found its way into even a few RCA models, but was short-lived.

A brief surge of tube design occurred in the mid-'50s on the arrival of automobiles with 12-V electrical systems, allowing for hybrid radios with tubes running with 12 V on their plates in the low-level stages and transistor output. This called for a special line, including space-charge triodes. Tung-Sol's 1956 line for this purpose was: 12AC6 RF-IF amp., 12AD6 converter, 12AE6 detector-audio, and 12K5 tetrode audio driver. There were numerous other types for this use, an application lost to solid-state in the early '60s.

END OF THE LINE

The mass market for tubes faded away under the force of transistorization: first hearing aids, then portable radios, followed by auto and home radios, then audio systems. Solid-state TV sets appeared in the late '60s. Industry leader RCA converted to 100% solid-state TV production in June 1974, signaling the doom of the Harrison receiving-tube plant, which closed in April 1976. Production of nuvistors, which was still commercially viable, was transferred to Sylvania. RCA continued for a time, distributing prestocked tubes and imports (from Brazil, England, Holland, Italy, and Japan), but eventually disappeared. RCA's Lancaster plant, which specialized in photo, transmitting, and television types, lasted into the '90s but is similarly gone, with its line of specialty products sold to a spin-off firm. Between a declining market and production overseas, Westinghouse's receiving-tube operation was an early casualty. Sylvania and GE's Ken-Rad were the final producers in the mass market, lasting into the early '80s; Sylvania, which at that point was a subsidiary of Philips, closed all its factories except the Altoona plant and the original Emporium facility, which was down to subminiatures and nuvistors. Ken-Rad, sold to MPD Inc. in 1987, lasted until 1993, after converting mainly to microwave components. Symbolic of audio as the last pillar of tube manufacture, its final tube was a 6550. Only specialty houses (Eimac, Richardson Electronics, AT&T/Lucent/Westrex) and the foreign makers remain.

The relative (and shifting) market shares of the various makers were private information. However, RCA announced the three-billionth tube made at its Harrison plant in 1969; Sylvania claimed its three-billionth in 1973. This suggests that, over a 50-year span, Sylvania was just behind RCA.

INTERPRETING TUBE DATA

1. HOUSEKEEPING DETAILS

Tube data presented in this book are organized with the following, necessarily lengthy, conventions.

- Bulb sizes are given, per industry practice, as "S" (spherical), "ST" (spherical-tubular), "GT" (globe-tubular), or "T" (tubular). The number following is the nominal diameter in eighths of an inch. Thus a "T12" bulb is tubular, 1-½" in diameter.
- Early bases are cited as follows. A "UX" base is the '20s four-pin style with locating pin; a "four-pin" base lacks the pin. A "large 4-pin" base is the size associated with '20s transmitting tubes, then called a "50-watt" base. An "industrial" 4-pin base has skewed, unequal spacing of the pins. A "UY" base is the '20s five-pin format.
- "Filament" refers to a directly heated electron emitter, whereas "heater" covers indirectly heated cathodes and "cold cathode" refers to a gas tube. "T-T" means thoriated tungsten. A voltage rating in "VCT" indicates a center-tapped element.
- Where a manufacturer is listed in a tube description, data from that maker was used in preparing the listing. This does not identify which manufacturer developed a given type, although the availability of a development number (see below) gives a hint. Indeed, it is quite difficult to distinguish companies that *made* a given tube from companies that *distributed* it - as in the case of the nuvistor, where all original production apparently was from RCA, whereas all later production was from Sylvania. Manufacturers' names are as abbreviated below.

AR: Arpin
AS: Arcturus
BL: Bomac Labs.
BR: Burroughs
BX: Bendix
CGT: Compagnie Generale de T. S. F.
CH: Chatham
CM: Canadian Marconi
EEV: English Electric Valve
EG&G: Edgerton, Germeshausen, & Grier
EL: Electrons, Inc.
EP: Electronic Products
GC: General Electrodynamics
GEC: Gen'l Electric Co. Ltd.
HI: Hitachi
HK: Heintz & Kaufman
HP: Hewlett-Packard
HU: Hughes
HY: Hytronic Labs., later Hytron, CBS Hytron, and CBS
IT: IT&T
JA: Jenkins & Adair
JB: Johnsonburg
KK: Kobe Kogyo
KL: Kuthe Labs.
LE: Lewis Electronics
LI: Litton Industries
LK: Lewis & Kaufman
LME: L. M. Ericsson
MA: Microwave Associates
MC: Marconi (Br.)
MET: Microwave Electron Tube Co.
MU: Mullard
NE: Northern Electric
NEC: Nippon Electric
NL: National Electronics
PH: Philips
PL: Penta Labs
PO: Polarad
RA: Raytheon
RE: Rogers Electronics
RL: Rauland
SFR: Societé Française Radioelectrique
SL: Slater
SP: Sperry
ST: Sonotone
SV: Svetlana
TF: Telefunken
TH: Thomson-Houston (Br.)
THF: Thomson-Houston (Fr.)
TL: Thermonic (sic) Labs
TSU: Toshiba
VA: Varian Associates
VN: Victoreen
VO: Valvo
WE: Western Electric
WT: Weltronic
W: Westinghouse

- Manufacturers' developmental or experimental numbers for tubes are listed in {brackets} where known. These numbers usually carry prefixes specific to the maker, as follows:

A: RCA (for amplifiers)
AGR-AX: Amperex
C: RCA (for photodevices)
DT: Tung-Sol
HD: Hytron
R: RCA (for rectifiers)
QF-QL-QM-QT-QY: Raytheon
SA-SB-SC: Sylvania
SD-SR: Sylvania
WX: Westinghouse
XQ (as suffix): WE
Z-ZG-ZP: GE

They help identify the designer of a given tube - except that RCA, say, in preparing to make a GE-designed tube, would assign a developmental number to cover its own manufacturing-standards work.

- For registered types, the date of release of the final registration (where known) is shown in parentheses.
- For base connections: most tubes have pinouts following one of the RMA/EIA standard basing codes, cited in <brackets>. For example, the 6AB4 triode uses pinout <5CE>. Base diagrams are not printed here, to avoid duplicating diagrams in the manufacturers' tube manuals, the H. W. Sams Tube Substitution Handbook, 1940-1978 editions of the ARRL Radio Amateur's Handbook, and several other present-day tube books.
- A letter prefix is often associated with a specific manufacturer - especially for nonregistered types - as follows:

BL: Bomac Laboratories
BW: United Electronics
C: Cunningham, Collins
CE: Continental Electric
CG: Navy, made by GE
CH: Chatham/Tung-Sol
CK: Raytheon
CV: United Electronics
CW: Navy, made by WE
D: WE special or temp. no.
DR: General Electronics
DRJ-DRO: Westinghouse
E: Arcturus (xmtg., 1932)
EE: Electronic Enterprises
EL: Electrons, Inc.
F: Federal
FA-FG-FJ-FP-FR-GL: GE
HF: Amperex
HK: Heintz & Kaufman
HV: United Electronics
HY: Hytron
K: Du Mont
KI-KX: Westinghouse
KU plus 2 digits: Kuthe Labs
KU plus 3 digits: Westinghouse
ML: Machlett Laboratories
NL: National Electronics
NU: National Union
PJ-PR-PT: GE
PM: Sylvania
QK-RK-RKR-RM: Raytheon
R: Sylvania
RJ-RO: Westinghouse
RX: Eitel-McCullough
SE: Navy (Bureau of Steam Eng.)
SR: Westinghouse
Sxx: Sperry
T-TT-TZ: Taylor Tubes
TE: Bendix, Red Bank Div.
TS: Tung-Sol
UE: United Electronics
UH: Eitel-McCullough
VC: Chatham/Tung-Sol
WL: Westinghouse
WT: Weltronic
ZB: Amperex

To the tube researcher, these may be simply "vanity" prefixes for common tubes (e. g., *GL*-807 vs. *NU*-807 vs. *RK*-807 vs. *WL*-807). Conversely, they may distinguish between two entirely different types under the same number (e. g., 951 vs. UE-951).

- For rectifiers, peak inverse voltage (PIV) is given. If no PIV rating is known, a voltage "in" is listed, this simply being the applied AC voltage. The rectified DC current is given where available; otherwise a peak value is cited.

▪ Triodes are associated with an amplification factor μ, and often a transconductance rating g_m. Tetrodes and pentodes generally carry only a transconductance rating.

▪ Transconductance is given in millisiemens (mS), the international standard unit of conductance. The old-time "micromho" is clumsy because most post-1930 tubes have g_ms above 1000 μmhos. For familiarization, 1.0 mS = 1000 μmhos.

▪ Tube power ratings, unless noted otherwise, are plate *dissipation*. The figure is for CCS (continuous commercial service), and is a total for multi-section tubes. Output power from a tube is a different matter; it depends on the mode of operation, and can vary up to two-to-one between Class C and Class A.

▪ Frequency rating, given for most transmitting types, is the limit for use at full power; operation above this frequency is feasible but at reduced power and/or with increased cooling.

▪ Technical parameters are quoted as the values most usually published; they may vary from catalog to catalog or from maker to maker. (Transconductance is particularly sensitive to specific operating voltage and current.) Occasionally a maker rerated a tube based on production experience - usually upward.

▪ Certain abbreviations pertain to microwave devices:

BWO: Backwave-wave osc.	IC: Integral cavity
CO: Coaxial output	IM: Integral magnet
EC: External cavity	TWA: Traveling-wave amp.
EM: External magnet	WG: Waveguide output

▪ Phototubes and photomultipliers are rated as to their spectral sensitivity per the following designations:

No.	Color	Peaks at (Å)
S-1	Red-IR	8000
S-3	Violet	4200
S-4	UV-Violet	4000
S-5	UV	3500
S-8	UV	3700
S-9	Blue	4700
S-10	Blue-Violet	4400
S-11	Blue-Violet	4400
S-13	Blue-UV	4500
S-17	Green-UV	4800
S-19	UV-Violet	3300
S-20	UV-Violet	4200

▪ Phosphors in cathode-ray tubes carried "P" numbers (now replaced by two-letter indicators). The following designations applied when tubes listed in the CRT chapter were registered:

No.	Color	Peak at (Å)	Persistence
P1 (GJ)	Green	5250	Medium
P2 (GL)	Blue-green	5450	Long
P3 (YB)	Yellow	6000	Medium
P4 (WW)	White	4100, 5400	Medium
P5 (BJ)	Bluish	4300	Short*
P7 (GM)	Blue/green-yellow	4350, 5550	Minutes
P10 (ZA)	(Dark trace)	-	Long
P11 (BE)	Bluish	4600	Short*
P12 (LB)	Orange	5900	Short
P14 (YC)	Purple-org.	4400, 6000	Short
P15 (GG)	Blue-green & near-UV	3900, 5050	Short
P16 (AA)	Near-UV	3900	Short
P22 (XX)	Color TV	4500, 5200, 6300	Medium

* For photo recording.

OTHER DEFINITIONS

▪ "AKA" ("also known as") refers to multiple designations of the same tube.

▪ "Remote cutoff" defines a tube that can be biased to handle strong signals without cutting off (and will handle automatic gain control). RCA literature of the '30s called this feature "super-control" or "variable-μ."

▪ A "dual-control" pentode (e.g. 6AS6} has its suppressor wired out separately for gate use, etc.

▪ "Beam power tubes" are receiving pentodes with beam-forming plates in place of suppressor grids. However, transmitting tubes include beam triodes and beam tetrodes with different constructions.

▪ "Radiator" (R) versions of water-cooled transmitting tubes are always forced-air cooled; they are sometimes derated to as low as half the power rating of a liquid-cooled prototype.

▪ "VHF" tubes are rated for full power to at least 30 MHz, while "UHF" types are rated to at least 300 MHz.

▪ "G-G" stands for "grounded-grid."

▪ "TR" refers to the transmit-receive switch in a radar which disconnects the receiver from the antenna during transmission. "ATR" is an anti-TR switch, which disconnects the transmitter from the antenna during reception.

▪ "NF" stands for "noise figure," a measure (in dB) of how noise an amplifier generates compared to a perfect amplifier.

▪ "Mobile-rated" means intended for use with varying battery voltage on the heater, with considerable on-off cycling, and with vibration resistance.

▪ "Gas" means argon, neon, xenon, etc., as opposed to mercury vapor.

▪ "Cermet" refers to ceramic-metal construction; RCA's trade term for it was "Cermolox."

▪ Numerous historic tube applications were in military systems. Examples have been cited where known, some of them Signal Corps "BC" components or "SCR" complete systems, or early Navy radars like SC or SG. Post-1942 military systems are denoted under the "AN" nomenclature system. The following subset of the "AN" code, paraphrased a bit, applies to the systems cited.

Installation	Equipment	Purpose
A: Airborne	L: Countermeasures	A: Auxiliary
C: Air transportable[1]	M: Meteorological[2]	C: Communications
D: Pilotless carrier	P: Radar	G: Gun directing
F: Ground, fixed	R: Radio	N: Navigation
G: Ground, gen'l use	S: Special	Q: Special[3]
M: Ground, mobile		R: Receiving
S: Shipboard		S: Search
T: Ground, transport.		T: Transmitting[4]
U: General utility		W: Remote control
		X: Identification, friend-or-foe

1. "C" obsoleted early, but used into the '50s.
2. Often used for radiosondes.
3. AN/MPQ- family is for mortar location.
4. Includes countermeasures transmitters like AN/APT-5.

▪ "Computer use" means intended for operation over long periods with plate current cut off.

- "Liquid-cooled" means cooling with Fluorocarbon FC-75, water, or ethylene-glycol/water - as opposed to "water-cooled." Non-water coolants were common in super-power military mobile transmitters, especially airborne.

2. A NOTE ON EUROPEAN TUBES

A quirk pertains to designations of European tubes. Receiving types of the '50s usually carry a letter indicating heater voltage, a letter(s) showing number of elements, a number indicating base style, and an arbitrary number - for example, ECC88 is a 6.3-V ("E") triode-triode ("CC") on a 9-pin base ("8"). However, many European ruggedized or special-purpose types use a revised sequence; for example, a special variant of the ECC88 is the E88CC. (These are familiar here as the 6DJ8 and its premium variant 6922 respectively.) This is not a hard-and-fast rule - the writer has not seen it elsewhere - but may help explain such cases as EL34 vs. E34L.

3. REGISTRATION THEORY

PREHISTORY

In the early years of the tube industry, numbering of tubes was a free-for-all, structured only by the development capabilities of GE and Westinghouse and the market power of RCA. Receiving tubes of early vintage (but late enough to have bases) are listed in the "Prehistory" chapter, including lists of uniquely numbered types from the smaller manufacturers.

Transmitting tubes of this era fell mainly into the 800-series, as numbered mainly (and randomly) by RCA. This did not prevent double use of identifiers - for example, 814s from Taylor vs. RCA, 825s from Taylor vs. RCA, or 889s from Du Mont vs. RCA. These types are detailed in the "800" chapter. Meanwhile GE, WE, Westinghouse, and the "independent" makers were free to devise their own tube types and give them their own nomenclature.

RECEIVING TUBES

The Radio Manufacturers' Association (today's Electronic Industries Association) began putting order into the numbering of receiving tubes in 1933. The first "release" of the Committee on Tube Numbering, chaired by David Grimes of Philco, was dated Feb. 5, 1934. It dealt with the 1A6 and the 1C6, as registered by Sylvania. Release 3 covered the 2A3 and 2A3H. Release 6 (Sept. 14, 1934) dealt with the 2A5, 6C6, and 6D6. The registration process continued throughout "tube history": for example, Release 5536 of June 26, 1967, defined the 6BW11.

Tube identifiers were assigned according to a simple code. An initial digit gives filament or heater voltage: "0" for cold-cathode, "1" for 1 to 2 volts, and on up. The code was gerrymandered slightly to accommodate the loktal series: "7" and "14" refer to 6.3- and 12.6-volt tubes in loktal format.

The letter(s) following the number were assigned (nominally) in sequence from "A," while rectifiers were lettered backward from "Z." As a special case, 6S( ) covered several single-ended versions of double-ended metal tubes, e. g. 6SK7 vs. 6K7.

The final character is a digit giving the number of "usable" elements, that is, elements wired out individually. Thus a simple cathode-type triode would be scored as (heater + cathode + grid + plate) or "4," as in 6J4. Shields having their own base pins counted; elements wired internally to other elements did not. Taps on heaters theoretically did not count, but there were exceptions like the 35Z5GT.

Suffixes later became common:

"A," "B," "C": improved, backward-compatible versions.
"E": "Export" version. It has never been clear what this implies, but numerous examples include 1LA6E, 6A7E, 6J8EG, and 7J7E.
"G": Large spherical-tubular (ST) glass bulb.
"GT": Small (T9) tubular glass bulb.
"GT/G": GT-size fits-all replacement for either "G" or "GT" type.
"S": Spray-shielded (zinc-clad) glass tube. In practice these were from Majestic in the U. S. (Raytheon for subminiatures) or Rogers in Canada.
"L": Loktal format, on an octal base with normal octal pinout.
"LM," "ML": "loktal-metal," RCA and GE's short-lived construction, ca. 1940.
"LT": RCA's briefly produced loktal-like tube with metal-ringed loktal-wafer base.
"MG": "Metal-glass" construction.
"W": Ruggedized.
"X": Ceramic-based variant (usually from HY, sometimes SY).
"Y": Micanol (mica-filled phenolic) base.

Variants suffixed "E," "L," "W," "X," and "Y" were not registered as such. Presumably the specifications for the "W" types were determined satisfactorily by military specifications like JAN-E-1A and its successors.

The chapter on registered receiving tubes lists types that are (A) from the 1934-1950 era, (B) prototypes of specials listed in other chapters, (C) obscure types not listed in regular tube manuals like RCA's RC-30, or (D) types that are simply odd or interesting.

CATHODE-RAY TUBES

The RMA began registering CRTs in 1939. The code used is simply (1) the nominal screen diameter in inches, (2) a letter or letters starting from "A," and (3) a "P" number giving the phosphor. Heater voltage was not a problem; the early tubes "all" used 2.5 V @ 2.1 A; later ones "all" used 6.3 V @ 600 mA. There was no attempt to encode the method of focusing or deflection.

SPECIAL-PURPOSE TYPES

The RMA took on assigning numbers for special-purpose and transmitting tubes in 1942, using a number-letter-number coding scheme. Under this system, about 480 tube types carry somewhat mysterious number-letter-number designations like "2E26" or "4D32." Among the first registrations, in September 1942, was Release 303 covering the 1C21. The code is given at the head of the chapter covering these types.

This system lumped together tubes of all sizes and uses having the same number of elements - "D" included both subminiatures and huge water-cooled items. It did not distinguish among vacuum, mercury-vapor, or gas-filled devices. Multiple-section tubes like the 2G21 were simply scored by the most complex section.

The coding plan didn't last long, being officially dropped ca. 1944 in favor of a non-system: simply assigning four-digit numbers beginning with "5500." Even so, registration releases continued to come out (e. g., for the 2J70) under the supposedly defunct 1942 system as late as 1955. The four-digit plan wasn't particularly easy on the user - the *7894* and *7984*, for example, were both small transmitting tubes, and it was easy to confuse them. The four-digit plan lasted until the end of new tube registrations, and occupies the longest chapter in this book. Somehow it includes germanium P-N junction devices (7223, 7224, 7467) and numerous cadmium-sulfide photoconductors (8100, et al.) - items that are hard to view as "tubes."

Besides the new types, most existing items in the 800 and 900-series were retroactively registered in the early '40s.

The exception to the four-digit plan was solid-state devices. The arrival of the transistor gave the superseded code new life. The first number was redefined: "1" had always applied to diodes, "2" now covered triode transistors, "3" described tetrode or dual-gate transistors, and "4" or "6" referred to multi-lead devices like diode-phototransistor optoisolators. The "1N" numbers reached about 1N6300; "2N," around 2N6800; "3N," about 3N260; "4N," ca. 4N50, "6N," around 6N140. Even so, a changeover to nonregistered (manufacturer-numbered) types slowly occurred, and was largely complete by the time integrated circuits arrived.

Many wartime and '50s tubes were "classified" military technology. Indeed, final registration documents on most radar devices (klystrons, magnetrons, TR tubes) were publicly released only several years after the war, after many of them had become obsolete. A couple of tubes remained secret: as late as 1967, the 2C27 and 2E27 were officially listed as classified - probably as the result of inattention rather than of any continuing military value. Correspondingly, the services maintained a list of types preferred for their use. The unclassified part of the list was widely published; the classified part was not, even (as far as is known) in later times.

Contrasting with the more obscure types, many registered tubes were highly successful and had long periods of popularity. The 2E26, for example, appeared in new equipment designs into the mid-'60s. The 2J42 magnetron was common in later years, both by itself and in improved variants numbered in the four-digit series. It is ironic that devices like the 2K22 klystron, developed to help shut down Germany and Japan, were made in the '60s by Telefunken and Toshiba!

The fact of being registered did not guarantee that a given type was made in quantity, or made at all. Some of the more obscure tubes - whether actually produced or not - had their registrations abandoned; the RMA/EIA registration list in later years simply omitted them and their release dates. Hence the lack of some dates in the lists.

Every process has exceptions. Some special-purpose tubes were designated under the receiving-tube plan. For example, RCA's special versions of the 12SA7, 12SN7, and 12SR7 for aircraft radios operating from 28-volt battery power were registered as the 12SY7, 12SX7, and 12SW7 respectively. Sylvania's 6BL6 and 6BM6 small klystrons were likewise numbered as if they were "receiving" tubes. When Sylvania and GE were hopeful of selling high-performance tubes for RF stages in UHF TV tuners, they registered the "receiving" 6BA4 and 6BY4 - variants respectively of the 2E26 "rocket" triode and the 6299-7077-7391 family of cermet triodes. Conversely, many late home-audio types were inexplicably registered as "special-purpose": the 7025, 7027, 7189, 7199, 7247, 7754, et al.

Even the simple four-digit numbering plan eventually faltered, while approaching 9000 in the 1970s. RCA, for one, eventually began assigning its own four-digit numbers - in the 2000- and 4000-series - for everything from TV transmitting tubes to cadmium-sulfide photocells.

WHETHER TO REGISTER

Tube makers were free to register their new designs or not. Whether to do so seems to have been a marketing decision. The military services strongly preferred registration, as consistent with their preferred-type lists and with competitive procurement. Registered, the tube might attract second-source makers, who could either (A) increase the salability of the tube or (B) take revenue away. Not registered, the tube might give control of a niche market - and certain industries like radar countermeasures were niches too small to support the administrative burdens of registration. The stakes could be high, as when Westinghouse won a competitive military contract to supply 3500 magnetrons of a registered type. The makers often registered tubes long after introduction, apparently as a result of changes in sales strategy. (The prime example was Eimac, which registered dozens of tubes in the 8000-9000 area when many of had been on the market for years.)

On a tube-box lid, evidence of a registration process out of control.

The RMA-EIA was a registrar, not a policeman. The manufacturers were quite happy to devise tubes only minimally different from those of their competitors. Pressure from transistors forced the development of tubes narrowly optimized for the color-TV market - usually in triple variants for series heater strings at 300, 450 and 600 mA. Toward the end of the tube era, especially with RCA's industry leadership eroding, this led to a glut of dual-, triple-, or even quadruple-numbered types barely distinguishable from each other. This remains a source of confusion today.

PREHISTORY

This section covers "prehistoric" tubes, that is, those that appeared before RMA-controlled registration began in 1934.

Besides the industry-standard types sponsored mainly by RCA, many small manufacturers offered lines of tubes of their own design and styling. This section gives most of the proprietary versions as well as the better known types. The scope is *based* tubes; earlier unbased types (tubular audions, AudioTrons, early Moorheads, etc.) are covered in Tyne's Saga et al.

The numbering of these early types is cause for confusion. Lacking coordinated numbering, the smaller manufacturers went their own ways. As a result, for example, an Arcturus 180 "is" an 80, but a Sparton 182 "is not" an 82. For best clarity, tubes under "standard" numbers (mostly RCA) are given first, then the uniquely numbered product lines of smaller makers.

GENERAL

00A [AKA (UX) 200A, q. v.]; detector triode.

01 [AKA C-301, PR-1-A, UV/UX-201, q. v.]; rcvg. triode.

01A [AKA 501A, 38001, C-301A, VT-30, PR-1-B, SE-4374, De Forest 401A, CV750, PJ-11]; rcvg. triode. Also see UX-201A.

1-V [AKA 6Z3]; diode rectifier, vac. version of KR1, 4-pin base; heater 6.3 V @ 300 mA, 1 kV PIV, 45 mA {R630} <4G>.

WD-11 Rcvg. triode, intro. 1921, spl. 4-pin base w/ enlarged plate pin, coated fil. 1.1 V @ 250 mA, μ 6.6, g_m 0.44 mS (W) <4F>. T10 and T8 bulbs, tipped and tipless, brass and bakelite bases. Variant: 864.

WD-12 Rcvg. triode, intro. 1922, T10- or T8-sized WD-11 on UV base (W) <4D>.

WX-12 [AKA VT-7]; rcvg. triode, WD-11 on UX base (W).

12A [AKA UX-112A, q. v.]; power triode.

14 Tetrode RF amp., w/ cap on UY base; 14 V @ 300 mA on heater; made by SY for Philco 110-volt DC receivers <5E>.

15 RF pentode, ST12 bulb w/ cap, UY base, *heater* 2.0 V @ 220 mA, g_m 0.7 mS, in some Philco 2-volt sets (SY, RCA) <5F>.

17 Triode sim. to 27, heater 14 V @ 300 mA, μ 9, g_m 1.0 mS <5A> (SY, Lansdale). Companion of 14 in Philco DC receivers.

18 [AKA KR18]; power pentode sim. to 42; heater 14 V @ 300 mA, sold by SY and NU for use in 110-V DC receivers <6B>.

19 Dual Class B power triode, ST12 bulb, 6-pin base, fil. 2.0 V @ 260 mA, 1J6G char. {A476} <6C>.

20 [AKA UX-120]; power triode, T8 bulb, UX base, fil. 3.3 V @ 132 mA, μ 3.3, 120 mW out <4D>.

WR-21 A, D Triodes, intro. 1921, tipped tubular bulb, special 4-pin brass base, for Aeriola Grand (W), pure-tung. fil. 800 mA @ ca. 4 V. "A" is amp. version; "D" is detector variant (green tip paint).

22 [AKA SE-3222, 38022, VT-26]; RF tetrode, ST14 bulb w/ cap, UX base, fil. 3.3 V @ 132 mA, g_m 0.5 mS {A287} <4K>.

24, A [AKA SE-3924, FM-500 (GE), 124, (UY-) 224]; RF tetrode, sim. to 35/51; S14 or ST14 bulb w/ cap on UY base; heater 2.5 V @ 1.75 A, g_m 1.05 mS. "A" [AKA VT-28, CV936, 38024] is quicker-heating version w/ revised construction and lower secondary emission from plate, not as well suited to dynatron oscs. {A279} <5E>.

WD-25 Triode equiv. to UV-201A, intro. 1924 (Can. W).

WX-25 Triode equiv. to UX-201A, intro. 1925 (Can. W).

25RE Export (Br.) full-wave rectifier, 6-pin base, heater 25 V @ 300 mA, 250 V in, 80 mA <6E>.

25/25S [AKA 1B5/25S, q. v.].

26 Rcvg. triode; 01A w/ fil. 1.5 VAC @ 1.05 A, w/ higher g_m of 1.1 mS, μ 8.3 {A189} <4D>.

27, A, HM [AKA SE-3373, 38027, DE1, VT-29, FR-503, CV-944]; rcvg. triode, S14, ST12, or (ultimately) tall T9 bulb on UY base, heater 2.5 V @ 1.75 A, μ 9, g_m 1 mS (sim. to 56) {A201} <5A>. "A" has low-current heater. "HM" is MG variant.

29 Wunderlich detector (SY); 2.5-V 1-A version of 69, S12 bulb on 6-pin base; μ 30, g_m 1.45 mS; uses separate cathodes connected internally.

30 [AKA 38030, FP-59, SE-3872, VT-27, CV604, RCA-230]; rcvg. triode; S14 (Maj.), ST12, or T9 bulb on UX base; fil. 2.0 V @ 60 mA, μ 9.3 {A205} <4D>. Variant: 1230.

31 [AKA 38031, FP-60, VT-31, CV931, RCA-231]; power triode, 30 format, fil. 2.0 V @ 130 mA, μ 3.8 {A206} <4D>.

32 [AKA 38032, FP-61, SE-3873, 1B4P, VT-44, CV711, RCA-232]; sharp-cutoff tetrode, ST14 bulb, fil. 2.0 V @ 60 mA, g_m 0.65 mS {A288} <4K>. Variant: 1229.

33 [AKA VT-33, CV949]; power pentode, ST14 bulb, UY base, fil. 2.0 V @ 260 mA, g_m 1.7 mS {A415} <5K>.

34 [AKA VT-54, CV1751]; remote-cutoff (variable-μ) pentode, ST14 bulb w/ cap, UX base, fil. 2.0 V @ 60 mA, g_m 0.6 mS {A458} <4M>.

35 [AKA 51, 4465, VT-35, 38035, CV1752]; remote-cutoff tetrode, variable-μ variant of 24A, heater 2.5 V @ 1.75 A, g_m 1.05 mS {A282} <5E>.

35/51 [AKA 35, q. v.]; tetrode.

36, A [AKA SE-4610, 36036, VT-36, CV1775]; tetrode, ST12 bulb w/ cap, UY base, heater 6.3 V @ 300 mA, g_m 1.05 mS {A284} <5E>. "A" has 400-mA heater.

37, A [AKA 4611, 38037, 37A, VT-37, CV606]; rcvg. triode, ST12 or (ultimately) T9 bulb on UY base, heater 6.3 V @ 300 mA, μ 9.2 (sim. to 76) {A196} <5A>. "A" has 400-mA heater.

38, A [AKA VT-38, CV712, SE-4612, 38038]; power pentode, heater 6.3 V @ 300 mA, g_m 1.05 mS {A442} <5F>. "A" has 400-mA heater.

39/44 [AKA 39, 44, 239, VT-49, CV1771, 38039]; remote-cutoff RF pentode, heater 6.3 V @ 300 mA, g_m 1.05 mS {A450} <5F>.

40 [AKA SE-3720, 38040, VT-40, CV2501]; high-μ triode, fil. 5.0 V @ 250 mA, μ 30 <4D>.

41 [AKA 38041, VT-48, CV608]; power pentode, ST12 or (ultimately) T9 bulb on 6-pin base, 6K6G-7B5 char. {A452} <6B>.

42 [AKA 10E/333]; power pentode, S17 or ST14 bulb on 6-pin base, 6F6G-2A5 char. {A453} <6B>.

43, MG power pentode, S17 or ST14 bulb on six-pin base, 25A6G char. {A457} <6B>. "MG" is octal metal-glass version. Also see 25A6 Coronet.

44 [AKA 39/44, q. v.]; pentode.

45, A [AKA SE-3921, 38045, UX-245, VT-45, FR-304, CV596]; power triode, S17 or ST14 bulb, fil. 2.5 V @ 1.5 A, μ 3.5, 800 mW out @ 250 V {A190} <4D>. "A" is uprated from 250 to 325 V; puts out 3 W instead of 1.6 (NU).

46 [AKA VT-63, CV2531]; dual-grid power triode (second grid connectable to first grid for Class A or to plate for Class B use), S17 or ST17 bulb on UY base, fil. 2.5 V @ 1.75 A, μ 5.6 {A283} <5C>.

47 [AKA 38047, VT-47, CV1772, PZ, KR48]; power pentode, fil. 2.5 V @ 1.75 A, g_m 2.5 mS, 2.5 W out @ 250 V {A441} <5B>. Variant: 1610.

48 Power tetrode for DC-operated receivers; heater 30 V @ 400 mA, g_m 3.8 mS, 2.5 W out {A461} <6A>.

49 Dual-grid power triode for Class A or B amps., battery-powered analog of 46, fil. 2.0 V @ 120 mA, g_m 1.125 mS, 3.5 W out {A290} <5C>.

50 [AKA 38050, 585, 586, SO-2, SE-3871, VT-50, CV-2533]; power triode, S21 or ST19 bulb, UX base, fil. 7.5 V @ 1.25 A, μ 3.8 {A218} <4D>.

51 [AKA 551, 35, q. v.]; pentode.

52 Dual-grid power triode, for car radios or DC home sets w/ 110-V B+; a 6-volt analog of 46; filament 6.3 V @ 300 mA, 1.2 W out @ 100 V. A pair could give six watts in Class B @ 180 volts. The double-grid feature used two grids tied together to get a high μ without using a delicate close spacing between the first grid and the filament - a technique also used in 46 and 49 (RA).

53 [AKA 38053]; dual power triode, heater 2.5 V @ 2 A, 6A6-6N7 char. {A456} <7B>.

55 Duodiode-triode, heater 2.5 V @ 1 A (85-6V7G char.) {A459} <6G>.

56, A [AKA 38056, VT-56, CV611, 27HM]; triode, heater 2.5 V @ 1 A, 76 char., sim. to 27 {A198} <5A>. "A" is 6.3-V, 400-mA variant (Maj.).

57, A [AKA 38057, VT-57, CV612]; sharp-cutoff pentode, heater 2.5 V @ 1.0 A, 6C6 char., sim. to 58 {A432} <6F>. "A" is 6.3-V, 400-mA variant (Maj.).

58, A, AS [AKA 38058, VT-58, CV613]; remote-cutoff RF pentode, heater 2.5 V @ 1.0 A, 6D6 char. {A433} <6F>. "A" is 6.3-V, 400-mA variant (Maj.). "AS" is "A" w/ sprayed shield.

59 [AKA 38059, VT-59, CV2538]; triple-grid power tube; sim. to 47 w/ dual cathodes. Suppressor was wired-out independently, so tube could be used as Class A or B triode, or Class A pentode. Heater 2.5 V @ 2 A {A436} <7A>.

64 RF tetrode for car radios; heater 6.3 V @ 400 mA (NU) <5E>. Replaceable by 36. Part of first line of car-radio tubes, 1931, w/ 65, 67, and 68; all considered obsolete as early as March 1933.

65 RF tetrode, remote-cutoff version of 64 (NU); repl. by 39/44 <5F>.

67 Power triode (NU), same heater as 64; rated at 650 mW output; replaceable by 37 <5A>.

68 Dual-grid power tube, same heater as 64; replaceable by 38 in parallel-heater receivers (NU) <5F>.

69 Wunderlich detector (SY); 6-volt version of 29, q. v.

70 Wunderlich detector (Cardon-Sparton), S12 bulb on 6-pin base, heater 6.3 volts.

71, A [AKA 271]; power triode, variant of 201A w/ T-T fil. 5.0 V @ 500 mA, 700 mW out @ 180 V <4D> "A" [AKA 171A, 271A, 38071, SE-4171] had oxide fil. cut to 250 mA, μ 3, g_m 1.7 mS, 800 mW out {A188}.

75 [AKA 38075, VT-75, CV614]; duodiode-triode, 2A6-6B6 char. {A467} <6G>.

76 [AKA 38076, VT-76, CV615]; triode, 6-V 56, sim. to 37-6P5G {A204} <5A>.

77 [AKA 38077, VT-77, CV616]; pentode sim. to 6C6 {A459} <6F>.

78 [AKA 38078, VT-78, CV2544]; remote-cutoff pentode (nearest equiv. is 6D6) {A460} <6F>.

79 Class B duo-triode power amp., 6Y7G char. {A447} <6H>.

80, M [AKA 38180, SE-3841, UX-280, VT-80, CV617]; full-wave rectifier, S17, ST14, or (ultimately) T9 bulb on 4-pin base, 5Y3G char. {R622} <4C>. "M" is mercury variant.

81 [AKA CG-2935A, 38181]; diode rectifier, S19 or ST19 bulb, fil. 7.5 V @ 1.25 A, 2 kV PIV, 85 mA <4B>.

82 [AKA 38182, AF]; full-wave mercury rectifier, fil. 2.5 V @ 3.0 A, 1550 V PIV, 115 mA {R624} <4C>.

83, V [AKA 38183, VT-83, CV1773]; full-wave mercury rectifier, higher ratings than 82; fil. 5 V @ 3 A, 1550 V PIV, 225 mA {R600} <4C>. "V" is cathode-type vacuum variant w/ 5V4G char.; slightly derated from 83 {R634A} <4A>.

84 [AKA 84/6Z4, 98, 38184, VT-84, CV619, P861]; full-wave rectifier, ST12 or (ultimately) T9 bulb, heater 6.3 V @ 500 mA, 1250 V PIV, 60 mA {R629} <5D>.

85, S [AKA 38085]; duodiode-triode, 6-V 55, 6V7G char. {A438} <6G>. "S" is spray-shielded.

88 Full-wave rectifier, like 83V but uprated to 125 mA and 450 V in - has also been attributed to Perryman 588, a mercury replacement for 80.

89, Y [AKA 38089, VT-89, CV833]; triple-grid power amp., heater 6.3 V @ 400 mA {A437} <6F>. See 59 re modes of use. "Y" has Micanol base.

91 [AKA FG-37]; mercury thyratron w/ 110-V heater, ST23 bulb, for use in pairs in inverters to power AC radios and phonos from a 110-V DC line; dev. by GE-RCA ca. 1931 but not made in qty. (*Electronics*, Oct. 1931, pp. 152-154). "Not listed for distributor sales" by RCA {R603} Also see 1904.

V99 [AKA UV-199, UV-299]; rcvg. triode, intro. 1923, T8 bulb on special nub-pin base, fil. 3.3 V @ 63 mA, μ 6.6, g_m 0.425 mS <4E>.

X99 [AKA CX-299, UX-199, UX-299]; rcvg. triode, conical UX base (V99 rebased) <4D>.

ZP-100 Rcvg. triode like UX-201A, fil. reduced to 100 mA (Zetka Laboratories).

UX-112A [AKA 12A, UX-112A, 38102, 38382, SE-3382]; power triode sim. to 71A, dev. from W prototype WL-112; fil. 5.0 V @ 250 mA, μ 8.5, g_m 1.8 mS {A219} <4D>.

UX-120 [AKA 20]; power triode.

UX-121B Power triode, low-current UX-201A, fil. 125 mA, μ 5 (between UX-201A and UX-171A) (Can.).

124 [AKA 24, q. v.]; tetrode (AS).

126 [AKA 26, q. v.]; triode (AS).

127 [AKA 27, q. v.]; triode (AS).

UX-171 [AKA 71, q. v.]; power triode.

UX-171A [AKA 71A, q. v.]; power triode.

180 [AKA 80, q. v.]; rectifier (AS).

UV-196 Full-wave "Rectron" rectifier, intro. 1924, for B-eliminators (W). Tipped, brass-based, has two fils., 4.25 V @ 550 mA each w/ common plate, 50 mA. Base shell is an element connection.

UV-199 [AKA V99, C299, q. v.]; triode, orig. on brass base w/ ceramic insulation.

UX-199 [AKA PT-199, UX-299, X99, q. v.]; triode.

UV-200, A Gas detector triode, intro. 1920, UV-201 with 50 microns of argon. "A" has T-T fil., current reduced to 250 mA.

UX-200, A Gas triode, UV-200 on UX base, intro. 1925. "A" has T-T fil., current reduced to 250 mA, gas changed to cesium vapor.

UV-201, A Rcvg. triode, pure-tung. fil. 5.0 V @ 1.0 A. "A" has T-T fil., reduced to 250 mA. Also see 01.

UX-201, A Rcvg. triode, UV-201 on UX base. "A" has T-T fil., reduced to 250 mA {A197}. Also see 01A. A rare variant, labeled as an RCA Radiotron, has a T9 tubular tipped bulb w/ WR-21-style fil. 4.0 V @ 800 mA (W) (fil.).

UX-201 AA, B, C Low-current UX-201As. "AA" has oxide-coated fil., current 125 mA (NU, Silvertone). "B" is same electrically, intro. 1927 (Can. W, CeCo, JB, Sonatron). "C" is cut further to 62.5 mA, intro. 1929 (Can. W).

UX-210 [AKA 10, 310, CG-2566A, CG-2010, PJ-13, PR-10-A]; power triode, anncd. 9-25 <4D>.

UV-212 Diode, unbased cylindrical, 3/4" dia. by 5-½" long; not released.

UX-213 Full-wave rectifier, intro. 1924, T-T fil. 5.0 V @ 2.0 A, 220 V in, 65 mA; lower-rated than, and superseded by, 80 <4C>.

UV-216, A, B Rectifier, intro. 1921; fil. 7.5 V @ 1.25 A, 550 V in, 65 mA (gridless 10?); superseded by 81. Designation "A" was originally assigned to UX-216B; changed to avoid confusion w/ WE 216A. "B" is T-T version, anncd. 9-25; dev. for RCA 104 speaker.

CX-**220** [AKA UX-120, 20, q. v.]; power triode (Cunningham).

UX-221 Rcvg. triode, 5.75-V, 60-mA export version of UX-201A, μ 9, intro. 1929 (Can. W).

UX-222 [AKA 22, q. v.]; tetrode.

UY-224, A [AKA 24, A, q. v.]; triode.

(UR) **225** GE and W forerunners of RCA 27.

UX-225 Triode, "Radiotron UX-225," as factory-engraved on base, S14 bulb, unique base w/ five pins in "W" formation, not in Tyne's Saga or Stokes' 70 Years but fourth example seen. Presumably a transition from orig. double-ended UX-225 and UY-227.

UX-226 [AKA 26, q. v.]; triode, UX-201A with 1.05-V, 1.5-A AC fil. <4D>.

UY-227, A [AKA 27, q. v.]; triode. "A" is Can. W variant.

230 [AKA 30, q. v.]; triode. First U. S. 2-V tube, and first tube made by RCA Radiotron Co. Orig. in T8 bulb, then S12, then ST12.

231-34 [AKA 31-34, q. v.]. Also see 230.

DX235 Triode, nonmicrophonic "Nor-Phonic" variant of UX-201A (235A) (NE).

235-39 [AKA 35-39, q. v.].

UX-240 [AKA 40, q. v.]; triode.

UX-245 [AKA 45, q. v.]; power triode.

247 [AKA 47, q. v.]; power pentode.

UX-250 [AKA 50, q. v.]; power triode.

257 Power pentode, S17 bulb on UY base, fil. 6.3 V (NU says 5.0 V) @ 300 mA, g_m 1.4 mS, 800 mW out, for DC line-operated receivers like Fada (NU, TS).

264 [AKA 864, q. v.]; triode.

UX-271 [AKA 71, q. v.]; power triode.

UX-280 [AKA 80, q. v.]; rectifier.

UX-281 [AKA 81, q. v.]; rectifier.

UX-288 Gas rectifier, W variant of BH, dev. 1925, not released commercially.

(UV) **299** [AKA V99, q. v.]; triode.

(UX) **299** [AKA X99, q. v.]; triode.

C-300 to **C-399** These are generally "industry" types as made by GE and W and marketed by Cunningham. The numbers parallel RCA's practice (CX-301A = UX-301A, etc. Only important exception is Cardon's C-373; see "373" under Sparton.

401 to **499** These are generally "industry" types as sold by De Forest. For exceptions, see the De Forest section.

551S [AKA 35S/51S, q. v.]; tetrode.

607 Gas (helium?) diode rectifier, tipped S14 bulb on UX base, in Atwater Kent B-eliminators. Suspected of having been made by SY.

AC608, R Marathon AC triode, detector-only, w/ blue label. Flanged UX-style base w/ clip-on nubs for heater, heater 5.5 V @ 1.0 A, μ 7.3, g_m 0.775 mS. "R" is for general use, has red label.

704 Power pentode (Champion), 1931.

WB-800 Ballast tube, intro. 1921, for W Aeriola Grand (W).

P861 [AKA 84, q. v.]; full-wave rectifier, UY base, heater 6.3 V @ 500 mA, 225 V in, 50 mA.

950 Power pentode (KR, NU), sim. to 1J5G and 33, replaceable by 1B4.

951 [AKA 1B4P, q. v.]; pentode (KR, NU, RA).

ARCTURUS

AD Mercury half-wave rectifier for car radios, 4-pin base, fil. 6.3 V @ 300 mA, 350 V in, 50 mA; replaceable by 1-V.

AE Mercury half-wave rectifier, heater 12.6 V.

AF Full-wave rectifier, "safety" type w/ internal fuse, rated slightly above 82 but replaceable by it.

AG Full-wave rectifier, "safety" type w/ internal fuse, rated somewhat above 83 but replaceable by it.

GA Rcvg. pentode, UY base, suppressor tied to center of fil., T-T fil. 5.0 V @ 250 mA, g_m 2 mS.

LA [AKA 6A4/LA, q. v.]; power triode.

PA Power pentode, six-pin base, heater 6.3 V.

PZ [AKA 47, q. v.]; power tetrode.

PZH [AKA 2A5, q. v.]; power pentode.

T1 Half-wave mercury rectifier, 4-pin base, fil. 6.3 V.

Wunderlich "Special detector," 1931. Contained a dual grid, acting simultaneously as the plates of a pair of diodes for detection and as a single grid for audio amplification. AS' tubes were "coplanar" (cocylindrical) with a single cathode and plate. One need not use a Wunderlich *tube* to get Wunderlich *detection*: the Majestic 25 used a pair of 27s. Related types: 29, 69, 70, 90, 92, KR-22, and KR-28. Variants:

- A — Two versions: in S-12 blue bulb on red six-pin base (least rare Wunderlich), or on UY base w/ top cap for the *cathode* connection; heater 2.5 V @ 1.0 A.
- A Auto — On 6-pin base, heater 6.3 V at 0.4 A <6N>.
- B — W/ top cap for an auxiliary plate, giving amplified AVC; 2.5-V heater. Actual production unknown.

AC-Series AC-heated triodes, S14 tipped bulbs (usually blue) on UX bases, 15-V carbon heater @ 350 mA, cathode wired to one side of heater <4Q>, 30-40 sec. warmup time. Negature temperature coefficient of carbon meant low start-up surges. Promoted for use w/ "common toy transformer." (A)28 and (A)30 first used in Hugo Gernsback's "Electrified Peridyne," *Radio News*, Feb. 1928.

- (A)22 — Rcvg. triode, g_m 0.6 mS.
- (A)26 — Detector triode.
- A28 — Rcvg. triode, μ 10.5, g, 1.16 $_{m}$S.
- (A)30 — Power triode, μ 3.8, 180 V.
- (A)31 — Rcvg. triode.
- (A)32 — Rcvg. triode, high μ, μ 30.
- (A)40 — [AKA SO1]; power triode, heater raised to 400 mA, μ 3, g_m 1.5 mS, 180 V <4Q>.
- (A)46 — Detector triode.
- (A)48 — [AKA RA1]; rcvg. triode, μ 11, g_m 1.17 mS <4Q>

CeCo

CeCo, like many others, sought "product differentiation" of ordinary tubes by unique identifications. For a table of equivalents [1], see the table at the end of this section. The company also developed a few unique offerings:

P-1 AC space-charge pentode, S15 bulb w/ cap and screw terminal on UY base for space-charge grid, heater 2.5 V @ 1.75 A, g_m 2.5 mS.

AC22 Rcvg. tetrode, 22 w/ *heater* 2.5 V @ 1.75 A.

S-27 Equivalent of Kellogg 401, q. v.

CROSLEY

Crosley "Deflectrons" were a line of gridless tubes using a control electrode placed near the filament [2]. Tubes are made like early De Forest DV-types, w/ tipped, heavily gettered bulbs on nickeled UV bases. Bases (and cartons) are stamped with the '20s Crosley Radio logo with a "D" superimposed on it. Plainly designed carton says "Notice to Purchaser: Insist on your dealer lighting the filament of this tube before purchase / The Crosley Vacuum Products Corporation." These are believed to be market-trial prototypes, possibly to gain leverage in Crosley's license negotiations with RCA. (For the corporate intertwining of Crosley and De Forest, see pp. 114 and 173 of Douglas' *Radio Manufacturers of the '20s*, Vol. I.) Known Deflectron variants are:

1. Sized like DV2.
2. Similar to DeF 20.
3. Like (2) but ¼" higher.
4. Like (2) but ½" higher.
5. In blue S14 bulb.
6. Like DV2 but w/ unplated base and not gettered.

DAVEN

MU-6 Power triode, fil. 6.0 V @ 500 mA, μ 6, g_m 1.4 mS.

MU-20 Rcvg. triode, fil. 6.0 V @ 250 mA, μ 20, g_m 0.5 mS.

De FOREST

The following are mid-'20s receiving types from the De Forest Radio Tel. & Tel Co., later the De Forest Radio Co. ("NB": nickeled base; "BkB": bakelite base; "IB": Isolantite base.) All are tipped; most are heavily gettered. Bulbs are T12 unless indicated. "DL" indicates UX (long-pin) base; "DV" indicates UV-style (short-pin) base. All were replaced in 1928 w/ "400"-series types [3]. For a 1930-32 view, see the De Forest chapter.

DR Rectifier, IB variant of UX-216B, tipped bulb, fil. 7.5 V @ 2.0 A, 550 V in, 75 mA.

DV-1 Rcvg. triode, NB, fil. 3.0 V @ 70 mA, μ 6.4, g_m 0.4 mS.

DL-2 Rcvg. triode, BkB or IB, fil. 5.0 V @ 250 mA, μ 8,

g_m 0.83 mS; sim. to UX-201A.

DV-2 Rcvg. triode; NB, BkB, or IB; see DL-2.

DL-3 Rcvg. triode, short T12 bulb, BkB or IB, fil. 3.0 V @ 60 mA, μ 6.6, g_m 0.42 mS; sim. to DV-1.

DV-3, A Rcvg. triode, short T12 bulb, BkB or IB, UV base. "A" has UV-199-style base of same materials.

DL-4 Rcvg. RF-amp. triode, IB, fil. 5.0 V @ 250 mA, μ 9.5, g_m 1.12 mS.

DL-5 Rcvg. triode, BkB or IB, fil. 5.0 V @ 250 mA, μ 9.7, g_m 1.0 mS

DV-5 Rcvg. triode, IB; see DL-5.

DV-6, A Rcvg. triode, short T12 bulb, NB, fil. claimed to be T-T but actually pure tung., claimed in the same ad to be both "metallic vapor" and vacuum type, fil. 4.0 V @ 740 mA, μ 8.1, g_m 0.27 mS. "A" has NB, T-T fil. reduced to 360 mA.

DL-7 Power triode, T12 or T14 bulb, BkB or IB, fil. 5.0 V @ 500 mA, μ 7, g_m 1.03 mS, 65 mW out.

DV-7 Rcvg. triode, IB; see DL-7.

DL-9 Power triode, T14 bulb, IB, fil. 7.5 V @ 1.6 A, μ 6.5, g_m 1.08 mS, 1.5 W out.

DL-14 Power triode, T14 bulb, IB, fil. 5.0 V @ 500 mS, μ 3, g_m 1.1 mS, sim. to UX-171, 600 mW out.

DL-15 Detector triode, possibly an alkali-vapor type, IB, fil. 5.0 V @ 250 mA, μ 17, g_m 0.725 mS.

A note: it has been pointed out [4] that the original De Forest Audion, based on published characteristic curves, had a μ of about 1.4 and a g_m of about 0.012 mS. This is not considered a high order of performance.

DONLE-BRISTOL

Early types made by Connecticut Tel. & Electric Co.; later by Donle-Bristol Corp. All are in short tipped bulbs (ca. T12) on UV bases. Most use a gridlike plate, a regular plate, and an external adjustable magnetic field.

DR-1V Full-wave vacuum rectifier, tipped S14 bulb on UX base sim. to two-plate version of UX-216B..

DA2 Rcvg. triode.

B6 Gas det. triode, fil. and gas heater 5.0 V @ 250 mA.

B8 Detector triode.

S-11 Sodium-vapor detector, tipped spherical bulb, brass UV base, used base-up.

S-13 Sodium-vapor detector, spl. base sim. to UV-199, "non-oscillating," fil. 3.8 V @ 240 mA shared w/ sodium-heater.

D-21 Sodium-vapor detector, UV base, able to oscillate. Bulb is inside nonevacuated, frosted outer bulb for thermal insulation. Heater, connected in series w/ filament, is wound on inner bulb; fil. and heater use 5.0 V @ 250 mA, μ 22 (*QST*, Dec. 1924).

DP71 Power triode.

JR-117 Gridless vacuum detector, tubular bulb w/ ring and pin contacts (*QST*, Dec. 1921).

EMERSON

Multi-valve Triple triode, S14 tipped bulb on flanged UX-style base w/ four auxiliary binding posts for grids and plates; fil. 5.0 V @ 250 mA total. Announced by Emerson Radval Corp. but traceable to Cleartron (here and in Britain) - used in Lodge "N" circuit as grid-leak detector and two audio stages.

JOHNSONBURG

The Johnsonburg Radio Corporation made mainly "industry" types. However, it did devise a couple of its own, as well as being an original source of the registered 2B6.

985 Full-wave mercury rect. sim. to 84, for car radios.

986 Full-wave mercury rect.; one-amp. version of 82, intended to repl. 80; rated at 150 mA.

KEN-RAD

KR1 Mercury single-diode rectifier, heater 6.3 V @ 300 mA, 350 V in, 50 mA, superseded by 1-V <4G>.

KR5 [AKA 6A4, q. v.]; power triode.

KR18 [AKA 18, q. v.]; power pentode.

KR20 [AKA 90]; Wunderlich detector, S12 bulb on 6-pin base, heater 2.5 V @ 1.0 A; like original AS design.

KR22 [AKA 92]; Wunderlich det., KR20 w/ heater 6.3 V @ 400 mA.

KR25 [AKA 2A5, q. v.]; power pentode.

KR26 Detector triode, htr. 2.5 V @ 1.0 A, μ 14, g_m 1.4 mS.

KR28 [AKA 84/6Z4, q. v.]; full-wave rectifier.

KR31 Single-diode mercury rectifier, heater 10.0 V @ 500 mA, 350 V in, 100 mA.

KR48 [AKA 47, q. v.]; power tetrode.

90 [AKA KR20]; Wunderlich detector.

92 [AKA KR22]; Wunderlich det., htr. 2.5 V @ 1.0 A.

95 Power pentode; see 2A5.

96 Diode mercury rectifier; heater 10 V @ 500 mA, 350 V in, 100 mA.

KR98 [AKA 84/6Z4, q. v.]; full-wave rect., equals KR28.

280M Rectifier, mercury variant of 80, uprated from 350 V input to 450.

281M Rectifier, mercury variant of 81.

LESTRON

These were to be AC tubes heated by 110 V, made by the Lestein Corp. of America and advertised in the Radio Trade Directory, Aug. 1925; then re-announced five years later (*Electronics*, Dec. 1930; *Radio-Craft*, April 1931). Had dual heaters wound so as to minimize hum, and UY bases. Plans included 32- and 220-V variants and a tetrode. Actual production unknown; these would be fine collectibles if they turned up.

1012 Rcvg. triode, intended to replace UX-112, μ 8.5, g_m 2.0 mS, 500 mW output @ 180 V.

1048 Power triode, μ 3.5, g_m 2.6-3.0 mS, 2.9 W output @ 250 V.

Unk. Full-wave rectifier.

LOEWE

While this is nominally an analysis of *North American* tubes, it may be of collector interest to show data on the German-British Loewe multi-section types with integral R-C coupling [5], which occasionally become available here.

2HF Dual space-charge tetrode, RC-coupled RF amp., 6-pin special base, fil. 4.0 V @ 160 mA, μ 17.5, g_m

0.5 mS (each of two sections).

3NF Bat — Triple-triode, grid-leak det. and two audio stages, 7-pin special base, fil. 4.0 V @ 320 mA; μ 57, g_m 0.2 mS (for each section). Seventh pin added to allow regenerative detector.

3NF Net — Like 3NF Bat, but w/ expanded output section; fils. 4.0 V @ 375 mA; output section has μ 8, g_m 0.7 mS.

3NFK — Triple-triode, like 3NF but w/ cathodes added to the low-level stages and the grid return of the second stage brought out to a base pin. Heaters 4.0 V @ 3.6 A total; output section gives 12 W @ 300 V.

3NFL — Triple-triode, 3NFK w/ higher g_m of 1.2 mS on detector triode.

3NFW — Triple-triode, 3NFK w/ heater current 3.0 A total, output section gives 6 W out @ 200 V.

WG33 — AC version of 3NFK, all three sections having cathodes, 11-pin base; output section is tetrode; heaters 50 V @ 180 mA.

WG34 — Dual tetrode, one of them an output stage; 12-pin base; heaters 50 V @ 180 mA.

WG35 — RF/IF tetrode, detector diode, and output tetrode; 12-pin base and cap; heaters 65 V @ 180 mA.

WG36 — Two remote-cutoff RF-IF pentodes and an osc. triode; 12-pin base and cap; heaters 65 V @ 180 mA.

MAGNAVOX

A — Gridless triode, w/ "unique control electrode," UX base, fil. 5.0 V @ 220-250 mA ("no tungsten whatever is used"), μ 7-10, g_m 0.7-0.8 mS.

150-D — "Quality" detector triode, odd tapered-tubular bulb on UX base, same ratings as UX-200A.

MAJESTIC

Grigsby-Grunow, as owner of a tube plant, had the ability to make unique tubes for its Majestic radios. In addition to RMA-registered types, and besides the types shown, the following were spray-shielded equivalents of industry types: 24S, 27S, 35S/51S, 55S-58S, 75S.

G2-G2S — Duodiode, heater 2.5 V @ 1.75 A, spray-shield.

G4-G4S — Duodiode, heater 2.5 V @ 1.0 A, spray-shield

4 — [AKA 2/4, q. v.].

56AS — Triode, 56A w/ spray shield.

57AS — Pentode, 57A w/ spray-shield.

G59B — Power tetrode, fil. version of 59.

G84 — [AKA 2Z2, q. v.]; rectifier. Not same as 84/6Z4.

85AS — Duodiode-triode; spray-shielded. Higher μ and g_m than 85; replaceable by 6C7.

McCULLOUGH-KELLOGG

The McCullough (later Kellogg and, in Canada, Rogers) triodes were early types for AC operation. Most of them used a unique tapered/conical bulb with two-pin heater connector at the top. Other small makers produced variants of the 401.

400 — Early triode, UV bakelite or brass base, little known and scarce. See 401.

401, C401 — AC triode, sim. to 201A w/ heater 3 V @ 10 A, μ 8.5 (or 10), g_m 0.8 mS, 165 mW out @ 180 V. Variants: 402, 403. Best known for use in Marti rcvrs. Marvin RX-401 used wires up the outside of the bulb to reach the top connector. Cardon C401, w/ heater 3 V @ 1.3 A, used studs on side of unusually tall base; was in Sparton sets. Sovereign AC tube was tubular variant w/ binding posts on top for heater. SY 401 had ST bulb and heater studs out sides of the UX base. There seems to have been a Speed "401," a variant on a UY base.

402 — Power triode, sim. to 71A w/ heater; heater 3.0 V @ 1.5 A, μ 3, g_m 1.0 mS. No known use.

403 — Power triode, sim. to 71A w/ heater 3.0 V @ 1.5 A; apparently electrically same as 402 but in tubular bulb. No known use.

MOORHEAD

Moorhead Laboratories made a group of early types up to mid-1922 [6]. Those for the U. S. market were on brass-tubing UV bases, w/ sealing tips, and generally carried serial numbers. Most were made under license from De Forest and/or American Marconi (and expiration of that license ended the company). Their base-type products are:

B — Triode made for British use in WW I, variant of "R" valve w/ vertical structure; spherical-bulbed, fil. 6.0 V @ 850 mA, g_m 0.2 mS.

R — Triode made for British use in WW I, horizontal structure; usually w/ Br. base but known w/ Shaw base; fil. 4.0 V @ 700 mA, g_m 0.16 mS.

VT32 — Triode made for British use in WW I; tubular bulb on Br. base; fil. 6.0 V @ 850 mA, g_m 0.2 mS.

SE1444 — U. S. Navy triode (WW I), tubular bulb, fil. 4.0 V @ 700 mA, g_m 0.16 mS.

VT — (Marconi); triodes, tubular bulbs, constructed same as SE1444, fil. 4.0 V @ 700 mA.

Class II (A) — Vacuum type, g_m 0.16 mS.

Class I (B) — Gas detector, variant of "A," fil. 4.0 V @ 700 mA.

VT — Triode, fil. 4.0 V @ 700 mA, g_m 0.16 mS. About 85% have gold-colored getters. Later ones carry Marconi-De Forest logo stamped into base.

ER — Gas triode "electron relay" detector, clear tubular bulb, spiral copper grid, aluminum cylindrical plate (later hairpin grid and nickel plate), fil. 4.0 V @ 700-1000 mA. "Tall plate" variant, made w/ and w/o mica, has 700-mA fil.

TT — Xmtg. triode, fil. 6.0 V @ 800 mA, 80-500 V B+ (distrib. by A-P Tube Co.).

Rectifier — Rectifier, spherical-tubular bulb, very sim. to De Forest "singer" triode less the grid; fil. ca. 9.5 V @ 3.25 A, 75 mA (distrib. by A-P Tube Co.)

QRS RED-TOP

The QRS Music Company of Chicago sold a line of tubes with red lacquer bulb coatings, generally in tubular bulbs on UX bases. "QRS" was marked on the tubes via a logo punched out of a gilt label pasted onto the glass. At least some tubes were on red bases. Internal electrode structure was two sets of elements in parallel, each being coaxial-cylindrical. Types offered included 112, 120, 171, 200A, 201A, and 199, plus four cold-cathode rectifiers sim. to Raytheon BA-BH, a 100-mA 500 V rectifier, and a VR tube.

RAYTHEON

The following cold-cathode rectifiers are in addition to a wide variety of "industry" types made by Raytheon, and to CK-RK-QK numbered types listed in their own section.

B [AKA B-60]; helium half-wave rect., 350 V in, 60 mA.
BA Full-wave rectifier, 320 V in, 350 mA, superseded by (CK) 1009.
BH [AKA BH-85]; full-wave rectifier, 350 V in, 85 mA.
BR Half-wave rectifier, 350 V in, 125 mA.

RADIO PRODUCTS CO.

The Radio Products Company of Newark, NJ made two special tubes in 1932 for a small-run miniaturized cathedral clock-radio sold locally [7]. Bulbs were about T6 size, on miniaturized five-pin bases w/ short pins. One was a triode, a mini-27 w/ 2.5-V heater. The other was a filament-type mercury diode rectifier.

RADIO SCIENTIFIC LABORATORY

These were full-wave mercury rectifiers, w/ tubular bulbs having dual downward arm extensions carrying plate caps (*Radio*, Jan. 1935). Fils. took 2.5 V; PIV was 10 kV.

RSL-9 On UX base, fil. 7 A, 1.2 A out.
RSL-14 On large 4-pin base, fil. 10 A, 2 A out.

ROGERS

Rogers Radio Tubes shared the sprayed-shield technique with Majestic. Besides the types shown, they made spray-shielded equivalents of the following "industry" types: 24, 30, 32-34, 41, 42, 46, 47, 59, 75, 77, 78, 84, 89, 6A7, and 6K7. (24 and 27 were also offered unshielded.) These used the usual suffix "S." "M" indicated "metal spray," an octal-based countermeasure to U. S. metal tubes. Rogers used corrugated grid caps as a distinguishing mark. The following are in addition to RMA-registered types.

R20 Power AC triode, heater 2.8 V @ 1.6 A, tapered bulb, sim. to McCullough 403.
R30 Triode, R32 w/ same tapered bulb, on UY base.
R32 AC triode, heater 2.8 V @ 1.0 A, tapered bulb ,sim. to McCullough 401.
R33 AC triode, heater 2.8 V @ 1.0 A.
85M "Metal spray" variant of 85.
86M Octals, tubular and ST14 bulbs w/ caps.
87S "Canadian type."
88S, M Duodiode-triode (?), ST12 bulb on 6-pin base, spray-shielded. "M" has tubular bulb on octal base.
89RS [AKA 6G7, q. v.]; rectifier-pentode.

SCHICKERLING

These types generally had odd triangular punched-out elements, of unknown functioning, as a trademark. Some had a "stabilizing element" encircling the plate. Later production had bakelite bases. (Note the inconsistent punctuation in the type numbers.) SX-227 was like regular UY-227.

MU-20 Rcvg. triode, fil. 5.0 V, sim. to Daven MU-20.
Z-50 Neon rect, 1925, tipped "S" bulb, brass UV base.
S. 200 Gas detector triode, brass UX base, fil. 5.0 V @ 250 mA, 16-45 V.
S. 300 Detector (?) triode, UV base, fil. 1.5 V @ 250 mA.
TVT 200 Detector triode, tubular tipped bulb on UV base, triple triangular plates.
S. 400 Rcvg. triode, UV base, fil. 3.0 V @ 0.1 A.
SX400 Rcvg. triode, tipped S14 bulb on UX base.
S. 500 Rcvg. triode, UV base, fil. 6.0 V @ 250 mA.
S-600 Rcvg. triode, UV-199 style tipped bulb and brass base, fil. 3-4 V @ 80 mA.
S. 700 Power triode, spherical tipped bulb on brass UV base, fil. 5.0 V @ 250 mA, w/ "stabilizing element."
S-1600 Rcvg. triode, T14 tipped bulb on brass UX base, fil. 5.0 V @ 160 mA.
S-4000 S14 tipped bulb, brass UV base, fil. 5.0 V @ 250 mA, w/ "stabilizing element."
S-8100 Tubular tipped bulb on UV base.

SEARS ROEBUCK "WLS"

Sears sold equivalents of standard types under its "WLS" brand, which celebrated its Chicago radio station of the same letters. For a table of equivalents [8], see p. 8 of this section.

SONATRON

MU-X 6 Power triode, blue bulb on UX base, fil. 5.0 V, μ 6, g_m 1.0 mS.
MU-X 20 Rcvg. triode, silver bulb on UX base, fil. 5.0 V, μ 20, g_m 0.675 (or 0.835) mS, for resistance-coupled amps. Parallels Daven and Schickerling MU-20s.
MU-X 30 Rcvg. triode, red bulb on UX base, fil. 5.0 V @ 250 mA, μ 32, g_m 0.8 mS, for resistance-coupled amps.
SP 60 Rcvg. triode, gold bulb on UX base, fil. 5.0 V, μ 8, g_m 1.0 mS.

SONORA (ARCTURUS)

D½	[AKA 81, q. v.]	**RE1**	[AKA 80, q. v.].
D1	[AKA 80, q. v.]	**RE2**	[AKA 81, q. v.].
DE1	[AKA 27, q. v.]	**SO1**	[AKA (AS) A40, q. v.].
RA1	[AKA (AS) A48, q. v.]	**SO2**	[AKA 50].

SPARTON

Like Grigsby-Grunow, Sparks-Withington had a tube factory under its Cardon subsidiary. The result was some special triodes close to, but not necessarily interchanging with, standard types. Some of these (e. g., 182A, 182B, 183) may be found with number prefixed by "C."

70 See "standard" listing.
181 Power triode, S14 bulb on UX base w/ heater on studs at sides of base, fil. 3.0 V @ 1.35 A, μ 3, 180 V.
182A Power triode, fil. 5.0 V @ 800 mA, μ 3, 200 V, replaceable by 71A.
182B/482B Power triode, variant of 182A, fil. 5.0 V @ 1.25 A, μ 5, 250 V, 1.8 W out (JB, Sparton).
183/483 [AKA 483]; power triode, fil. 5 V @ 1.25 A, μ 3.2, 250 V, sim. to 5-V 45 (JB, Sparton).
373 Rcvg. triode, heater 3 V @ 1.3 A to side pins on base.
482A [AKA 182/482, q. v.].

SEARS "WLS," CECO, AND VAN HORNE EQUIVALENTS OF STANDARD TYPES

RCA/ Industry	Sears WLS	CeCo	Van Horne
CX (WX)-12	4615		*
UV-112	2977/WLS512		
UX-112	2953, 4612	F	5VC*
UX-120		E	*
UX-171	2968, WLS 571	J	
UX-171A	4616	J-71-A	*
UV-199	2951, 4610	B	3VB
UX-199^		C	
UX-199	2952, 4611		3VBX*
UX-199 (med.-size)		BX	
UV-200A "Spl."@		H	
UX-200	2950, 4602		
UX-200A			5VD
UV-201A	2964	A	5VA*
UX-201A BkB	2976, 4600	AX+	5VAX
UX-201A (RF-amp.)		K	
UX-201B		01-B	
UX-210	4618/WLS510	L-10	6P*
UX-222		RF-22	*
UY-224	4603		*
UX-226	4630/WLS526	M	*
UY-227	4631/WLS527	N	*
UX-240	2969 Hi-Mu#	G	*
UX-250	4617/WLS550	L	*
UX-280	4620/WLS580	R-80	*
UX-281	4619/WLS581	R-81	*
Multi-Valve	4613		
Raytheon BH	4621	D-G	BH-125

* Later used "industry" numbering. # Probable.
^ WX-12 style base. + Arcturus also.
@ Non-microphonic construction.

483 Power triode, fil. 5.0 V @ 1.25 A, 250 V, 2.0 W out.

484, A Rcvg. triode, heater 3 V @ 1.3 A, μ 12.5, for sets operated from 32-V DC; superseded by 485.

485 Rcvg. triode, variant of 27, heater 3.0 V @ 1.25 A (JB, Sparton).

486 Power triode, blue bulb, typically unmarked, nominally on 4-pin base but has been observed on UY base w/ one pin unused, fil. 3.0 V @ 250 mA.

585 Power triode, 50 w/ oxide fil. and μ reduced to 3.8.

586 [AKA 585, q. v.]; electrically same.

686 Battery triode, fil. 3 V @ 250 mA, supersed. by 486.

SPEED "TRIPLE-TWIN"

These are two-section types: a triode cathode-follower as power-detector or plain amp., direct-coupled to an output triode. The sections are built independently; input section has heater while output section has filament (RMA base diagram 5G notwithstanding). All are on UY bases w/ caps. Promoted for disc-TV use on the basis of flat freq. response. "Triple-Twin" refers to claims of triple the 45's output and twice the 47's.

291 DC-line version, S14 bulb, fil. 12.6 V @ 300 mA, input section μ 6.8, g_m 0.78 mS; output μ 11.2, g_m 2.55 mS <5G>.

293 Automobile version, S12 bulb, heater 6.3 V @ 600 mA, 1.25 W out @ 250 V <5G>.

295 AC-line version, S21 bulb; total power 2.5 V @ 4 A, input section μ 14.4, g_m 1.2 mS; output μ 13, g_m 4.35 mS, 4.5 W out @ 250 V <5G>. Led to 2B6, 6N5, et al.

VAN HORNE

The Van Horne Tube Co. made "standard" products under its own numbering [9]. Equivalents are tabulated at left.

REFERENCES

1. L. A. Sibley, "A Glimpse into CeCo," *Radio Age*, Vol. 20 No. 3 (March 1995), p. 3.
2. B. Kelley, "Crosley Deflectron Tubes," *Old Timer's Bulletin*, Vol. 31 No. 3 (Oct. 1990), p. 48.
3. W. Condon, "The De Forest DV-DL Series," *Old Timer's Bulletin*, Vol. 23 No. 4 (March 1983), pp. 12-16.
4. G. Symonds, "The Audion Revealed," *MAARC Newsletter*, Vol. 8 No. 4 (April 1991), p. 6.
5. L. A. Sibley, "The Loewe Multi-Section Tubes," *Old Timer's Bulletin*, Vol. 29 No. 4 (Feb. 1989), p. 28.
6. W. Condon, "Moorhead and the Electron Relay," *Old Timer's Bulletin*, Vol. 29 Nos. 3, 4 (Nov. 1988, Feb. 1989), pp. 12-14, 14-16.
7. A. S. Douglas, "Radio Products Co.," *Radio Age*, July 1990, pp. 3-4.
8. D. Craft, "WLS Tubes," *Arizona Antique Radio News*, Vol. XIII No. 2 (Spring 1996), pp. 14-17.
9. J. Cross, "The Van Horne Tube Company," *Old Timer's Bulletin*, Vol. 35 No. 2 (May 1994), pp. 8-13.

House-brand tubes are interesting collectibles. These are from the Bamberger and Macy department-store chains.

RECEIVING TYPES UNDER RMA-EIA SYSTEM

0A2, WA [AKA WT-210-0051, 150C2, STV150/30]; min. voltage regulator, 150 V @ 5-30 mA (HY) {S856, HD51} (6-8-45) <5BO>. "WA" is rated at 148 V. Variants: 6073, 6626, 6830.

0A3 [AKA VR-75/30, 38275, VT-260]; octal voltage regulator, 75 V @ 5-30 mA (HY) (3-6-44) <4AJ>.

0A4G Cold-cathode octal triode, promoted for use in calculating machines and RF remote-control systems operating via AC power lines; 225 V, 25 mA (4-25-38) <4V>. Variant: 1267.

0A5 Min. cold-cathode relay pentode, for triggering photoflash tubes; 1 kV max. (SY) (1-15-47). Used with flash tubes like SY R4330.

0B2, WA Min. voltage regulator, 108 V @ 5-30 mA (HY) {HD52, S860} (5-18-45) <5BO>. Variants: 5787, 6074, 6627, 6831.

0B3 [AKA VR-90, VR-90/30, 38290, VT-184, WT-210-0016]; octal voltage regulator, 90 V @ 5-40 mA (HY) (9-14-42) <4AJ>. Variant: 1265.

0C2 Min. voltage regulator, 75 V @ 5-30 mA (SY) {S859} (3-31-58) <5BO>.

0C3, A, W [AKA VR-105, VR-105/30, 38205, VT-200, CV686]; octal voltage regulator, "105" (108) V @ 5-30 mA (HY) (3-6-44) <4AJ>. "A" has T9 bulb. "W" has T9 bulb inside deep Micanol cup base.

0D3, A, W [AKA VR-150, VR-150/30, 38250, VT-139, CV216, WT-210-0018]; octal voltage regulator, 150 V @ 5-30 mA (HY) (9-14-42) <4AJ>. For "A" and "W," see 0C3.

0G3 [AKA 85A2, STV85/10]; min. voltage regulator, 85 V @ 1-10 mA (5-28-52).

0Y4, G Gas rectifier, ionically heated cathode w/ starter electrode, 300 V PIV; gives 40-75 mA from 117 V; promoted to replace 117Z6GT (RA) (9-20-45) <4BU>. "G" is in 0Z4G format.

0Z3 Gas rectifier, ionically heated cathode, 350 V applied, 75 mA, intro. 1933 (RA) (6-17-35) <5N>. No known use.

0Z4, A, G Full-wave gas rectifier, ionically heated cathode, 1.6 kV PIV, 30-75 mA (RA) (10-24-35). Uprated to 90 mA in 1945. "G" is small (T7) size. "A" [AKA (CK) 1003] is uprated to 110 mA (RA) (10-15-56) <4R>.

1A1 Ballast tube, 1.0 V @ 500 mA (SY) (10-26-34).

1A3 Min. HF diode, fil. 1.4 V @ 150 mA, 330 V PIV, 5 mA (RCA, W) (5-9-42) <5AP>.

1A4, P, T Rcvg. remote-cutoff pentode, ST12 bulb w/ cap, 4-pin base, fil. 2.0 V @ 60 mA, 1D5 char. (RCA) {A4028} (5-21-35) . "P" and "T" lettering derives from industry confusion: SY introduced 1A4 as a tetrode <4K>; others as a pentode <4M>. The pentode version eventually won out. The two should interchange today.

1A5GT/G [AKA VT-124, CV756]; power pentode, fil. 1.4 V @ 50 mA, g_m 0.8 mS, 100 mW out (1LA4 char.); needs 90 V B+; sim. to 1Q5GT, 1T5GT (SY) (6-10-38) <6X>. 1A5G was in the first mass-market series of 1.5-V 50-mA tubes, announced by SY in mid-1938; the others: 1A7G, 1C5G, 1H5G, and 1N5G.

1A6 Pentagrid conv., same as 1C6 except for capacitance (RCA) {A564} <6L>.

1A7GT/G [AKA VT-147, CV2]; pentagrid conv. sim. to 1B7GT, 1LA6 (6-10-38) <7Z>.

1AB5 [AKA SD719]; loktal HF pentode, "high" g_m of 1.35 mS (SY, W) (3-20-44) <5BF>.

1AB6 [AKA DK96]; min. pentagrid conv., fil. 1.4 V @ 25 mA (HY, TF) (10-16-51) <7DH>. Also see 1AC6.

1AC5 T3 submin. pentode, part of early SY-RCA line for AM radios including 1AD6, 1E8, and 1T6; fil. 1.25 V @ 40 mA, 50 mW out (12-31-48) <8CP>.

1AC6 [AKA DK92]; min. pentagrid conv., fil. 1.4 V @ 50 mA (HY, TF) (10-16-51) <7DH>. Appears to be re-filamented 1AB6.

1AD4 [AKA DF62]; T2X3 submin. RF-AF pentode, spray-shielded, 1.25 V @ 100 mA, g_m 2.0 mS (RA, SY, TF) {QF-408} (1-27-49). Variant: 5875.

1AD5 T3 submin. sharp-cutoff pentode, fil. 1.25 V @ 40 mA (12-31-48) <8CP>.

1AD6 Submin. pentode, fil. 1.25 V @ 40 mA; see 1AC5.

1AE4 Min. pentode, sharp cutoff, fil. 1.25 V @ 100 mA, g_m 1.6 mS (HY) (3-21-50) <6AR>.

1AE5 Submin. mixer, fil. 1.25 V @ 60 mA {QF-409} (1-27-49).

1AF4 [AKA 1AJ4]; min. remote-cutoff RF pentode, fil. 1.4 V @ 25 mA, g_m 1.0 mS (HY, SY) (2-8-50) <6AR>.

1AF5 [AKA 1AH5]; min. diode-pentode, fil. 1.4 V @ 25 mA, g_m 1 mS (HY, SY) (2-8-50) <6AU>.

1AG4 T2X3 submin. power pentode, fil. 1.25 V @ 40 mA, g_m 1.0 mS, 35 mW out (RA, SY, W) (8-16-54).

1AG5 T2X3 submin. diode-pentode, fil. 1.25 V @ 40 mA, g_m 0.25 mS (SY, W) (9-12-55).

1AH4 T2X3 submin. flat-press RF pentode, spray-shielded, fil. 1.25 V @ 40 mA, g_m 0.75 mS (4-23-51).

1AH5 [AKA DAF96, 1AF5, q. v.]; min. diode - AF pentode, fil. 1.4 V @ 25 mA, g_m 0.55 mS (HY, TF) (10-16-51) <7DJ>.

1AJ4 [AKA DF98]; remote-cutoff RF pentode, fil. 1.4 V @ 25 mA, g_m 0.75 mS (HY, TF) (5-29-53) <6AU>. Also see 1AF4.

1AJ5 T2X3 submin. diode-pentode, fil. 1.25 V @ 40 mA, g_m 0.425 mS (TS, W) (8-16-54).

1AK4 T2X3 submin. RF pentode, fil. 1.25 V @ 20 mA, g_m 0.75 mS (RA, SY) (8-16-54).

1AK5 T2X3 submin. diode-pentode, fil. 1.25 V @ 20 mA, g_m 0.28 mS, sim. to 5804 (RA, SY) (8-16-54).

1B1 Ballast tube, 1 V @ 360 mA (NU, SY) (5-21-35).

1B3GT HV rectifier, T9 bulb w/ cap on octal base, fil. 1.25 V @ 200 mA; 8016 uprated to 30 kV, later derated to 26 kV, and from 2 mA to 0.5 mA (RCA) {R6158A} (6-28-46). Variant: 6215. History: developed from prewar 8016. Announced in 1946 w/ 1" 8-lead button stem. Had silver type numbers on side until 1947; these were found to contribute to electrolysis of glass; base-brand only thereafter. Changed to 4-lead pressed stem, 1949. Got lead-glass bulb in 1951 to reduce X-rays, new flatter stem shield, and removal of mica spacers and two stem leads. White-banded variant appeared, 1952, using alundum coat to avoid gas from bulb. Red-etched (1955-58) w/ non-metallic ink on hydrogen-fired bulb. 1B3GT/-1G3GT is shortened version, derated on PIV, introduced later (3-25-57) <3C>.

1B4P, T Sharp-cutoff pentode, 1E5GP char., 2.0 V @ 60 mA, g_m 0.65 mS, repl. 951 (RCA) {A4027} (8-28-35) <4M>. "T" was tetrode version <4K>. See 1A4 re "P" vs. "T."

1B5/25S Duodiode-triode, 1H6G char. (RCA) {A483} <6M>.

1B7GT/G Pentagrid conv. sim. to 1A7GT, fil. 1.4 V @ 100 mA (7-28-39) <7Z>.

1B8GT Diode-triode-pentode (9-15-39) <8AW>.

1C1 Ballast tube, 1 V @ 745 mA (SY) (10-26-34).

1C3 Min. triode based on 1LE3, "export" type, fil. 1.4 V @ 50 mA, μ 14.5 (HY, SY) (8-15-49) <5CF>.

1C4 Pentode, sim. to 1A4P (Aust.) (12-10-35) <4K>.

1C5GT/G [AKA VT-125, CV1805]; pentode, 100 mA; more output than 1A5 but requires 90 V; sim. to 1Q5GT, 1T5GT (TS) (6-10-38) <6X>.

1C6 Pentagrid conv., 1C7G char. (RCA) {A593} (2-5-34) <6L>.

1C7G Pentagrid conv., octal 1C6 (RCA) {A5507} (5-4-36) <7Z>.

1C8 T3 submin. fil. triode-heptode conv. - part of an AM radio complement with 1Q6, 1V5, and 1W5 (SY) (1-25-46) <8CN>.

1D1 Ballast tube, 1.0 V @ 240 mA (SY) (10-26-34).

1D2 Ballast tube (9-21-36).

1D3 T3 submin. low-μ RF triode, fil. 1.25 V @ 30 mA, μ 8.7 (SY) {SD1066A} (9-21-50) <8DN>.

1D4GT/G Power pentode, fil. 2.0 V @ 240 mA (Aust.) (12-10-35) <5B>.

1D5G, GP, GT Remote-cutoff pentode, ST12 bulb w/ cap, octal 1A4P; fil. 2.0 V @ 60 mA, g_m 750 mS, sim. to 1E5G (RCA) {A4042} (5-4-36) <5Y>. "GT" is tetrode version <5R>; see 1A4 re "GP" vs. "GT." In this case, "GT" does not imply a T9 bulb!

1D7G Pentagrid conv., octal 1A6 (RCA) {A5508} (5-4-36) <7Z>.

1D8GT [AKA VT-148, CV1811]; diode-triode-pentode, fil. 1.4 V @ 100 mA, triode μ 25, pentode g_m 0.875 mS, 200 mW out (RCA) (6-23-39) <8AW>.

1DN5 Min. diode-pentode, 1S5 w/ pinout scrambled (GE, RCA, W) (4-6-57) <6BW>.

1E1 Ballast tube, 1.0 V @ 480 mA, supersedes 1P1 (NU, SY) (5-21-35).

1E2 Ballast tube (9-21-36).

1E3 [AKA DC80]; min. 9-pin UHF power triode, fil. 1.25 V @ 220 mA, g_m 3.5 mS (HY) (10-16-51) <9BG>.

1E4G Triode sim. to 1H4G and 1LE3, fil. 1.5 V @ 50 mA, μ 14.5 (SY) (2-10-39) <5S>.

1E5G, GP, GT [AKA VT-170, CV45]; RF pentode, octal 1B4P, ST12 bulb w/ cap, sim. to 1D5GP; repl. 1E5G and 1E5GT (RCA) {A4027} (5-4-36) <5Y>. "GT" is tetrode version <5R>; another non-T9 "GT." See 1A4 re "GP" vs. "GT."

1E7G [AKA 38717E]; twin power pentode, ST12 bulb, fil. 2.0 V @ 240 mA, g_m 1.6 mS, 575 mW out (RCA) {A5510} (3-26-36) <8C>.

1E8 T3 submin. conv., fil. 1.25 V @ 40 mA (SY) (12-31-48) <8CN>. Also see 1AC5.

1F1 Ballast tube, 1.0 V @ 720 mA (SY) (7-9-35).

1F4 Power pentode, 5-pin base, 1F5G char. (RCA) {A4029} (2-1-36) <5K>.

1F5G Power pentode, octal 1F4, ST14 bulb, fil. 2.0 V @ 120 mA, g_m 1.7 mS, 300 mW out (RCA) {A4044} (5-4-36) <6X>.

1F6 Duodiode-pentode, ST12 bulb w/ cap, 6-pin base, fil. 2.0 V @ 60 mA, g_m 0.65 mS (RCA) {A518} (4-14-36) <6W>.

1F7G, GH, GV Duodiode-pentode, octal 1F6 w/ cap. Versions are: "G" (RCA) {A523} (6-12-36) <7AF>; "GH" (TS) (3-18-38); "GV" (RCA) (6-7-37). "GV" has one diode above the other at the (-) end of the fil.

1G1 Ballast tube, 1.0 V @ 420 mA (SY) (7-9-35).

1G4GT/G Low-μ octal triode, fil. 1.4 V @ 50 mA, μ 8.8, sim. to 1E4G, 1H4G (1-20-39) <5S>.

1G5G Power pentode, ST14 bulb, octal base, sim. to 1J5G, fil. 2.0 V @ 120 mA, g_m 1.5 mS, usable with low B+ voltage (SY) (3-9-37) <6X>.

1G6GT/G Twin Class B power triode, T9 bulb on octal base, fil. 1.4 V @ 100 mA, g_m 0.675 mS, 675 mW out @ 90 V (1-20-39) 7AB>.

1H4G Triode, octal 30, ST12 bulb (RCA) {A2041} (5-4-36) <5S>.

1H5GT/G [AKA VT-223, CV104]; diode-triode, fil. 1.4 V @ 50 mA, μ 65; triode has 1LH4 char. (RCA) (6-10-38) <5Z>.

1H6G Duodiode-triode, octal 1B5, fil. 2.0 V @ 60 mA, μ 20 (RCA) {A4045} (5-4-36) <7AA>.

1J1 Ballast tube (9-25-35).

1J5G Power pentode, octal 950 (3-31-37) <6X>.

1J6GT/G, High-μ power duotriode, octal 19 (RCA) {A4046}

GX (5-4-36) <7AB>.

1K1 Ballast tube, 1.0 V @ 550 mA, repl. 5H1, 10A3, LLL25 (SY) (5-4-36).

1K4 Duodiode-pentode, 1K5G char. (Aust.) (1-20-37).

1K5G Duodiode-pentode (Aust.), octal w/ cap (3-22-38). In rcvr. and xmtr. of W. S. No. 101 of 1938.

1K6 "Aust. type" (9-30-36).

1K7G Aust. pentagrid conv., octal w/ cap, same use as 1K5G (3-20-38); also in W. S. No. 11 (Aust.).

1L4 [AKA DF92]; min. sharp-cutoff RF-IF pentode, for FM rcvrs, based on 1N5, fil. 1.4 V @ 50 mA, g_m 0.93 mS (SY, TF) (5-20-42) <6AR>.

1L5G "Aust. type." In xmtr. of Aust. FS6 xcvr. (3-22-38).

1L6 Min. pentagrid conv., fil. 1.4 V @ 50 mA, based on 1LA6. Favored over 1R5 based on SW performance (4-5-49) <7DC>. Variant: 1U6.

1LA4, E Power pentode, loktal 1A5GT (NU) <5AD>.

1LA6 Loktal pentagrid conv. sim. to 1A7G, 1C7G, 1LC6 (5-16-39) <7AK>.

1LB4 Loktal power pentode, fil. 1.4 V @ 50 mA, g_m 0.875 mS (char. of pentode of 1D8), suited to B+ as low as 60 V (9-4-39) <5AD>.

1LB6 Loktal pentagrid converter, fil. 1.4 V @ 50 mA (SY) (1-24-40) <8AX>.

1LC5 Loktal sharp-cutoff RF-IF pentode, suited to B+ as low as 45 V, fil. 1.4 V @ 50 mA, g_m 0.75 mS (5-8-40) <7AO>.

1LC6 [AKA VT-178, CV3]; loktal pentagrid conv. sim. to 1LA6, suited to B+ as low as 45 V, fil. 1.4 V @ 50 mA (5-8-40) <7AK>.

1LD5 Loktal diode - audio pentode, pentode sim. to 1LN5, suited to B+ as low as 45 V (5-8-40) <6AX>.

1LE3, Spl [AKA VT-239, CV46]; loktal med.-μ triode, fil. 1.4 V @ 50 mA, μ 14.5 (4-15-40) <4AA>. "Special," used in Army I-107F VTVM, is supersed. by 1LF3.

1LF3 Loktal triode, 1LE3 w/ low grid-current limits (4-10-48) <4AA>.

1LG5 Loktal remote-cutoff pentode, fil. 1.4 V @ 50 mA, g_m 1.0 mS (SY) (12-20-45) <7AO>.

1LG6 Loktal, postwar (SY), otherwise unknown.

1LH4 [AKA VT-177, CV780]; diode-triode, loktal 1H5G (SY) (5-16-39) <5AG>.

1LN5, E [AKA VT-179, CV781]; loktal sharp-cutoff RF-IF pentode sim. to 1N5, fil. 1.4 V @ 50 mA, g_m 0.8 mS (NU) (5-16-39) <7AO>.

1M5G Aust. type; RF pentode, octal w/ cap, fil. 2.0 V @ 120 mA, used in rcvr. of W. S. No. 11 (Aust.) (3-22-38) <5Y>.

1N1 Ballast tube (3-18-36).

1N5GT/G [AKA VT-146, CV37]; sharp-cutoff RF-IF pentode, fil. 1.4 V @ 50 mA, g_m 0.75 mS (SY) (6-10-38) <5Y>.

1N6GT Diode - power pentode, fil. 1.4 V @ 50 mA, g_m 0.8 mS, 100 mW out (SY) (4-12-39) <7AM>.

1P1 Ballast tube replaceable by 1E1 (3-18-36).

1P5GT/G RF remote-cutoff pentode, fil. 1.4 V @ 50 mA, g_m 0.75 mS, eff. same as 1N5 (3-28-39) <5Y>.

1Q1 Ballast tube (3-18-36).

1Q5GT/G Power pentode, fil. 1.4 V @ 100 mA, g_m 2.2 mS, 400 mW out, sim. to 1C5GT, 1T5GT, 1A5GT (2-17-39) <6AF>.

1Q6 T3 submin. diode-pentode, 1.25 V @ 40 mA, g_m 0.6 mS (SY) (1-25-46) <8CO>.

1R1G Ballast tube, 1 V @ 540 mA (SY) (5-26-36).

1R4 [AKA 1294, VT-183, CV91]; loktal VHF diode, fil. 1.4 V @ 150 mA, 365 V PIV, 1 mA (SY) (2-1-43) <4AH>.

1R5 [AKA VT-171, CV782, DK91]; min. pentagrid conv. based on 1A7, fil. 1.4 V @ 50 mA (RCA) (11-13-39) <7AT>; part of the first complement of min. AM radio tubes, with 1S4, 1S5, and 1T4.

1R6 [AKA RTS-98, VC1568]; "min. fil. reactance modulator" (unverified).

1S1G Ballast tube (6-17-36).

1S4 [AKA VT-210, CV181]; min. power pentode based on 1Q5, fil. 1.4 V @ 100 mA, g_m 1.575 mS, 270 mW out (11-13-39) <7AV>. Also see 1R5.

1S5 [AKA VT-172, CV124, DAF91]; min. diode - sharp-cutoff pentode, fil. 1.4 V @ 50 mA, g_m 0.63 mS, sim. to 1LD5, 1LH4, 1U5 (11-13-39) <6AU>. Also see 1R5.

1S6 T3 submin. diode-pentode, fil. 1.25 V @ 40 mA, g_m 0.3-0.6 mS (SY) (1-18-49) <8DA>.

1SA6GT Octal med.-cutoff RF pentode, fil. 1.4 V @ 50 mA, g_m 0.95 mS (TS) (4-22-41) <6BD>.

1SB6GT Octal diode-pentode, fil. 1.4 V @ 50 mA, g_m 0.65 mS (SY) (4-22-41) <6BE>.

1T1G Ballast tube, 1.0 V @ 560 mA (SY) (5-26-36).

1T4, WA [AKA VT-173, CV35, DF91]; min. remote-cutoff pentode based on 1P5, sim. to 1LC5, fil. 1.4 V @ 50 mA, g_m 0.8 mS (SY) (11-13-39) <6AR>. Also see 1R5.

1T5GT Beam power tube, fil. 1.4 V @ 50 mA, g_m 1.1 mS, 170 mW out, sim. to 1A5GT, 1C5GT, 1Q5GT (SY) (5-19-39) <6X>.

1T6 T3 submin. diode-pentode, fil. 1.25 V @ 40 mA, g_m 0.6 mS (SY) (12-31-48) <8DA>. Also see 1AC5.

1U1 Ballast tube (11-13-36).

1U4, WA [AKA M16-T-51970, DF904]; min. sharp-cutoff RF-IF pentode based on 1LN5 (or 1N5). Same fil. and pinout as 1L4, different curves, g_m 0.9 mS (TS) {DT153} (8-15-45) <6AR>. Variant: 5910.

1U5 Min. diode-pentode; 1S5 w/ pinout scrambled (SY) (10-25-45) <6BW>.

1U6 Min. pentagrid conv., fil. 1.4 V @ 25 mA, like half-current 1L6 (2-8-50) <7DC>.

1V5 T3 submin. pentode, fil. 1.25 V @ 40 mA, g_m 0.75 mS (SY) (1-25-46) <8CP>; see 1C8.

1V6, S T2X3 submin. triode-pentode osc.-mixer, fil. 1.25

V @ 40 mA, triode section has g_m 0.55 mS; pentode has 0.2 mS. Also see 1C8 (GE, SY, W) (4-23-51). "S" has short leads for socket use.

1W1 Ballast tube (1-17-38).

1W4 Min. power pentode, based on 1LB4, fil. 1.4 V @ 50 mA, g_m 0.925 mS, 200 mW out (SY, TS) (8-19-49) <9U>.

1W5 T3 submin. fil. pentode, fil. 1.25 V @ 40 mA, g_m 0.735 mS (SY) {SD828F} (1-25-46) <8CP>. Also see 1C8.

1Y1 Ballast tube, 1.0 V @ 540 mA (SY) (9-11-36).

1Y2 HV rectifier, ST14 bulb w/ cap on 4-pin base, fil. 1.25 V @ 265 mA, 50 kV PIV, 2 mA (CH, SY) (9-6-46) <4P>.

1Z1 Ballast tube, 1.0 V @ 900 mA (SY) (9-11-36).

1Z2 Min. rectifier, w/ cap, fil. 1.25 V @ 265 mA, 20 kV PIV, 2 mA (RCA) {R1045} (8-10-45) (CH, SY, TS) <7CB>.

2A3, A, H, W [AKA 38213, VT-95, CV1831]; power triode, ST-16 bulb, UX base, fil. 2.5 V @ 2.5 A, μ 4.2, 15 W (RCA, SY, TS) {A181} <4D>. Variants: 6A3, 6A5G, 6B4G, 5930. "A" and "W" are ruggedized. "H" is heater version (AS, NU, RA) (5-6-41).

2A4G Argon triode octal thyratron, ST12 bulb, fil. 2.5 V @ 2.5 A, 200 PIV, 100 mA (CH, HY, TS) (9-27-39) <5S>. Stepping-switch controller in Philco "Mystery Control" receiver; also promoted as grid-controlled rectifier.

2A5 [AKA KR25, 38215]; power pentode, heater 2.5 V @ 1.75 A, 11 W (42 char.) (RCA) {A451} <6B>.

2A6 Duodiode-triode, 75 & 6B6G char., heater 2.5 V @ 800 mA, μ 100 (RCA) {A4006} <6G>. Variant: 1659.

2A7, S Pentagrid conv., 2.5-V 6A7 (RCA) {A508} (11-22-34) <7C>. "S" has spray-shield.

2B4 Thyratron, heater 2.5 V @ 1.5 A, replaceable by 885 (DuM) (1-2-43) <5A>.

2B6 Power amp., direct-coupled duotriode, htr. 2.5 V @ 2.25 A, μ 18, 4.0 W out (AS, JB) (5-6-41) <7J>.

2B7 [AKA 38227]; duodiode-pentode, 2.5-V 6B7 (RCA, TS) {A557} <7D>.

2C4 Min. gas triode thyratron, heater 2.5 V @ 600 mA, 350 V PIV, 5 mA (HY) (7-14-43) <5AS>.

2E5 Electron-ray tuning ind., 2.5-V 6E5 (4-24-36).

2F7 Triode-pentode, 2.5-V 6F7 (KR).

2G5 Electron-ray tuning indicator, 2.5-V 6G5 (TS) (4-24-36).

2V3G Diode rectifier, octal w/ cap, for CRT HV, fil. 2.5 V @ 5 A, 16.5 kV PIV, 2 mA (RCA, TS) (1-9-39) <4Y>.

2W3, GT Rectifier, metal, fil. 2.5 V @ 1.5 A, 350 V applied, 55 mA (NU) (5-2-38) <4X>. "GT" has T9 bulb (6-6-39).

2X2, A, W [AKA 143D, 879, VT-119, CV248]; HV rectifier, UX base w/ cap, for CRT HV and radar "keep-alive" supplies, fil. 2.5 V @ 1.75 A, PIV 12.5 kV, 7.5 mA (RCA) (7-6-39) <4AB>. "A" [AKA 2B21] is ruggedized (GE, HY, RCA, SY) {R612G} (8-23-46). "W" is ruggedized further.

2Y2 HV rectifier sim. to 2X2, octal, heater 2.5 V @ 1.75 A, 4.4 kV in, 5 mA (NU) (12-5-38) <4P>.

2Y3, 2Y4 "Canadian type."

2Z2 [AKA G84]; diode rectifier, fil. 2.5 V @ 1.5 A, 350 V applied, 50 mA (Maj., NU) <4B>.

3A4 [AKA DL93]; min. power pentode, fil. 2.8 VCT @ 100 mA, g_m 1.9 mS, 2 W (5-9-42) <7BB>. Variant: 1662.

3A5 [AKA DCC90]; min. dual HF power triode, fil. 2.8 VCT @ 110 mA, μ 15, 2 W total (RCA, SY) (5-9-42) <7BC>.

3A8GT [AKA VT-149]; diode-triode-pentode, octal w/ cap, fil. 2.8 VCT @ 50 mA, equals 1H5 + 1N5 (TS) (8-21-39) <8AS>.

3B4, WA [AKA DL98]; min. VHF power tube, fil. 2.5 VCT @ 165 mA, g_m 4.3 mS, 2.7 W (HY) {HD30} (4-30-48) <7CY>. Final amp. in AN/PRC-6.

3B5GT Beam power tube, fil. 2.8 VCT @ 50 mA, g_m 1.5 mS, for 67½-V B+, 200 mW out (TS) (4-22-41) <7AO>.

3B7, Spl [AKA 1291, VT-182, CV228]; loktal UHF power duotriode, fil. 2.8 VCT @ 50 mA, μ 20, 5.4 W total, repl. 1292 (2-1-43) <7BE>. "Special" is selected. Variants: 1288, 1291.

3C4 [AKA DL96]; min. power pentode, fil. 2.8 VCT @ 50 mA, g_m 1.25 mA, replaceable by 3V4 (HY) (10-6-51) <6BX>.

3C5GT Beam power tube, fil. 2.8 VCT @ 50 mA, g_m 1.55 ma, 260 mW out (6-6-39) <7AQ>.

3C6 [AKA XXB]; loktal duotriode, fil. 2.8 VCT @ 50 mA, μ 14.5 (11-30-44) <7BW>.

3D6 [AKA 1299, VT-185, CV187]; loktal beam power tube., fil. 2.8 VCT @ 110 mA, 4.5 W, usable to 50 MHz (SY) (2-1-43) <6BA>.

3E5 Min. beam power tube, fil. 2.8 VCT @ 25 mA, g_m 1.3 mS, 175 mW out (HY, TS) (2-5-50) <6BX>.

3E6 Loktal sharp cut-off pentode, fil. 2.8 VCT @ 50 mA, g_m 2 mS (SY, TS, W) (2-18-46) <7CJ>.

3LE4 Loktal beam power tube, fil. 2.8 VCT @ 50 mA, g_m 1.6 mS, 300 mW out (SY) (7-18-40) <6BA>.

3LF4 [AKA VT-180, CV187]; loktal beam power tube, loktal 3Q5 (SY) (10-22-41) <6BB>.

3Q4 [AKA VT-264, CV182, DL95]; min. power pentode, min. 3Q5 (4-1-41) <7BA>.

3Q5GT/G [AKA VT-221, CV185]; beam power tube, fil. 2.8 VCT @ 50 mA, g_m 2.2 mS, 270 mW out (SY) (6-30-39) <7AP>.

3S4 [AKA VT-174, CV183, DL92]; min. power pentode based on 3Q5 - or 1S4 w/ dual fil. - fil. 2.8 VCT @ 50 mA, g_m 1.6 mS, 270 mW out (11-22-40).

3V4 [AKA DL94]; power pentode, 3Q4 w/ pinout scrambled (SY) (10-25-45) <7BA>.

4A6G Power duotriode, for Class B use, fil. 4.0 VCT @ 60 mA, μ 20, 1 W out (NU) (5-2-38) <8L>.

5A6 Min. 9-pin xmtg beam power tube, fil. 5.0 VCT @ 230 mA, 5 W, 100 MHz (HY, SY, TS) {DT275} (4-11-49) <9L>.

5AR4 [AKA GZ34]; full-wave rectifier, T11 bulb, heater 5.0 V @ 1.9 A, 1.7 kV PIV, 130 mA (SY) (1-16-56) <5L>.

5AS4, A Full-wave rectifier, variant of 5U4GB in ST-16 bulb (GE, SY) (1-3-55). "A" has T12 bulb.

5AT4 Full-wave cathode-type rectifier, T16 bulb on octal base, heater 5.0 V @ 5.5 A, 1.55 kV PIV, 800 mA (SY, TS) (2-16-61) <5L>.

5AU4 Full-wave rectifier, T12 bulb, fil. 5.0 V @ 3.75 A, 1.4 kV PIV, 200 mA (GE, SY) (4-16-54) <5T>.

5AZ4 Full-wave rectifier, loktal 5Y3GT (GE, SY) (11-25-46) <5T>.

5R4GY, GYA, GYB Full-wave rectifier, ST-16 bulb on Micanol octal base, fil. 5.0 V @ 2.0 A, high PIV at high altitude (RCA) {R6175B} (9-10-42) <5T>. PIV is graded: 2.8 kV @ 150 mA, or 2.4 kV @ 175 mA, or 2.1 kV @ 250 mA. "WGY" is ruggedized "GY" w/ conventional base. "GYA" (5-2-55) has T12 bulb, is uprated to 2.8 kV @ 165 mA. "GYB" ((3-16-59) is uprated to 3.1 kV (@ 20,000 ft.). "WGA" is ruggedized w/ T14 hard-glass bulb and skirt base. "WGB" is ruggedized "GYA" w/ hard glass, shock-isolated skirt base, and special processing. Common in 1943-56 military equipment but largely unknown to radio restorers; where the large bulb will fit, makes a long-lived replacement for 5Y3G.

5T4 [AKA VT-114, CV276]; full-wave rectifier, 5U4G ratings (RCA) {R6022} (2-4-37) <5T>.

5U4G, GA, GB, WG [AKA VT-244, CV276]; full-wave rectifier, octal 5Z3, fil. 5.0 V @ 3.0 A, 1.55 kV PIV, 225 mA, sim. to Br. VU71 (RCA, TF) (7-27-36) <5T>. "WG" is ruggedized "G." "GA" (4-23-54) has T11 bulb and is uprated to 250 mA; "GB" (4-30-54, only a week after "GA") has T12 bulb w/ button stem and radiating fins on plate; is uprated to 275 mA. Variant: 5931.

5V4G, GA, GY [AKA VT-206A, CV729]; full-wave cathode-type rectifier, octal 83-V, ST14 bulb, heater 5.0 V @ 2.0 A, 1.54 kV PIV, 175 mA (RCA) {R6030} (4-28-36) <5L>. "GA" (4-16-54) has T12 bulb.

5W4, GT/G [AKA VT-97, CV1849]; metal full-wave rectifier, fil. 5.0 V @ 1.5 A, 1.4 kV PIV, 100 mA (RCA) {R6009} (5-1-36) <5T>. "GT/G" (6-13-40) is in T9 bulb.

5X3 Full-wave rectifier, fil. 5.0 V @ 2.0 A, 1275 V in, 30 mA, called a "TV" type (AS) (12-15-38) <4C>.

5X4G, GA Full-wave rectifier, 5U4G w/ pinout scrambled, reportedly to prevent substitution of a (lower-rated) 5Y3. Common in (unique to?) '30s Philcos (RCA) {A6031} (5-26-36) <5Q>.

5Y3, G, GT/G, WGT, WGTA, WGTB [AKA VT-197A, CV1268]; full-wave rectifier, octal 80, fil. 5.0 V @ 2.0 A, 1.4 kV PIV, 125 mA (6-2-41) <5T>. "GT/G" is in T9 bulb; "WGT" and "WGTA" are ruggedized; "WGTB" is "WGTA" tested with added parameters. A "5Y3" (6-6-35) may have been intended as a metal version; actual production unknown. Variants: 6004, 6087, 6106, 6853.

5Y4GT/G, GA Full-wave rectifier, 5Y3GT/G w/ pinout scrambled: fil. on pins 7 & 8 (5-26-36) <5Q>. "GA" is in T12 bulb (SY, TS) (6-11-54).

5Z3 [AKA VT-145, CV275]; full-wave rectifier, ST16 bulb, 4-pin 5U4G (RCA) {R626} <4C>. Variant: 1275.

5Z4, GT/G [AKA VT-74, CV1864] (RCA) {R6008, ZP-315} (5-7-35) <5L>; full-wave cathode-type rectifier, 5Y3 ratings First member of GE-RCA "original nine" metal-octal line. "GT/G" (9-27-40) is in T9 bulb.

6A3, Spl Power triode, 2A3 w/ fil. 6.3 V @ 1.0 A (5-10-35). "Spl" [AKA FNH] is special 6A3, not further identified.

6A4 [AKA LA, KR-5]; fil.-type power pentode for car radios sim. to 52, fil. 6.3 V @ 300 mA, g_m 2.2 mS, 1.4 W out. Introduced by AS as type LA; RCA dev. no. {A468} <5K>.

6A5G Power triode, octal 6B4, cathode connected to center of heater (1-5-37).

6A6, X [AKA 38616]; duotriode, 6-V variant of 53 (6N7 char.) (RCA) {A493} (8-14-34) <6T>.

6A7, M, S Pentagrid conv., ST12 bulb w/ cap, 6-V variant of 2A7, heater 6.3 V @ 300 mA, conversion g_m 0.36-0.55 mS (RCA) {A506} (11-22-34) <7C>. "M" (Rogers) and "S" (Maj.) have spray-shield.

6A8, G, GT, GTX Pentagrid conv., metal 6A7, sim. to 6D8G (GE, RCA) {ZP-310, A5500} (8-15-35) <8A>. "G" [AKA VT-151, CV16] is in ST12 bulb (RCA) {A5505} (6-6-35); regis. actually issued earlier than metal 6A8. "GT" [AKA VT-151B, CV580] is in T9 bulb (3-11-38). "GTX" has ceramic base.

6AB4 [AKA EC92]; min. high-μ RF triode, for G-G amps., heater 6.3 V @ 300 mA, μ 62 (SY) (4-25-49) <5CE>. Variant: 6664.

6AB5 [AKA 6N5]; electron-ray tuning ind., heater 6.3 V @ 150 mA, sim. to 6E5 and 6U5/6G5 (4-19-37).

6AB6G Direct-coupled duotriode, heater 6.3 V @ 500 mA, g_m 1.8 mS, 3.5 W out (SY) (12-28-36) <7AU>.

6AB7 [AKA 1853, VT-176, CV66]; "television" pentode; remote-cutoff version of 6AC7, heater 6.3 V @ 450 mA, g_m 5 mS (3-18-39) <8N>. "Y (3-18-39 has Micanol base.

6AB8 [AKA ECL80}; min. 9-pin triode-pentode, heater 6.3 V @ 300 mA, triode μ 18, pentode g_m 3.4 mS (HY) (2-15-51) <9AT>.

6AC5GT/G High-μ power triode for direct coupling from

6AE5, 6P5, or 37 driver, heater 6.3 V @ 400 mA, μ 58 (RCA) {A2058} (9-29-37) <6Q>.

6AC6G, GT Direct-coupled duotriode, heater 6.3 V @ 1.3 A, μ 54, g_m 3.0 mS, 3.6 W out (SY) (1-8-37) <7W>.

6AC7, A, W, Y [AKA 1852, VT-112, REL39, CV660]; sharp-cutoff metal "television" pentode, IF and video amp. for radar and TV; heater 6.3 V @ 450 mA, g_m 9.0 mS (3-18-39) <8N>. Variant: 1617. "A" is early ruggedized version; "W" (RCA) {A4073F} is later. Variants: 6AB7, 6AJ7, 1649, 6134.

6AD4 [AKA 5898]; T3 submin. triode, heater 6.3 V @ 150 mA, μ 70, 330 mW (SY) {SN980D} (8-8-49) <8DK>. Variant: 5719.

6AD5G, GT Triode sim. to 6F5, heater 6.3 V @ 300 mA, μ 100, called a "TV" type (AS, SY) (12-15-38) <6Q>.

6AD6G Dual electron-ray tuning indicator, to be driven by 6AE6, replaceable by 6AF6G (NU, TS) (5-26-38) <7AG>.

6AD7G Low-μ triode - power pentode w/ common cathode; heater 6.3 V @ 850 mA, triode μ 6, pentode equals 6F6G (4-1-40) <8AY>. In (unique to?) Zenith 7S598 and 7S681, 1941. Should directly replace 6F6G by cutting off pins 1 and 6!

6AD8 [AKA EBF89, 6DC8]; min. 9-pin duodiode-pentode, heater 6.3 V @ 300 mA, g_m 1.1 mS (HY) (5-28-51) <9HE>.

6AE5GT/G Triode, driver for 6AC5, heater 6.3 V @ 300 mS, μ 4.2 (TS) (10-18-38) <6Q>.

6AE6G Single-grid twin-plate triode to drive 6AD6; plates have remote- and sharp-cutoff char., heater 6.3 V @ 150 mA, μs 25 and 33 (NU, TS) (6-14-38) <7AH>.

6AE7GT Twin-grid triode; two grids and cathodes, one plate; heater 6.3 V @ 500 mA (RCA) (5-9-40) <7AX>. In (unique to?) Zenith 12S550, 1941.

6AE8 Min. 9-pin triode-hexode conv., heater 6.3 V @ 300 mA (HY) (11-21-50) <8DU>.

6AF5G Triode sim. to 6P5GT, heater 6.3 V @ 300 mA, μ 7.4 (SY) (4-11-39) <6Q>. In (unique to?) Zenith 7S487, 8S432, etc., 1941.

6AF6G Twin electron-ray tuning indicator, like shortened 6AD6G; heater 6.3 V @ 150 mA (RCA, SY) (6-7-38) <7AG>.

6AF7G Twin electron-ray tuning indicator (French), heater 6.3 V @ 200 mA (4-17-39) <8AG>.

6AG5, /S, WA Min. VHF pentode, sim. to 6AK5, based on 6SH7, heater 6.3 V @ 300 ma, g_m 5.0 mS (RCA) (9-10-42) <7BD>. Variant: 6186. "/S" is selected for high g_m for TV use (SY).

6AG6G [AKA EL33}; power pentode sim. to scaled-up 6V6G, heater 6.3 V @ 1.25 A, g_m 9 mS, 5 W out (Brimar) (11-10-38) <7AC>.

6AG7, W, Y [AKA VT-247, CV207]; metal power/video pentode, scaled-up 6AC7, heater 6.3 V @ 650 mA, g_m 11 mS, 9.0 W (5-24-39) <8Y>. "W" and "Y" are ruggedized and Micanol-based respectively. Variant: 6AK7.

6AH5G Beam power tube, heater 6.3 V @ 900 mA, g_m 5.2 mS, 11 W out (SY) (1-9-42) <6AP>.

6AH6 Min. sharp-cutoff RF pentode and gate tube, suppressor wired out for control use, heater 6.3 V @ 450 mA, g_m 9.0 mS (RA) (12-5-46) <7BK>. Variant: 6485.

6AH7GT Duotriode, octal, heater 6.3 V @ 300 mA, μ 16, 1.5 W (2-5-41) <8BE>. Variant: 12AH7GT.

6AJ4 Min. 9-pin UHF triode, for G-G amps., 225 mA, μ 42, 2 W (GE, HY) {Z2103} (3-5-42) <9BX>.

6AJ5 Pentode, 6AK5 optimized for 28-V plate supply, g_m 2.75 mS (SY, WE) (7-5-44) <7BD>. Variant: 7755.

6AJ6 Tetrode, heater 6.3 V @ 450 mA; rumored to be quick-heating, possibly beam type, used in WW II lifeboat xmtr.; known samples were made for U. S. Navy in 1944.

6AJ7 Pentode, 6AC7 w/ ring contact for direct grounding of shell, w/ specified cutoff action via suppressor grid (Can.) (1-25-44) <8N>.

6AJ8 [AKA ECH81]; min. 9-pin triode-heptode conv., heater 6.3 V @ 300 mA (HY) (6-27-52) <9CA>.

6AK4 T3 UHF submin. triode, 6.3 V @ 0.225 A, μ 20, 3.3 W (SY) (3-17-52) <8DK>.

6AK5, W [AKA (WE) 403A, (WE) 731A, EF95, M8180, DP61]; min. sharp-cutoff VHF pentode, heater 6.3 V @ 175 mA, g_m 4.3-5.1 mS (WE) (5-27-43) <7BD>. "W" is ruggedized. Variants: 5591, 5654, 5656, 5702, 6028, 6096, 6968.

6AK6 Power pentode, min. 6G6G, 2.75 W (RCA) {A4361} (8-27-43) <7BK>.

6AK7 Pentode, 6AG7 w/ grounding ring (2-8-44) <8Y>. Also see 6AJ7.

6AK8 [AKA EABC80]; min. 9-pin triple diode - triode, one diode on own cath., heater 6.3 V @ 450 mA, μ 70, replaceable by 6T8 (HY) (6-7-52) <9E>.

6AL5, W [AKA D27, EAA91, EB91]; min. duodiode, heater 6.3 V @ 300 mA (RCA) {R6269} (8-11-44) <6BT>. "W" is ruggedized . Variants: 5704, 5726, 5829, 6058, 6097, 6663, 6887, 7055, 7631.

6AL6G Beam power tube, 6L6G w/ plate cap; sweep amp. in prewar GE TV (RA, SY) (8-21-39) <6AM>.

6AL7GT Twin electron-ray indicator, T9 bulb on octal base, heater 6.3 V @ 150 mA (SY) (6-20-46) <8CH>.

6AM5 [AKA EL91]; min. power pentode, heater 6.3 V @ 200 mA, g_m 2.6 mS, 1.4 W out (1-28-49) <7BD>.

6AM6 [AKA EF91]; min. power pentode, heater 6.3 V @ 300 mA, g_m 7.5 mS (1-28-49) <7BD>. Variants: 6064, 6516.

6AN4 Min. UHF triode, for G-G amps., heater 6.3 V @ 225 mA, μ 70 (RCA) {R1553D} (5-27-52) <7DK>. Variants: 7738, 1218A/8334.

6AN5 [AKA CK612]; pentode, min. 6AG7, for video and computer use - "first" type processed for low interface resistance (RA) (11-10-48) <7BD>.

6AN6 Quadruple diode, common cathode, heater 6.3 V @ 200 mA, 117 v in, 3 mA per plate (HY, SY, T) (8-12-46) <7BJ>.

6AN7 Min. 9-pin triode-hexode conv., heater 6.3 V @ 300 mA (HY) (4-29-49) <9Q>.

6AN8, A Min. 9-pin triode-pentode, heater 6.3 V @ 450 mA, triode μ 19, pentode g_m 7.8 mS (SY) (1-15-54) <9DA>. Variants: 7258, 8489.

6AQ4 [AKA EC91]; min. UHF triode, for G-G amps., heater 6.3 V @ 300 mA, μ 100 (HY) (3-20-53) <7DT>.

6AQ5, A, W [AKA EL90]; min. beam power tube; based on 6V6GT/G derated from 14W to 12 W (RCA) {A4450} (2-14-46) <7BZ>. Variants: 6005, 6094, 6095, 6669, 6928. "A" has controlled warm-up time.

6AQ6 [AKA D28]; min. duodiode-triode, 150 mA, based on 6T7G; min. 6SZ7 (RCA) {A4375A} (4-29-44) <6BT>.

6AQ7GT Duo-diode - high-μ triode; for FM det. use; diodes have own cathode; heater 6.3 V @ 300 mS, μ 70 (6-20-46) <8CK>.

6AQ8 [AKA ECC85]; min. 9-pin duotriode, heater 6.3 V @ 435 mA, μ 57, g_m 5.7 mS (SY) (8-15-55) <9AJ>.

6AR5 Min. power pentode based on 6K6, heater 6.3 V @ 400 mA, g_m 2.4 mS, 3.2 W out (8-8-47) <6CC>.

6AR6 Power tetrode, radar-scope sweep amp., T11 bulb, octal base, heater 6.3 V @ 1.2 A, 21 W when forced-air cooled; designed by Bell Labs and made by TS et al. (8-10-45) <6BQ>. Variants: 6098, 6384, 7756.

6AR7GT Duodiode-pentode (Aust.), typically used in reflex IF-AF stages (8-15-49) <7DE>.

6AR8 Min. 9-pin beam-deflection tube, sheet-beam balanced modulator, heater 6.3 V @ 300 mA (TS) (4-16-54) <9DP>. Also see 7360, 6JH8, 6ME8.

6AR11 Compactron duopentode, dual TV IF amp., heater 6.3 V @ 800 mA, g_m 10.5 mS (SY) (12-4-61) <12DM>.

6AS5 Min. beam power tube based on 7A5, heater 6.3 V @ 800 mA (2-16-48) <7CV>.

6AS6, W [AKA (WE) 409A]; min. VHF pentode, w/ suppressor brought out for use as gate electrode in phantastron dividers, etc., heater 6.3 V @ 175 mA, g_m 1.9-3.2 mS (WE) (3-3-47) <7CM>. Variants: 5725, 5784, 5916, 6187, 6486, 7752.

6AS7G, GA, GYB Dual low-μ power triode originally intended for TV use but usually found as "pass" tube in voltage regulators, heater 6.3 V @ 2.5 A, μ 2.0, 26 W total (RCA, SV) {A1834 (1-10-46) <8BD>. "GA" is in T12 bulb (RCA) {A4475} (7-11-55). Variants: 5998, 6080, 6082, 6520, 7105.

6AS8 Min. 9-pin diode - sharp-cutoff pentode, heater 6.3 V @ 450 mA, g_m 6.2 mS (HY, SY) (2-19-54) <9DS>.

6AT6 [AKA EBC90]; min. duodiode-triode based on 6Q7 (SY) (10-25-46) <7BT>. Variant: 6066.

6AT8, A Min. med.-μ triode - sharp-cutoff pentode based on 6X8, for VHF converters (SY) (6-11-54) <9DW>.

6AU5GT Beam tetrode, heater 6.3 V @ 1.25 A, 10 W, special pinout (RCA) (10-21-49) <6CK>.

6AU6, A Min. sharp-cutoff power/video pentode based on 6SH7 (10-25-45) <7BK>. Variants: 6136, 7543, 8425.

6AU8, A Min. 9-pin med.-μ triode - sharp-cutoff video pentode, heater 6.3 V @ 600 mS, triode μ 40, pentode g_m 8 mS (SY) (11-15-54) <9DX>. Variant: 7060.

6AV4 [AKA EZ91]; min. duodiode, heater 6.3 V @ 950 mA, 1250 V PIV, 90 mA; uprated 6X4 (HY) (10-10-52) <5BS>.

6AV6 [AKA EBC91]; min. duodiode - high-μ triode based on 6Q7 (3-8-47) <7BT>.

6AW7GT Duodiode-triode, heater 6.3 V @ 300 mA, μ 8.0 mS (SY) (11-15-48) <8CQ>.

6AW8, A [AKA 2081]; min. 9-pin triode - video pentode, heater 6.3 V @ 600 mA, triode μ 70, pentode g_m 9.5 mS (SY) (7-16-54) <9DX>. "A" differs in curve-knee characteristics.

6AX5GT Full-wave rectifier, heater 6.3 V @ 1.2 A, 1250 V PIV, 125 mA (RCA) {R6272A} (1-9-50) <6S>.

6AX7 Min. duotriode, 12AX7 w/ heater 6.3 VCT @ 300 mA (SY) (6-11-54) <9A>.

6AY11 Compactron duodiode-duotriode, heater 6.3 V @ 690 mA, μ 100 - equals 12AX7 and two diodes (GE) (6-18-62) <12DA>.

6AZ5 T3 submin. duodiode, heater 6.3 V @ 150 mA, 420 V PIV, 8 mA (GE, SY) (4-9-51) <8DF>.

6AZ6 T3 submin. duodiode, heater 6.3 V @ 150 mA, 450 V PIV, 20 mA (GE) (5-4-51) <8EH>.

6B4G Power triode, octal 6A3 (TS) (5-26-36) <5S>.

6B5 Direct-coupled duotriode, 6N6 char. (TS) (1-29-35) <6D>. Base diagram from 1938 (TS) is labeled "6Db"; shows a built-in cathode resistor on the input triode.

6B6, G, M Direct-coupled power duotriode, 6.3-V 2B6 (JB) <7J>. "G" and "M" involve an apparent duplication of numbers on an entirely different tube: "G" is a duodiode-triode, octal 75, sim. to 6T7G and replaceable by 6Q7G (GE, RCA) (-17-35) <7V>. "M" is Rogers "metal-spray" 6B6G.

6B7, S [AKA 10E/340, VT162 (Br.), VT-68, CV1891]; duodiode-pentode, heater 6.3 V @ 300 mA, g_m 1.3 mS, 2B7 and 6B8 char. (RCA) {A558} <7D>. "S" has spray-shield.

6B8, G, GT [AKA VT-93, CV1894]; duodiode-pentode, metal 6B7 (RCA) {A521} (6-15-36) <8E>. "G" [AKA VT-93A, CV1893, q. v.] has ST12 bulb (7-17-36).

"GT" has T9 bulb (5-22-40).

6BA4 "Rocket" disc-seal low-noise SHF triode, 2E36 format (q. v.), for UHF TV preamp use, heater 6.3 V @ 400 mA (SY) (9-4-53).

6BA5 T3 submin. pentode, heater 6.3 V @ 150 mA (SY) (8-8-49). Per SY in 1951, 6BA5 "old" is an RF amplifier, g_m 3.3 mS, with unique pinout; 6BA5 "new" is an AF amplifier, g_m 2.15 mS, w/ different capacitance values and slightly changed pinout <8DY>.

6BA6, W Pentode, min. 6SG7; effectively a remote-cutoff 6AU6; heater 6.3 V @ 300 mA, g_m 4.7 mS (10-25-45) <7BK>. Variants: 5749, 6660.

6BA7 Min. pentode conv. based on 6SB7, intended for FM use (SY) (6-9-48) <8CT>.

6BC4 Min. 9-pin med-μ UHF triode, heater 6.3 V @ 225 mA, μ 48 (HY, SY) (4-9-54) <9DR>.

6BC5 Min. pentode, 6AG5 w/ higher g_m of 6.1 mS (SY) (6-8-49) <7BD>.

6BC7 Min. 9-pin triple diode, heater 6.3 V @ 450 mA (HY, SY) (2-14-50) <9AX>.

6BD6 Min. remote-cutoff RF pentode, min. 6K7 (SY) (5-15-46) <7BK>.

6BD7, A [AKA EBC81]; min. 9-pin duodiode-triode, heater 6.3 V @ 230 mA, μ 70 (3-15-50) <9Z>.

6BE6 Min. pentagrid conv., min. 6SA7 (SY) (10-25-45) <7CH>. Variants: 5750, 5915, 7036.

6BE7 [AKA EQ80]; min. 9-pin heptode conv., heater 6.3 V @ 200 mA (HY) (3-15-50) <9AA>.

6BF5 Min. power pentode, heater 6.3 V @ 1.2 A, 5.5 W (SY) (2-14-50) <7BZ>.

6BF6 Min. duodiode-triode, min. 6SR7 (RCA, SY) {A4442} (6-28-46) <7BT>.

6BF7, A, W T3 submin. flat-press VHF med.-μ power duotriode; heater 6.3 V @ 300 mA, individual cathode leads; μ 35, g_m 4.8 mS, 2.0 W total (SY) (4-3-50) <8DG>. "W" is ruggedized." Variant: 6021.

6BG7 T3 submin. med.-μ duotriode, heater 6.3 V @ 300 mA, μ 35 (GE) (4-3-50) <8DG>.

6BH5 [AKA EF81]; min. 9-pin variable-μ remote-cutoff pentode (5-28-51) <9AZ>.

6BH6 Min. sharp-cutoff RF-IF pentode, heater 6.3 V @ 150 mA (RCA) {A4524A} (11-15-47) <7CM>. Variants: 6065, 6265, 6661, 7693.

6BJ5 Min. power pentode, heater 6.3 V @ 640 mA, g_m 10.5 mS, 9.9 W (HY, SY) (11-21-50) <6CH>.

6BJ6, A Min. pentode, heater 6.3 V @ 150 mA, based on 6SS7 (SY) (12-10-46) <7CM>. Variants: 6662, 7694.

6BJ7 Min. 9-pin triple-diode, heater 6.3 V @ 450 mA, 330 V PIV, 1.0 mA (HY, SY) (2-5-54) <9AX>.

6BK6 Min. duodiode-triode, heater 6.3 V @ 300 mA, μ 100 (TS) (1-17-50) <7BT>.

6BK7, A, B Min. VHF duotriode based on 12AV7, heater 6.3 V @ 450 mA, μ 43 (SY) (8-10-51) <9AJ>.

6BL6 Reflex klystron, 1.4-6.5 GHz, EC, heater 6.3 V @ 750 mA, 325 V, 250 mW out (RA, SY) {SD-835H} (10-5-46). Variants: 6BM6, 5836, 5837.

6BL8 [AKA ECF80]; min. triode-pentode, heater 6.3 V @ 450 mA, triode μ 20, pentode g_m 6.2 mS (RCA, SY) (9-4-56) <9DC>. Variant: 7643.

6BM5 [AKA 6P9]; min. power pentode, heater 6.3 V @ 450 mA, g_m 7.0 mS, 3.5 W out (HY) (6-30-52) <7BZ>.

6BM6, A Reflex klystron, 0.55-3.8 GHz, EC, 3-pin min. base, swaged rings to match cavity, heater 6.3 V @ 650 mA, 325 V, 175 mW CW out (BX, LI, RA, SFR, SY) {SD835K} (10-4-48). Variant: 5837.

6BM8 [AKA ECL82]; min. 9-pin triode-pentode, heater 6.3 V @ 780 mA, triode μ 70, pentode g_m 6.4 mS, 3.5 W out (GE, SV, SY) <9EX>.

6BN6 Min. gated-beam FM detector, heater 6.3 V @ 300 mA (SY) (9-26-49) <7DF>.

6BQ5 [AKA EL84]; min. 9-pin audio non-beam power pentode, heater 6.3 V @ 760 mA, g_m 11.3 mS, 12 W (SY) (7-9-56) <9CV>. Variants: 7189, 7320.

6BQ7, A Min. VHF duotriode, 6 V @ 400 mA, μ 38, g_m 6.4 mS, interchanges with 6BZ7 and 6DJ8 (SY) (8-11-50) <9AJ>. Variant: 7062.

6BR7 Min. sharp-cutoff RF pentode based on 6J7, heater 6.3 V @ 150 mA, g_m 0.9 mS (HY) (10-25-51) <9BC>. Variant: 6059.

6BS5 Min. 9-pin beam pentode, heater 6.3 V @ 750 mA, g_m 7.0 mS, 4.5 W out (HY) (10-30-53).

6BT6 Min. duo-diode - high-μ triode, heater 6.3 V @ 300 mA, μ 70 (SY) (1-17-50) <7BT>.

6BU6 Min. duo-diode - low-μ triode, heater 6.3 V @ 300 mA, μ 16.5 (HY) (1-17-50) <7BT>.

6BV7 Min. 9-pin duodiode - power pentode, heater 6.3 V @ 800 mA, g_m 8.0 mS, 2.0 W out (HY) (8-27-51) <9BU>.

6BW4 Min. 9-pin full-wave rectifier, heater 6.3 V @ 900 mA, 900 V in, 100 mA (HY, SY) (7-11-55) <9DJ>.

6BW6 Min. 9-pin beam power tube based on 6V6, beam plates wired out separately, heater 6.3 V @ 450 mA, 12 W (GE) (4-17-50) <9AM>. Variant: 6061.

6BW7 Min. 9-pin sharp-cutoff RF pentode, heater 6.3 V @ 150 mA, g_m 9.0 mS (HY) (7-10-53).

6BX4 Min. full-wave rectifier, heater 600 mA, 1350 V PIV, 90 mA out (HY) (10-30-53) <5BS>.

6BX6 [AKA EF80]; min. 9-pin sharp-cutoff RF pentode, heater 6.3 V @ 300 mA, g_m 7.2 mS, 2.75 W (HY, SY) (2-15-51) <9AQ>.

6BY4 Cermet UHF triode, 5/16" dia. by 3/8" high, for UHF TV preamps, heater 6.3 V @ 200 mA, μ 100, g_m 6 mS, NF 8 dB @ 900 MHz (GE) (10-17-55).

6BZ7 Min. 9-pin VHF med.-μ duotriode, 6 V @ 400 mA, based on 6BQ7, μ 36; interchanges with 6BQ7A (SY) (8-22-52) <9AJ>. Variant: 7057.

6C4, W Min. VHF med.-μ power triode based on 6J5, heater 6.3 V @ 150 mA, μ 17, 3.5 W (originally 5 W) (RCA, SY) (5-9-42) <6BG>. Variants: 6100, 6135.

6C5, G, GT/G [AKA VT-65, CV582]; metal triode; contains a set of 6J7 pentode elements internally triode-connected (!); heater 6.3 V @ 300 mA, μ 20 (RCA) {A2021} (8-15-35) <6Q>. "G" [AKA VT-65A, Br. CV581, NR78] is in ST12 bulb (RCA) {A2040} (6-6-35). "GT/G" is in T9 bulb.

6C6 [AKA 38636, 10E/337, VT161 (Br.)]; pentode, ST12 bulb w/ cap on 6-pin base, 6J7 char. (RCA) {A478A} <6F>. Variant: 1221.

6C7 Duodiode-triode, spray-shielded 85A, sim. to 6R7 (Maj., NU, RCA, TS) <7G>.

6C8G [AKA VT-163, CV1896]; duotriode, ST12 bulb w/ cap, heater 6.3 V @ 300 mA, μ 36 (1-21-37) <8G>.

6C10 Compactron triple-triode, heater 6.3 V @ 400 mA, equals 1-½ 12AX7s (GE, SY) (8-1-60) <12BQ>. Variant: 7689.

6CA4 [AKA EZ81]; min. 9-pin full-wave rectifier, heater 6.3 V @ 1.0 A, 1 kV PIV, 150 mA (SY) (8-27-56) <9M>.

6CA7 [AKA EL34]; octal power pentode (SY: "beam pentode"), tubular bulb, heater 6.3 V @ 1.5 A, g_m 11 mS (MU, SV, SY) (12-12-52) <8ET>.

6CB6, A Min. pentode, for TV IF amps., heater 6.3 V @ 300 mA, g_m 6.2 mS (SY) (10-21-49) <7CM>. "A" has g_m raised to 8 mS. Variants: 6676, 7056, 7732.

6CG6 Min. remote-cutoff pentode, heater 6.3 V @ 300 mA, g_m 2.0 mS (HY) (10-10-50) <7BK>.

6CH6 [AKA EL821]; min. 9-pin RF/video pentode, heater 6.3 V @ 750 mA, g_m 11.0 mS (HY) (4-18-51) <9BA>. Variants: 6132, 7499.

6CJ6 [AKA EL81]; min. 9-pin pentode, audio or horiz. sweep, heater 6.3 V @ 1.05 A, g_m 4.6 mS (HY) (10-16-51) <9AS>.

6CK6 [AKA EL83]; min. 9-pin RF/video pentode, heater 6.3 V @ 710 mA, g_m 10.0 mS (HY) (10-16-51) <9AR>.

6CL6 Min. 9-pin RF/video pentode, heater 6.3 V @ 650 mA, g_m 11 mS, 7.5 W (RCA, SY) (6-18-52) <9BV>. Variants: 6197, 6297, 6677.

6CM6 Beam power tube, min. 9-pin 6V6GT (GE) (10-9-53) <9CK>.

6CQ6 [AKA EF92]; min. variable-μ pentode, heater 6.3 V @ 200 mA, g_m 2.5 mS (HY) (3-20-53) <7DR>.

6CW4 Nuvistor high-μ VHF triode, for G-G amps., heater 6.3 V @ 130 mA, μ 62, g_m 9.8 mS, 1.5 W (RCA) (7-4-60) <12AQ>. Gave NF of 5.5 dB at Ch. 13 in RCA KRK-90A TV tuner, or about 1 dB better than the 6GK5, 1961's best conventional tube. Variants: 2CW4, 6DS4, 13CW4.

6CW5 Min. 9-pin power pentode, heater 6.3 V @ 760 mA, 12 W (GE, SY) (7-9-56) <9CV>.

6CY5 Min. sharp-cutoff RF tetrode, heater 6.3 V @ 200 mA, g_m 8 mS (SY) (9-10-56) <7EW>. Variants: 7167, 7717.

6CZ5 Min. beam power tube, heater 6.3 V @ 450 mA, 11 W (SY) (11-5-56) <9HN>. Variant: 6973.

6D4 [AKA VC17]; min. argon triode thyratron, heater 6.3 V @ 250 mA, 320 V PIV, 25 mA, usable for usual thyratron functions but best known as an RF noise generator; w/ tube diode-connected and mounted inside a magnet, noise spectrum peaks at 500 kHz and is still strong at 1 MHz (HY, SY, TS) (3-10-44) <5AY>.

6D5, G Metal power triode, announced 1935 by RCA and SY, but not made in qty.; heater 6.3 V @ 700 mA, μ 4.7 (NU) (5-6-41). "G" is glass variant (6-6-35); neither is used in any known radio.

6D6 [AKA 38646, VT-69, (Br.) VT160, CV1900]; pentode, heater 6.3 V @ 300 mA, g_m 1.6 mS (58 and 6U7 char.) (RCA) {A479A} <6F>.

6D7 Pentode, spray-shielded 6C6 on 7-pin base (Maj., NU) <7H>.

6D8G [AKA VR57 (Br.)]; pentagrid conv., 150-mA version of 6A8G (RCA) {A5517} <8A>. "G" is in ST12 bulb.

6DB6 Min. sharp-cutoff RF pentode, heater 6.3 V @ 300 mA, g_m 2.1 mS (HY, SY) (2-26-54) <7CM>.

6DC6 Min. semi-remote-cutoff RF-IF pentode, heater 6.3 V @ 300 mA, g_m 5.5 mS (SY) (3-12-54) <7CM>, intended for TV IFs but used more in comm. gear. Hard to find today; "hotter" substitutes include 6BZ6, 6CB6, 6CF6, 6DE6; lower-g_m substitutes include 6BH6, 6BJ6, 6JF6.

6DG6GT [AKA 6W6GT, q. v.]; beam power tube (4-25-55).

6DJ8 [AKA ECC88]; min. 9-pin VHF duotriode, designed for TV tuners but better known as audio preamp, heater 6.3 V @ 365 mA, μ 33, g_m 12.5 mS (SY) (5-12-58) <9AJ>. Very sim. to 6BQ7A despite difference in g_m, and officially replaceable by it in Tektronix 502 and 545 scopes. Variants: 6922, 7308.

6DL4 Min. high-μ triode, heater 6.3 V @ 165 mA, μ 65, 2.0 W (GE) (1-17-66) <9NY>. Variant: 8255.

6DS4 Nuvistor vari-μ triode, variant of 6CW4 w/ extended cutoff, for G-G amps., heater 6.3 V @ 135 mA, μ 63, g_m 9.0 mS (GE, RCA) (8-21-61) <12AQ>. Variant: 2DS4.

6DV4 Nuvistor med.-μ triode, UHF osc., gold-plated shell and pins for stability, heater 6.3 V @ 135 mA, μ 35, 1 W, usable to 950 MHz (GE, RCA) {A15239G} (9-10-62) <12EA>. Variant: 2DV4.

6E5 [AKA VT-215, 38665E, CV239]; electron-ray tuning ind. (sim. to 6AB5/6N5), T9 bulb, 6-pin base, heater 6.3 V @ 300 mA (RCA) {CA4} 6-27-35) <6R>. Variants: 2E5, 1629.

6E6 Dual power triode, heater 6.3 V @ 600 mA, g_m

1.6 mS, 1.6 W out (NU) (2-15-35) <7B>.

6E7 Pentode, spray-shielded 6D6 on 7-pin base (Maj., NU, T) <7H>.

6E8, G, MG Triode-hexode conv. (French), octal w/ cap, heater 6.3 V @ 300 mA (6-2-39) <8K>.

6EU7 Min. duotriode, 12AX7 w/ nontapped heater of 6.3 V @ 300 mA, wired to pins 1 and 2 to reduce hum; claimed to be quieter than 12AX7, but has same published hum level as 12AX7A (GE, RCA, SY) (3-10-60) <9LS>.

6EW7 Duotriode, "9-T9" bulb, for TV vert. osc. and output - in first group of 9-T9 types along with 6HC8 (SY) (3-7-60) <9HF>.

Acorn UHF osc. triode, heater 6.3 V @ 225 mA, μ 17, 2 W, usable to 1.2 GHz, dual pins for grid and plate (RCA) {A2227} (8-11-44) <7BR>.

6F5, G, GT, M Metal triode, 7B4 char., μ 100 (RCA) {A2015} (5-28-35) <5M>. "G" (6-6-35) and "GT" {A2039} (4-11-38) are in ST12 and T9 bulbs respectively. "M" is Rogers "metal-spray" version.

6F6, G, GT [AKA VT-66, CV1186]; metal power pentode, 42 char., heater 6.3 V @ 700 mA, 11 W (GE, RCA) {A-4016, ZP-308} (5-28-35) <7S>. Variants: 1611, 1613, 1621. "G" [AKA 10E/346, KT63, NR85, ARP17, W1535, VT-66A, CV1911] is in ST14 bulb (RCA) {A4035} (6-6-35). "GT" is in T9 bulb (1-2-40).

6F7, M, S [AKA 38667, VT-70, CV1915]; triode-pentode, heater 6.3 V @ 300 mA (RCA) {A560} <7E>. "M" (Rogers) and "S" are spray-shielded.

6F8G [AKA VT-99, 38768F, CV1917]; duotriode, ST12 bulb w/ cap, equals two 6J5s (6SN7GT and 7N7 char.) (RCA) (9-24-37) <8G>.

6FV8, A Min. 9-pin triode-pentode, heater 6.3 V @ 450 mA, triode μ 40, pentode g_m 6.5 mS (SY) (2-16-59) <9FA>. Variant: 8446.

6FW7 T3 submin. VHF duotriode, heater 6.3 V @ 300 mA, μ 36, 1.7 W (SY) (10-22-62) <8LM>. Variant: 3FW7.

6FW8 Min. 9-pin VHF duotriode, heater 6.3 V @ 400 mA, μ 33 (RCA) (6-22-59) <9AJ>. Variant: 7803.

6FX7 T3 submin. VHF duotriode, heater 6.3 V @ 300 mA, μ 36, 1.7 W (SY) (7-9-62) <8LK>. Variant: 3FX7.

6G5 [AKA 6U5]; electron-ray tuning indicator, remote-cutoff 6E5 (RCA) {CA4A} (2-21-36) <6RA>.

6G6G [AKA VT-198A, CV1926]; power pentode, heater 6.3 V @ 150 mA, g_m 2,3 mS, 2.75 W (SY) (11-29-37) <7AC>. Derivative: 6AK6.

6G7 [AKA 89R]; duodiode rectifier - pentode (Can.) sim. to 6B7.

6G8G Duodiode-pentode (Aust.) <8E>. In reflex IF-AF stage in AWA Radiola 517-M.

6GE8 Min. 9-pin triode-pentode, heater 6.3 V @ 900 mA, triode μ 5.4, pentode g_m 3.2 mS (SY) (10-5-59) <9LC>. Variant: 7734.

6GM5 Beam power tube, 7591 in 9-T9 bulb (SY) (3-13-61) <9MQ>. (Compactron equiv.: 7868.)

6H4GT Single octal diode, heater 6.3 V @ 150 mA (TS) (5-11-39) <5AF>.

6H5 [AKA 6G5]; electron-ray tuning ind., ring display, replaceable by 6U5/6G5 (RA) (10-24-36) <6R>.

6H6, G, GT, WGT [AKA VT-90, CV1301, 38566H]; metal duodiode, heater 6.3 V @ 300 mA, 420 V PIV, 16 mA total (RCA) {A6016, ZP-314} (8-15-35) <7Q>. "G" (6-6-35) is in ST12 bulb. "GT" [AKA VT-90A, CV1931, q. v.] is in T9 bulb (4-11-38). "WGT" is ruggedized "GT.".

6H7, M, S High-μ triode - power pentode (Can.). "M" and "S" (8-21-34) are spray-shielded.

6H8G Duodiode-pentode, octal w/ cap (French), heater 6.3 V @ 300 mA, g_m 2.4 mS (6-27-39) <8E>.

6J4, WA Min. UHF triode, for G-G amps., heater 6.3 V @ 400 mA, μ 55, g_m 12 mS (RCA, SY) {A2209A} (3-6-44) <7BQ>. Variants: 7137, 7245, 8532.

6J5, Sel., G, G Sel., GX, GT, GTX, M, WGT [AKA 38565J, VT-94, CV1067]; med.-μ metal triode, heater 6.3 V @ 300 mA, μ 20 (RCA) (6-24-37) <6Q>. "Selected" is AKA VT-94B. "G" [AKA VT-94A, 38765J; Br. NR78A, VR67, L63, CV1932] is in ST12 bulb (RCA) {A2042} (5-26-36). "G Selected" is AKA VT-94C. "GX" and "GTX" have ceramic bases ((HY, SY). "GT" [AKA VT-94D, CV1934] is in T9 bulb (4-11-38). "WGT" is ruggedized "GT." "M" is Rogers "metal-spray" version.

6J6, A, W Min. UHF med.-μ duotriode, heater 6.3 V @ 450 mA, μ 38, 3 W total, common cathode (RCA) {A4434} (9-10-42) <7BF>. Variants: 5844, 5920, 5964, 6030, 6045, 6099, 6101, 6211, 6535, 6927, 7244.

6J7, G, GT, GTX, MG [AKA VT-91; (Br.) VR56, CV1074, 10E/11402, EF36, NR49]; sharp-cutoff pentode, metal 6C6 (RCA) {A4014} (8-15-35). Variants: 1620, 6059, 7000. "G" is in ST12 bulb; has internal shield cage on Pin 1; is sim. to Br. NR83 (RCA) {A4040} (6-6-35) <7R>. "GT" [AKA VT-91A, CV1937] is in T9 bulb (4-11-38). "GTX" is "GT" w/ ceramic base (HY). "M" is Rogers "metal-spray" version.

6J8G, EG, M Triode-heptode conv., sim. to 7J7 (KR, SY, TS) (2-21-38) <8H>. "M" is Rogers "metal-spray" version.

6JH8 Min. 9-pin beam-deflection tube, sheet-beam balanced modulator, heater 6.3 V @ 300 mA (GE) (7-10-61) <9DP>. Also see 7360, 6AR8, 6ME8.

6K4 T3 submin. triode, heater 6.3 V @ 150 mA, μ 20, 2.25 W, oscillates to 1.5 GHz; effectively a submin. 6J5 {SD834} (GE, SY) (11-23-45). Variants: 5703, 5897, 5977, 5892, 6778.

6K5, G, GT Octal triode equiv. to triode section of 6Q7 (TS) (5-26-36) <5U>.

6K6G, GT [AKA VT-152A, CV1938]; power pentode, octal 41, 7B5 char. (RCA) {A4047} (5-26-36) <7S>. "GT" [AKA VT-152, CV207] has T9 bulb (4-11-38).

6K7, G, GT, GTX, M [AKA VT-86, 38567K; Br. CV1942, ARP34, EF39, VR53]; remote-cutoff pentode, metal 78, heater 6.3 V @ 300 mA, g_m 1.7 mS (RCA) {A4015} (8-15-35) <7R>. "G" [AKA VT-86A, Br. NR81, CV1941] has ST12 bulb (RCA) {A4036} (6-6-35). "GT" [AKA VT-86B, CV1943] has T9 bulb (4-11-38). "GTX" is "GT" w/ ceramic base (HY). "M" is Rogers "metal-spray" version. Variant: 5732.

6K8, G, GT, GTX [AKA VT-167, 38568K, CV18]; metal triode-hexode conv., heater 6.3 v @ 300 mA (RCA) (1-31-38) <8K>. "G" [AKA VT-167A, CV1944] has ST12 bulb (6-15-38). "GT" has T9 bulb (3-31-39). "GTX" is "GT" w/ ceramic base.

6L4 Acorn triode, sim. to 6F4 but derated to 1.7 W, slightly lower capacitances, rated at higher voltage (RCA) {A2297} (5-23-49) <7BR>.

6L5G [AKA VT-213A, CV151]; triode, 6C5G w/ 150-mA heater (RCA) {A2045} (6-12-36) <6Q>.

6L6, A, G, GA, W, Y, WGA, GAY, GB, GC [AKA VT-115, CV208]; metal beam power tube, heater 6.3 V @ 900 mA, 19 W (RCA) {A494} (3-24-36) <7S>. Preceded, and electrically relat ed to, 807. Grandfather of all beam power tubes. Introduced 1936 with eyeletted base seal. Variants of 6L6: 1614, 1619, 1622, 1631, 1632, 7581. "A" and "W" are first and final ruggedized versions. "Y" has Micanol base. "G" [AKA VT-115A, CV1947] has ST-16 bulb; sim. to Br. NR77 (RCA) {A4048} (6-8-36). Variants of "G": 6AL6G, 6BG6G, T21, 1223. "GA" has ST-14 bulb, still rated at 19 W (SY) (3-16-43). Variants of 6L6GA: 1222, 5932. "WGA" is ruggedized "GA." "GAY" is "GA" w/ Micanol base. "GX" has ceramic base (HY, SY). "GB" [AKA 5881] has T12 bulb (4-16-54). "GC" is "GB" uprated to 30 W by use of button stem and 11-mil. plate material in place of 7-mil. (RCA) (12-1-58). Variants of "GC": 7027A, 7581.

6L7, G [AKA VT-87, CV1951]; metal pentagrid mixer-amp., two control grids, for separate osc. or for use in volume expanders, etc. (RCA) {A599} (4-29-35) <7T>. Variant: 1612. "G" [AKA VT-87A, CV1951] has ST12 bulb (RCA) {A5506} (6-6-35).

6M5 Min. 9-pin power pentode, heater 6.3 V @ 710 mA, g_m 10.0 mS, 3.9 W out (HY) (4-29-49) <9N>.

6M6G Power pentode (French) sim. to EL11, heater 6.3 V @ 1.2 A (12-5-38) <7AC>.

6M7G Triple-grid amp. (French), heater 6.3 V @ 300 mA, g_m 3.4 mS (1-8-40) <7R>.

6M8G, GT Diode-triode-pentode, heater 6.3 V @ 600 mA, μ 100, g_m 1.9 mS (8-1-39) <8AU>.

6ME8 Min. 9-pin beam-deflection tube, sheet-beam balanced modulator, heater 6.3 V @ 300 mA (GE) (7-10-61) <9RV>. Also see 7360, 6AR8, 6JH8.

6MJ8 Compactron triple-triode, heater 6.3 V @ 900 mA, μ 17; equals 1-½ 12AU7s (GE) (6-5-67) <12HG>. Also see 7688.

6N4 [AKA CK611]; min. UHF triode, heater 200 mA, μ 32, 3 W; char. sim. to 2C35 (HY, RA, TS) (9-25-45) <7CA>.

6N5 [AKA 6AB5, q. v.]; electron-ray tuning ind. (RCA, TS) {CA13} <6R>.

6N6, GT/G, MG Direct-coupled duotriode, metal 6B5, sim. to 6AB6G, heater 6.3 V @ 800 mA. Triad successor to Speed 293 "Triple Twin" (7-9-36) <7AU>. Makers licensed to make 6N6 were AS, HY, KR, NU, RA, SY, STC, and Triad. "GT" (9-27-40) has T9 bulb.

6N7, Sel, G, GT/G [AKA VT-96, CV1957]; power duotriode, metal 6A6, 53 char. (RCA) {A4017} (3-21-36) <8B>. "Selected" is AKA VT-96B. "G" (octal 6A6) has ST16 bulb.; is sim. to Br. NR73 (RCA) {A4039}. "GT/G" has T9 bulb (9-27-40).

6N8 [AKA EBF80]; min. 9-pin duodiode-pentode, heater 6.3 V @ 300 mA, g_m 2.2 mS (HY) (8-19-49) <9HE>.

6P5GT/G Triode, T9 bulb, octal 76 (TS) (4-11-36) <6Q>.

6P6 Aust. power pentode, heater 6.3 V @ 700 mA (3-25-36) <6AC>.

6P7G Triode-pentode, octal 6F7 (6-6-35) <7U>.

6P8G Triode-hexode conv. (Br.), heater 6.3 V @ 800 mA (11-10-38) <8K>.

6Q4 [AKA EC80]; min. 9-pin triode, usable to 500 MHz, heater 6.3 V @ 450 mA, 4 W, g_m 12 mS (AX, HY, RCA) {A2298, A2304} (7-15-49) <9S>. RCA: "export" type replaceable by 6BC4.

6Q5G [AKA 884]; thyratron for scope-sweep generation (DuM) (1-2-43) <6Q>.

6Q6G Diode-triode, 6Q7G w/ 150-mA heater; nearest equiv. is 6T7G (6-12-36) <6Y>.

6Q7, G, GT [AKA VT-92, CV588]; duodiode - high-μ triode, metal 75, μ 70 (RCA) {A4018} (12-9-35) <7V>. Variant: 6118. "G" [AKA VT-92A; Br. NR68, DH63, CV589] has ST12 bulb, can replace 6B6G (RCA) {A4038}. "GT" has T9 bulb (4-11-38).

6Q8 Foreign equiv. of 6A8 <8A>.

6R4 [AKA EC81]; min. 9-pin triode, oscillates to 1.2 GHz, heater 6.3 V @ 240 mA, μ 16, 5 W (AX, HY) (7-15-49).

6R6G Octal pentode, heater 6.3 V @ 300 mA, g_m 1.45 mS, called a "TV" type (AS) (12-15-38) <6AW>.

6R7, G, GT, M [AKA VT-88, 38567R; Br. VR55, 10E/11401, AR-21, EBC33, NR48, 1639, CV1963]; duodiode - med.-μ triode, metal 85, μ 16; 7E6 and 12BD6 char. (RCA) {A4022} (1-21-36) <7V>. "G" [AKA VT-88A, CV1962] has ST12 bulb (RCA) {A4041} (5-26-36). "GT" [AKA VT-88B,

CV1964] has T9 bulb (6-6-39). "M" is Rogers "metal-spray" version.

6R8 Min. 9-pin triple-diode - triode, one diode has own cathode; heater 450 mA, μ 16 (HY, TS) (10-14-49) <9E>.

6S5 Electron-ray tuning ind. (Can.) (4-1-37) <6R>.

6S6GT Power pentode, heater 6.3 V @ 450 mA, g_m 4.0 mS (1-12-40) <5AK>.

6S7, G Remote-cutoff pentode, metal 6D6 w/ 150-mA heater (TS) (2-23-38) <7R>. "G" (6K7G w/ 150-mA heater) has ST12 bulb (RCA) {A4064} (6-12-36).

6S8GT Triple-diode - 6SQ7 plus independent diode w/ its own cathode, μ 100 (TS) (9-28-45) <8CB>.

6SA7, Y, GT, GTX, GTY WGT [AKA VT-150, CV1966]; metal pentagrid conv. (RCA) (11-18-38) <8R>. "Y" has Micanol base. "GT" [AKA VT-150A, CV1967] has T9 bulb (3-17-39) <8AD>. "GTX" and "GTY" are "GT" w/ ceramic and Micanol bases. "WGT" is ruggedized "GT."

6SB7Y Metal pentagrid conv. w/ Micanol base, intended for FM use; has higher current ratings than 6SA7; heater 6.3 V @ 300 mA, high conversion g_m of 0.88 mS (RCA) {A5581A} (1-10-46) <8R>.

6SC7, GT, GTY [AKA VT-105, CV1969]; metal high-μ duotriode, heater 6.3 V @ 300 mA, μ 70, g_m 1.3 mS (RCA, TS) (11-18-38) <8S>. Metal version of 6SL7GT w/ grid leads pinned-out away from heater. Variant: 1655. "GT" has T9 bulb (1-29-42). "GTY" is "GT" w/ Micanol base.

6SD7GT Remote-cutoff pentode, heater 6.3 V @ 300 mA, g_m 3.4 mS, sim. to 6SG7GT (RA, TS) (4-15-40) <8N>.

6SE7GT Triple-grid amp. sim. to 6SG7GT, 6SH7GT, heater 6.3 V @ 300 maA, g_m 3.1 mS (4-30-40).

6SF5, GT Metal triode, single-ended 6F5, 7B4 char. (RCA) (10-7-38) <6AB>.

6SF7, GT Metal diode - remote-cutoff pentode, 6B8 less one diode and single-ended, unrelated to 6F7 (3-3-41) <7AZ>.

6SG7, GT, Y [AKA VT-211, CV1978]; metal remote-cutoff RF pentode, heater 6.3 V @ 300 mA, g_m 4.7 mS; effectively a remote-cutoff 6SH7 (RCA, SY) (2-7-41) <8BK>. Variant: 6006. "GT" has T9 bulb (1-15-46). "GTY" is "GT" w/ Micanol base.

6SH7, GT, L Metal sharp-cutoff HF pentode sim. to 7L7, heater 6.3 V @ 300 mA, g_m 4.9 mS (RCA, SY) (8-1-41) <8BK>. "GT" has T9 bulb (11-20-42). "L" is "GT" shortened by 0.32" (SY) (1-17-46).

6SJ7, Y, W, GT, GTX, GTY, WGT [AKA VT-116, CV70]; metal sharp-cutoff pentode; not specifically a single-ended 6J7; g_m 2.5 mS (RCA, SY) (10-7-38) <8N>. Variant: 5693. "Y" [AKA 6SJ7 Spl, VT-116B, CV866] has Micanol base. "W" is ruggedized. "GT" [AKA VT-116A, CV590] has T9 bulb (5-1-39). "GTX" and "GTY" are "GT" w/ ceramic and Micanol bases. "WGT" is ruggedized "GT."

6SK7, Y, W, GT, GTX, GTY, WGT [AKA VT-117, CV1981]; metal remote-cutoff pentode, heater 6.3 V @ 300 mA, g_m 2.3 mS (RCA, SY) (10-7-38) <8N>; effectively a remote-cutoff 6SJ7. "GT" [AKA VT-117A, CV-1982] has T9 bulb (5-1-39). For "W," "Y," "GTX," and "GTY," see 6SJ7.

6SL7GT, A, L, WGT [AKA VT-229, CV171]; high-μ duotriode, 7F7 char., sim. to 6SC7GT (RCA, SY) (7-2-41) <8BD>. Variant: 5691. Also see 6SC7. "A" and "L" are early ruggedized versions; "WGT" is final. Variant: 6113.

6SN7GT, A, L, WGT [AKA VT-231, CV170]; med.-μ duotriode, dual 6J5GT, or single-ended GT version of 6F8G (RCA, SY) (3-3-41) <8BD>. Variants: 1633, 5692, 6042, 6180. "A" and "L" are early ruggedized versions; "WGT" is final ruggedized type, w/ heater-to-cathode voltage uprated from 100 V to 250 (RCA) {A4273B}. "GTA" is improved, uprated to 7.5 W total (4-6-50). "GTB" is controlled-warmup version of "GTA" for series-string receivers (6-8-54).

6SQ7, W, GT/G [AKA VT-103, CV1990]; metal duodiode - high-μ triode, sim. to 75 and 2A6; not precisely a single-ended 6Q7, μ 100 (RCA) (10-7-38) <8Q>. "W" is ruggedized. "GT/G" has T9 bulb (5-8-41).

6SR7, GT [AKA VT-233, CV109]; metal duodiode - med.-μ triode, single-ended 6R7 (RCA) (4-1-40) <8Q>. "GT" has T9 bulb (9-21-42).

6SS7, GT [AKA VT-199, CV70]; metal remote-cutoff pentode, 6SK7 w/ 150 mA heater (TS) (4-22-41) <8N>. "GT" has T9 bulb (1-5-46).

6ST7 [AKA VT-205, CV109]; metal duodiode - med.-μ triode, 6SR7 w/ 150-mA heater (12-1-41) <8Q>.

6SU7GTX, GTY Duotriode; sel. 6SL7GT w/ Micanol base, low gas and leakage; tested for balance (TS) (3-15-44) <8BD>. "6SU7 matching specifications are passed by about half of production 6SL7s." Variant: 6188. "GTX" is "GTY" w/ ceramic base; "GTY" has Micanol base.

6SV7 Metal diode - sharp-cutoff pentode, heater 6.3 V @ 300 mA, g_m 3.6 mS (SY, TS) (6-20-46) <7AZ>.

6SZ7 Metal duodiode - high-μ triode; single-ended, 150-mA 6Q7 (TS) (11-20-45) <8Q>.

6T4 Min. low-μ UHF triode, heater 6.3 V @ 225 mA, μ 13 (HY, TS) (9-19-52) <7DK>.

6T5 Tubular electron-ray tuning indicator w/ ring display (4-12-37) <6R>, disc. in 1939. Replaceable by 6U5/6G5.

6T6 Triple-grid amp. (Can.; Rogers?), heater 6.3 V @ 450 mS, g_m 5.5 mS (9-29-39) <6Z>, "metal-spray" type.

6T7G [AKA 6Q6G]; duodiode-triode, 150-mA version of 6Q7G (RCA, TS) {A4064} (10-22-36) <7V>.

6T8, A [AKA EABC80]; min. 9-pin triple-diode-triode; one diode has own cathode; heater 6.3 V @ 450 mA, μ 70; sim. to min. 6S8GT (SY) (5-20-47) <9E>.

6T9 Compactron, equiv. to ½ 12AX7 + 6AQ5 (GE, SY) (9-9-63) <12FM>

6U5 [AKA 6G5, VT-98, CV504]; electron-ray tuning indicator, T9 tubular 6G5 (RCA) {CA18} (4-19-37) <6R>.

6U6GT Beam power tube, optimized for low supply voltage, heater 6.3 V @ 750 mA, 11 W, sim. to 6W6GT (RA, SY) (6-6-40) <7S>.

6U7G Remote-cutoff pentode, octal 6D6, sim. to 6S7G, 6W7G, g_m 1.6 mS (RCA, TS) {A4071} (8-11-36) <7R>.

6U8, A [AKA ECF82]; min. 9-pin med.-μ triode - sharp-cutoff pentode, heater 6.3 V @ 450 mA, triode μ 40, pentode g_m 5 mS (SY, TS) (5-15-51) <9AE>. Variants: 6678, 7059, 7731, 8445.

6V4 [AKA EZ80]; min. 9-pin full-wave rectifier, heater 6.3 V @ 600 mA, 350 V PIV, 90 mA (HY, TF) (4-29-49) <9M>.

6V5GT Beam power tube, slightly derated 6V6GT w/ heater on pins 7 & 8 (GE) (11-15-48) <6AO>.

6V6, G, GTA, GTX, GTY [AKA VT-107, CV510]; metal octal beam power tube sim. to 7C5, heater 6.3 V @ 450 mA, 12 W (RCA) {A4055A} (1-3-38) <7S>. Variant: 7184. "G" [AKA VT-107B, CV509] has ST14 bulb (RCA) {A4055} (11-9-36). "GT" [AKA VT-107A, CV511] has T9 bulb (7-10-39). "GTA" is "GT" up-rated to 14 W, w/ controlled warmup time. Variants: 5871, 5992, 7408. "GX" and "GTX" have ceramic bases. "Y" and "GTY" have Micanol bases.

6V7G Duodiode-triode, octal 85 (TS) (8-14-36) <7V>.

6V8 Min. 9-pin triple diode - triode, 6T8 w/ pinout scrambled (HY, SY) (9-6-50) <9AH>.

6W5G [AKA VT-196, CV279]; duodiode rectifier, ST12 octal bulb, heater 6.3 V @ 900 mA, 1250 V PIV, 100 mA (TS) (1-4-37) <6S>.

6W6GT [AKA 6DG6GT]; beam power tube, heater 6.3 V @ 1.2 A, 10 W; sim. to 6U6GT, 6Y6G (TS) (6-8-39) <7S>.

6W7G Sharp-cutoff pentode, 150-mA version of 6J7G (TS) (1-17-38) <7R>.

6X4, W [AKA EZ90]; min. full-wave cathode-type rectifier, min. 6X5GT, heater 6.3 V @ 450 mA, 1250 V PIV, 70 mA (RCA, SY) {R6277} (4-2-46) <5BS>. Variants: 5993, 6063, 6202, 6203.

6X5, G, GT, L, WGT [AKA VT-126, CV279]; metal full-wave rectifier, heater 6.3 V @ 600 mA, 1250 V PIV, 70 mA (RA, RCA, SY) {R6019} (2-13-36) <6S>. Variant: 5852. "G" [AKA VT-126A, CV572] is octal 84 in ST12 bulb (RCA) {R6028}. "GT" [AKA VT-126B, CV574] has T9 bulb (4-11-38). Variants: 1274, 5838, 5839. "L" and "WGT" are ruggedized.

6X6G Electron-ray tuning indicator, octal (Rogers) (4-1-37) <8EV>.

6X8 Min. 9-pin triode-pentode, heater 6.3 V @ 450 mA, triode μ 70, pentode g_m 5.5 mS (SY) (4-2-51) <9AK>.

6Y3G Diode rectifier (Can.), heater 6.3 V @ 700 mA (1-9-42) <4AC>.

6Y5 Spray-shielded full-wave 6-pin rectifier, heater 6.3 V @ 800 mA, 350 V in, 50 mA <6J>. A mercury type as made by Majestic, but a vacuum type as produced by RCA and TS.

6Y6G, GA, GT [AKA VT-168A, 38766Y, CV205]; beam power tube for low B+ voltage, ST14 octal bulb, heater 6.3 V @ 1.25 A, 12.5 W, sim. to 6U6GT, 6W6GT (RCA) {A4072} (7-15-37) <7S>. "GA" has T12 bulb; "GT," T9 bulb.

6Y7G Class B duotriode, octal 79 (TS) (1-5-37) <8B.

6Z3 [AKA 1-V, q. v.]; rectifier, w/drawn.

6Z4 [AKA 98, 84/6Z4, q. v.]; full-wave rectifier (8-15-35) <5D>.

6Z5 [AKA 12Z5]; full-wave rect, 84 char., heater 12.6 VCT @ 300 mA, 230 V in, 60 mA (Maj., NU) <6K>.

6Z7G Dual power triode for Class B amps, 300-mA heater, 16 W total (TS) (5-24-37) <8B>.

6ZY5G Full-wave rectifier, heater 6.3 V @ 300 mA, 1250 V PIV, 40 mA, like low-htr.-cur. 6X5G (RCA, TS) {R6052} (5-25-37) <6S>.

7A4 [AKA VT-192, CV1770, XXL]; loktal triode, 6J5 char. w/ short-lead VHF construction (SY) (3-8-39) <5AC>.

7A5 Loktal beam power tube, heater 6.3 V @ 750 mA, 35L6 char. but derated to 5.5 W (SY) (5-4-39) <6AA>.

7A6 Duodiode, 150-mA loktal 6H6 (SY) (6-10-38) <7DX>. Variant: 5679.

7A7, LM Pentode, loktal 6SK7 (SY) (12-9-38) <8V>. "LM" is rare "loktal-metal" version in metal shell with "octalox" pin wafer (RCA) (9-21-39).

7A8 Octode conv., 150-mA loktal 7B8, sim. to 6D8G (SY)) (6-10-38) <8U>.

7AB7 [AKA 1204]; short loktal VHF pentode, heater 6.3 V @ 150 mA, g_m 1.8 mS (SY, TS) {SC651} (11-5-45) <8BO).

7AD7 Loktal power/video pentode, heater 6.3 V @ 600 mA, g_m 9.5 mS (SY) (4-2-47) <8V>.

7AF7 [AKA M1039]; loktal med.-μ duotriode, heater 6.3 V @ 300 mA, μ 16, 2.75 W (NU, SY) (11-30-44) <8AC>.

7AG7 Loktal sharp-cutoff RF pentode, heater 6.3 V @ 150 mA, g_m 4.2 mS (SY) (11-15-45) <8V>.

7AH7 Loktal semi-remote-cutoff pentode, heater 6.3 V @ 150 mA, g_m 3.3 mS (TS) (1-30-47) <8V>.

7AJ7 Loktal sharp-cutoff pentode for preamps, heater 6.3 V @ 300 mA, g_m 1.6-2.3 mS (SY) (8-15-47) <8V>.

7AK7 Loktal pentode w/ suppressor brought out as gating electrode; heater 6.3 V @ 800 mA, g_m 6 mS (SY) (3-20-48) <8V>. Claimed by SY as the first tube developed specifically for computer use.

7AN7 [AKA PCC84]; min. 9-pin med.-μ duotriode, heater 7.0 V @ 300 mA, μ 24, for VHF cascode amps. (HY) (8-15-55) <9DD>.

7AU7 Min. duotriode, 12AU7 w/ heater 7.0 VCT @ 300 mA (HY, SY) (11-15-54) <9A>.

7B4 High-μ triode, loktal 6F5 (SY) (1-30-40) <5AC>.

7B5, LT Power pentode, loktal 41 or 6K6 (SY) (2-16-39) <6AE>. "LT" (GE) is rare soldered-seam metal version (3-13-40); for RCA "LT," see 35A5LT.

7B6, LM Loktal duodiode - high-μ triode, 300-mA variant of 7C6 (SY) (2-16-39) <8W>. For "LM," see 7A7LM (RCA) (3-13-40).

7B7 Remote-cutoff pentode, 150 mA, g_m 1.8 mS (SY) (12-8-38) <8V>.

7B8, LM [AKA VT-208, CV22]; pentagrid conv., loktal 6A8 (SY) (12-16-39) <8X>. For "LM" [AKA VT-208A], see 7A7LM (RCA) (3-13-40).

7C4 [AKA 1203A]; loktal UHF diode, heater 6.3 V @ 150 mA, 420 V PIV, 9 mA (3-26-43) <4AH>.

7C5, LT Beam power tube, loktal 6V6GT (SY) (2-16-39) <6AA>. Also see 2C48. "LT" is rare "soldered-seam" version (GE); for RCA "LT," see 35A5LT (3-13-40).

7C6 Loktal duodiode-triode, 150-mA 7B6 (SY) (12-9-38) <8W>.

7C7 [AKA VT-193, CV76]; loktal sharp-cutoff pentode, heater 6.3 V @ 150 mA, g_m 1.3 mS, sim. to 6W7 (SY) (3-31-39) <8V>. Variant: 1273.

7D4 [AKA NU-1018]; never registered? (NU).

7D7 Loktal diode-pentode, heater 6.3 V @ 450 mA (7-6-39) <8AR>.

7E5 [AKA VT-241, 1201, CV890]; loktal UHF triode, heater 6.3 V @ 150 mA, μ 36, g_m 3.0 mS, 4 W; unusually short leads; oscillates to 750 mHz (TS) (3-26-43) <8BN>.

7E6 [AKA VT-188, CV891]; duodiode - med.-μ triode, loktal 6R7 (TS) (4-12-39) <8W>.

7E7 Duodiode-pentode, loktal 6B8 (TS) (6-12-39) <8AE>.

7F7, W [AKA VT-189, CV893]; duotriode, loktal 6SL7 (8AC) (SY) (6-12-39) <8AC>. "W" is ruggedized.

7F8, W Short loktal VHF duotriode, 300 mA, μ 48 (SY) (7-31-44) <8BW>. "W" is ruggedized.

7G7 [AKA 1232]; loktal sharp-cutoff "television" pentode, heater 6.3 V @ 450 mA, g_m 4.5 mS, sim. to 7L7 (TS) (1-30-40) <8V>.

7G8 [AKA 1206]; short loktal VHF twin tetrode, heater 6.3 V @ 300 mA, 1.65 W, usable to 400 MHz (SY, TS) {SD706} (3-22-44) <8BV>. Variant: 1296.

7H7 [AKA VT-190, CV75]; loktal remote-cutoff TV-FM IF pentode, 300 mA, g_m 3.8 mS; sim. to 6AB7, 6SG7 (SY) (3-20-40) <8V>.

7J7 [AKA VT-194, CV22]; triode-hexode conv., loktal 6J8G, sim. to 7S7 (SY) (7-12-39) <8BL>.

7K7 Loktal duodiode-triode; diodes have own cathode; triode equals ½ 7F7 (SY) (8-15-41) <8BF>.

7L7 Loktal sharp-cutoff pentode, sim. to 7G7 w/ 300 mA heater current; also to 6SH7, 7H7; g_m 3.0 mS (RA, SY) (3-1-40) <8V>.

7N7 Duotriode, loktal 6SN7GT (RA, SY) (4-15-40) <8AC>.

7Q7 Pentagrid conv., loktal 6SA7 w/ higher conversion g_m of 0.55 mS (SY) (8-21-39) <8AL>.

7R7 Loktal duodiode - remote-cutoff pentode, like 6SF7 w/ g_m raised to 3.0 mS and a second diode (SY) (3-7-41) <8AE>.

7S7 Loktal triode-heptode conv. sim. to 6J8, 7J7; higher conversion g_m (0.525 mS) than 7J7 (SY) (5-19-41) <8BL>.

7T7 Loktal triple-grid amplifier sim. to 7L7, 7G7, 7H7; heater 6.3 V @ 300 mA, g_m 4.9 mS (GE, TS) (3-4-42).

7V7 [AKA 1231]; loktal sharp-cutoff power/video pentode, heater 6.3 V @ 450 mA, g_m 5.8 mS, 4.0 W (SY) (1-21-41) <8V>. 1231 was the first loktal tube released, in May 1938. Actually has four grids, nos. 2 and 3 being connected together as screen. Also see 7W7.

7W7 Loktal variable-μ RF-IF pentode, essentially same as 7V7 except for capacitances (SY) (11-7-41) <8BJ>. Variant: 1282.

7X6 Full-wave rectifier, heater 6.3 V @ 1.2 A, 700 V PIV, 75 mA (TS) (7-6-49) <7AJ>.

7X7 [AKA XXFM]; loktal duodiode - hi-μ triode; one diode has own cathode; heater 6.3 V @ 300 mA, μ 100 (SY) (11-30-44) <8BZ>.

7Y4 Loktal full-wave rectifier, sim. to 6X5 w/ heater cut to 500 mA and limited to 70 mA out (SY) (12-9-38) <5AB>.

7Z4 [AKA VT-181, CV252]; full-wave loktal rectifier, robust 7Y4, heater 6.3 V @ 900 mA, 1250 V PIV, 100 mA out (SY) (12-30-40) <5AB>.

12A5 Power pentode, ST12 bulb, 7 pins, heater 12.6 VCT @ 300 mA, 8 W (*not* a 12-V 6A5) (JB, RA, TS) <7F>.

12A6, Y, GT, GTY [AKA VT-134, CV525]; metal power pentode, heater 12.6 V @ 150 mA, 7.5 W (RCA) (10-16-39) <7S>. Variant: 5659. "GT" has T9 bulb (5-21-43). "Y" and "GTY" have Micanol bases.

12A7GT/G Diode rectifier - power pentode (*not* a 12-V 6A7 or 25A7), ST12 bulb w/ cap, 7-pin base, heater 12.6 V @ 300 mA, pentode 550 mA out; rectifier 30 mA (RCA, TS) {A512} <7K>.

12A8G, GT Pentagrid conv., 12-V versions of 6A8G and 6A8GT (SY) (1-7-39) <8A>.

12AB5 Min. beam power tube, heater 12.6 V @ 200 mA, 12 W, sim. electrically to 12V6 (SY) (7-5-55) <9EU>. Variant: 7061.

12AD7 Min. 9-pin high-μ duotriode, 12AX7 w/ heater raised to 225 mA, promoted on basis of low hum (HY, SY) (10-10-55) <9A>.

12AH7GT [AKA VT-207, CV529]; duotriode, 12-V 6AH7GT (TS) (2-5-41) <8BE>.

12AH8 Min. 9-pin triode-heptode conv., heater 12.6 VCT @ 150 mA (HY) (10-25-51) <9BP>.

12AJ7 Min. 9-pin triode-heptode, heater 12.6 V @ 150 mA (HY) (12-19-55) <9CA>.

12AL5 Min. duodiode, 12-V 6AL5 (SY) (6-6-47) <6BT>. Variant: 7055.

12AT6 Min. duodiode-triode, sim. to min. 12Q7 (RCA, SY) (10-25-45) <7BT>.

12AT7, A, WA, WB, WC [AKA ECC81, B152, B309]; min. 9-pin high-μ duotriode, dual 6AB4, heater 12.6 VCT @ 150 mA, μ 62 (SY) (5-20-47) <9A>. "WA," "WB," and "WC" are ruggedized versions w/ progressively tighter process controls. Variants: 6060, 6201, 6677, 6679, 7728.

12AU6 Min. pentode, 12-V 6AU6 (SY) (12-20-46) <7BK>. Variant: 8426.

12AU7, A, W [AKA ECC81]; min. 9-pin med.-μ duo-triode, separate cathodes (dual 6C4), heater 12.6 VCT @ 150 mA, μ 17 (RCA, SY) {A4498} (10-18-46) <9A>. "A" is low-microphonics version. Variants: 5814, 5963, 6067, 6189, 6670, 6680, 7316, 7318, 7489, 7730.

12AV6 Min. duodiode - high-μ triode, 12-V 6AV6 (RCA) (12-20-47) <7BT>.

12AV7 Min. low-noise sharp-cutoff duotriode, dual 6BC4, heater 12.6 VCT @ 225 mA, μ 48 (SY) (2-14-50) <9A>. Variants: 5965, 6829.

12AW6 Min. sharp-cutoff pentode, 12-V 6AG5 (SY) (2-3-47).

12AX7, A, WA [AKA ECC83]; min. 9-pin high-μ duotriode, min. 6SL7GT, heater 12.6 VCT @ 150 mA μ 100 (SY) (9-25-47) <9A>. "A" has lower hum and microphonics. Variants: 6EU7, 12AD7, 5721, 5751, 6057, 6681, 7025, 7058, 7494, 7729.

12AY7 Min. 9-pin high-μ duotriode, for audio preamps, heater 12.6 VCT @ 150 mA, μ 48 (GE) (12-7-48) <9A>. Variants: 2082, 6072.

12AZ7, A Min. 9-pin high-μ duotriode, heater 12.6 VCT @ 225 mA, μ 60 (SY) (3-2-51) <9A>. Variants: 6085, 7898.

12B6M Diode-triode (Can.), octal w/ cap, heater 12.6 V @ 150 mA, μ 100, spray-shielded (7-1-40) <6Y>.

12B7, ML [AKA 14A7, q. v.]; loktal pentode (*not* a 12-V 6B7) (6-6-39). "ML" is rare soldered-can metal version (GE) (7-7-39).

12B8GT Triode-pentode, heater 12.6 V @ 300 mA, triode μ 97, pentode g_m 1.8 mS (*not* a 12-V 6B8) (GE) (2-15-39) <8T>.

12BA6, A Pentode, min. 12SG7 (GE) (10-25-45) <7BK>.

12BA7 Pentagrid VHF conv., 12-V 6BA7 (GE) (6-9-48) <8CT>.

12BD6 Min. RF pentode, 12-V 6BD6 (GE) (5-15-46) <7BK>.

12BE6, A Pentagrid conv., min. 12SA7 (GE) (10-25-45) <7BK>.

12BF6 Min. rcvg. duodiode -triode, min. 12SR7 (GE) (10-25-45) <7BT>.

12BH7, A Min. 9-pin med.-μ duotriode based on 12SN7, heater 12.6 V @ 300 mA, μ 16.5 (GE) (5-19-50) <9A>. Variant: 6913.

12BK6 Duodiode-triode, different from 12AV6 only in being 7/8" taller (GE) (1-17-50) <7BT>.

12BR7, A Min. 9-pin duodiode - high-μ triode; diodes have own cathode; heater 12.6 VCT @ 225 mA, μ 60 (GE) (1-24-55) <9CF>. Variant: 8447.

12BT6 Min. duodiode - high-μ triode; differs from 12AT6 only in being ½" taller (1-17-50) <7BT>.

12BU6 Min. duodiode - low-μ triode; 12BT6 w/ μ of 16 (GE) (1-17-50) <7BT>.

12BV7 Pentode, 12BY7 w/ g_m 12 mS instead of 13 (GE) (2-21-55) <9BF>.

12BY7, A [AKA 12BV7]; min. 9-pin sharp-cutoff RF/video pentode, heater 12.6 VCT @ 300 mA, g_m 13 mS, 6.25 W (6-11-51) <9BF>. Variants: 7054, 7733, 8448.

12C8, Y, -K9, Spl. [AKA VT-169]; metal duodiode - semi-remote-cutoff pentode; a 12-V 6B8, *not* a 12-V 6C8; heater 12.6 V @ 150 mA, g_m 1.3 mS;(GE) (2-2-39) <8E>. Variant: 5660. "Y," "-K9," and "Spl" [AKA VT-153, 1664, CV128] have Micanol base and shell connected to Pin 1. Variant: 1664.

12DW7 Min. 9-pin asymmetric duotriode, equals ½ 12AU7 + ½ 7025, heater 12.6 VCT V @ 150 mA, μs 17 and 100 (1-26-59). Variant: 7247.

12E5GT/G Triode, 12-V 6P5GT/G (GE) (7-26-39) <6Q>.

12F5GT Triode, 12-V 6F5GT (TS) (1-7-39) <5M>.

12G4 Min. med.-μ triode, min. 12J5GT (GE, HY) (4-10-53) <6BG>.

12G7G Duodiode-triode, octal base, heater 12.6 V @ 150 mS, μ 70 <7V>.

12H4 Min. med.-μ triode, heater 12.6 VCT @ 150 mA, μ 20 (GE, HY) (11-20-53) <7DW>.

12H6 [AKA VT-214, CV916]; metal duodiode, 12-V 6H6 (GE) (5-5-41) <7Q>.

12HG7 Ruggedized frame-grid version of 12BY7A, uprated to 10 W and g_m 32 mS; heater cut to 260 mA (GE, RCA) (12-7-64) <9BF>.

12J5, GT, WGT [AKA VT-135A, CV534]; metal triode, 12-V 6J5 (GE) (4-6-50) <6Q>. "GT" [AKA VT-135, CV535] has T9 bulb (3-28-39). "WGT" is ruggedized "GT."

12J7G, GT Pentode, 12-V 6J7G (GE) (2-6-40) <7R>.

12K7G, GT Pentode, 12-V 6K7G (GE) (5-22-39) <7R>.

12K8, Y, Spl., GT Metal triode-hexode conv., 12-V 6K8 (RCA) (10-16-39) <8K>. "Y" and "Spl." [AKA VT-132, CV24] have Micanol bases. "GT" has T9 bulb.

12L6GT Beam power tube, 35L6GT w/ heater 12.6 V @ 600 mA (GE) (5-7-54) <7AC>. Variant: 1632.

12L8GT Dual power pentode, dual 6G6, heater 12.6 V @ 150 mA, 2.5 W (GE) (8-27-43) <8BU>. Variant: 1644.

12Q7G, GT Duodiode-triode, 12-V 6Q7G (GE) (1-7-39) <7V>.

12S8GT Triple-diode-triode, 12-V 6S8GT (GE) (1-30-47) <8CB>.

12SA7, Y, GT/G, GTY [AKA VT-161, CV537]; metal pentagrid conv., 12-V 6SA7 (RCA) (2-2-39) <8R>. "Y" has Micanol base. "GT/G" has T9 bulb (12-1-41) <8AD>.

12SC7 [AKA VT-268, CV540]; metal duotriode, 12-V 6SC7 (GE) (2-2-39) <8S>. Variant: 1634.

12SF5, GT Metal triode, 12-V 6SF5 (GE) (9-21-39) <6AB>.

12SF7, Y Metal diode-pentode, 12-V 6SF7 (GE) (3-3-41) <7AZ>. "Y" has Micanol base. "GT" has T9 bulb (2-1-46).

12SG7, Y, GT [AKA VT-209, CV694];metal pentode, 12-V 6SG7 (GE) (2-17-41) <8BK>. "Y" has Micanol base. "GT" has T9 bulb.

12SH7, GT [AKA VT-288, CV922]; metal pentode, 12-V 6SH7 (GE) (8-15-41) <8BK>. "GT" has T9 bulb.

12SJ7 [AKA VT-162, CV697]; metal pentode, 12-V 6SJ7 (GE) (2-2-39) <8N>.

12SK7, Y, GT/G, GTY [AKA VT-131, CV543]; metal pentode, 12-V 6SK7 (GE) (2-2-39) <8N>. Variant: 5661. "GT/G" has T9 bulb (12-1-41). "Y" and "GTY" have Micanol bases.

12SL7GT, WGT [AKA VT-289, CV924]; duotriode, 12-V 6SL7GT (GE) (6-2-41) <8BD>. "WGT" is ruggedized.

12SN7GT, GTA, WGT Duotriode, 12-V variant of 6SN7GT (GE) (4-22-41) <8BD>. "GTA" has controlled warm-up time.

12SQ7, GT [AKA VT-104, CV546]; metal duodiode-triode, 12-V 6SQ7 (GE) (2-2-39) <8Q>. "GT" has T9 bulb (5-1-39).

12SR7, GT [AKA VT-133, CV700]; metal duodiode-triode, 12-V 6SR7 (GE) (10-16-39) <8Q>. "GT" has T9 bulb (5-21-43).

12SW7, GT Metal duodiode-triode; version of 12SR7 for operation from aircraft DC power (GE) (6-20-46) <8Q>. "GT" (also 6-20-46) was disc. in favor of 12SW7.

12SX7GT Duotriode, version of 12SN7GT for operation from aircraft DC power (GE) (6-20-46) <8BD>.

12SY7, GT Metal pentagrid conv., version of 12SA7(GT) for operation from aircraft DC power (9-25-45) <8R>. "GT" has T9 bulb <8AD>.

12X3 Diode, bulb w/ cap (Can.) (1-31-44). HV rectifier in AN/VVX-1.

12Y4 Duodiode (Can.), heater type. In W. S. (Cdn.) No. 9 (4-15-44) as det.-AVC tube.

12Z3 [AKA 14Z3]; rectifier, ST12 bulb, 4-pin base, heater 12.6 V @ 300 mA, 700 V PIV, 55 mA (GE, RCA, TS) {R6001} <4G>.

12Z5 [AKA 6Z5, q. v.]. (JB).

13CW4 Nuvistor triode, 6CW4 w/ heater 13.5 V @ 60 mA, promoted for antenna-mounted (remote-powered) TV preamps (RCA) (11-26-62) <12AQ>.

14A4 Loktal UHF triode, 12-V 7A4 (GE) (10-24-40) <5AC>.

14A5 Loktal power pentode, 12-V 7A5 (GE) (10-24-40) <6AA>.

14A7 [AKA 12B7]; loktal pentode, 12-V 7A7 (SY) (10-28-40) <8V>. Intro. by SY as 12B7.

14AF7 [AKA XXD]; duotriode, 12-V 7AF7 (SY) (7-21-42) <8AC>.

14B6 Loktal duodiode-triode, 12-V 7B6 (GE) (3-1-40) <8W>.

14B8 Loktal pentagrid conv., 12-V 7B8 (GE) (10-24-40) <8X>.

14C5 Loktal power pentode, 12-V 7C5 (SY) (10-14-40) <6AA>.

14C7 Loktal sharp-cutoff pentode, slightly different from 7C7; optimized for 120-V B+ (SY) (10-14-40) <8V>. Variant: 1280.

14E6 Loktal duodiode-triode, 12-V 7E6 (GE) (11-18-40) <8W>.

14E7 Loktal duodiode-pentode, 12-V 7E7 (GE) (4-22-41) <8AE>.

14F7 Loktal duotriode, 12-V 7F7 (SY) (10-24-40) <8AC>.

14F8 Rcvg, loktal duotriode, 12-V 7F8 (GE) (7-29-46) <8BW>.

14GT8, A Min. 9-pin rcvg. duodiode - high-μ triode, heater 14.0 V @ 150 mA, μ 72, separate cathodes for diodes (GE) (12-14-59) <9KR>. Variant: 7724.

14H7 Loktal pentode, 12-V 7H7 w/ g_m reduced from 4.0 to 3.8 mS (GE) (8-9-40) <8V>.

14J7 Loktal triode-heptode conv., 12-V 7J7, sim. to 14S7 (GE) (8-23-39) <8BL>.

14K7 [AKA UCH42]; octal triode-hexode, *not* a 12-V 7K7, heater 14.0 V @ 100 mA, part of Euro. 100-mA AC-DC series (10-10-55) <8GY>.

14L7 [AKA UBC41]; octal duodiode-triode, *not* a 12-V 7L7, heater as for 14K7 (10-10-55) <8GZ>.

14N7 Loktal duotriode, 12-V 7N7 (GE) (10-14-40) <8AC>.

14Q7 Loktal pentagrid conv., 12-V 7Q7 (SY) (3-1-40) <8AL>.

14R7 Loktal duodiode-pentode, 12-V 7R7 (SY) (3-7-41) <8AE>.

14S7 Loktal triode-heptode conv., 12-V 7S7, sim. to 14J7 (GE) (11-7-41) <8BL>.

14W7 Pentode, 12-V 7W7 (GE) (11-7-41) <8BJ>.

14X7 Loktal duodiode - high-μ triode, 12-V 7X7 (GE) (4-4-47) <8BZ>.

14Y4 Loktal rectifier, 12-V 7Y4 (GE) (10-7-40) <5AB>.

14Y7 (3-4-57). No further information.

14Z3 Diode rectifier, 4-pin base, heater 14.0 V @ 300 mA, 250 V in, 60 mA <4G>.

18FW6, A Min. remote-cutoff pentode, 12BA6 w/ heater 18 V @ 100 mA, part of SY 100-mA AC-DC lineup w/ 18FW6, 18FY6, 34GD5, and 36AM3 (SY, RCA) (12-15-58) <7BK>.

18FX6 Min. pentagrid converter, 12BE6 w/ heater 18 V @ 100 mA (SY, RCA) (12-15-58) <7CH>. Also see 18FW6.

18FY6 Min. duodiode - high-μ triode, 12AV6 w/ heater 18 V @ 100 mA (RCA, SY) (12-15-58) <7BT>. Also see 18FW6.

20A5 Gas trigger tube (SY).

20J8GM Triode-heptode conv. (Can.), 6J8 w/ heater 20.0 V @ 150 mA, spray-shielded (Rogers?) (9-18-39) <8H>.

25A6, GT Metal power pentode, octal 43, heater 25 V @ 300 mA, 5.3 W (RA, RCA, TS) {A4011} (12-9-35) <7S>. "GT" (4-11-38) has T9 bulb (RCA) {A4037}.

25A7GT/G Octal diode rectifier - pentode (*not* a 25-V 6A7 or 12A7), heater 25 V @ 300 mA; pentode is small, at 2.25 W (TS) (6-23-36) <8F>.

25AC5GT/G Triode, 25-V 6AC5; same suggested driver tubes (TS) (10-18-38) <6Q>.

25B5 Duotriode direct-coupled amp., sim. to 6B5 but derated; heater 25 V @ 300 mA (TS) (8-17-36) <6D>.

25B6, G, GT Metal power pentode, heater 25 V @ 300 mA, 12.5 W (*not* a 25-V 6B6) (TS) (8-5-36) <7S>. Variant: 5824. "GT" has T9 bulb (3-28-39).

25B8GT Triode-pentode, 25-V 12B8GT, but *not* a 25-V 6B8GT (GE, TS) (3-28-39) <8T>.

25C6G, GA Beam power tube, 6Y6G w/ heater 25 V @ 300 mA (*not* a 25-V 6C6) (GE, TS) (5-15-39) <7AC>. "GA" has T12 bulb.

25D8GT Diode-triode-pentode (*not* a 25-V 6D8), heater 25 V @ 150 mA, triode μ 100, pentode g_m 1.9 mS (TS) (3-31-39) <8AF>.

25L6, G, GT [AKA VT-201, CV220]; metal beam power tube, 35L6GT w/ heater 25 V @ 300 mA, 10 W (RCA) {A4050} (11-4-36) <7AC>. "G" has ST14 bulb (RCA, TS) {A4050} (3-30-37). Variant: 1285. "GT" [AKA VT-201C, CV551] has T9 bulb (4-11-38). Variant: 6046.

25N6G Duotriode, octal 25B5 (TS) (8-17-36) <7W>.

25X6GT Full-wave rect./doubler, heater 25 V @ 150 mA, 125 V in, 60 mA (3-31-39) <7Q>.

25Y4, GT [AKA PY31]; metal diode rectifier, heater 25.0 V @ 150 mA, 125 V in, 75 mA (6-6-39) <5AA>.

25Y5 Full-wave rectifier/doubler, heater 25 V @ 300 mA, 700 V PIV, 85 mA per plate (GE) (11-7-34) <6E>.

25Z3 Rectifier, heater 25.0 V @ 300 mA (5-6-41) <4G>.

25Z4, GT Metal rectifier sim. to 25Y4GT, same heater (12-12-38) <5AA>.

25Z5 [AKA 38255]; full-wave rectifier/doubler, heater 25 V @ 300 mA, 700 V PIV, 75 mA per plate (RCA, TS) {R604} (8-30-35) <6E>.

25Z6, GT, WGT Metal full-wave rectifier/doubler, octal 25Z5 (RCA) {R6014} (12-9-35) <7Q>. "GT" is in T9 bulb (RCA) {R6027} (4-11-38).

26A5 Min. beam power tube, heater 26.5 V @ 300 mA, for 26.5 V B+; char. sim. to half of 26A7 (RCA) {A4449}.

26A6 Pentode, 6BA6 w/ heater 26.5-V @ 70 mA (HY, RCA) {A4465E} (6-20-46) <7BK>.

26A7GT Twin beam power tube (*not* a 25-V 6A7), heater 26.5 V @ 600 mA, optimized for 26.5-V B+, 4.0 W total (RCA, TS) {A5637E} (6-8-45) <8BU>.

26BK6 Min. duotriode-triode, 6BK6 w/ heater 26.5 V @ 70 mA (HY) (1-17-50) <7BT>.

26C6 Min. duodiode - low-μ triode; 6AT6 w/ heater 26.5 V @ 70 mA, optimized for 26-V B+ (HY, RCA) {A4454} (6-20-46) <7BT>.

26CG6 Min. remote-cutoff pentode, heater 26.5 V @ 70 mA, g_m 2.0 mS (HY) (10-10-50) <7BK>.

26D6 Pentagrid conv., 6BE6 w/ heater 26.5-V @ 70 mA (RCA) (6-20-46) <7CH>.

26E6WG Octal beam power tube, heater 26.5 V @ 300 mA, 12.5 W (TS) (10-3-55) <7S>.

26Z5W Min. 9-pin full-wave rectifier, heater 26.5 VCT @ 200 mA, 100 mA out (TS) (3-15-51) <9BS>.

28D7, W Loktal dual beam power tube, for 26.5 V on heater and plates, heater 400 mA, 6 W total (SY) (10-1-42) <8BS>. In BC-1206A, AN/ARN-5, etc. Variant: 1238.

28Z5 Full-wave loktal rectifier, 7Z4 w/ heater 28 V @ 240 mA (10-1-42) <6BJ>.

32L7GT Diode rectifier - beam power tube, heater 32.5 V @ 300 mA, 1.5 W out (TS) (2-15-39) <8Z>.

34GD5, A Min. beam power tube, 35C5 w/ heater 34 V @ 100 mA, 5 W (RCA, SY) (6-13-60) <7CV>. Also see 18FX6.

35A5, LT Beam power tube, loktal 35L6GT (TS) (12-9-38) <6AA>. "LT" is rare RCA version w/ "octalox" metal-banded base.

35B5 Min. beam power tube, min. 35L6GT (TS) (11-20-46) <7BZ>.

35C5, A Min. beam power tube (35B5 w/ revised pinout so that AC-DC receivers would not have 117 V on chassis even w/ 35C5 in wrong socket) (TS) (2-16-48) <7CV>.

35D5 (3-24-58). No further information.

35L6GT/G Beam power tube, heater 35 V @ 300 mA, 8.5 W (TS) (1-6-39) <7AC>.

35W4, A Rectifier, min. 35Z5GT/G (TS) (10-25-45) <5BQ>.

35Y4 Rectifier, loktal 35Z5 (TS) (11-8-40) <5AL>.

35Z3, LT Rectifier, loktal 35Z4 (TS) (12-9-38) <4Z>. For "LT," see 35A5LT.

35Z4, GT Rectifier, no lamp tap (1-6-39) <5AA>.

35Z5GT/G Rectifier, w/ lamp tap, heater 35 V @ 150 mA, 700 V PIV, 100 mA (TS) (2-17-39) <6AD>.

35Z6G Full-wave rectifier, 25Z6 uprated to 110 mA, heater 35.0 v @ 300 mA (TS) (3-15-40) <7Q>.

36AM3, A, B — Min. rectifier, w/ lamp tap, heater 36 V @ 100 mA, 365 V PIV, 82 mA (RCA, SY, TS) (12-15-58) <5BQ>. Also see 18FX6.

40A1 — Ballast tube for use in cathode ckt. of TV sweep tubes like 6BQ6GT (SY) (6-20-52).

40Z5GT — [AKA 45Z5GT, q. v.]; rectifier (12-18-39) <6AD>.

45Z3 — Min. rectifier, no lamp tap; heater 45 V @ 75 mA, introduced for use in 3-way powered portable radios (RCA, TS) {R6130} (4-1-41) <5AM>.

45Z5GT — [AKA 40Z5GT]; rectifier, 45-V 35Z5GT (TS) (3-24-39) <6AD>.

50A1 — Min. 9-pin ballast, 30 V @ 52 mA to 65 V @ 56 mA (Amperite, HY) (5-1-53) <9CM>. Apparently unique to late-model tube Zenith Trans-Oceanics.

50A2 — [AKA 165R8, q. v.]; ballast (5-28-35).

50A5 — Beam power tube, loktal 50L6GT (TS) (11-8-40) <6AA>.

50B2 — [AKA 165R4, q. v.]; ballast (5-28-55).

50B5 — Min. beam power tube, min. 50L6GT (RCA, TS) {A4388} (10-25-45) <7BZ>.

50C5, A — Min. beam power tube, 50-V 35C5 (TS) (2-16-48) <7CV>.

50C6G, GA — Beam power tube, 50-V 25C6G (TS) (10-9-39) <7AC>.

50L6GT/G — Beam power tube, 50-V 35L6GT, uprated to 10 W (TS) (3-31-39) <7AC>.

50X6 — Rectifier, loktal 50Y6 (TS) (10-29-46) <7AJ>.

50Y6GT/G — Full-wave rectifier/doubler, 50-V 25Z6 (TS) (12-2-40) <7Q>.

50Y7GT — Full-wave rectifier ("G" 50Y6 w/ lamp tap) (TS) (4-30-48) <8AN>.

50Z6G — Full-wave rectifier/doubler, sim. to 50Y6GT/G, heater 50 V @ 300 mA (4-18-38) <7Q>.

50Z7G — Full-wave rect./doubler, sim. to 50Y6GT/G, heater 50 V @ 150 mA w/ lamp tap (TS) (6-2-39) <8AN>.

70A7GT — Diode rectifier - beam power tube w/ lamp tap, heater 70 V @ 150 mA, 1.5 W out, rectifier 60 mA (TS) (6-30-39) <8AB>.

70L7GT — Diode rectifier - beam power tube, no lamp tap, heater 70 V @ 150 mA, 1.8 W out, rect. 70 mA, sim. to 25L6 + 117Z4 (TS) (3-28-39) <8AA>.

117L7/ M7GT — Diode rectifier - beam power tube, heater 117 V @ 90 mA, 6.0 W; rectifier 350 V PIV, 75 mA (TS) (9-25-39) <8AO>.

117N7GT — Diode rectifier - pentode, sim. to 117P7GT except for pinout; more power than 117L/M7 (TS) (5-15-40) <8AV>.

117P7GT — Diode rectifier - pentode, heater 117 V @ 90 mA, 1.0 W, rectifier 350 V PIV, 117N7GT w/ pinout scrambled (TS) (1-7-41) <8AV>.

117Z3 — Rectifier, min. 117Z4GT, heater 117 V @ 40 mA, 330 V PIV, 90 mA (TS) (10-31-45) <4CB>.

117Z4GT — Rectifier, heater 117 V @ 40 mA, 350 V PIV, 90 mA (TS) (4-22-41) <5AA>.

117Z6 GT/G, GC — Full-wave rectifier/doubler, heater 117 V @ 75 mA, 700 V PIV, 60 mA per plate (TS) (8-5-39) <7Q>. Introduced with center-tapped heater; tap abandoned ca. 1940. "GC" (SY) had center-tapped heater like origonal 117Z6G; produced temporarily.

SPECIAL-PURPOSE TYPES: 800-SERIES

800 [AKA RK-30, VT-64, CV2657]; xmtg. triode, S21 bulb w/ grid and plate caps, UX base, T-T fil. 7.5 V @ 3.25 A, μ 15, 35 W, 60 MHz, intro. 10-33 (AX, GE, RCA) {A165} <2D>. In BC-320A.

801 [AKA 305D, q. v.]; xmtg. triode, graphite-anode, ceramic-based version of 10, uprated to 20 W (AX, GE, RCA) {A220} <4D>. Also see 801A. Variant: 1608.

801A/801 [AKA 38101, UE-310]; xmtg. triode, ST16 bulb, UX base, T-T fil. 7.5 V @ 1.25 A, μ 8, 20 W, 60 MHz (C, GE, HY, RCA, W) (9-24-43). Two in Navy WEA 25-kHz sonar.

802 Xmtg. general-use pentode, ST16 bulb w/ cap, 6-pin base, third grid wired out for suppressor modulation, heater 6.3 V @ 0.9 A, 10 W, 30 MHz (GE, RCA, W) {A4000} (9-24-43) <6BM>. Intro. 1935 under RCA-DeF brand.

803, W [AKA 38803, VT-106, CV623, (WE) 322A]; xmtg. pentode, T20 stepped-tubular bulb w/ cap, "giant" 5-pin ceramic base, graphite anode, third grid wired out for suppressor modulation, T-T fil. 10 V @ 5 A, 125 W, 20 MHz (GE, NU, RCA, T, W) {A492} (9-24-43) <5J>. Two in QCJ 17-26 kHz sonar.

804 [AKA RK-20]; xmtg. pentode, stepped T16 bulb w/ cap, med. 5-pin base, third grid wired out for suppressor modulation, T-T fil. 7.5 V @ 3.0 A, 40 W, 15 MHz (GE, RCA) {A4001} (9-24-43) <5J>. Intro. 1936.

805 [AKA (WE) 331A, RK-57, VT-143, CV625, UE-905]; high-μ xmtg. triode, T18 bulb w/ cap, jumbo 4-pin base, graphite anode, T-T fil. 10 V @ 3.25 A, variable μ for Class B modulator use, 125 W, 30 MHz (AX, GE, NU, RCA, T, W) {A312} (9-24-43) <3N>. Intro. 1936.

806 Xmtg. triode, GT30 spherical-tubular bulb w/ cap, jumbo 4-pin base, grid-out-side, Ta plate, T-T fil. 5 V @ 9.5 A, μ 12.6, 150 W, 30 MHz, "copy of Eimac 150T" (GE, RA, RCA, W) {A2043} (10-31-44) <2N>.

807, A, W, Modified [AKA HY61, 38807, VT-100, CV124]; xmtg. beam tetrode, ST16 bulb w/ cap, 5-pin Micanol base (originally ceramic), heater 6 V @ 900 mA, 25 W, 60 MHz, sim. to Br. VT60 and ATS25, intro. 10-37 (GE, HY, NU, RCA, RA, W) {A4051} (9-24-43) <5AW>. RK-39 is non-beam version. "A" and "W" are ruggedized. "807 modified" is undefined but probably relates to internal construction; later 807s were uprated from 600 V to 750 V B+. Variants: RK-66, 1624, 1625, 5933, 8007.

808 [AKA 38808]; xmtg. triode, G22 bulb w/ cap, grid-out-side, UX base, Ta plate, fil. 7.5 V @ 4 A, 50 W, 30 MHz, possibly copy of RK-35 (EE, GE, RCA, W) {A2048} (9-24-43) <2D>. One each in QBF 17-25 kHz sonar and TBS xcvr.

809 [AKA (HY, NU) 30Z]; xmtg. triode, ST19 bulb w/ cap, UX base (originally ceramic), T-T fil. 6.3 V @ 2.5 A, μ 50, 25 W, 60 MHz; sim to RK-11, -12; intro. 10-37 (GE, NU, RCA, W) {A2059} (9-24-43) <3G>. Variant: 1623.

810 Xmtg. triode, T20 bulb w/ cap, grid-out-side, jumbo 4-pin base, T-T fil. 10 V @ 4.5 A, μ 36, 125 W, 30 MHz (GE, RCA, W) (9-24-43) <2N>. Variant: 1627. Modulators and finals in Collins 300G and Gates BC-250T 250-W xmtrs; finals in RCA BTA-250L.

811, A 811-3, -10 [AKA VT-217, CV628]; xmtg. triode for Class B modulators, ST19 bulb w/ cap, UX base, T-T fil. 6 V @ 4 A, μ 160, 40 W, 60 MHz, intro. 9-39 (EE, GE, GN, NU, RCA, W) (9-24-43) <3G>. "A" has redesigned plate, uprated to 45 W (W: 65 W) but derated to 30 MHz (GE, SV, W) (4-25-49). Two as modulators in T-47/ART-13; four each in Collins 30L-1, early Gonset GSB-201, and Heathkit HA-10 linear amps. Variant: 572B. SV811-3 and -10 are present-day SV offerings w/ differing μ.

812, A Xmtg. triode for Class C amps., low-μ version of 811 (μ 29), intro. 9-39 (EE, GE, GN, NU, RCA, W) (9-24-43) <3G>. "A" has redesigned plate, uprated to 45 W (W: 65 W) but derated to 30 MHz (GE, RCA, W) (6-29-48).

813 [AKA VT-144, CV26]; xmtg. beam pentode, T20 bulb w/ cap, "giant" 7-pin base, T-T fil. 10 V @ 5 A, 100 W, 30 MHz, intro. 10-38 (GE, ML, NU, RA, RCA, T, UE, W) (9-24-43) <5BA>. Modulator in SJ; final in T-47/ART-13.

814 [AKA 4C24, HV12, q. v.]; RF-heating/xmtg. triode (T). *Unrelated to 814 below.*

814 [AKA 38814, VT-154, CV629]; xmtg. beam tetrode, stepped T16 bulb w/ cap, 5-pin base (originally ceramic), fil. 10 V @ 3.25 A, 50 (or 65) W, 30 MHz (GE, RCA, W) {A326A} (9-24-43). One in ATD; two in BC-652 of SCR-506. RK-48A is similar but higher-rated.

815 [AKA VT-287, CV2663]; dual xmtg. HF beam tetrode, T16 bulb w/ plate caps, button stem, metal-shell Micanol octal base, common screen, heater 12.6 VCT @ 800 mA, 20 W (originally 13.5 W) (W: 25 W), 150 MHz (GE, RCA, W) (4-5-45). Designed as low-cost 829. Variant: 3E22.

816 [AKA VT-216, CV724]; mercury rectifier, sim. to 866A w/ ½ ratings; ST12 bulb w/ cap, UX base, fil. 2.5 V @ 2.0 A; 7.5 kV PIV, 125 mA (uprated 2B26 w/ plate out top) (GE, HY, RCA, W) (9-24-43) <4P>. Variant: 6288.

820B [AKA 520B, F3652A, 1652, q. v.]; xmtg. triode, rated at 2.5 kW for broadcast xmtrs (SY).

822, S [AKA HV-27, q. v.]; graphite-anode high-μ variant of 814 in hard glass; μ 27 (T) <3N>.

825 Xmtg. UHF inductive-output amp., all-glass stepped-tubular bulb, seven pins at small end for htr. etc.; flying leads for collector, etc. at large end, for use w/ external cavity and electromagnet; heater 6.3 V @ 750 mA, 50 W, 500 MHz, ca. 6-40 (GE, NU, RCA). Uses quasi-klystron output coupling but does not employ velocity modulation. *Unrelated to 825 below.* No known commercial use, but TO 08-1-25 says "825" is used in BC-699.

825 Xmtg. triode, UX base, fil. 7.5 V @ 2 A, μ 20, 40 W (T) <4D>.

826 Xmtg. VHF triode, T16 all-glass construction, all leads out "dish" base, fil. 7.5 V @ 4 A, μ 31, 60 W w/ forced-air cooling, 250 MHz (GE, RCA, W) (9-24-43) <7BO>.

827R Xmtg. VHF beam tetrode, ring-seal construction, flying ribbon leads for fil. and grid, ring for screen, ext. anode w/ radiator; fil. 7.5 V @ 25 A, 800 W, 110 MHz, intro. 8-40 (GE, RCA) (6-15-45). (No evidence that there ever was a *plain* 827.) Driver in RCA 50-kW SW BC xmtr. Also see 7C24.

828 Xmtg. beam tetrode, variant of 814 for AB_1 audio amps, higher g_m, uprated to 70 W (GE, RCA) or 80 W (W) (8-1-45) <5J>. Modulators in RCA BTA-250L.

829, A, B [AKA VT-259, CV632]; dual VHF beam tetrode, all-glass/hard-glass T16 bulb w/ plate pins and "dish" base; scaled-up 832 w/ common screen, heater 12.6 VCT @ 1.125 A, 28 W (GE, NU, RA, RCA) (9-24-43). "A" [AKA 3E29, q. v.] supersedes 829 in, for example, AN/MPG-1 (RCA). "B" has Pt-plated Mo grid wire, uprated to 40 W, 200 MHz, intro. 12-43 (AX, EEV, GE, ML, MC, RA, RCA, SY, W) (10-26-43) <7BP>. In T-14/TRC-1. Variants: 3E21, 3E29, 5894.

830, B [AKA 530, 930]; xmtg. triode, stepped T16 bulb w/ cap, grid-out-side, UX base, T-T fil. 10 V @ 2 A, μ 8, 40 W, 6 MHz (AX). "B" [AKA UE-930B, HY-51B] is uprated to 60 W, 15 MHz (C, GE, RCA, T) {A2018B} <3G>.

831, A [AKA WL-674]; xmtg. triode, intro. 10-32, GT56 bulb w/ side arm and grid cap, blade-and-pins base, flying leads out side-arm for plate, fil. 11 V @ 10 A, μ 14.5, 400 W, 20 MHz (AX, RCA, W) {A245}. "A" has T-T fil. and graphite anode (SY) (1-34). Also see 861.

832, A [AKA VT-118; Br. VT88, CV634]; xmtg. VHF dual beam pentode, common screen, intro. 6-38, all-glass cylindrical bulb w/ 2 plate pins, "dish" base, heater 12.6 VCT @ 800 mA, 15 W, 200 MHz (GE, NU, RCA) (9-24-43) <7BP>. Includes internal screen-bypass cap. "A" [AKA VT-286, Br. CV634] has Pt-plated Mo grid wire (AX, EEV, GE, RA, RCA, W) (9-24-43). In SCR-522 and RT-18/ARC-1. Variants: 6252, 6907, 8024.

833, A Xmtg. and RF-heating triode, all-glass tubular bulb

833C w/ two prongs at each end, Ta plate, T-T fil. 10 V @ 10 A, μ 35, 300 W, 30 MHz, intro. 9-37 (C, RCA) {A2029B} (8-10-45, *after* 833A). Two in RCA ET-8017 xmtr., 350-500 kHz. "A" [AKA 38833, WT-210-0086] has Zr-coated plate, introduced ca. 10-40; uprated to 400 W w/ forced air (C, GE, GN, ML, RCA, SV, T, UE) or 450 W (W) (9-24-43). Finals and modulators in Gates BC-1D, 1946. "C" is present-day SV offering.

834 [AKA BW-11, (WE) 304B, RK-32, UH50]; xmtg. UHF triode, S21 bulb w/ grid and plate pins out top, ceramic UX base, T-T fil. 7.5 V @ 3.1 A, μ 10.5, 50 W, 100 MHz, intro. 1-36 (EI, GE, RA, UE) {A2009} <2D>. Variants: 8019 and Br. types NT58, NT58A, VT62 and DET12 are same except for Br. 4-pin base.

835 [AKA CG-1984]; xmtg. triode, low-capacitance variant of 211, C_{gp} reduced from 14.5 to 9.3 pF, uprated from 15 to 20 MHz (GE, RCA) <4E>; "for renewal use only" as of 1942; replaceable by Navy 38111A.

836, A [AKA 38266A, VT-236, CV636]; xmtg. vacuum rectifier, ST19 bulb w/ cap, two parallel cathodes, heater 2.5 V @ 5 A, 5 kV PIV, 250 mA (EE, GE, HY, RA, RCA, W) {A6003} <4P>. Variant: 3B27.

837 RK-44, 38837, VT-101, CV637]; xmtg. pentode, ST16 bulb w/ cap, seven-pin base (originally ceramic), third grid wired out for suppressor modulation, heater 12.6 V @ 700 mA, 12 W, 20 MHz (GE, HY, RA, RCA, W) {A4002} (9-24-43) <6BM>.

838, W [AKA HF-130, UE-938, RK-58, 38138]; xmtg. triode, variable-μ variant of 203A for use in zero-bias Class B modulators, lower capacitances, 100 W (AX: 120 W), 30 MHz (AX, C, GE, NU, RCA, T, W) {A300} <4E>.

839 Xmtg. tetrode, 2.5-V heater, med. 5-pin Isolantite base, "not listed for distributor sales" by RCA as of 4-1-34 {A254}.

840 [AKA NL840, 8754]; rcvg. RF pentode, ST12 bulb w/ cap, 5-pin base, fil 2.0 V @ 135 mA, gm 0.4 mS, "not listed for distributor sales" by RCA as of 4-1-34 (GE, RCA) {A440} <5J>.

841 [AKA VT-51, (UE) (DE) 941]; xmtg. triode, 10 w/ higher μ of 30 and uprated to 15 W (AX: 20 W), 6 MHz, number assigned 1927 (AX, C, GE, HY, RA, RCA) (6-28-45) <4D>. *Unrelated to 841A below.*

841A, SW RF-heating triode, "S" bulb w/ cap, UX base, fil. 10 V @ 2 Aμ 14.6, 50 W, 1.25 kV (T) <3G>. "SW" [AKA UX-CV11, 211H] is like 841A but rated at only 1 kV (T).

842 [AKA UE-942, 38842, VT-72, PJ-9]; xmtg. audio triode, number assigned 1927, S16 bulb (later ST16), for Class A amps, variant of 10 w/ lower μ of 3 (AX, GE, RCA) (6-28-45) <4A>. "Not as desirable as a 2A3, which will provide more output at a lower plate voltage." AX uprated to 20 W.

843 [AKA 38143, VT-73, CV639]; xmtg. triode, S17 or ST16 bulb, 5-pin base, heater 2.5 V @ 2.5 A, μ 8, 12 W, 6 MHz (GE, RCA) {A155} (4-26-43) <5A>.

844 Xmtg. tetrode, ST-16 bulb w/ cap, 5-pin base, heater 2.5 V @ 2.5 A, g_m 0.6 mS, 15 W (RCA) {A253} <5AW>.

845, B, W [AKA 545, 845B, SE-3945, 38145, VT-43, WL-412, CV735]; xmtg. audio triode for Class A and AB1 amps, 203A format, 75 W later uprated to 100 W; "will replace [WE] 284D in all but a few RF applications," number assigned 4-27 (AX, C, GE, NU, RCA, T, UE, W) (6-28-45). Two in Navy QCL 20-kHz sonar.

846 [AKA 38146, F346A, (WE) 284D]; xmtg. triode, number assigned 8-32, fil. 11.0 V @ 51 A, μ 40, 2.5 kW, 50 MHz, water-cooled via ext. cooler (GE, ML, RCA) {ZP-70}. "Purchased outside and merchandized by RCA Mfg. Co." as of 1-20-38.

848 [AKA F348A]; xmtg. triode, number assigned 7-27; 207 w/ lower μ (8, vs. 20), derated to 6 kW; 891 char. "Purchased outside and merchandized by RCA Mfg. Co." as of 1-20-38. "Replaced by 891" as of 1935.

849, A, H [AKA 549, UE-949, 38149]; xmtg. triode for Class A amps, 204A format, T-T fil. 11 V @ 5 A, μ 19, 400 W, 3 MHz. Number assigned 1927; "purchased outside and merchandized by RCA Mfg. Co." as of 1-20-38 (AX, C, F, GE, RCA, UE, W) (6-28-45). Interchanges w/ WE 270A and 296A. "A" is redesigned; fil. raised to 7.7 A, uprated to 500 W (AX, C). "H" is low-capacitance variant, grid-out-side, for HF RF heating; rated at 500 W to 30 MHz (AX, C).

850 [AKA 38150, VT-60]; tetrode variant of 203A w/ grid cap, g_m 2.75 mS, 13 MHz; number assigned 1930 (GE, RCA, W). Two in W Type EB 50-W police xmtr.

851 [AKA 551, F351A, UE-951, 38151, CG-2172, VT-41, PR-51-A, CV2671]; xmtg. triode, number assigned 1924, "expanded 204A format: stepped T48 bulb w/ cap, blade-and-pins base; T-T fil. 11 V @ 15.5 A, μ 20.5, 750 W, 3 MHz (AX, C, GE, RCA, UE, W). "Purchased outside and merchandized by RCA Mfg. Co." as of 1-20-38 (4-26-43).

852, A [AKA 552, F352A, FP-1, UE-952, 38152]; xmtg. triode, GT-30 bulb w/ arm, plate-out-side, UX base, flying leads for grid and plate; fil. 10 V @ 3.25 A, μ 12, 100 W, 30 MHz, intro. 3-27 (AX, C, GE, RCA, W) <2D>. Also see 860.

853, 854 "Obsolete" (unverified).

855 Rectifier, "obsolete" (unverified).

856 Vacuum rectifier, 4-1/16" dia. by 15-½" long, fil. 11 V @ 16.75 A, 50 kV PIV, 270 mA (W). (Also reported as having 14.5-V fil.)

857 [AKA FG-15, 38157]; mercury rect., number assigned 3-29, superseded by 857B (GE, RCA) {ZG-15}. "Purchased outside and merchandized by RCA Mfg. Co." as of 1-20-38.

857A, B [AKA 38157A]; mercury rect., fil. 5 V @ 40 A, 22 kV PIV, forced-air cooled (AX, F, GE). "B" [AKA F357A, SE-3072, 38157B] has GT56 spherical-tubular bulb, flying leads at bottom, plate cap at top, fil. reduced to 30 A, 22 kV PIV @ 10 A (at 30-40 °C) or 10 kV @ 10 A (at 25-60 °C) (GE, HY, RCA, W). Sim. to [WE] 266C. "Purchased outside and merchandized by RCA Mfg. Co." as of 1-20-38 (11-10-45) (F, GE, HY, W).

858 [AKA SE-3717, F358A, 38158]; xmtg. MF triode, number assigned 9-29, grid-out-side, fil. 22 V @ 52 A, μ 42, 20 kW, 1.5 MHz, water-cooled via external cooler (GE, RCA). "Purchased outside and merchandized by RCA Mfg. Co." as of 1-20-38 (F, GE, RCA) {ZT-6} (4-26-43). In '30s exp. TV xmtrs.

859 Rcvg. triode, UX-199 w/ μ raised to 30, for microphone preamps, mid-'20s (RCA).

860 [AKA (WE) 260A, 560, 38160, F311A, SE-3119, VT-17, CV640]; tetrode version of 852, intro. 3-29, same format and basic ratings, g_m 1.1 mS (GE, RCA, W) {ZF-19} (4-26-43).

861, A [AKA 561, 38161, F331A, SE-3124A, VT-19, PR-861, WL-407, CV641]; tetrode version of 831, same format but w/ cap added on bulb for screen, same basic ratings, number assigned 12-27 (GE, ML, RCA, W) {A302, ZF-24} (4-26-43).

862, A [AKA FT-1-B, 38162, SE-3675]; xmtg. and RF-heating triode, three feet of copper pipe w/ a tube inside, grid-out-side, flying leads for fil., 4-1/8" dia. by 60-3/8" overall, T-T fil. 33 V @ 207 A, μ 45, 100 kW, 1.6 MHz, water-cooled via ext. cooler; superseded by 862A, number assigned 7-28 (F, GE, RCA, W) (4-26-43). "Purchased outside and merchandized by RCA Mfg. Co." as of 1-20-38. GE claimed to have sold "60 million watts" of 862s as of 4-41, or 600 tubes. Two in RCA 50-kW xmtr. at WWL. 12 in WLW 500-kW system as RF amps, eight more as modulators. Variant: 898. "A" (F, GE) has 898A as variant.

863 [AKA F363A]; xmtg. triode, 207 w/ higher μ (50 vs. 20), number assigned 8-28, superseded by 892 (F, RCA). "Purchased outside and merchandized by RCA Mfg. Co." as of 1-20-38.

864 [AKA 264, 38064, SE-3864A, VT-24, FR-300, CV2675]; rcvg. triode, T9 bulb, small 4-pin base, fil. 1.1 V @ 250 mA, μ 8.2, g_m 0.65 mS (low-microphonic, rebased, improved WD-11) (AX, GE, HY, RCA) {A214} (9-24-43) <4D>.

865, A [AKA 565, 38165, SE-3294, VT-55, PJ-27, CV2676]; xmtg. tetrode, ST-16 bulb w/ cap, UX base, intro. 3-29, T-T fil. 7.5 V @ 2 A, g_m 0.75 mS, 15 W (De F: 7.5), 15 MHz (GE, RCA, W) {A301} (9-24-43).

866, A [AKA VT-46, PJ-28, SE-3069, CV32, UE-966];

xmtg. mercury rect., intro. 1-29, S19 bulb w/ cap, UX base, simple unshielded internal structure, fil. 2.5 V @ 5 A, 7.5 kV PIV (AX, C, EE, HY, RA, RCA, T, TS, W) (9-24-43) <4P>. "A" [AKA UE-966A, F366A, 38166, SE-3572, VT-46A, WT-210-0008, Br. CV32], intro. 4-33, has shielded element structure in S19 or ST19 bulb, uprated to 10 kV PIV @ 250 mA (30-40 °C) or 5 kV @ 250 mA (25-70 °C) (AX, C, CE, CH, EE, EI, GE, GN, HY, ML, NU, RA, RCA, T, UE, W) (9-24-43). Superseded for mil. use by 3B28. Taylor had a "shielded" 866 claimed to equal 866A. Thyratron variant: 5557.

866A/866 Mercury rect., final version, intro. 1-41; claimed high perveance of 866 and high PIV of 866A (GE, RA, RCA). Also see 3B28.

866B Mercury rectifier, 866 w/ fil. 5 V @ 5 A, uprated to 8.5 kV PIV and 1 A peak (T).

866JR [AKA 2B26, q. v.] (HY, NU, T).

867 [AKA PJ-15]; argon phototube, 868 format, 20 μA/lumen, 2-½" dia. <2K> Superseded by 868.

868 [AKA WL-735, CE-1D, NU1, 38268, SE-3982, PJ-23, 59A4]; argon-cesium phototube, dwarf (UX-199-style) 4-pin base, T9 tubular bulb 3-½" above socket, side-on cathode, for sound repro., S-1 response, 90 μA/lumen; used in early RCA Photophone projectors (GE, HY, RCA, W) (6-16-47) <2K>. Variant: 1P23. Format used in 1P29, 1P30, 1P36, 1P37.

869, A, B [AKA VT-39, PJ-26, SE-3071, 38169]; xmtg. mercury rect., number assigned 12-28, fil. 5 V @ 20 A, 20 kV PIV, forced-air cooled (AX, GE, RCA) {ZG-8}. "A" [AKA VT-39A, CV2723, 38169A] had number assigned 12-31, GT40 spherical-tubular bulb w/ cap, blade-and-pins base, fil. reduced to 18 A, 2.5 A (AX, HK, RCA, W). "Purchased outside and merchandized by RCA Mfg. Co." as of 1-20-38. "B" was improved, w/ tube drop of 10 V vs. 11 (sim. to WL-424) (AR, CE, EE, F, GE, HY, RCA, W) (4-26-43).

870, A [AKA 38170]; glass xmtg. mercury rect., spherical-tubular bulb 24-½" high w/ cap, flying leads at base for fil., number assigned 4-33, heater 5.0 V @ 65 A, 16 kV PIV; 75 A, forced-air cooled. "Purchased outside and merchandized by RCA Mfg. Co." as of 1-20-38 (GE, RCA) {ZG-42}. "Developed for" WLW 500-kW system, 1934, which used six, but also used in Navy TAW 300-kW xmtr. w/ 862s, 1931. "A" [AKA 38171, SE-4102] was redesigned into 25" tubular steel case; same ratings (GE, RCA). Six in Bethany, OH, 200-kW SW xmtr., 1945. Also see FG-166.

871 [AKA (WE) 253A, UE-971, SE-4102]; xmtg. mercury rect., T9 bulb w/ cap, UX base, fil. 2.5 V @ 2.0 A, 5 kV PIV, 125 mA (GE, RCA, TS, W) {R602} <4P>. "Not listed for distributor sales" per RCA, 4-1-34.

872, A, "Spl. fil." [AKA 572, VT-42, FG-19, SE-3070, CV642, 38172]; xmtg. mercury rect., T18 bulb w/ cap, jumbo 4-pin base, simple unshielded elements; number assigned early 1929, fil. 5 V @ 10 A, 7.5 kV PIV, 1.25 A (AX, RCA, T, W) {R648} (4-26-43) <4AT>. "A" [AKA 572A, 38172, F353A, F872B, UE-972, WT-210-0027] intro. 4-33, has shielded element structure, fil. reduced to 6.75 A, uprated to 10 kV PIV; similar to (WE) 319A (AX, C, CE, CH, EE, EI, F, GE, GN, HK, HY, ML, NU, RA, RCA, T, TL, UE, W) {R649} (9-24-43). 872A "Spl. fil." is AKA VT-42. Replaceable by any-temperature 4B32. Variant: 8008.

872A/872 Mercury rect., final version, intro. 12-44 (RCA).

873, A [AKA UE-973, WT230]; mercury triode thyratron, fil. 5.0 V @ 6.75 A, 3 kV PIV, 2.5 A; 872A w/ grid, derated on PIV (EE, F, HY, T).

874 [AKA 38274, SE-3918]; voltage regulator, 150 V @ 10-50 mA, S16 (early) or ST16 bulb, UX base; dev. 1925 for intial use in RCA 104 amplified speaker (GE, RCA) <4H>.

875, A [AKA 575A, q. v.]; mercury rectifier (HY, T, TS).

876 [AKA 38276, UV-876, (GE) B-27]; ballast tube, T16 bulb, mogul screw base, 1.7 A @ 40-60 V, to be used in series w/ 120-V line to receiver (GE, RCA). In RCA 104 amplified spkr. when used w/ 50-60-Hz power; also in power unit for Navy RAK/RAL rcvrs. Also see 886, 896.

877 Protective ballast tube, double-contact base, 2.5 V @ 20 mA to 45 V @ 90 mA, or 5 V @ 20 mA to 90 V @ 90 mA depending on wiring (RCA).

878, A [AKA 38278 (?)]; vacuum HV rect., T14 bulb w/ cap, UX base, fil. 2.5 V @ 5.0 A, 20 kV PIV, 5 mA (GE, RCA) {R613} (9-24-43) <4P>. "A" is AKA REL52.

879 [AKA 2X2, q. v.]; repl. H2-10, 143D (7-6-39).

880 Xmtg. VHF triode, 7" dia. all-glass construction w/ reentrant ext. anode, pure-tung. fil. 12.6 V @ 320 A, μ 20, 20 kW, 25 MHz, water-cooled via ext. cooler. Intro. 8-40 (EI, GE, ML, RCA, W) {R612} (1-25-50). Used in HF broadcast xmtrs (two in GE 50-kW FM xmtr., two in RCA 50-kW SW xmtr.) and high-power radar jammers. Variants: 5658, 5771, 7085, 8009.

881 Xmtg. gas rectifier (W) (1-27-43).

884, W [AKA 38884, 6Q5G, VT-222, CV647]; triode gas thyratron rect. and relaxation osc., ST bulb, octal base, heater 6.3 V @ 600 mA, 350 V PIV, 2 mA (CH, GE, HY, RCA, W) {R6045A} (3-21-45) <6Q>. Common time-base generator in early oscilloscopes. Variant: 885.

885 Thyratron, five-pin 884, heater 2.5 V @ 1.5 A, sim. to WL-629 and 2B4 (GE, RCA, W) {R631} (3-21-45) <5A>.

886 [AKA 38286, (GE) B-41]; ballast, 876 to run at 2.05 A; same voltage range (GE, RCA). In RCA 104

A; same voltage range (GE, RCA). In RCA 104 amplified spkr. when used w/ 40-Hz power.

887 Xmtg. VHF triode, tubular construction w/ stud-mounted anode at center, flying ribbon leads for fil. at one end, grid cap at the other; fil. 11 V @ 24 A, μ 10, 600 W, 250 MHz, water-cooled via integral jacket, intro. 5-37 (GE, RCA) {A2026B}. Variant: 888.

888 Xmtg. VHF triode, high-μ variant of 887, μ 30, intro. 5-37 (GE, RCA) {A217B}.

889 Electron-microscope demonstrator, ST-21 bulb on 7-pin base, heater 1.5 V @ 1.5 A, 1 kV. Dome is fluorescent screen; image of cathode is enlarged 25 X and projected onto screen (DuM). [*Electronics*, 12-36, p. 52.] *Unrelated to 889 below.*

889, A, M, RA Xmtg. triode, 10-5/8" overall, pure-tung. fil. 11 V @ 125 A, μ 21, 5 kW, 50 MHz, water-cooled via ext. cooler (EI, GE, RCA, W). "A" is improved version, for xmtrs or RF heating (GE, ML, RCA, W) (10-31-45). Variant: 5666. "M" is RF-heating variant (EP) (1947). "AR" {A2251B} is radiator version of "A."

889R, RA Xmtg. triode, radiator variant of 889, derated to 25 MHz (super-seded by 889RA), intro. 8-40 (GE, W) {A2251A} (4-26-43). Pair used in GE 10-kW FM xmtr. "RA" is improved version, freq. derated to 40 MHz (GE, ML, RCA) (10-31-45). In RCA ET-4750 15-kW HF CW xmtr. Variant: 5667.

891, M Xmtg. triode, 848 w/ dual fils., 11 V @ 60 A each, for single- or dual-phase operation, apparently same as 892 except for μ (F, GE, ML, RCA, W) (1-27-43). "Purchased outside and merchandized by RCA Mfg. Co." as of 1-20-38. "M" is RF-heating variant (EP) (1947).

891R Xmtg. triode, radiator version of 891, same ratings except derated to 4 kW and plate voltage limited to 10 kV instead of 12 (AX, F, GE, ML, RCA, W), sim. to WE 341AA (1-27-43). "Purchased outside and merchandized by RCA Mfg. Co." as of 1-20-38. (AX also offered a "radiator" 891 as of 1938.) Two as modulators in W 5-kW AM xmtr., 1941.

892, A, M [AKA WT-210-0045, 38192]; xmtg. triode, 863 w/ dual pure-tung. fils., 11 V @ 60 A each, for single- or dual-phase operation, for RF heating as well as xmtrs (superseded GE-452), same as 891 except for μ of 50 (AX, F, GE, ML, RCA, W) (1-27-43). "Purchased outside and merchandized by RCA Mfg. Co." as of 1-20-38. Intro. 1936. Variants: 5668, 5705, 6333, 6446. "A is later version; four as drivers in Navy Jim Creek megawatt station (see 5831) (F). "M" is RF-heating variant (EP) (1947).

892R Xmtg. triode, radiator version of 892, same ratings except derated to 4 kW, and plate voltage limited to 12.5 kv instead of 15 (F, GE, ML, RCA, W) (1-27-43) "Purchased outside and merchandized by RCA Mfg. Co." as of 1-20-38. Finals and modulators in RCA BTA-10F xmtr, 1946. One in Collins 21B 5-kW xmtr; two in 21L 10-kW; one in W 5-kW AM xmtr, 1941. Variants: 5669, 5706.

893, A, AR [AKA 38193]; xmtg. triode, for BC and HF xmtrs and RF heating, 28-3/4" overall, grid-out-side, wingnuts for fil. connections, six-unit fil., 20 V @ 183 A in single-phase use, μ 35, 20 kW, 5 MHz, water-cooled via external cooler (F, GE, ML, RCA, UE, W) (4-26-43). "AR" is radiator version of 893A, same ratings (GE, ML, RCA, W).

893R Xmtg. triode, radiator version of 893 (superseded by 893AR), intro. 8-40 (F, GE, WL) (4-26-43).

895, R Xmtg. triode, general use, 4-½" dia. by 25" long, triple Y-connected fils., 19 V @ 138 A each, μ 37, 20 kW (W: 40 kW), 6 MHz, water-cooled (GE, ML, W) (1941). "R" is radiator version, same ratings (GE, ML) (W: 20 kW).

896 [AKA (GE) FB-50, 38296]; ballast tube, T9 bulb, med. screw base, 5-8 V @ 250 mA; "inactive" as of 4-1-34 but still catalogued as of 1945 (GE, RCA, W).

897 [AKA (GE) FB-49]; ballast tube, 1.15 A (GE).

898, A Xmtg. triode, 862 with six-unit fil., 16.5 V @ 70 A per unit, and pins for filaments instead of flying leads, 1936. "Purchased outside and merchandized by RCA Mfg. Co." as of 1-20-38 (F, GE, RCA, W). WLW 500-kW xmtr. was refitted from 862s to 898s, but 898 was not a universal replacement for 862: both were offered, at same price of $750, as of 1945.

899, A Xmtg. triode, "purchased outside and merchandized by RCA Mfg. Co." as of 1-20-38. "A" ratings are fil. 14.5 V @ 180 A, 30 kW (W) (1-10-47).

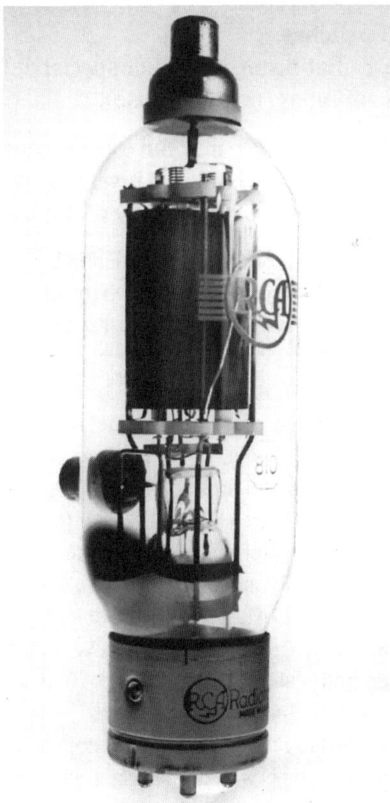

An 810 with prewar RCA logo. *Photo: Bob Dennison, W2HBE.*

SPECIAL-PURPOSE TYPES: RMA 1942-1944 SYSTEM

CODING SYSTEM

FIRST NUMBER: HEATER POWER

1: 0 (no filament/heater; cold-cathode device, etc.)
2: Up to 10 W
3: > 10 to 20 W
4: > 20 to 50 W
5: > 50 to 100 W
6: > 100 to 200 W
7: > 200 to 500 W
8: > 500 to 1000 W
9: > 1000 W

LETTER: TYPE OF DEVICE

A: Single-element (ballasts, vacuum-type resistors).
B: Diode (including protective tubes, spark gaps, voltage regulators). Radar TR tubes were considered diodes, with or without a third or "keep-alive" electrode.
C: Triode.
D: Tetrode.
E: Pentode.
F: Hexode.
G: Heptode.
H: Octode.
J: Magnetically controlled, usually incorporating a resonator (read: "magnetron").
K: Electrostatically controlled, incorporating a resonator (velocity-modulated tubes, klystrons, inductive-output tubes) - in effect, klystrons.
L: Vacuum capacitors.
N: Crystal rectifiers; later, **solid-state devices.**
P: Photoemissive (phototubes, photomultipliers, pickup tubes).
Q: Cavities (microwave resonant cavities for frequency control).
R: Ignitrons, pool tubes.
S: Vacuum switches.
T: Storage, radial-beam, and other special deflection control tubes (no known use).

The letters V through Y may have been considered for other types of tube, but, on investigation, were not officially part of the code and were not used.

SECOND NUMBER: SERIAL

Assigned sequentially, beginning with 21 to avoid confusion with receiving or cathode-ray tubes.

1B21, A [AKA (GL) 471, A]; TR tube.

1B22 Argon-hydrogen spark-gap diode radar modulator, glass body 1-1/8" dia. by 2-11/16" w/ large end caps, aluminum cathode, fires at 2.7 kV, 75 A peak current, used in series pairs (WE) {1378X} (6-7-48). Cathode included internal corona points to aid firing. Two, at 10.5 kW pulse power, in AN/APS-4.

1B23 [AKA (WE) 729A]; TR cell, disc-seal, for use w/ external cavity, 0.9-1.2 GHz, 450 kW, filled with hydrogen and water vapor; replaced 702B w/ mod. kit (BL, WE) (6-7-48). In RF-1/TPS-1.

1B24, A [AKA X6011, REL30]; TR tube, copper tunable integral-cavity type w/ external glass gas reservoir, radium-activated, covers 8.5-9.6 GHz; RG-52/U WG in and out (BL, MA, SY, VA, W) {WX3199} (2-11-46). In GE "A" version, gas reservoir is of steel. In AN/APS-1, RT-15/APS-15, AN/MPN-1, and AN/-MPG-1. Variants: 1B26, 1B60.

1B25 [AKA S853]; gas switching tube (11-3-44).

1B26 [AKA X6010]; TR tube, 1B24 scaled-down to 1.2 cm; covers 23.4-24.6 GHz, 35 kW (BL, SY, W) {WX-3210} (2-11-46).

1B27 [AKA X6013]; disc-seal TR cell, tunes 2.3-3.8 GHz, for use w/ external cavity, has tuning screw to change internal gap spacing, 450 kW (BL, SY) (1-26-48).

1B28 [AKA X6012]; gas switching tube.

1B29 Argon-hydrogen spark-gap diode radar modulator (scaled-down 1B22), glass body 1" dia. by 2" w/ end caps of differing diameters, aluminum cathode, fires at 1.9 kV, 30 A peak current, used in series pairs (WE) {1437GT} (6-7-48). Cathode includes internal corona points to aid firing.

1B30 Gas switching tube {ZG530}.

1B31 Argon-hydrogen spark-gap diode radar modulator, for airborne use, glass body ca. dia. by 4" w/ end caps of different diameters, aluminum cathode, used in series pairs, 300 A peak, delivers 230 kW pulse to 2J53; not made in quantity (WE) {1391Z} (9-10-48).

1B32 [AKA (GL, WL) 532A; (SigC) GA-4, GA-5, GA-5A]; diode TR tube (w/o keep-alive electrode) {ZG532} (2-11-46). In AN/TPS-2.

1B33 Gas switching tube.

1B34 Spark-gap diode radar modulator, large (ca. 1" dia. by 5" long, 200 A pulse {WX3226} (2-11-46).

1B35, A [AKA X6031]; ATR tube, fixed-tuned, for 9.0-9.6 GHz, 5 to 250 kW pk., loaded Q 6.5, used w/ 1B24 (BL, GE, SY) {ZG626} (8-20-45). Variants: 1B36, 1B37.

1B36 [AKA X6015]; ATR tube, 1B35 for 24 GHz, 75 kW (BL, GE) {ZG627} (8-20-45).

1B37, A ATR tube, 1B35 for 8.5-9.0 GHz (BL, GE, SY) {ZG629} (8-20-45).

1B38 Pre-TR tube, for 2.7-2.91 GHz, 1 MW, typically used in tandem w/ 1B27; filled with argon and water vapor (BL, GE) {ZG645} (8-20-45).

1B39 [AKA WL-539A]; gas switching tube.

1B40 TR tube, 1075-1095 MHz, 1 kW, for use w/ external cavity; comprises a grounded base electrode and a keep-alive electrode; the actual TR discharge is "electrodeless" (BL, SY) {SD847A} (10-15-47).

1B41 Argon-hydrogen spark-gap radar modulator, ca. 3" dia. by 6" long, 90 cm H_2 pressure, 450 A pulse (W) {WX3240D} (2-11-46). Variants: 1B45, 1B49.

1B42 [AKA D-170135]; series-gap diode radar modulator, sintered "iron-sponge" mercury cathode, glass body w/ anode cap and large cathode assembly w/ fins and threaded mounting; for long-pulse ship radars, 300 A pulse, three used in series for 1.4 MW (WE) {1438-GT} (12-15-47).

1B43 Gas switching tube {SD829} (10-30-47).

1B44 ATR tube, in waveguide package, for 2680-2820 MHz, 20 kW min. pk, 1 MW max. pk. (BL, GE, SP, SY) {ZG610} (10-4-45). Variant: 1B56.

1B45 Spark-gap radar modulator, 1B41 w/ H_2 pressure raised to 150 cm, 450 A pulse (W) {WX3240B} (2-11-46).

1B46 Min. volt. reg., T4½ black-painted bulb on polarized bayonet base, 80 V, 1-2 mA (SY, CH) {R1160A} (3-11-46). Variant: 1B47.

1B47 Min. voltage regulator, 1B46 for greater range of source voltage (3-11-46).

1B48 [AKA CK1011A]; min. cold-cathode HV rect., in T-6½ bulb w/ four pins, one of them elongated to ca. 1", 2.7 kV PIV, 6 mA (RA) (9-6-45). Also see 5517.

1B49 Spark-gap radar modulator, 1B41 w/ H_2 pressure raised to 110 cm, 450 A pulse (W) {WX3240G} (2-11-46).

1B50 [AKA X6032]; TR tube, plated-steel integral-cavity type w/ external glass gas reservoir, tunes 6.0-7.1 GHz (BL, W) {ZG626} (10-30-47).

1B51 [AKA X6033]; ATR tube, fixed-tuned to 6425 MHz, 500 kW (BL) (6-15-46).

1B52 ATR tube, fixed-tuned, for 3.55-3.70 GHz, 750 kW (BL, SY) {ZG621} (6-15-46). Variant: 1B53.

1B53 ATR tube, 1B52 for 3.40-3.55 GHz (BL, SY) (7-10-46).

1B54 [AKA X6036]; pre-TR tube, 3.4-3.7 GHz bandpass type, fixed-tuned, 750 kW (BL) (1-26-48).

1B55 [AKA X6034]; TR tube, triple-gap bandpass unit sealed into waveguide, RG-48/U WG in and out, 3.5 GHz fixed-tuned, 750 kW (BL, GE) {ZG625} (2-26-54). Variant: 1B58.

1B56 ATR tube, 1B44 for 2775-2925 MHz (BL, GE, SP, SY) (10-5-45).

1B57 ATR tube, fixed-tuned, for 3250-3400 MHz, 750 kW (BL, SP) (10-5-45).

1B58, A TR tube, 1B55 for 2800 MHz (BL, SY) (10-5-45).

1B59 [AKA R1130B]; cold-cathode crater-glow mod. tube, actinic point light source for fax and half-tone scanners. Crater dia. 0.05", 140 V, 5-35 mA, takes modulation to 15 kHz (SY) (3-11-46). Also see R1131.

1B60 TR tube, lightweight 1B24 (BL) (10-30-47).

1B62 TR tube, modified (WE) 721B, q. v. (3-25-46).

1B63, A, B [AKA X-6044]; TR tube, two-gap fixed-tuned bandpass type in waveguide package, RG-52/U WG in and out, for 8.5-9.6 GHz, 250 kW pk. The two gaps provide pre-TR and TR functions (BL, GE, EI, SY) (4-25-46). SY ad, 12-52, claims "the accepted wideband TR for X-band use." "B" (SY) has 1000-hr. claimed life. Variants: 6646, 6647.

1B64 Voltage regulator (3-11-46).

1B67 [AKA VG10A]; mica-window beta-gamma counter, 1.2 kV, in stepped machined cylindrical body (VN) (10-28-48).

1B68 [AKA VG-13, G-11, K-11]; beta-gamma counter (10-28-48).

1B69 [AKA 1E, 62005]; gamma counter (4-30-48).

1B70 [AKA 1M, 62006]; gamma counter (4-30-48).

1B71 [AKA 4E, 62007]; gamma counter (4-30-48).

1B72 [AKA 4M, 62008]; gamma counter (4-30-48).

1B73 [AKA 10E, 62009]; gamma counter (4-30-48).

1B74 [AKA 10M, 62010]; gamma counter (4-30-48).

1B75 [AKA BS-6, 62011, 100C, T-61003]; argon self-quenched beta- and X-ray counter (4-30-48).

1B76 [AKA 100N, 62012]; neon self-quenched beta counter (4-30-48).

1B77 [AKA 120C, 62013]; argon self-quenched beta-and X-ray counter (4-30-48).

1B78 [AKA 150C, 62014]; beta-gamma counter (4-30-48).

1B79 [AKA 150M, 62015]; beta-gamma counter (4-30-48).

1B80 [AKA BS-1, 150N, 150NA, 62016]; neon self-quenched beta-gamma counter (4-30-48).

1B81 [AKA 200C, 62017]; argon self-quenched alpha-beta-gamma counter (4-30-48).

1B83 [AKA TGC-1]; beta-gamma counter, 1-½" dial by 3-5/8" long, 4-pin base (Tracerlab) (5-5-48).

1B84 [AKA TGC-2]; general-purpose beta-gamma counter, aluminum body 5-1/64" dia. by 4-1/8" long, 950 V, self-quenched (VN) (5-5-48). Also see 6306.

1B85 Beta-gamma counter, self-quenched "thyrode" in ribbed aluminum cylinder, 900 v (VN) (11-15-48).

1B86 Counter (7-11-50).

1B87 [AKA VXG-11]; submin. glass "thyrode" gamma counter, 900 V (VN) (11-4-48). Variant: 1B88.

1B88 Gamma counter, 300-volt variant of 1B87 (VN).

1B90 Beta-gamma counter (10-26-49).

1B97-99 Counters (all 8-18-49).

1B105 Counter (8-18-49).

1C21 Cold-cathode triode, sim. to 0A4G in GT bulb but for DC use only, 25 mA (HY, NU, RCA) {R6160} (9-10-42) <4CK>.

1D21 [AKA 631P1, SN4]; cold-cathode neon "strobotron" flash tube, tubular bulb, 4-pin base, 350 V (SY) {R4100, R4101} (1-26-48).

1L21-38 Glass-bulb vacuum capacitors, axial electrodes sized to fit 30-A fuse clips (all GE). Types rated at 7.5 kV are in tubular bulbs; types rated at 16 kV are in bulged-tubular bulbs. Dimensions given below exclude electrodes. Voltages are peak values.

1L21 12 pF, 7.5 kV, 1-5/8" dia. by 1-3/4" long (3-1-44).

1L22 25 pF, 16 kV, 2" dia. by 3" long (12-10-43).

1L23 50 pF, 16 kV, 2" dia. by 3" long (12-10-43).

1L24 100 pF, 16 kV, 2-¼" dia. by 3" long (8-20-45).

1L25 12 pF, 16 kV, 2" dia. by 3" long (3-1-44).

1L31 6 pF, 16 kV, 2" dia. by 3" long (8-20-45).

1L32 6 pF, 7.5 kV, 1-5/8" dia. by 1-3/4"' long (8-20-45).

1L33 100 pF, 7.5 kV, 1-5/8" dia. by 3" long (12-10-43).

1L36 25 pF, 7.5 kV, 1-5/8" dia. by 1-3/4"' long (12-10-43).

1L38 50 pF, 7.5 kV, 1-5/8" dia. by 1-3/4"' long (12-10-43). Appears to be same unit as in BC-442 and RE-2/ARC-5 antenna relay units, coded GE ZA-38.

1Nxx Crystal diodes; all listed here are silicon or germanium point-contact types.

1N21 A-C [AKA D-163366, R7011]; 3-GHz silicon radar mixer n ceramic cartridge; typ. NF 16.4 dB @ 3 GHz. In TN-54/APR-4. Lettered variants have progressively lower loss and noise. "A" [AKA D-168355] has typ. NF 14.6 dB @ 3 GHz (SY, WE). In BC-1241A. "B" [AKA D-169113] has typ. NF 12.7 dB @ 3 GHz (SY, WE). "C" has typ. NF 11.1 dB @ 3 GHz (SY, WE).

1N22 [AKA D-164389]; radar mixer (WE). In I-126 test set.

1N23 A-D [AKA D-168356]; 10-GHz mixer sim. to 1N21; typ. NF 17.1 dB @ 9375 MHz (SY, WE). Lettered variants have progressively lower loss and noise. "A" [AKA D-169304] has typ. NF 14.9 dB @ 9375 MHz (SY, WE). In AN/MPG-1. "B" [AKA D-169305] has typ. NF 13.4 dB @ 9375 MHz (SY, WE). "C" is from BL, SY; "D" is from SY.

1N24 [AKA X6020]; silicon mixer in ceramic cartridge.

1N25, A, AR [AKA D-170247]; 1-GHz silicon radar mixer in ceramic cartridge; typ. NF 14.7 dB @ 1 GHz (SY, WE).

1N26, A-C [AKA X6021, D-168707]; 24-GHz silicon radar mixer in coaxial package; typ. NF 15.2 dB (SY).

1N27, A Silicon radar direct video detector in ceramic cartridge. In AN/UPN-2 and AN/CPN-8 beacons.

1N28 [AKA D-168030]; silicon 3-GHz radar mixer in ceramic cartridge; typ. NF 13.2 dB @ 3 GHz (WE).

1N29 Silicon 3-GHz radar direct video detector or mixer, in ceramic cartridge.

1N30 Silicon 9-GHz radar direct video detector in ceramic cartridge.

1N31, A [AKA D-171936]; silicon 9-GHz direct video detector in coaxial package (SY). In AN/UPN-4 beacon.

1N32, A Silicon radar 3-GHz, direct video detector in ceramic cartridge; nearest equiv. 1N27 (MA, SY).

1N33 [AKA V-1000, R7013]; silicon 3-GHz radar direct video detector.

1N34, A Germanium general-purpose diode, 50 V PIV (AX, RA, SY). "A" is sealed in glass.

1N35 Dual matched 1N34s, mounted together (SY).

1N38, A, B [AKA R7113]; Ge high-conductivity diode, high PIV (100 V) (AX,SY).

1N39 Ge xtal diode, high PIV (200 V) (SY).

1N40-42 Quad Ge diode bridges, for bal. modulators (SY).

1N40 [AKA V301]; in 6H6 tube format, 50 PIV.

1N41 1N40 potted in case w/ solder lugs.

1N42 1N40 uprated to 100 PIV.

1N43-47 Ge general-purpose diodes, 400-type (WE), 1948.

1N43 [AKA 400A]; 60 V PIV.

1N44 [AKA 400B]; 115 V PIV.

1N45 [AKA 400C]; 75 V PIV.

1N46 [AKA 400D]; 50 V PIV.

1N47 125 V PIV, WE equiv. of 1N38.

1N48-52 Ge general-purpose diodes (GE).

1N48 [AKA 32C406G5]; 50 mA, 85 V PIV.

1N49 [AKA 32C406G5A]; 50 V PIV.

1N50 [AKA 32C406G5B]; 50 V PIV.

1N51 [AKA 32C406G5C]; 25 mA, 50 V PIV.

1N52 [AKA 32C406G5D]; 85 V PIV.

1N53, A-D [AKA X7103]; silicon 35-GHz mixer, coaxial case, NF 13 dB (MA, SY).

1N54, A Ge general-purpose diode, 35 V PIV (AX, HY, SY). "A" has high back resistance.

1N55, A, B Ge diode, 150 V PIV (HY, SY). "A" and "B" have higher conductivity.

1N56, A Ge diode, 30 V PIV (HY, SY). "A" has higher conductivity and 40 V PIV.

1N57, A Ge diode, 80 V PIV (SY).

1N58, A Ge diode, 100 PIV (AX, HY, SY). "A" is glass-sealed.

1N59 Ge diode, 250 V PIV (SY).

1N60, A Ge diode, 30 V PIV (AX, HY, SY). "A" is uprated to 40 V.

1N61 Ge diode, 130 V PIV (SY).

1N62 Ge diode, 110 V PIV (Ohmite).

1N63, A [AKA 32C406G5E]; Ge diode, 50 mA, 125 V (AX, GE, HY, RA, SY).

1N64, A Ge video detector, 20 V PIV (GE, HY).

1N65 Ge video restorer, 50 mA, 85 V PIV (GE, HY).

1N66, A Ge general-purpose diode, 60 V PIV (RA).

1N67, A Ge video restorer, 50 V PIV (AX, RA).

1N68, A Ge video restorer, 100 V PIV (AX, RA).

1N69, A Ge diode, 60 V PIV (GE, HY).

1N70, A Ge diode, 100 PIV (GE, HY).

1P21 Photomultiplier, for photometers, 931 format and response ("selected 931A") w/ sensitivity between 931 and 931A (RCA) (11-10-43) <11K>. Variant: 4473.

1P22 Photomultiplier, for photometers and scintillation counters, S-8 response, 931 format (RCA) {C7104} (9-28-45) <11K>.

1P23 [AKA CE-1, GL-553]; gas phototube for sound repro., 868 format w/ painted envelope, 190 μA/lumen, S-1 response (7-6-43) <2K>.

1P24 [AKA GL-564, VT-280, 936, (GL, ZP) 516]; vac. phototube {C7063} (8-20-45). No CV-no.; VT-no. cancelled; no known use despite the many type numbers.

1P25, A, G [AKA 1-B13]; IR-sensitive image tube, ca. 4" dia. cathode, 1" viewing screen, 4 kV, cond-cathode, three accelerating electrodes, response peaks at 800 nm {C7493, C73011} (11-3-44).

1P26-27 Phototubes (3-1-44).

1P28, A, B, V1 Photomultiplier, for photometers, S5 response via special glass in bulb, 931A format (RCA) (4-29-44) <11K>. "A" has doubled sensitivity. "B" was originally numbered 4818. "V1" has modified response.

1P29 [AKA (GL, FJ) 401]; gas phototube, 868 format, S-3 response, for photometers (GE, RCA, W) (6-15-44) <2K>.

1P30 [AKA CE2C, 71A3]; gas phototube, for sound repro., S-1 response, 868 format, 150 μA/lumen (2-10-45) <2G>.

1P31 [AKA CE3D, (WE) 3A, 79A4, G10]; gas phototube, flying leads, spherical bulb, S-1 response, 135 μA/e-lumen (2-10-45) <2AL>.

1P32 [AKA CE25C, 51A3]; gas phototube, 927 format, S-1 response, 135 μA/lumen (2-10-45) <2F>. 927 replaces.

1P33 [AKA CE26D]; gas phototube, med. 4-pin base, short tubular bulb, S-1 response, 75 μA/lumen (2-10-45) <2G>.

1P34 [AKA CE48]; gas phototube (2-10-45).

1P35 [AKA CE-5, 52A3]; gas phototube, 927 format, S-1 response, 125 μA/lumen (2-10-45) <2F>.

1P36 [AKA 58A3, CE4C]; gas phototube, 868 format, S-1 response, 125 μA/lumen (4-20-45) <2G>.

1P37 [AKA CE-91]; gas phototube, for sound repro from dye-image film track w/ incandescent light, S-4 response, 868 format, 124 μA/lumen (RCA) {C7106} (3-11-46) <2K>.

1P39 [AKA CE-29]; phototube, for photometers and relay use, 929 w/ Micanol base and lower dark current (RCA, W) (3-11-46) <3J>.

1P40 Phototube, for sound repro and relay use, 930 w/ Micanol base and lower dark current (RCA, W) (3-11-46) <3J>.

1P41 [AKA CE-22]; gas phototube, relay control, "peewee" 3-pin min. base, flat-ended tubular bulb 2-1/16" long, head-on cathode, S-1 response, 90 μA/lumen (repl. 924 in new designs) (RCA, W) (4-25-46) <2AR>.

1P42 Vac. phototube, relay use, S-9 response, head-on tubular bulb ¼" dia. by 1-11/32" long, ring for cathode, pin for anode, 25 μulatorA/lumen (RCA) (10-18-46) <2AT>.

1Q21-25 Beacon cavities: vacuum reference-cavity radar freq. markers, WG input-output; unloaded Q 1900-2400, loaded Q 4-6; resonate at the correct local osc. freq. to let the radar be tuned to the desired beacon freq.; usually mounted in WG just ahead of the mixer unit. Frequency tolerance is ± 0.5 MHz (W) (2-11-46).

1Q22 For 9250 MHz

1Q23 For 9280 MHz

1Q24 For 9310 MHz.

1Q25 For 9375 MHz.

1Q26, A Reference cavity, for 9280 MHz (GE) (8-20-45). "A" has temp. drift below 400 kHz.

1R21 [AKA GL-415]; ignitron, for welder control, steel case, air- or water-cooled, handles 22.4 A @ 600 V. Also see 5550 (2-11-46).

1S21 [AKA GL-34A]; SPDT vacuum switch, T-shaped glass bulb w/ short actuating rod, 20 A @ 7.5 kV (GE) (4-1-44). Also see FA-6 and FA-15.

1S22 SPDT vacuum switch, 10 kV, rated at 1 million operations at 2 A and 10 kV (CH, TS) (8-20-45).

2A21 Ballast tube, octal, 1.0 A @ 6.5-9.5 V (WE) (9-10-48).

2A22 Ballast tube (9-10-48).

2B21 HV rectifier, improved 2X2/879 {R612G}.

2B22 Lighthouse planar diode; modified 559, used as TR switch or detector, heater 6.3 V @ 750 mA, 300 V PIV, 20 mA, 1.5 GHz (GE, NEC) {R6279} (2-1-46).

2B23 Magnetically controlled diode, octal, heater 6.3 V @ 300 mA, 150 V out @ 30 mA (GE) {ZP650} (5-3-46).

2B24, A [AKA VT-284]; submin. diode {QF-197} (9-25-48). Never made?

2B25 [AKA CK-1091]; min. vacuum rectifier, fil. 1.4 V @ 110 mA, 2.8 kV PIV, 1.5 mA (HY, RA) (6-15-45).

2B26 [AKA 866JR]; mercury rect., sim. to 816 w/ plate lead out base, fil. 2.5 V @ 2.5 A, 5 kV PIV, 125 mA (HY).

2B27 [AKA REL35]; UHF lighthouse diode, dev. from ZP464 (GE) {ZP447}, ca. 3-41.

2B35 [AKA EA50, 10E/105, VR92 (Br.), q. v.]; submin. UHF diode, heater 6.3 V @ 150 mA {QF-195}. "2B35" cancelled by RMA.

2B36 [AKA VR78 (Br.), q. v., et al.]; submin. diode. "2B36" cancelled by RMA.

2C21 [AKA RK-33, 1642, 38233]; duotriode, 7 pins w/ cap, heater 6.3 V @ 600 mA, μ 10.4 (NU, RA, SY) (11-6-42) <7BH>. In (unique to?) Navy RU-18 and -19 aircraft rcvrs.

2C22 [AKA 7193]; VHF triode, GT octal w/ grid and plate caps, heater 6.3 V @ 300 mA, μ 20, 3.3 W; very nearly a double-ended 6J5GT (NU, SY) (1-15-43) <4AM>.

2C23 [AKA 10Y, q. v.]; power triode.

2C24 [AKA HY-114B]; VHF triode, ceramic octal base and dual caps, fil. 1.4 V @ 120 mA, μ 12, 2.5 W, usable to 300 MHz.

2C25 [AKA VT-25A, 10 Spl, 10Y, q. v.]; xmtg. triode (HY) <4D>.

2C26, A [AKA D10]; xmtg. triode, T9 octal bulb w/ grid and plate caps, heater 6.3 V @ 1.1 A, μ 16.3, 10 W (HY, NU) (4-10-43) <4BB>. "A" is rated for high-altitude use. In BC-1267A IFF interrogator and AN/APN-12 IFF transponder, giving 500-W pulses at 160-237 MHz. Two in AN/CPN-2 beacon as 1200-V 6-mA reg.

2C27 [AKA VT-285]; submin. triode {QF-200C} (9-25-48). No known use; "classified" as of 1967; never made?.

2C28 Submin. triode {SA780}.

2C29 [AKA X102G]; xmtg. disc-seal VHF triode, glass dish base, g_m 5.5 mS, 30 W, 7 W CW out @ 1250 MHz.

2C31 [AKA X102L]; xmtg. VHF triode.

2C32 Xmtg VHF triode.

2C33 [AKA RX233, RX233A]; triode thyratron, heater 2.5 V @ 2.5 A (RL) (4-22-43).

2C34 [AKA RK-34, VT-224; Br. CV18, DET 19, VT61B]; VHF xmtg. duotriode, heater 6.3 V @ 800 mA, μ 13, 10 W (RA). Originally promoted for VHF model-control xmtrs but used widely in IFF transponders like BC-647 (2-1-43).

2C35 Submin. UHF triode, heater 6.3 V @ 200 mA, μ 32, for use to 1.2 GHz {QF-215}.

2C36 Xmtg. UHF glass disc-seal coplanar triode, first of a line of "rocket" triodes, for pulsed-osc. use up to 1.2 GHz, plate on disc, grid on ring, heater 6.3 V @ 400 mA, 125 W pulse out (LI, SY) {SB846} (7-10-46). Variants: 2C37, 5764-5768, 6018, 6481, 6503, 6BA4.

2C37 Planar CW osc. triode, 2C36 format, 2.0 W output to 3.3 GHz (LI, SY) {SB846A} (6-20-46).

2C38 [AKA L-30-C]; rcvg UHF planar triode, physically same as, and replaceable by, 2C39.

2C39 A, WA, WA/C, WA/G, B, BA [AKA (GL, ZP) 572, 3X100A11]; UHF xmtg. triode, "oil can" planar design, heater 6.3 V @ 1.1 A, μ 100, g_m 17 mS, 25 W, 500 MHz (7-10-46). One of those "classic" designs that last indefinitely: varian ts include 5648, 6897, 7211, 7289, 7698, 7815, 8250, 8403, 8533, 8745, ML-381, 3CX100A5, and 3CX100F5. In AN/APT-9 jammer (200-2400 MHz) and AN/GPA-8 IFF interrogator, giving 1.5 kW pulses at 990-1130 MHz. Two in Motorola 460-MHz CB base station and mobile unit, "first [FCC] type-approved" unit, 5-53. In REL 757C 900-MHz relay xmtr. "A" and "WA" have glass or cermet construction depending on maker and vintage; uprated to 100 W and 2.5 GHz w/ forced-air cooling (GE, EI, ML, NEC, RCA) (7-10-46). Replaces 2C38. "WA/G" is glass; "WA/C" is cermet. In AN/APX-67 IFF set, giving 1-kW pulses at 950-1150 MHz, and RT-265/UPX-6. "B" and "BA" [AKA 6897, q. v.]; are cermet 2C39As (EI) (7-23-56).

2C40, A Rcvg lighthouse planar triode; replaced 446-series; heater 6.3 V @ 750 mA; μ 35, 6.5 W, usable to 3 GHz or more (GE, GEC, NEC, RCA, SY, TF) {ZP620, A2229A} (4-20-46) <6BY>. Rcvr preamp in radars, et al. Transmitter in AN/APN-7 S-band beacon, giving 200 W peak at 3.3 GHz; likewise in AN/APN-60 S-band beacon. In receiver of BP shipboard IFF interrogator. Variant: 2C49.

2C41 [AKA L-30-P]; xmtg. UHF triode, 2C39 format, not interchangeable phyically, heater 6.3 V @ 1.03 a, μ 100, uprated from 1 kV to 3.5 kV pulse, 3 GHz {ZP602} (GE, ML, SY).

2C42 Lighthouse triode, "special 2C43" {A2293C} (2-1-46).

2C43 [AKA GL-464A]; xmtg. UHF lighthouse triode; replaces 464, 464A, and 2C44; heater 6.3 V @ 900 mA, μ 48, 12 W, 3370 MHz (GE, NEC, RCA) {ZP-613, A2222} (4-20-46). In AN/UPX-5 and AN/UPX-12 IFF transponders, giving 1 kW pulses at 1090-1100 MHz. Variants: 2C42, 2C46, 2C47.

2C44 [AKA GL-464]; xmtg. lighthouse triode, superseded by 2C43 {ZP570}.

2C45 [AKA 38142, 45 Spl, VT-52]; xmtg. triode, 45 w/ fil. 7-V, 1.18 A, made by NU under WE contract for BC-230/430 aircraft radios; also distributed by HY <4D>.

2C46 Xmtg. lighthouse triode, selected 2C43 {A2294C} (2-1-46). In AN/APX-25 IFF transponder, giving 1 kW at 1090-1130 MHz.

2C47 Xmtg. lighthouse triode, selected 2C43.

2C48 [AKA QL134]; dual loktal low-μ power triode (to supersede 7C5 in some servo amps.), heater 6.3 V @ 600 mA, μ 9.5, 7.7 W. Variant: 2C50.

2C49 Lighthouse triode, pulse-rated 2C40 {Z(P?)682}.

2C50 [AKA CK1800]; dual low-μ power triode, for aircraft control systems, bantal 2C48 w/ triode conns. brought out independently; heater 12.6 V @ 0.3 A, μ 9.5, g_m 1.75 mS (GE, RA) (3-28-55) <8BD>.

2C51, W [AKA (WE) 396A]; min. 9-pin VHF med.-μ remote-cutoff duotriode, usable to 800 MHz, heater 6.3 V @ 300 mA, μ 35 (SY, WE) {1491} (11-25-46) <8CJ>. Variants: 5670, 6185, 6385, 6386, 7861.

2C52 High-μ octal duotriode, for aircraft controls; heater 12.6 V @ 300 mA, μ 100 (RA) (8-16-48) <8BD>.

2C53 [AKA M1060]; HV triode, shunt regulator, tall GT bulb w/ cap, heater 6.3 V @ 300 mA (11-25-46).

2D21, W [AKA EN91, WT-210-0001]; min. tetrode thyratron, min. 2050, heater 6.3 V @ 600 mA, 1.3 kV PIV, 100 mA, for relay, grid-controlled rectifier, and pulse-modulator use; also usable as broadband noise gen. (GE, HY, MU, RCA, RA, SY, W) {R6200} (4-29-44) <7BN>. Variants: 5727, 5757, R6288. "W" was pulser in AN/APN-19 and AN/APN-60 transponders.

2D28 Rcvg tetrode.

2D29 [AKA GE-N4]; submin. thyratron {SA-782B}.

2E21 Rcvg tetrode.

2E22 [AKA TS-70]; xmtg. pentode, quick-heating fil., heater 6.3 V @ 1.5 A, 30 W {DT70} (1-11-43) <5J>. In BC-1306, AN/GRC-9 and later TBW tranceivers.

2E24 Xmtg. VHF beam tetrode, tubular octal w/ cap, quick-heating fil., 6 V @ 650 mA, 10 W, 125 MHz (GE, RCA) (12-15-45) <7CL>. Variant: 2E26.

2E25, A Xmtg. beam tetrode, quick-heating T-T fil., 6 V @ 1.0 A, 15 W; replaced HY-65 (HY) {HD25} (9-27-46) <7CL>.

2E26 Xmtg. beam tetrode, heater version of 2E24, heater 6.3 V @ 0.8 A, 10 W, 125 MHz (GE, RCA) (3-11-46) <7CK>. Widely used in '50s mobile radios. Used into the mid-'60s in Clegg '22er and '66er, Knight-Kit TR-106, etc. Variant: 6893.

2E27 [AKA VT-283]; submin. pentode superseded by 2E29) {QF-206} (9-25-48). "Classified" as of 1967; never made?

2E28 [AKA HY-145ZT, q. v.]; quasi-submin. pentode.

2E29 Submin. pentode {SA781A}.

2E30 [AKA 5830]; xmtg. min. audio-VHF beam pentode, quick-heating fils. 3.0 V @ 650 mA each, 10 W, 165 MHz (GE, HY) {HD59} (3-15-47) <7CQ>. Mfgrs. also suggested use as receiver output tube, with fil. controlled by squelch circuit to save power.

2E31 [AKA CK553AX]; submin. sharp-cutoff RF pentode, flat-bodied, spray-shield, long leads, fil. 1.25 V @ 50 mA, g_m 0.5 mS (11-10-45) <2E31>. RA promoted, as a set for pocket-radio use, the 2E31/32, 2G21/22, 2E41/42, and 2E35/36. In the Belmont Boulevard personal set. Variant: 2E32.

2E32 [AKA CK553AY]; submin. pentode, 2E31 w/ short plug-in leads (RA) (11-10-45) <2E31>.

2E35 [AKA CK502AX]; submin. power pentode, flat-bodied, long leads, fil. 1.25 V @ 30 mA, g_m 0.39 mS (RA) (11-10-45) <2E31>. Variant: 2E36.

2E36 [AKA CK502AY]; submin. pentode, 2E35 w/ short plug-in leads (RA) (11-10-45) <2E31>.

2E41 [AKA CK551AX]; submin. diode-pentode, flat-bodied, long leads, fil. 1.25 V @ 30 mA, g_m 0.4 mS (RA) (11-10-45) <2E41>. Variant: 2E42.

2E42 [AKA CK551AY]; submin. diode-pentode, 2E41 w/ short plug-in leads (RA) (11-10-45) <2E41>.

2F21 Monoscope, 5", 40° magnetic-deflection, modernized version of 1899, same Indian-head test pattern, shortened to 11-9/16" long, pattern-electrode connector recessed into faceplate instead of cap, heater changed to 6.3 V @ 600 mA, derated to 1.5 kV (RCA) {C-7364} (3-11-46). Variant: 1699.

2G21 [AKA CK555AX]; submin. triode-heptode, flat-bodied, 1.25 V @ 50 mA, long leads, conversion g_m 0.06 mS (11-10-45) <2G21>. Variant: 2G22.

2G22 [AKA CK555AY]; submin. triode-heptode, 2G21 w/ short plug-in leads (RA) (11-10-45) <2G21>.

2H21 "Phasitron" magnetic-deflection phase-mod. tube in '40s GE and Gates FM xmtrs; domed tubular bulb, 11-pin magnal base, heater 6.3 V @ 300 mA (GE) (4-25-46). Also see 5593.

2J21, A [AKA WL-442, D2RN1, Y2, D2]; magnetron, fixed at 9375 MHz, interchangeable w/ 725A (2-11-46). In RT-10/APS-3.

2J22-29 Magnetron, fixed-freq., EM, CO, heater 6.3 V @ 1.5 A, 22 kV @ 30 A, 265 kW pulse output, forced-air cooled (RA) (5-20-44).

2J22 [AKA J1RF, J-714]; for 3300 ± 33 MHz. Has own anode-block design (see 2J23). Variant: 2J52.

2J23 [AKA K-0]; for 3085.5 ± 14.5 MHz (RA) (5-20-44). Shares anode-block design w/ 2J24-2J29. In BC-1224-A.

2J24 For 3059 ± 14 MHz.

2J25 For 3033 ± 14 MHz.

2J26 For 3005.5 ± 13.5 MHz (replaceable by 2J58).

2J27 For 2978.5 ± 13.5 MHz.

2J28 For 2952 ± 13 MHz.

2J29 For 2926.5 ± 12.5 MHz.

2J30-34 Magnetrons, fixed-freq., part of overall 2J22-2J34 series, CO, EM, 20 kV, 285-kW pulse out, have own anode-block design. Freq. tolerance ± 20 MHz.

2J30 [AKA K-7, VT-257]; for 2880 MHz (EEV, RA, TH) (5-20-44).

2J31 [AKA 110E/790]; for 2840 MHz (EEV, F, RA, TH). In SCR-784 and AN/MPQ-10A.

2J32 [AKA K-7-RF4, VT-251 (WL-441 family)]; for 2800 MHz (EEV, F, RA, TH). In SCR-584, RF-7/MPN-1, AN/MPQ-12, AN/MPQ-10A, and Gilfillan ASR-1.

2J33 For 2760 MHz (EEV, F, RA, TH). In AN/MPQ-10A.

2J34 For 2720 MHz (EEV, RA, TH). In AN/MPQ-10A.

2J35 Split-anode magnetron, T10 bulb, UX base, two plate pins, fil. 1.8 V @ 2.0 A; w/ 1 kV on plates @ 5 mA, puts out 1 W @ 3 GHz; has stabilizer electrodes taking 650 V @ 10 mA (RCA) (2-12-43) <4AP>.

2J36, A Magnetron, fixed at 9085.5 ± 82.5 MHz, EM, WG, heater 6.3 V @ 1.5 A, 14 kV @ 12 A, 14 kW pulse output (EEV, F, RA) (10-20-45). "A" has 28 kW output. In RF-6/MPN-1.

2J37, A Magnetron (10-20-45). "A" is tunable. In T-8/CPN-3.

2J38 Magnetron, fixed at 3256 ± 7 MHz, IM, CO (7/8" dia.), heater 6.3 V @ 1.3 A, 6 kV @ 8 A, 5 kW pulse output, forced-air cooled; radiator has mounting site for a thermostat for frequency stabilization (RA) (10-20-45). Variant: 2J39.

2J39 Magnetron, 2J38 for 3300 ± 33 MHz, 8.7 kW output; has smaller radiator and no thermostat mounting (RA) (11-23-45). In AN/DPN-1 beacon.

2J40 Magnetron (11-23-45).

2J41 [AKA A121]; magnetron, "coupled-cavity" stabilized design (see 6865), tunes 9290-9330 MHz, IM, WG, low-power (200-1250 W), forced-air cooling not needed (RCA) {A2116} (9-19-55). Typical beacon magnetron, as in AN/UPN-3.

2J42, A, BH, F [AKA LVX, X6001, A116]; magnetron, fixed at 9375 ± 30 MHz, IM, WG, heater 6.3 V @ 600 mA, 6 kV @ 5.5 A, 8 kW output (AX, EEV, KK, LI, MA, NEC, RA, SY, TSU, VO) {LD2} (5-15-46). In AN/APS-10, AN/APG-30 tail-warning radar and AN/SPS-35. "A" can operate at 7.4 kV @ 7.5 A, giving 20 kW output (MA). "F" has heater reduced to 480 mA (SFR). "H" suited to unpressurized operation to 60,000 feet (MA, RA, SY, TSU). Variant: 6027.

2J47 [AKA DRH505]; magnetron.

2J48-50 Magnetrons, variants of 725A, for use w/ external stabilizing cavity, EM, WG, heater 6.3 V @ 1.0 A, 16 kV @ 16 A, 50 kW pulse output {1392M, 1393M} (10-25-48). Freq. tolerance on all is ± 75 MHz.

2J48 For 9315 MHz (AX, RA, WE). In AN/CPN-6.

2J49 For 9080 MHz (AX, F, PH, TSU, RA, VO, WE). In AN/MPG-1.

2J50 For 8825 MHz (AX, PH, RA, RCA, TSU, VO, WE) (10-25-48). In AN/SPS-23.

2J51, A Magnetron, tunable IM variant of 725A, covers 8.5-9.6 GHz, heater 6.3 V @ 1.0 A, 16 kV @ 16 A, 60 kW pulse out, interchanges w/ 725A by removing 725A's magnet (AX, PH, RA, VO, WE) {1424MQ} (10-6-48). In AN/UPW-1. "A" from AX, THF, PH, RA, VO.

2J52 Magnetron, tunable 2J22, covers 3.3-3.4 GHz, EM, heater 6.3 V @ 1.5 A, 15 kV @ 15 A, 70 kW pulse out (RA).

2J53 Magnetron, variant of 725A for long-pulse use, fixed at 9375 MHz, 16 kV @ 16 A, 40 kW pulse out, repl. early by 4J50 and 4J52 (10-25-48) (RA, WE).

2J54, A, B Magnetron, tunes 3123-3259 MHz, EM, heater 6.3 V @ 1.5 A, 14 kV @ 15 A, 45 kW pulse out (RA) (11-23-45).

2J55 [AKA QK-47]; magnetron, packaged 725A-style, fixed at 9375 ± 30 MHz, IM, WG, heater 6.3 V @ 1.0 A, 16 kV @ 16 A, 50 kW pulse output, derived from 2J51 (AX, EEV, F, NEC, PH, RA, SY, TSU, WE) (11-23-45). In Mk. 12 (SP), 1660X (RA-Selenia), and AVQ-10 (RCA) small-aircraft weather radar. Variant: 2J56.

2J56, A Magnetron, 2J55 for 9245 ± 55 MHz (AX, F, RA, WE) (11-23-45). In AN/APG-32 for B-47E aircraft.

"A" is derated to 12 kV @ 12 A and 40 kW out (AX, EEV, PH, RA, VA).

2J57 Magnetron.

2J58 Magnetron, tunable 2J26, covers 2992-3100 MHz, EM, heater 6.3 V @ 1.5 A, 22 kV @ 15 A, 50 kW pulse output (RA) (11-23-45).

2J59 Magnetron, EM, CO, for use w/ external stabilizing cavity (1-10-46).

2J60 Magnetron.

2J61, A [AKA 706T1]; magnetron, tunes 3000-3100 MHz, EM, heater 6.3 V @ 1.5 A, 15 kV @ 15 A, 35 kW, replaces 706AY-DY inclusive (RA) (10-20-45).

2J62, A Magnetron, 2J61 for 2914-3010 MHz (RA) (10-20-45). Repl. 706DY-GY inclusive. In Mk 28 Mod 3.

2J64 [AKA A-124]; tunable magnetron, forced-air-cooled (10-5-45).

2J65 Magnetron, 2J41 w/o external tuner.

2J66-69 Magnetrons, sim. to 2J30, fixed-freq., EM, heater 6.3 V @ 1.5 A, 20 kV @ 25 A, 150 kW pulse out (RA) (1-10-46).

2J66 [AKA K7T-1]; for 2890 ± 15 MHz

2J67 [AKA K7T-2]; for 2825 ± 30 MHz.

2J68 [AKA K7T-3]; for 2775 ± 30 MHz.

2J69 [AKA K7T-4]; for 2725 ± 30 MHz.

2J70, B Magnetron, fixed at 3070 ± 40 MHz, IM, CO, heater 6.3 V @ 1.4 A, 7.5 kV @ 15 A, 20 kW pulse output (RA) (12-12-55). "B" in RA-Selenia 1660-12S radar.

2J71 Magnetron, fixed at 3195.5 ± 55 MHz, IM, CO, heater 6.3 V @ 1.4 A, 5 kV @ 5 A, 6 kW pulse out (RA).

2K21 [AKA WL-421, 421C]; klystron.

2K22 Reflex klystron, 723A/B format, tunes 4.24-4.91 GHz, heater 6.3 V @ 440 mA, 330 V, 115 mW output (AX, HI, NEC, THF, PH, RA, SP, TF, TSU, VO, WE) {1400XQ} (12-20-48). In AN/TRC-6 rcvr.

2K23 Pulsed reflex klystron, tunes 4280-4880 MHz, 723A/B format, dev. from 2K29, 400 mW out (WE) {1428-XQ} (12-20-48). Original tube in AN/TRC-6 xmtr. Also see 2K54-55.

2K24 Reflex klystron (WE) {1409XQ}.

2K25 Reflex klystron, up-speced 723A/B w/ redesigned electron gun, tunes 8.5-9.7 GHz, heater 6.3 V @ 440 mA, 300 V, 25 mW out (AX, HK, KR, RA, WE) {1410XQ} (10-6-48). Interchanges w/ 723A/B; some were marked "2K25/723A/B." In AN/APS-4, AN/-APQ-7, AN/APQ-13, AN/APS-15, AN/APS-31, etc.

2K26 Reflex klystron, 723A/B format, tunes 6.25-7.06 GHz, heater 6.3 V @ 440 mA, 330 V, 120 mW output (HI, NEC, RA, RCA, TSU, WE) {1411XQ-1412XQ} (12-20-48).

2K27 Klystron {1413XQ} (12-20-48).

2K28, A [AKA QK28]; reflex klystron (shortened 707B), 1200-3750 MHz, EC, CO, heater 6.3 V @ 660 mA, 330 V, 125 mW output (RA, THF, WE) (2-16-44).

2K29 Reflex klystron, 3.4-3.96 GHz, 723A/B format but with larger cavity and redesigned electron gun, heater 6.3 V @ 440 mA, 330 V, 100 mW (RA, THF, WE) {1349XQ, 1414XQ-1415XQ} (10-6-48).

2K30 Klystron.

2K31 [AKA (SP) 411]; klystron (10-5-45).

2K32 [AKA 412]; klystron.

2K33, A, B [AKA QK33]; disc-seal reflex klystron, 23.5-24.5 GHz, octal base, cap for reflector, IC, WG, hand-tuning knob, heater 6.3 V @ 650 mA, 1.8 kV, 40 mW out (RA) (3-20-46). British design, further dev. by RA.

2K34 Klystron amplifier, 2.73-3.30 GHz, heater 6.3 V @ 1.6 A, 1.9 kV, 12 W output, forced-air cooled (10-30-47).

2K35 Klystron amplifier, 2.73-3.30 GHz, heater 6.3 V @ 1.6 A, 3 kV, 25 W out, forced-air cooled (SP) (1-26-48).

2K36 [AKA 416]; klystron frequency multiplier, 300 MHz in, 3 GHz out, octal base, threaded coax fittings for input and output, internal cavities for both freqs.

2K37 Klystron.

2K38 [AKA 418]; klystron (11-10-45).

2K39 [AKA 419B]; reflex klystron, 7.50-10.3 GHz, heater 6.3 V @ 1.0 A, 1250 V, 250 mW output, forced-air cooled (SP) (11-10-45).

2K40 [AKA D-164694]; klystron, EC, 9 GHz (WE) {1280-CT} (2-11-46).

2K41-44 Reflex klystrons, improved 417B, octal base, IC, CO, heater 6.3 V @ 1.3 A, 1250 V, 250 mW out, forced-air cooled (11-10-45)

2K41 For 2.66-3.31 GHz (KK, SP).

2K42 [AKA XG-8071]; for 3.3-4.2 GHz (SP).

2K43 [AKA XG-7893]; for 4.2-5.7 GHz (SP).

2K44 [AKA XG-8072]; for 5.7-7.5 GHz (HI, SP).

2K45 Reflex klystron, thermally tuned, covers 8.50-9.66 GHz, shortened-6L6 case w/ octal base, IC, coax. line out base as with 723A/B, heater 6.3 V, 300 V, 32 mW out, tuner plate current 0-25 mA (WE) {1407XG} (12-20-48). Thermal tuning uses an internal triode whose plate is a wall of the klystron cavity; changing grid voltage affects the plate temperature, and thus the shape of the cavity wall. For use in frequency-agile (jamming-resistant) radars; tunes the band end-to-end in 6 sec. Variant: 6116.

2K46 Klystron frequency tripler, 2.7-3.3 GHz in and 8.1-10 GHz out, heater 6.3 V @ 1.3 A, 1.5 kV, 70 mW output, forced-air cooled.

2K47 Klystron freq. multiplier, 250-280 MHz in and 2250-3360 MHz out, heater 6.3 V @ 1.3 A, 1.0 kV, 150 mW output, forced-air cooled.

2K48 Reflex klystron, 4.0-11.0 GHz, short-6L6 body w/ waveguide out top, EC, CO, heater 6.3 V @ 700 mA, 1.5 kV, 20 mW output (WE) {1429CT8} (12-20-48).

2K49 Klystron (WE) {1429CT9} (12-20-48).

2K50 Reflex klystron, IC, WG, thermally tuned, covers 23,216-24,751 MHz, short-6L6 case w/ octal base, heater 6.3 V @ 775 mA, 330 V, 10 mW output (BX, WE) {1462} (12-20-48).

2K51 Klystron (W) {WX3243A} (2-11-46).

2K52 Klystron (W) {WX3243B} (2-11-46).

2K53 Klystron (W) {WX3243C} (2-11-46).

2K54, A, B, C, DA Reflex klystron, 723A/B format, for pulsed use, tunes 4290-4560 MHz, heater 6.3 V @ 450 mA, 400 V, 900 mW pulse output, replaces 2K23 (WE) {1468} (12-20-48). In AN/TRC-6 xmtr. "A" covers 3.85-4.1 GHz. "B" tunes 3.6-3.9 GHz. "DA" (NEC) covers 4.25-4.35 GHz.

2K55 Klystron, variant of 2K54 for 4590-4860 MHz (WE) {1469} (12-20-48).

2K56 Reflex klystron, 723A/B format, tunes 3840-4460 MHz, heater 6.3 V @ 440 mA, 380 V, 100 mW output, dev. from 2K29 (RA, RCA, WE) (12-20-48).

2K57 [AKA SRC-2]; reflex klystron, for xmtr use, freq. not stated (9 GHz?), IC, WG, mechanically tuned, octal base, cap for reflector, 700 V, 1.5 w output, IC, WG (SP) (3-2-50).

2P21 [AKA LM15]; "television pick-up tube" {C7557A} (2-1-50).

2P22 "Television pick-up tube," smaller 2P21 {C73009} (7-30-54).

2P23 Image orthicon for broadcast TV remote pickups, 5826 format; more sensitive than iconoscopes but w/ excessive IR response (GE, RCA) (5-10-46). Related to 5655; replaceable by 5820.

3B21 [AKA EL-302.5, 38401]; argon full-wave rectifier, graphite anode, fil. 2.5 V @ 5.5 A, 500V PIV, 1 A, replaceable by 3B22 (10-28-42) <4C>.

3B22 [AKA (EL) 1C, 38405, CE222]; xmtg. full-wave xenon rect., tantalum anodes, fil. 2.5 V @ 6.25 A, 725 V PIV, 1 A (EL, HY, W) (10-28-42) <4C>.

3B23 [AKA RK-22, 38222]; xmtg. full-wave vac. rect., heater 2.5 V @ 4 A, 3.5 kV, 200 mA (11-2-42).

3B24, W [AKA (WE) 732A, CE230]; xmtg. vac. rect., T12 bulb w/ cap on UX base, fil. 2.5/5.0 V @ 3.0 A (half or full fil. may be used), 30/60 mA, 20 kV (LK, RA, UE, WE) {428T, 1299XQ} (5-27-43). Also see 72. In BN IFF interrogator. Common in radars; in RT-5/APS-4 et al.

3B25 Xmtg. xenon rect., stepped tubular bulb w/ cap, UX base, fil. 2.5 V @ 5 A, 4.5 kV PIV (orig. 4 kV), 500 mA (HY, RCA) {R6187} (8-21-44) <4P>.

3B26 [AKA QK36]; radar clipper diode, 73 w/ cathode, octal base w/ cap, heater 2.5 V @ 4.75 A, 15 kV PIV, 20 mA (RA) (1-10-46). Variant: QK95.

3B27 Xmtg. rectifier, improved 836, PIV raised from 5 to 8.5 kV but current derated from 250 mA to 150 (EE, GN) (12-20-44) <4B>.

3B28 Xmtg. xenon rectifier, stepped tubular bulb w/ ceramic-ringed cap, UX base, fil. 2.5 V @ 5 A, 10 kV PIV, 250 mA (5 kV @ 500 mA) (CH, GE, HY, RCA, UE, W) (6-25-45) <4P>. An any-temperature replacement for 866A.

3B29 [AKA QK95]; vacuum rectifier and clipper diode, improved 3B26, T12 bulb w/ cap, 4-pin base, same heater, uprated to 16 kV PIV and 65 mA (TS, UE) (7-30-51) <4AB>.

3C21 [AKA 120, ZB-120, 38220, C120, CEO220]; xmtg. zero-bias triode, 4-pin jumbo base, fil. 10 V @ 2 A, μ 90, 120 W (3-15-43) <4E>.

3C22 [AKA (GL) 522]; octal xmtg. lighthouse triode, heater 6.3 V @ 2.0 A, 125 W, 1.4 GHz, transverse forced-air-cooled (GE) (1-5-46). In AN/APT-5.

3C23 [AKA WT-210-0047]; argon-mercury triode thyratron, for motor controls, etc., ST bulb w/ cap, UX base, fil. 2.5 V @ 7 A, 1250 V PIV, 1.25 A, sim. to CE311, 393A (CE, GE, HY, NL, RCA, T, TS, W) (7-28-43) <3G>.

3C24 [AKA 25TG, 3-25D3, VT-204, (DR, HK) 24G, PE-130A, CV745]; xmtg. VHF triode, 25T w/ grid-outside; T-T fil. 6.3 V @ 3 A, μ 25, 25 W (EEV, EI, GN, HK, LE, LK) (9-6-45) <2D>. Variants: 3C28, 3C34. In BC-640-A.

3C25 Xmtg. UHF triode.

3C26 [AKA X129D]; xmtg. UHF triode.

3C27, A, B [AKA CV178]; xmtg. UHF pulse triode, fil. 6.3 V @ 2.5 A, sim. to Br. CV55 and E1458. "A" is rated for higher duty cycle. "B" has grid radiator; superseded by 3C37 (9-12-47).

3C28 [AKA PE130B]; xmtg. VHF triode, 3C24 w/ two grid leads out opposite sides (LE) (4-25-46).

3C29 Xmtg. VHF triode {ZP-449}, superseded by 3C22.

3C30 Xmtg. triode, HY30Z w/ fil. 6.0 V @ 2.25 A (HY) {HD50} (10-25-46) <3G>.

3C31, A [AKA (EL) C1B, CE303]; thyratron, superseded by 5664, q. v. (EL).

3C32 Xmtg. UHF triode {A2214B}.

3C33 Xmtg. duotriode, 826 format, heater 12.6 V @ 1.125 A, μ 11, 30 W total; built in VHF format but intended as a servo amp. (10-18-46) <7CG>.

3C34 [AKA 25T, 3-25A3, HK-24, PE130C]; xmtg. VHF triode, 3C24 w/ grid lead in base (LE) (9-6-45) <3G>.

3C36 [AKA R1001]; xmtg. UHF coaxial triode (2-15-46).

3C37 [AKA 3X150A3]; xmtg. radar triode, mod. 3C27, ring-seal design with large center anode flange, fil. 6.3 V @ 2.4 A, giving 10 kW pulses up to 1150 MHz (EI, NU) (2-15-46).

3C45, W [AKA H45, 5137]; triode hydrogen thyratron radar modulator, T12 bulb w/ cap on glass stalk, 4-pin base, heater 6.3 V @ 2.3 A, 3 kV, 35 A pk., 45 mA avg. (BL, KL, RCA) {ZG547} (10-20-47) <4BL>. "W" is all-glass variant w/o stalk; includes hydrogen reservoir (KL). In AN/APS-10. Variant: 6130.

3D21, A, B, WB [AKA H5015]; vacuum pulser tetrode, short ST14 bulb w/ cap, octal base, dual heaters, 6.3 V @ 850 mA each, 15 W (HY, IT) (5-25-45) <8JU>. Pulser for 2J41 in AN/UPN-3. Variant: 7403.

3D22, A Tetrode gas thyratron rect., short T16 bulb, giant 7-pin base, heater 6.3 V @ 2.6 A, 1350 V PIV, 750 mA (NL, RCA) {R6241} (1-15-45) <7BV>. "A" is uprated to 1500 V and 800 mA.

3D23 [AKA TB35]; xmtg. tetrode, 35T format, fil. 6.3 V @ 3.0 A, g_m 2.75 mS, 35 W, 250 MHz - seems to be a tetrode version of 35T (LE) (9-10-45).

3D24 Xmtg. beam tetrode, hard-glass loktal w/ cap, graphite anode, fil. 6.3 V @ 3.0 A, 45 W, 125 MHz (SY) (4-2-47). Appears to be the US' only capped loktal tube. In TEMCO amateur xmtr., ca. 1948.

3D29 Xmtg. tetrode.

3E21 Xmtg. twin pentode, equiv. to 829B.

3E22 Xmtg. dual beam tetrode, 815 uprated from 25 W to 35 W in intermittent "mobile service," but derated to 15 MHz (RCA) (4-25-46) <8BY>.

3E29 Xmtg. beam tetrode, pulser version of 829B, same heater and g_m (GE, RCA) (7-3-53) <7BP>. In T-25/TPS-2, AN/APG-30, AN/APN-3, AN/MPG-1, and SCR-584. Four in parallel in O-18/CPN-2A.

3J21 Magnetron, 4J50 format, fixed at 23,986 MHz, IM, WG, 15 kV @ 15 A, 60 kW pulse output (WE) {1456M} (10-19-48). Similar to 3J31 w/ IM and stabilizing cavity in output waveguide.

3J22 Magnetron {SD849B} (9-16-46).

3J24 [AKA (GL, ZP) 484]; magnetron.

3J25 [AKA (GL, ZP) 485]; magnetron.

3J26 [AKA (GL, ZP) 486, VT-256]; magnetron. RMA registration dropped; specs not found; not seen in surplus; no CV no. Never made?

3J30 [AKA X6000]; magnetron, replaceable by 3J31.

3J31 [AKA X6002, X6003, E5A]; magnetron, fixed at 24 GHz, EM, WG, heater 6.0 V @ 1.7 A, 50 kW pulse output, forced-air cooled (5-15-46). Also see 3J21.

3J35 [AKA GL-505]; magnetron.

3K21 [AKA 410-1L]; klystron amplifier, variant of 410R, 2300-2725 MHz, heater 6.3 V @ 1.6 A, 2.0 kV, 15 W out, forced-air cooled (11-10-45). Variants: 3K21, 3K24.

3K22 [AKA XZA-8073]; 3K21 for 3320-4000 MHz, variant of 410T (11-10-45). Variant: 3K25.

3K23 [AKA R5660201]; reflex klystron, 950-1150 MHz, heater 6.3 V @ 1.6 A, 1.0 kV, 5 W output, forced-air cooled (SP) (11-10-45). Variant: 3K27.

3K24 Klystron, 3K21 w/ tuner.

3K25 Klystron, 3K22 w/ tuner.

3K26 Klystron, 3K30 w/ tuner.

3K27 Klystron, 3K23 for 750-960 MHz (7-2-47).

3K30 Klystron amplifier, variant of 410R, 2.0-3.3 GHz, heater 6.3 V @ 1.6 A, 2 kV, 15 W output, forced-air cooled (11-10-45). Variant: 3K22.

4A21 Ballast, octal, 1.62 A @ 11-15 V (NU) (11-15-43).

4B22 [AKA EL5B2.5, EL5B, 38402]; full-wave argon rectifier ("standard duty"), pear-shaped bulb w/ dual plate pins, screw base, graphite anodes, fil. 2.5 V @ 12 A, 340 V PIV, 5 A (CH) (10-28-42). Also see 4B23.

4B23 [AKA EL5BHD, 38403]; full-wave argon rect ("heavy duty"), fil. 2.5 V @ 17 A, 425 V PIV, 5 A (10-28-42). Also see 4B22.

4B24 [AKA (EL) 3C, 38404, CE224]; full-wave xenon rect., lugged base, fil. 2.5 V @ 11.5 A, 725 V PIV, 2.5 A (EL, HY) (10-28-42). Variant: 6484.

4B25 [AKA WL-670A, CE221, 38674, 38674A]; full-wave xenon rectifier, fil. 2.5 V @ 17 A, 725 V PIV, 6.4 A. EL6C is electrically same but has different base (10-28-42).

4B26 [AKA 2000, F60 (Allen), (WE) 289A, R6A, 38116, WL-289416D, J213AAA, CE226, GE 189049]; argon rectifier, pear-shaped bulb w/ plate pin, mogul screw base, fil. 2.2 V @ 6 A, 375 V PIV, 6 A (HY, RCA, W) (1-5-43).

4B27 [AKA (GE) 16X897, 38897, 1297, CE201]; full-wave gas rectifier., tubular stepped bulb w/ anode lead and medium screw base, fil. 2.5 V @ 10 A; replaceable w/ CE-201A (5-5-45).

4B28 [AKA (GE) 189048, CE225, WL-289414]; argon rectifier, 4B26 derated to 300 V PIV but w/ wider temp. limits; replaceable by CE-225 (HY, W) (11-30-44).

4B30 [AKA 8022, 15R, 2-15A]; xmtg. vac. HV rectifier (gridless 15E), all-glass, T-T fil. 5.0 V @ 4.0 A, 20 kV PIV, 30 mA (EI).

4B31 [AKA QK98]; vac. diode clipper-rectifier, heater 5 V @ 5.25 A, 16 kV PIV, 60 mA (HY, RA) (1-10-46).

4B32 Xenon xmtg. rectifier, any-temperature replacement for 872A, stepped tubular hard-glass bulb w/ ceramic-ringed cap, jumbo 4-pin base, fil. 5.0 V, 10 kV PIV, 1.25 A (CH, UE) (9-6-45).

4B33 [AKA HK-953B]; xmtg. vacuum rectifier, T16 bulb w/ cap, med. 4-pin base, fil. 7.5 V @ 6.5 A, 30 kV PIV, 40 mA (HK) (10-20-43).

4B35 [AKA 12X825, WL-966626, (WE) 327A]; gas rectifier, fil. 2.5 V @ 12 A, 200 V PIV, 2.0 A (W) (5-25-45).

4B36 [AKA (GE) 20X672, Tungar 5A, WL-859483]; mercury rectifier, fil. 2.0 V @ 12 A, 30 V PIV, 5 A (W) (5-25-45).

4C21 [AKA 211 Spl, Army VT-4C]; xmtg. triode, 211 w/ "low grid-current limit" probably from reduced grid-plate capacitance (10-25-43).

4C22 [AKA RK-51, HF-100, C101, P150]; VHF RF-heating triode, tubular bulb w/ cap, grid-out-side, fil. 10 V @ 2 A, μ 23, 75 W (3-15-43).

4C23 Xmtg. triode.

4C24 [AKA T814, HV12, q. v.]; RF-heating triode, T21 bulb w/ cap, 4-pin jumbo base, graphite anode, T-T fil. 10 V @ 4 A, μ 12, 200 W, 30 MHz (UE).

4C25 [AKA (HK) (HY)-54, CV49, 3B/501A] xmtg. VHF triode, UX base, grid and plate pins, fil. 5 V @ 5 A, μ 27, 50 W; nearest equiv. 35TG (9-6-45) <2D>.

4C26 [AKA 54B]; xmtg. triode.

4C27 [AKA 8026; Br. CV92, CV199, REL7, NT99, E1232]; Br. "micropup" disc-seal xmtg. triode; body 3-½" long, w/ turned brass plate radiator, finned for forced-air cooling, at mid-body, flying leads for fil., grid socket at other end (like shortened version of REL1). Oxide-coated fil. 6 V @ 6.7 A; a pair could give 100-150 kW pulse output in grounded-plate ckt. (10-21-46). Two in BC-1084 xmtr of SCR-602A 200-MHz air-warning radar; also in BC-1186A xmtr of SCR-636A.

(220-260 MHz); two in O-18/CPN-2 SHORAN set.

4C29 Xmtg. triode (5-20-44).

4C30 [AKA 15E, 8023, 3-15A3]; xmtg. UHF triode, T-T fil. 5.0 V @ 4.0 A, μ 25, 20 W. Four in 500-MHz ring osc. in Navy ASB radar. Two in BP, and one in BN, shipboard IFF interrogators. One in YJ 500-MHz beacon transponder.

4C32 RF-heating triode, tubular bulb w/ cap, grid-out-side, 4-pin jumbo base, graphite anode, fil. 10.0 V @ 4.5 A, μ 30, 700 W, 60 MHz (LE) (10-29-46) <2N>.

4C33 Xmtg. UHF pulse triode, coaxial glass ring-seal construction, heater 5.0 V @ 9.1 A, μ 25, 250 W, 625 MHz, 130 kW pulse output at 600 MHz, forced-air-cooled (RCA) {A2212B} (9-15-49).

4C34 Xmtg. triode sim. to HF-300 (9-5-47).

4C35, A [AKA GL-529, H350]; hydrogen thyratron modulator; tubular bulb w/ ceramic-ringed cap or cap on glass stalk (depending on maker), large 4-pin base; fil. 6V @ 6 A, 8 kV and 90 A peak, 100 mA avg. (BL, EEV, GE, KL, SY) {R6210, R6238, ZG529} (9-25-44). In AN/APS-1 and AN/SPN-18. Variant: 6268.

4C36 Xmtg. triode, 4-pin base w/ plate cap and two grid caps, fil. 5.0 V @ 7.5 A (9-5-47).

4D21, A [AKA X103, 4-125A, (AT, PE) 340, QY3-125]; xmtg. VHF beam tetrode, T-T fil. 5.0 V @ 6.5 A, 125 W, 120 MHz, forced-air-cooled (AX, EI, GE, LE, LK, MU, PL, RCA, W̲) (7-20-46) <5BK>. "A" (PL) is uprated to 175 W. "A" (AX) is all-glass, graphite anode. First of Eimac postwar tetrode line, w/ "Pyrovac" plate material and non-emitting grid. Variant: 6155.

4D22 Xmtg. VHF beam tetrode, fils. 12.6 @ 800 mA each, 50 W (RA) (12-5-50).

4D23 Xmtg. beam tetrode, repl. by 4D21 (10-29-46) <5BK>.

4D32 Xmtg. VHF beam tetrode, fil. 6.3 V @ 3.75 A, 50 W (RA) (12-5-50). In Collins 32V-2, Hallicrafters HT-20, and Johnson Viking I.

4E27, A [AKA 8001, HK-257, RK-48A]; xmtg. VHF pentode, jumbo 5-pin base w/ plate pin, fil. 5.0 V @ 7.5 A, 125 W, 75 MHz (ICE, LE, RCA) {A4229} (6-8-50) <7BM>. In BC-751 and AM-14/APT-2. "A" [AKA 5-75A, 5-125B, HK-257B, RK-48A] has same ratings but 3/4-inch shorter bulb (EI, RCA, W̲) (10-27-50). In AM-8E/TRA-1 power amp. for AN/TRC-1.

4J21-30 family, cavity magnetrons fixed at 1220-1350 MHz, EM, CO, 23.5 V @ 2 A, 27 kV @ 46 A, 600 kW pulse output, forced-air cooled, replacing 5J23 (WE) {1402} (all 10-19-48). 4J28 is fixed at 1280 MHz; 4J30, at 1226 ± 6 MHz. 4J26-30 cover 1220-1280 MHz. 4J25 in AN/CPS-5.

4J31-35 Magnetrons, fixed-freq., EM, CO, 16.0 V @ 3.1 A, 30 kV @ 70 A, 685 kW pulse output. Same anode block and cathode structure as 4J76-77; interchanges w/ 4J74-77. (EEV, RA) (10-20-45). Freq. tolerance on all is ± 20 MHz. This family used in AN/APN-19 beacon transponder.

4J31 [AKA HK7); for 2880 MHz.

4J32 For 2840 MHz.

4J33 For 2800 MHz.

4J34 For 2760 MHz.

4J35 For 2720 MHz.

4J36-41 Magnetrons, fixed-freq., EM, WG, 16.0 V @ 3.1 A, 30 kV @ 70 A, 850 kW pulse output. Same anode block and cathode structure as 4J70-71; interchanges w/ 4J70-73 (RA) (11-23-45). Freq. tolerance on all is ± 25 MHz. 4J36 is AKA H15.

4J36 For 3675 MHz.

4J37 For 3625 MHz.

4J38 For 3575 MHz.

4J39 For 3525 MHz.

4J40 For 3475 MHz.

4J41 For 3425 MHz.

4J42 [AKA 700T]; tunable 700-type magnetron (replaces 700-series), covers 660-730 MHz, 12 kV @ 9 A, 40 kW output (WE) {1440M} (10-25-48).

4J43 Magnetron, fixed at 3005.5 ± 13.5 MHz, EM, CO, 16 V @ 3.1 A, 30 kV @ 70 A, 900 kW pulse out (EEV, RA) (11-23-45). Variant: 4J53.

4J44 Magnetron, 4J43 for 2952 ± 13 MHz (EEV, RA) (11-23-45).

4J45-47 Magnetrons, long-pulse versions of 720-series, 21 kV @ 45 A, 1.1 MW pulse out (WE) {1379M} (10-25-48).

4J45 Variant of 720B (WE).

4J46 Variant of 720C (WE).

4J47 Variant of 720D (AX, WE).

4J50, A, TO, TR Magnetron, fixed at 9375 ± 30 MHz, IM, WG, heater 13.8 V @ 3.4 A, 23 kV @ 28 A, 250 kW pulse output, forced-air cooled (AX, EEV, LI, PH, RCA, SY, VO, WE) {1379M} (11-8-48). "A" from LI and THF. Also see 4J52 and 4J78. "TO" and "TR" tune 8.5-9.6 GHz, 200 kW output.

4J51 [AKA 728T]; magnetron, tunable 728, EM, CO, covers 900-970 MHz, 23 kV @ 20 A, 285 kW pulse out; supersedes 728-type (WE) {1443M} (10-25-48).

4J52, A, F, T Magnetron, fixed at 9375 ± 30 MHz, IM, WG, heater 12.6 V @ 2.1 A, 15 kV @ 15 A, 80 kW pulse output, sim. to 4J50 w/ lighter magnet and smaller cathode (AX, F, LI, MA, RCA, SY, VO, WE) {1471} (10-19-48). 4J52 in AN/TPS-21 (7 kW) and AN/TPS-25 (40 kW) anti-intrusion radars; 4J52A in AN/APN-59. "F" is uprated to 16 kV @ 16 A (SFR). "T" is tunable over 8.5-9.6 GHz (CGT).

4J53 Magnetron, 4J43 for 2803 ± 10 MHz (GEC, RA) (11-23-45).

4J54-59 Magnetrons, fixed-freq., IM, WG, 12.6 V @ 3.75 A, 25 kV @ 35 A, 250 kW pulse output (RA) (all 1-10-46). Freq. tolerance on all is ± 50 MHz

4J54 For 6825 MHz.

4J55 For 6725 MHz.

4J56 For 6625 MHz.

4J57 [AKA QK915C]; for 6525 MHz (AX, PH, RA). In AN/SPS-5B.

4J58 [AKA QK915B]; for 6425 MHz (AX, PH, RA).

4J59 [AKA QK915A]; for 6325 MHz (AX, PH, RA).

4J60-65 CW magnetrons, tunable, IM, CO, 18-vane anode, octal base, heater 6.3 V @ 3.8 A, 1.2 kV @ 125 mA, 50 W out, for jammers (all RA).

4J60 [AKA QK59]; tunes 2230-2470 MHz.

4J61 [AKA QK60]; tunes 2450-2720 MHz.
4J62 [AKA QK61]; tunes 2700-3010 MHz.
4J63 [AKA QK62]; tunes 2990-3330 MHz.
4J64 [AKA QK63]; tunes 3310-3670 MHz.
4J65 [AKA QK64]; tunes 3650-4030 MHz.

4J70-74 [AKA 3064]; magnetrons, tunable by means of attached cavity, EM, WG, 300-1000 kW pulse output (2-27-47). Also see 4J36.
4J70 Tunes 3540-3710 MHz.
4J71 Tunes 3390-3560 MHz.
4J72 Tunes 3240-3410 MHz.
4J73 Tunes 3090-3260 MHz.

4J74-77 Magnetron, coaxial-output variant of 4J70, 1-5/8" dia. coax line (2-11-46).
4J74 Tunes 2990-3110 MHz.
4J75 Tunes 2890-3010 MHz.
4J76 Tunes 2790-2910 MHz. Also see 4J31.
4J77 Tunes 2690-2810 MHz. Also see 4J31.

4J78 Magnetron, 4J50 fixed at 9085 ± 82.5 MHz (AX, EEV, SY, VO, WE) {1397M, 1471} (10-19-48).

4K34 [AKA 414]; klystron.

5B21 [AKA G683, R15A, 2183, (CE, GE) 235, GE 217283, WL-766776]; argon rectifier, mogul base, fil. 2.5 V @ 25 A, 225 V PIV, 15 A (HY, W) (9-20-45).

5B24 [AKA 38674, 45X674, 1674]; full-wave gas rectifier, replaceable by 4B25 (4-10-48).

5B33 [AKA HK-953D]; xmtg. vacuum rectifier, GT30 bulb 12-½" high w/ cap, large 4-pin base, pure-tung. fil. 7.5 V @ 11 A, 75 kV PIV, 60 mA (HK) (10-20-43).

5C21 [AKA (EL) C6J, CE306]; xenon triode thyratron rectifier, T16 bulb w/ cap, super-jumbo 4-pin base, fil. 2.5 V @ 18 A, 1250 V, 6.4 A (CE, EL, RCA) (9-28-45). Also see 5878.

5C22, A [AKA QK50, H1000, HT415]; hydrogen thyratron, tubular bulb w/ plate pin on glass stalk, large 4-pin base, fil. 6 V @ 10.5 A, 16 kV, 325 A peak, 200 mA avg. (BL, EEV, GE, KL, PL, RA, RCA, SY) {R6243, ZG569} (5-4-46). At least four internal constructions, depending on maker. Pulser in AN/APS-45, AN/-MPQ-10A, AN/MPS-19, AN/SPS-5B, AN/SPS-6C, and AN/TPS-1D. Variants: 6279, 6587, 7872.

5C23 [AKA 53AX]; xmtg. triode.

5C24 Xmtg. triode, 5-pin base w/ cap, fil. 10 V @ 5.2 A, μ 8, 160 W, radiation-cooled, for Class A and AB1 amps. (GE) (6-20-46) <6B>.

5C27 [AKA (EI) 227A, 3-100E3]; xmtg. UHF triode, VT-127A format.

5C30 [AKA 1552, CE315, C5B, q. v.]; thyratron (11-30-44).

5C37 [AKA (EI) 327A, 3-100G3]; xmtg. UHF triode, VT-127 format but only one plate pin, pin out top; plate looks same size as in VT-127; fil. 10.5 V @ 10.6 A.

5D21 Pulse tetrode, 715B upgraded w/ corona grading ring on plate cap, quadruple cathodes, grid cooling via base pins instead of radiating fins; heater 26.0 V @ 2.1 A, 20 kV, 15 A, 60 W, replaceable by 715C (9-25-47). In SCR-717, MD-8/APQ-13, AN/MPG-1, and AN/CPN-6.

5D22 [AKA QY4-250, X117, 4-250A]; xmtg. VHF tetrode, fil. 5.0 V @ 14.5 A, 250 W, 110 MHz (GE: 75 MHz), forced-air-cooled, sim. to 5D24 (AX, EI, GE, MU, PL, RCA, W) (10-5-46) <5BK>. In Collins 20V 1-kW AM xmtr. Variant: 6156.

5D23 [AKA RK-65]; VHF xmtg. tetrode, spherical-tubular hard-glass bulb w/ plate pin, two grid pins out side, 4-pin jumbo base, fil. 5 V @ 14 A, 215 W, intro. ca. 4-40 (1-10-46).

5D24 Xmtg. tetrode, replaceable by 5D22 (6-20-46).

5J21-25 Magnetrons, fixed-frequency, EM, CO, 24.5 kV @ 33 A, 450 kW pulse output (WE) {1382M} (10-19-48)
5J21 Freq. not known. In RF-1/TPS-1.
5J22 Freq. not known (not made?).
5J23 For 1050 ± 6 MHz. In RF-1/TPS-1.
5J24 Freq. not known (not made?).
5J25 Freq. not known (not made?).

5J26 Magnetron, tunable replacement for 4J21-30, covers 1220-1350 MHz, EM, CO, 23.5 V @ 2.2 A, 34 kV @ 55 A, 550 kW output, forced-air cooled (AX, F, THF, PH, RA, TSU, VO, WE) {1475M} (10-19-48). Widely used: in AN/FPS-71, AN/GSS-7, AN/SPS-6C, AN/-TPS-1D, etc.

5J29 [AKA (F, ZP) 579]; split-anode CW magnetron, tubular single-ended all-glass bulb, EM, w/ loop between anodes to increase max. freq., water-cooled via tubular pins and the interanode loop, superseded by Z675 (GE) (1-5-46).

5J30 [AKA (F, ZP) 590]; split-anode CW magnetron, all-glass single-ended bulb, for external cavity or tuned circuit, EM, 350-750 MHz, fil. curr. 30 A, 1.5 kV @ 300 mA typical, 150-500 W out, water-cooled via concentric tubing into tubular pins, superseded by 5J32 (1-5-46). In AN/APT-6, -7, -8.

5J31 [AKA (F, ZP) 584]; CW magnetron, water-cooled, superseded by 5J33.

5J32 CW magnetron, double-ended tubular glass bulb, water-cooled as for 5J30 (GE) {ZP646} (1-5-46).

5J33 CW magnetron {ZP676} (1-5-46).

5K40 [AKA 420]; klystron amplifier.

6B21 Xmtg. diode vacuum rectifier (LE) (2-12-46).

6C21 [AKA 1000 Spec.]; xmtg. triode radar modulator; fil. 8.2 V @ 16.8 A, μ 30, 30 kV, 15 A, 300 W; roughly sim. to 1000T; suited to long-pulse use (EI, GE, ML) {ZP527} (7-25-44) <4AQ>. Pulser in AN/MPS-9, AN/MPQ-18, and SCR-584.

6C22 Xmtg. UHF triode, glass ring-seal construction, fil. 6.5 V @ 18.5 A, g_m 13 mS, 1 kW, 600 MHz, screws into external cooler, water- or air-cooled, for UHF TV xmtrs. Gives 50 W out @ 1 GHz (F) {L-600N, F-600N} (4-25-46).

6C23 [AKA X-126H]; xmtg. UHF triode, fil. 7 V @ 26 A (1-10-47).

6C24 Xmtg. and RF-heating triode, tubular ring-seal construction, flying leads for fil., finned anode radiator at mid-body, grid pin at other end, fil. 11.0 VCT @ 12.1 A, μ 30, 600 W, 160 MHz, 1.1 kW out (ML, RCA) {A2255} (11-20-45). Also see 8014A. Variant: 5786.

6D21 Xmtg. tetrode pulse amp. (RA) (3-20-46), fil. 8.2 V @ 20 A, 400 W (3-20-46).

6D22 Xmtg. beam tetrode, fil. 5.0 V @ 28.5 A (RA).

6J21 CW magnetron, tunes 2460-3610 MHz, EM, IC, WG, flying leads, heater 6.3 V @ 16 A, 4-5 kV, 1 kW out, for jammers, limited production {ZP612} (11-25-47).

7B33 [AKA HK-953E]; xmtg. vacuum rectifier, GT40 bulb 22-½" high w/ cap, large 4-pin base, pure-tung. fil. 15 V @ 15 A, 150 kV PIV, 200 mA (HK) (10-20-43).

7C21 Xmtg. triode, dev. from ZP655.

7C22 [AKA D-164714]; xmtg. duo-triode VHF osc.; appears to be a Navy metal-body packaged radar oscillator roughly comparable to SigC VT-158, but not widely produced; put out 600-1000 kW pulses at 200 MHz (12-20-48).

7C23 [AKA "Mod. F-5303," 5680]; pulse xmtg. triode, corona ring about anode, pins out crown, fil. 13.0 V @ 36 A, μ 25, 1.2 kW, 5 MHz, forced-air cooled (F) (1-26-48).

7C24 Xmtg. UHF triode, ring-seal construction, for G-G use in FM and VHF TV xmtrs, flying ribbon leads for fil., grid ring, anode radiator, T-T fil. 12.6 V @ 29 A, μ 25, 2 kW, 110 MHz (ML, RCA) {A2261B, WX-3074} (4-25-46). Like a triode version of 827R, refilamented and uprated. Variant: 5762.

7C25 [AKA F5303]; RF-heating/xmtg. triode, glass body w/ external anode, flying ribbon leads, fil. 11.0 V @ 29 A, μ 25, 1.2 kW, 50 MHz, forced-air-cooled (F) (5-17-46).

7C26 Xmtg. triode, fil. 9.0 V, μ 17, 1 kW, 150 MHz, forced-air-cooled (F) (5-17-46). Promoted for FM xmtrs.

7C27 Xmtg. triode, glass ring-seal construction, grid on ring, fil. 10.5 V @ 28 A, μ 21, 3 kW, 150 MHz, forced-air-cooled (F) (10-10-46). Promoted for FM xmtrs.

7C29 Xmtg. triode, T-T fil. 10.5 V @ 28 A, μ 29, 500 W, 110 MHz, forced-air-cooled (GE) (9-5-46).

7C30 Xmtg. triode, fil. 10.5 V @ 28 A, forced-air-cooled (12-5-46).

7D21 Xmtg. tetrode, glass ring-seal construction, T-T fil. 6.3 V @ 30 A, 1200 W, 110 MHz, forced-air-cooled, for FM xmtrs. (GE) (7-10-46).

8C22 [AKA X111B]; xmtg. UHF triode, fil. 4.1 V @ 320 A (1-10-47).

8C23 [AKA X111AD]; xmtg. UHF triode, fil. 4.1 V @ 320 A (1-10-47).

8C25 Xmtg. audio triode, fil. 7.0 V @ 110 A, μ 6, 5 kW, forced-air-cooled (F) (1-26-48).

8D21 Xmtg. twin tetrode, glass ring-seal construction, dual anode rods out top, mounting flange, tubular pins out bottom for fil. and grids; fil. 4.2 V @ 135 A, 6 kW total, 300 MHz, water- and forced-air-cooled; tubular pins carry cooling water, in early postwar VHF TV transmitters (GE, RCA) (4-25-46).

9C21 Xmtg. and RF-heating triode, fil. 19.5 V @ 415 A, μ 38, 40 kW, 15 MHz, water-cooled via external cooler (AX, GE, ML, RCA) (12-30-43). Also see 5770.

9C22 Xmtg. triode, radiator version of 9C21, derated to 20 kW and 5 MHz (AX, GE, ML, RCA) (12-30-43). Four in parallel in final of RCA BTA-50F, mod. by two more (1946).

9C23 Xmtg. triode, fil. 22.0 V @ 82 A, μ 32, 15 kW, 20 MHz, water-cooled (F) (4-25-46).

9C24 Xmtg. VHF triode, glass ring-seal construction, fil. 6.3 V @ 240 A, μ 21, 5 kW, 220 MHz, for G-G Class B TV and Class C FM use; forced-air- and water-cooled via external cooler, dev. from ZP-655 (GE) (11-20-46). Variant: 6039.

9C25 Xmtg. triode, radiator version of 9C27, derated to 17.5 kW (GE, ML, RCA) (7-20-46).

9C26 Xmtg. triode, for 10-kW FM xmtrs, T-T fil., 7.5 kW, 30 MHz, forced-air-cooled (RCA) (1-3-47).

9C27 Xmtg. triode, for G-G amps., fil. 6.0 V @ 285 A, μ 32, 25 kW, 30 MHz, forced-air- and water-cooled via external cooler (RCA) (7-20-46).

9C28 Xmtg. audio triode, dual grid-out-side pins, T-T fil. 7.5 V @ 135 A, μ 4.8, 12 kV, 40 kW, for 50-kW xmtrs., water-cooled via external cooler (F) (4-25-47).

9C29 Xmtg. audio triode, radiator version of 9C28, derated to 20 kW (F) (4-25-47) (F).

9C30 Xmtg. triode, dual grid-out-side pins, T-T fil. 7.5 V @ 135 A, μ 42, 15 kV, 40 kW, 20 MHz, water-cooled via external cooler, for 50-kW xmtrs. (F) (10-15-46).

9C31 Xmtg. triode, radiator variant of 9C30, derated to 20 kW (F) (10-15-46).

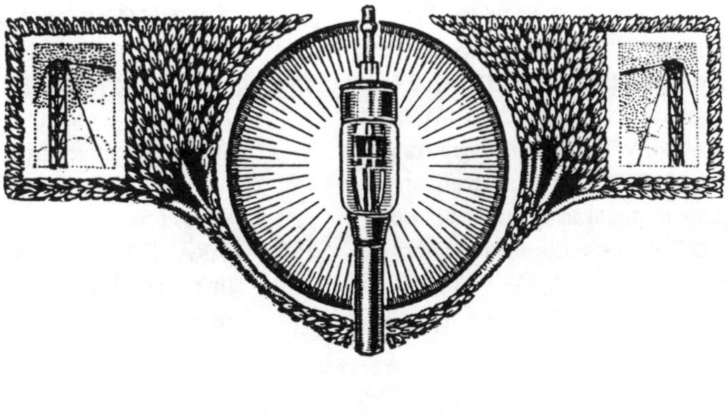

SPECIAL-PURPOSE TYPES: RMA-EIA 5500 SERIES

5500-11 Unknown; no evidence of introduction or use.

5512 Xmtg. triode, fil. 6.3 V @ 435 A (9-20-49).

5513 Xmtg. air-cooled triode, glass ring-seal construction, T-T fil. 6.3 V @ 30 A, μ 49, 1200 W, 220 MHz, forced-air cooled (GE) (2-5-47).

5514 High-μ power triode, heater 7.5 V @ 3.0 A, μ 145, 65 W, 60 MHz; supersedes HY30Z, HY40, HY40Z, HY51-series (HY) {HD69} (5-26-47) <4BO>.

5516 Xmtg. beam tetrode, shortened HY65, T11 bulb w/ cap on octal base, quick-heating fil., 6.0 VCT @ 700 mA, 15 W, 80 MHz (HY) {HD57} (5-26-47) <7CL>.

5517 [AKA CK1013]; min. cold-cathode rectifier w/ cap and starter electrode on own cap out side, for photoflash units; 2.8 kV PIV, 12 mA; replaces 1B48 (SY, HY, TS) {QM296} (4-30-47).

5518 Xmtg. air-cooled triode, glass ring-seal construction, pure-tung. fil. 6.3 V @ 235 A, μ 21, 4 kW, 110 MHz, forced-air cooled (GE) (9-5-46). Promoted for use in G-G pairs in 10-kW FM xmtrs for postwar FM band.

5519-23 X-ray sources (all 8-2-48).

5527 Iconoscope, electrostatic focus & deflection, tubular body 2" dia. by 9" long, 1.4" mosaic, 250-line resolution, diheptal base, heater 6.3 V @ 600 mA, supersedes 1847 (RCA) (2-10-47).

5528 [AKA CE326, C6L]; xenon triode thyratron, same base and bulb as C6J/5685; fil. 2.5 V @ 20 A, 500 V PIV, 6.4 A (HY) (9-27-46). Variant: 6044.

5530, H Xmtg. VHF triode, fil. 5.0 V @ 55 A, μ 26, 4 kW, 110 MHz, forced-air-cooled (WE, ML) (11-15-46). "H" (ML) is uprated from 5 to 8.5 kV, from 8.8 to 10 kW, but down from 110 to 30 MHz.

5531 RF-heating/xmtg. VHF triode, T-T fil. 6.3 V @ 92 A, μ 24, 10 kW, 30 MHz, forced-air-cooled (ML) (2-10-49).

5532-40 X-ray sources (all 2-25-48).

5541 Xmtg. triode, fil. 7.5 v @ 55 A, μ 26, 10 kW, 110 MHz, forced-air-cooled (ML) (11-15-46). In Continental Electronics 5-kW AM xmtr.

5542 Xmtg. triode, forced-air-cooled (2-10-49).

5544 [AKA XR1-3200, NL730]; xenon triode thyratron, tubular bulb w/ cap, jumbo 4-pin base, fil. 2.5 V @ 12.0 A, 1500 V PIV, 3.2 A (AX, GE, HY, MU) (2-28-48) <4BZ>. Also see 5877.

5545 [AKA XR1-6400]; xenon triode thyratron, for motor controls, etc., stepped tubular bulb w/ ceramic-ringed cap, large 4-pin base, fil. 2.5 V @ 21.0 A, 1.5 kV PIV, 6.4 A (AX, GE, MU) (1-20-47). Also see 5878, 6807.

5549 RF-heating triode, pure-tung. fil. 12.6 V @ 57 A, μ 23, 4 kW, 50 MHz, forced-air-cooled (GE) (12-20-46).

5550 [AKA FG-415, WL-681/686, WT-210-0070]; ignitron, "Size A" stainless steel case 2-3/4" dia. by 5" high, anode lead out top, ignitor terminal at other, cathode on shell, 900 V PIV, 12.2 A, cooled by air or clamp-on water cooler (GE, HY, NL, RCA, W) (9-18-46). Also see 1R21.

5551, A, A-PC [AKA FG-271, WL-652/657, WT-210-071, WT-210-0149, WT-210-0156]; ignitron, "Size B" stainless-steel case, 2-3/4" dia. by 9" high, anode lead out top, terminals at bottom, 500 V PIV, 18 A, water-cooled via integral jacket, for rectifier or control use (AX, GE, HY, RCA, W) (9-30-46). "PC" has thermostat control. Variant: 6346.

5552, A, A-PC [AKA WL-651/656, FG-235A, WT-210-072, WT-210-0157]; ignitron, "Size C" stainless steel case, 4-¼" dia. by 12" high, anode lead at top, terminals at bottom, 500 V PIV, 100 A, water-cooled via integral jacket, for rectifier or control use (AX, GE, HY, RCA, W) (9-18-46). "PC" has thermostat control. Variant: 6347.

5553, A, B, B-PC [AKA WL-655/658, FG-258A, WT-210-0073]; ignitron, "Size D" stainless steel case 5-5/8" dia. by 19-½" high, anode lead at top, terminals at bottom, 1.5 kV PIV, 112 A, water-cooled via integral jacket, for rectifier or control use (AX, GE, HY, RCA, W) (9-30-46). "B" and "B-PC" have thermostat control. Variant: 6348.

5554 [AKA FG-259B, WL-679]; ignitron, 5552 format, dual ignitors, 2.1 kV PIV, 75 A, water-cooled via integral jacket, for rectifier or control use (GE, RCA, W) (9-28-46).

5555, A [AKA FG-238B, WL-653B]; ignitron, 5553 format, in welder-control use, 2.4 kV applied, 207 A; in rect. use, 2.1 kV PIV, 150 A; water-cooled via integral jacket (AX, GE, HY, RCA, W) (9-30-46). Also see 5788. Five used in series in crowbar circuit for RF heating system for C-Stellarator controlled-fusion experiment, 1961, handling 60,000 A peak current.

5556 [AKA PJ-8]; xmtg. power triode, S-17 bulb (later T12), UX base, T-T fil., 4.5 V @ 1.1 A, μ 8.5, 10 W, 6 MHz (GE, RCA) (10-25-46); promoted for use in relay applications as well as RF.

5557 [AKA FG-17, (UE) 967, 1701, WT-210-0015, WT-210-0062, WT-210-0069, XGS-500]; grid-controlled version of 866A: mercury triode thyratron, S19 or ST bulb w/ cap, UX base, fil. 2.5 V @ 5.0 A, 5 kV PIV, 500 mA (AX, GE, HY, MU, RCA, W) (12-12-46) <3G>. Also see FG-17.

5558 [AKA FG-32, WL-5558/32]; mercury rect., ST24 bulb w/ cap, UX base, heater 5.0 V @ 4.5 A, 5 kV PIV, 2.5 A (GE, HY, RCA, W) (11-5-46) (GE rates FG-32 at only 1 kV PIV).

5559 [AKA FG-57, WL-5559/57, WL-631, WT-210-0056, XG1-2500]; mercury triode thyratron, 5558 w/ grid added, negative grid char., derated to 1.5 kV PIV (AX, HY, MU, RCA, W) (11-26-46) <4BL>. Also see FG-57.

5560 [AKA FG-95, WL-632A, 1295]; mercury tetrode thyratron, grid-out-side variant of 5559, high grid impedance (AX, GE, HY, NL, RCA) (12-12-46)

<4CD>. ML-5560 has grid cap mounted on base. Variant: 5720.

5561 [AKA FG-104, WL-5561/104]; mercury thyratron rect., conical or tubular ST30 bulb w/ cap, special 4-pin base, heater 5.0 V @ 10 A, 3 kV PIV, 6.4 A or, in welder-control use, 10 kV PIV @ 4 A (GE, HY, RCA, W) (11-5-46).

5562 Xmtg. beam tetrode, 4-pin base w/ cap, graphite anode, T-T fil. 6.3 V CT @ 3.0 A, ca. 50 W, 120 MHz (UE) (1946).

5563, A Mercury triode thyratron, T24 stepped bulb w/ cap, 4-pin jumbo base, fil. 5 V @ 10 A, 15 kV PIV, 1.6 A (RCA) (3-25-47). "A" is uprated to 20 kV; has T20 bulb.

5564 [AKA GL-507]; ignitron, steel case, 2.1 kV PIV, 300 A, water-cooled, for rectifier or control use, dual igniters (GE) (11-25-47).

5566 [AKA ML-101]; vac. rect., fil. 22 V @ 52 A.

5567-70 Vacuum rectifiers, fil. 10.0 V @ 11.5 A.

5571 Vacuum rectifier, fil. 13.0 V @ 12.4 A.

5572 Vacuum rectifier, fil. 10.0 V @ 17.8 A.

5573-74 Vacuum rectifiers, fil. 13.0 V @ 12.4 A.

5575 [AKA ML-100]; vacuum rect., fil. 20 V @ 24 A, 150 kV PIV, 500 mA, convection cooled (ML) (W rates at 320 mA) (3-10-48). Also see 5625.

5576 [AKA ML-200]; vac. rect., 20 V @ 32 A, 150 kV PIV, 1.25 A, convection cooled (ML) (3-10-48).

5577-78 Vacuum rectifiers, fil. 13.0 V @ 12.4 A.

5581 [AKA CE-59]; phototube, 930 w/ S-4 response, for sound from dye image track w/ incandescent light, 135 μA/lumen (RCA, W) (2-10-47).

5582 Phototube, 921 w/ S-4 response, same use as 5581, 120 μA/lumen (RCA) (2-10-47).

5583 [AKA CE-64]; phototube, 927 w/ S-4 response, same use as 5581, 135 μA/lumen (RCA) (2-10-47).

5584 Phototube, 920 w/ S-4 response, 120 μA/lumen, same use as 5581 (RCA) (2-10-47).

5586 Magnetron (tunable 4J31-35), 2.7-2.9 GHz, EM, CO, 16.0 V @ 3.1 A, 30 kV @ 70 A, 800 kW pulse out {HK7T} (AX, PH, RA, THF) (9-25-48). In AN/ASR-2, AN/FPS-14, AN/MPQ-18, AN/MPS-9, and AN/MPS-19.

5588 Xmtg. triode, glass ring-seal, coaxial electrodes, heater 6.3 V @ 2.5 A, μ 16, 200 W, 1.2 GHz, forced-air-cooled (GE, RCA) (2-25-47). Also see 6161.

5589 [AKA (WE) 376B]; cold-cathode triode, T9 painted bulb, octal base, 200 V PIV, 70 mA (WE) (2-10-47).

5590 [AKA (WE) 401A]; min. pentode, heater 6.3 V @ 150 mA, g_m 2.09-2.2 mS (GE, HY) (1-10-47) <7BD>.

5591 [AKA (WE) 403B]; min. pentode, heater 6.3 V @ 150 mA, g_m 5 mS; also see 6AK5 (GE, HY, WE) (2-4-47) <7BD>.

5592 Xmtg. triode, for G-G use in FM xmtrs, radiator-type, two fils., 11 V @ 412 A each, μ 32, 17.5 kW, 110 MHz (RCA) (4-1-47). Two in RCA BTF-50A 50-kW xmtr.

5593 Phasitron, improved 2H21; same format and pinout, ratings different (9-2-47).

5594 Xenon thyratron rectifier sim. to FG-17, T16 stepped-tubular bulb, 4-pin base, fil. 2.5 V @ 5.0 A, 5 kV PIV, 500 mA (CH, TS) (3-12-47).

5599 X-ray source (7-15-50).

5600-02 X-ray sources (all 7-15-50). (5602 is also reported to be a premium 12AT7.)

5603 [AKA (WE) 398A]; rcvg. power pentode, short T11 bulb on octal base, quick-heating fil. 6.3 V @ 500 mA, g_m 5.2-5.5 mS, 8 W (2-26-47).

5604, A RF-heating triode, fil. 11.0 V @ 176 A, μ 19.5, 10 kW, 25 MHz, radiator-type (AX, F, ML, RCA, W) (3-10-48). "A" has fluted radiator requiring less air flow.

5605 Crystal diode (!). (No sign of actual production.)

5606, A Xmtg. triode, fil. 22.0 V @ 60 A, μ 50, 10 kW, 1.6 MHz, water-cooled (F, GE, W) (11-19-56).

5607 CW magnetron, tunable 6J21, covers 2.5-3.6 GHz, EM, WG, 5.5 V @ 17.0 A, 6 kV @ 1 A, 1 kW output (LI) (11-25-47).

5608, A Med.-μ duotriode for "special control equipment," 7-pin base, fil. 2.5 V @ 2 A, μ 32 (GE, RA) (7-15-48) <7B>.

5609, A CW magnetron, for RF heating, fixed at 2450 ± 25 MHz, IM, CO, heater 6.3 V @ 18 A, 1.45 kV @ 135 mA, 70 W output (AX, RA, VO) (2-6-53).

5610 Min. triode, general industrial use, heater 6.3 V @ 150 mA, μ 14 (GE) (9-20-48) <6CG>.

5611 Klystron osc. (12-20-48).

5612-17 Gas phototubes, S-4 response (all 7-10-47).

5618 Min. rcvg. VHF power pentode, heater 6.0 VCT @ 230 mA, 5 W, 40 MHz (HY, RCA) (10-2-47).

5619 RF-heating triode, HF variant of 5604 water-cooled via external cooler, same ratings (W) or uprated to 20 kW (AX, F, ML) (3-10-48).

5620-24 Ballast tubes (all GE) (all 7-15-47).

5620 [AKA FB-50, 896, q. v.].

5621 [AKA B-6]; tubular bulb, med. screw base, 0.98 A @ 15-21 V.

5622 [AKA B-25]; "S" bulb, med. screw base, 7-16 V @ 1.12 A.

5623 [AKA B-47]; 5622 format; 8-18 V @ 2.2 A.

5624 [AKA B-46, E82M]; 5622 format, 8-18 V @ 2.98 A.

5625 [AKA KC4, WL-616]; vacuum HV rectifier sim. to F606, stepped tubular bulb w/ cap, bipin base, fil. 20 V @ 24.5 A, 150 kV PIV, 750 mA peak (GE) (6-30-48). Also see 5575.

5626 [AKA FA-15]; SPDT vacuum switch, in metal can w/ long actuating rod. Also see 5627 and 1S21 (GE) (7-15-48).

5627 [AKA FA-6]; SPDT vacuum switch, glass body w/ long actuating rod. Also see 5626 and 1S21 (GE) (7-15-48).

5628 Pressure-indicator tube (Pirani gauge?) (9-15-48).

5629 Pressure-measuring compensator tube (9-15-48).

5630 Ignitron, stainless steel case 5-3/4" dia. by 22-3/16" long, 35 kV PIV, 50 A avg. or 20 kA peak, water-cooled, for capacitor-discharge use or for rectifier use in transmitters (six will give 2.5 MW at 17 kV) (GE) (7-18-47). Used as crowbar in RCA BTH-250A 250-kW HF broadcast xmtr - could stop a flash arc in final tube in 5-10 μsec.

5632 [AKA C3J]; xenon triode thyratron, for welder controls, etc., T12 bulb w/ cap, UX base, fil. 2.5 V CT @ 9 A, 1250 V PIV, 2.5 A; appears to be uprated 5684 (AX, EL, GE, NL, RCA) (6-6-47). Variant: 7216.

5633 T3 submin. remote-cutoff RF pentode, grid out top, heater 6.3 V @ 150 mA, g_m 3.4 mS (GE) {SN-944} (2-25-48) <5633>.

5634 [AKA ME1500]; T3 submin. sharp-cutoff RF pentode, grid out top, heater 6.3 V @ 150 mA, g_m 3.5 mS (GE, SY) {SN-828E} (2-25-48) <5633>. Also see 5638.

5635 T3 submin. med.-μ duo-triode, heater 6.3 V @ 450 mA, μ 38 {SN-955B} (3-28-49) <8DB>.

5636, A [AKA Ef730]; T3 submin. dual-control pentode mixer, heater 6.3 V @ 150 mA; for phantastron dividers, mixers up to 400 MHz, etc. (GE, RCA, MU, SY, TS) {SN-1007A} (3-28-49) <8CD>. Variant: 8522.

5637 T3 submin. high-μ triode, heater 6.3 V @ 150 mA, μ 70, g_m 3.5 mS (SY) {SD917A} (12-15-47) <5637>.

5638 T3 submin. high-μ triode, heater 6.3 V @ 150 mA, g_m 3.3 mS, 600 mW (GE) (2-25-48) <5638>.

5639, WA T3 submin. video power pentode, heater 6.3 V @ 450 mA, g_m 10 mS, 3.8 W (RA, SY, TS) {SN953A} (4-12-50) <8DL>.

5640 T3 submin. audio beam power tube, heater 6.3 V @ 450 mA, g_m 4.4 mS, 1 W out (SY) {SN-947} (3-20-48) <5640>.

5641 [AKA EY70]; T3 submin. low-voltage rectifier, heater 6.3 V @ 450 mA, 1 kV PIV, 40 mA (SY) {SN954} (2-25-48) <6CJ>.

5642 [AKA DY70]; T3 submin. double-ended HV rectifier, fil. 1.25 V @ 200 mA (originally 1.1 V @ 130 mA), 10 kV PIV, 250 μA (GE, MU, RCA, SY, TS, W) {SN-956A} (3-20-48) <5642>. In TV sets and oscilloscopes like Tektronics 524D. 1BG2/DY51 is highly similar.

5643 [AKA EN70, TD17]; T3 submin. xenon "shield-grid" (high-impedance) thyratron sim. to 5696, heater 6.3 V @ 150 mA, 500 V PIV, 16 mA (BX, MU, SY, TS) {SN-949} (3-28-49) <8DD>.

5644 T3 submin. voltage reg., 95 V (MU, SY) {SN-948} (4-12-50) <4CN>.

5645 T2 submin. med.-μ triode, heater 6.3 V @ 150 mA, μ 20, g_m 2.7 mS (GE) {SN957A} (3-20-48) <5645>.

5646 T2 submin. high-μ rcvg. triode, heater 6.3 V @ 150 mA, μ 70, g_m 2.6 mS (GE, SY) {SN1006} (6-29-48) <5645>.

5647 [AKA EA21]; T1 sub-submin. diode, 1-¼" long, heater 6.3 V @ 150 mA, 460 V PIV, 10 mA (GE, SY) {SN946} (12-31-48) <5647>.

5648 [AKA 2C39A]; glass-metal xmtg. triode, indistinguishable from 2C39A (GE) (11-25-47).

5649 [AKA (WE) D-170573]; L-band reflex klystron, superseded by 5981 (WE) {1466} (1-30-48).

5650 [AKA (WE) D-176627]; reflex klystron, 1.2-1.4 GHz, IC, CO, heater 6.3 V @ 450 mA, 450 V, 165 mW output, superseded by 5981 (WE) {1596} (1-30-48).

5651, A, WA [AKA STV85/10]; min. neon voltage reference, 87 V, 3.5 mA (GE, MU, RA, RCA, SY, TS) {S901C} (1-26-48) <5BO>. "A" operates at 85.5 V.

5652 Vacuum phototube, S-4 response, T9 bulb on octal base, 2-5/16" seated height, side-on, two complete tubes built back-to-back internally for use w/ audio-tone excitation, 45 μA/lumen (RCA) (10-20-47). For use in fax machines as combined scan sensor and balanced modulator: wired into a bridge circuit and supplied w/ carrier tone, generates double-sideband line signal directly. Electrodes serve alternately as anodes and cathodes. Includes capacitance-balancing electrode. Also see 1645.

5653 Vacuum phototube, S-4 response; indistinguishable from 929; replaceable by CE-29 (RCA) (9-25-47).

5654 [AKA 6AK5W, E95F, M8100, TS-51, etc.]; ruggedized 6AK5 (GE, MU, RA, RCA, SY, TS) (7-15-48) <7BD>.

5655 Image orthicon, variant of 2P23, for B & W broadcast use, 3" dia. bulb stepped to 2", 15-¼" long, 7-pin jumbo base at large end, diheptal base at small end; replaceable by 5820 or 5826 (GE, RCA) (8-22-47).

5656 Min. 9-pin rcvg. duotetrode (twin 6AK5) w/ common screen lead, heater 6.3 V @ 400 mA (HY, RA) {QM322} (6-5-50).

5657 Magnetron, tunes 2.9-3.1 GHz, EM, CO, 16.0 V @ 3.1 A, 32 kV @ 70 A, 800 kW pulse output (sim. to HK7T) (AX, EEV, RA, THF) (2-6-53).

5658 RF-heating triode, improved 880, used Kovar seals in place of feather-edge, new fil. design, etc; same ratings except fil. voltage of 12.0 vice 12.6. Directly interchangeable (AX, EI, F, ML, W) (3-10-48).

5659 Mechanically improved 12A6 (11-15-47) <7S>.

5660 Mechanically improved 12C8; lower g_m and screen-voltage rating (11-15-47) <8E>.

5661 [AKA VP12D]; pentode, ruggedized 12SK7 (11-15-47) <8N>.

5662 Min. gas thyratron for use as "fuse," promoted for use in electric-blanket controllers, heater 6.3 V @ 150 mA, 200 V PIV, cathode current to fuse: 150 A of 60-Hz half-wave (GE, HY) (4-22-49) <5662>.

5663 Min. tetrode xenon thyratron, heater 6.3 V @ 150 mA, 500 V PIV, 20 mA (GE, HY, RCA) (11-25-47) <6CE>.

5664 [AKA C1B/A]; triode gas thyratron rectifier, improved 3C31, fil. 2.5 V @ 6.3 A, 1250 V PIV, 1 A (EL, W) (9-25-47).

5665 [AKA C16V, C16J, WT-210-0066]; xenon triode thyratron rectifier, T20 bulb, bracket mount, anode lead at top, other leads at base, fil. 2.5 V @ 31 A, 1250 V,

18 A (EL, GE, NL, RCA, W) (7-30-48).

5666 [AKA ML-309]; RF-heating triode, substantial redesign of 889A; uses special water jacket (AX); replaces w/o eqpt mod. (ML), same ratings (fil. 11.0 V @ 120 A, 12.5 kW, 20 MHz) (AX, F, ML) (3-10-48).

5667 [AKA ML-309R]; RF-heating triode, redesigned 889RA, replaces it directly, uprated (by longer radiator) to 7.5 kW and 50 MHz (F, W: derated to 22.5 MHz) (AX, EI, ML) (3-10-48).

5668 [AKA ML-308, WT-210-0093]; RF-heating triode, redesigned 892, replaces it directly, all leads out crown, uprated to 20 kW and 5 MHz (F, ML, W) (3-10-48).

5669 [AKA ML-308R]; RF-heating triode, redesigned 892R, replaces it directly, same ratings except uprated to 10 kW and 5 MHz (F, ML) (3-10-48).

5670, WA Min. duotriode, ruggedized 2C51, heater raised to 350 mA, computer-rated (GE, RA, RCA, SY, TS) (9-15-48) <8CJ>.

5671 RF-heating triode, T-T fil. 11 V @ 285 A, μ 39, 25 kW, 10 MHz, radiator-type; radiator has multiple layers of fins (RCA) (11-25-47).

5672 [AKA DL69, DL620, XFY14]; T2X3 submin. audio power pentode, 1.25 V @ 50 mA, g_m 0.65 mS, 65 mW out (GE, MU, RA, RCA, TS) (11-26-48).

5674 Electrometer duotriode (11-15-47).

5675 Glass pencil UHF triode, for G-G use, heater 6.3 V @ 135 mA, μ 20, 5 W, 1.7 GHz (GE, RCA, SY, TS) {A2302B} (11-15-49) <5675>. Variants: 5794, 5876, 5893, 6263, 6264.

5676 [AKA CK556AX, XFR2]; T2X3 submin. med.-μ UHF osc. triode, fil. 1.25 V @ 120 mA, μ 15 (GE, RA, TF, TS) (5-5-48) <5676>. Variants: 6050, 6121, 7246.

5677 [AKA CK568AX]; T2X3 submin. med.-μ UHF osc. triode, fil. 1.25 V @ 60 mA, μ 13.5 (RA) (7-15-48) <5676>.

5678 [AKA CK569AX, DF60, DF634, XFR2]; T2X3 submin. RF pentode, spray-shielded, 1.25 V @ 50 mA (GE, MU, RA, RCA, TF, TS) (7-15-48) <1AD4>.

5679 Duodiode, 7A6 w/ CT heater (GE) (12-1-47) <7CX>.

5680 Xmtg. triode, 7C23 rerated to 13 V, 2.5 kW, 30 MHz (F) (2-25-48). Two in Federal 196-A AM xmtr.

5681 Xmtg. glass ring-seal triode, bayonet 60° twist-in quick-change mounting; T-T fil. 12.0 V @ 220 A, μ 25, 75 kW, 30 MHz, water-cooled by integral water jacket (EI, ML) (11-21-50). Dev. jointly by BTL and ML to replace WE 298B, for FM broadcast xmtrs (incl. Armstrong KE2XCC). In postwar WLW 50-kW xmtr, Continental Electronics 50-kW AM xmtr, WCBS-TV Ch. 2 '50s xmtr, VOA xmtrs, RF heaters, REL "Project Diana" 50-kW 151-MHz moon-radar xmtr., etc.

5682 Xmtg. glass ring-seal triode, larger 5682, bayonet-type mount, fil. 16.5 V @ 325 A, μ 30, 120 kW, 30 MHz (EI, ML) (1-15-51). In Continental 105B 1-MW AM-LF xmtr, as installed in '50s VOA megawatt station on 173 kHz: one as driver, four as carrier amp., four as peak amp.

5683 [AKA C1J-A]; thyratron rectifier, imp. C1J (W) (12-3-47). Also see 6014.

5684 [AKA C3J-A]; 5632 with forward anode voltage uprated from 900 V to 1000 V (AX, EL, HY) or 1250 V (NL, W) (12-3-47).

5685 [AKA C6J-A, WT-210-0063]; 5C21 with forward anode voltage uprated from 750 V to 1000 V (AX, EL, HY, RCA) or 1250 V (W) (12-3-47).

5686 [AKA M328]; min. 9-pin ruggedized AF-VHF beam power tube, heater 6.3 V @ 350 mA, 7.5 W, 160 MHz, mobile-rated, quite like 5763 except for pinout (GE, HY, RA, RCA, SY, TS) (3-16-50) <9G>.

5687, WA, WB [AKA TS-229]; min. 9-pin duo-triode, med.-μ power type for computer use, heater 12.6 VCT @ 450 mA, μ 17, 8.4 W total, unusual pinout (RCA, SY, TS) (2-18-48) <9H>. Variants: 6900, 7370.

5690-93 Long-life "Special Red" line intro. by RCA.

5690 Full-wave rectifier, GT octal, heaters 6.3 V @ 1.2 A each, 1120 V PIV, 125 mA; heaters and cathodes pinned-out individually (GE, HY, RCA) {R6362} (1-30-53) <5690>. Added to the line after 5691-5693.

5691 Duotriode, 6SL7GT; heaters in series so that one failure will disable both sections in bridge ckts. Heater raised to 600 mA (GE, RCA, SY, TS, W) (2-10-48) <8BD>.

5692 Duotriode, 6SN7GT; derated to 1.7 W per section, heaters in series as in 5691 (GE, RCA, SY, W) {A4273B} (4-30-48) <8BD>.

5693 Pentode, 6SJ7 (GE, RCA, SY, TS, W) (4-30-48) <8N>.

5694 Dual AF power triode, premium 6N7GT w/ separate cathode leads (RA) (11-26-48) <8CS>.

5695 Min. shield-grid (high-impedance) xenon thyratron, possibly a 2B26/866JR w/ gas fill and grid added (1-26-48).

5696, A [AKA EN92]; min. xenon tetrode thyratron, heater 6.3 V @ 150 mA, 500 V PIV, 25 mA avg, fast deionization of 10 μsec (GE, HY, MU, RCA, TS, W) (9-20-48) <7BN>. GL-546 uprated on cathode current, w/ different pinout. "A" is uprated to 28 mA.

5697 [AKA CK570AX]; flat submin. electrometer triode, 0.625 V @ 20 mA (RA) (4-16-48).

5702, WA, WB [AKA CK605CX]; T3 flat-press submin. sharp-cutoff VHF pentode, 6AK5 char., heater raised to 200 mA, suppressor wired out separately (GE, RA, SY, TS) (7-15-48) <5702>. Variants: 6148, 6245, 6540, 7083.

5703, WA, WB [AKA CK608CX]; T3 flat-press submin. UHF triode, heater 6 V @ 200 mA, μ 25, 3.3 W; produces 750 mW @ 500 MHz; like ruggedized 6K4 (GE, RA, SY, TS) (7-15-48) <5703>. Variant: 6149.

5704 [AKA CK606BX]; T2 flat-press submin. diode (½ 6AL5), heater 6.3 V @ 150 mA, 420 V PIV, 9 mA

(GE, RA, SY) (7-15-48) <5704>.

5705 Xmtg. triode sim. to 892, fil. 22.0 V @ 60 A (6-28-48).

5706 Xmtg. triode sim. to 892R, fil. 22.0 V @ 60 A (6-28-48).

5710 [AKA Z1450]; xmtg. triode, forced-air-cooled (5-5-48).

5711-12 X-ray sources (7-15-50).

5713 Xmtg. triode, ring-seal design, for G-G use, heater 3.3 V @ 11.5 A, μ 25, 250 W, 220 MHz, forced-air-cooled (EEV, RCA) (4-30-48).

5718 [AKA EC70, 6778]; T3 submin. UHF triode, heater 6.3 V @ 150 mA, μ 27, 900 mW, 500 MHz (GE, MU, RCA, ST, SY, TS) {A2394A, SN979} (8-16-48) <8DK>.

5719 T3 submin. audio triode, 6AD4 w/ same μ but g_m raised to 1.7 mS, computer-rated (GE, RCA, ST, SY, TS) {A2428, SN980} (8-16-48) <8DK>.

5720 [AKA FG-33]; gas-mercury triode thyratron, triode variant of 5560, for motor controls, etc., ST bulb w/ cap, UX base, heater 5.0 V @ 4.5 A, 1 kV PIV, 2.5 A (GE, HY, MU, W̲) (4-30-48).

5721 [AKA QK205]; reflex klystron, 2.0-12.0 GHz, EC, 4-pin min. base, heater 6.3 V @ 580 mA, 1.25 kV, 250 mW output (RA) (10-19-48).

5722 Min. temperature-limited tungsten-filament diode for use in noise generators, fil. 2-5.5 V, ca. 1.5 A, depending on noise needed, 150 V, 500 MHz (SY, HY) {SD1033} (6-8-48 <5CB>. Also see 5845.

5725 [AKA M8196, 6AS6W]; ruggedized 6AS6, computer-rated (GE, MU, RA, RCA, SY, TF, TS) (3-15-50) <7CM>.

5726 [AKA 6AL5W, 6097, M8079, M8212, E91AA, EAA901, QA2404, QD77]; ruggedized 6AL5 (GE, MU, RA, RCA, TS) {R6411} (3-15-50) <6BT>.

5727 [AKA E91N, M8204, PL21, WT-210-0077]; thyratron, rugged 2D21 (AX, GE, MU, RA, RCA, SY, W̲) {QM502} (10-25-51) <7BN>.

5728 [AKA FG-67]; mercury triode thyratron, for regulated rectifiers, ST23 bulb w/ cap, UX base, heater 5 V @ 4.5 A, 1 kV PIV, 2.5 A avg. Appears to be a variant of 5559 (GE, RCA) (6-25-48) <4BL>.

5729 Beam-switching tube, collector/commutator for 30-channel multiplex systems, short tubular bulb w/ 42 pins or flying leads (to order), heater 6.3 V @ 450 mA, has 30 grid leads, 1 lead for 29 anodes, and 1 lead for the 30th anode. Uses electrostatic focusing from an external six-phase 60-Hz supply (NU) (4-9-51). Also see 5730.

5730 Beam-switching tube, separator/decommutator inverse of 5729, same bulb, heater, and focusing system, has 30 anode leads, one lead for 29 grids, and 1 lead for the 30th grid. (4-9-51).

5731 Acorn triode, premium 955 (GE) (8-3-48) <5BC>.

5732 Rcvg. pentode, ruggedized 6K7 (Can.) (9-13-48) <7R>.

5734, A Submin. triode "vibrotron" transducer tube, stepped metal anode shell 0.33" dia by 3/4" long, plate shaft out top, flying leads at bottom, heater 6.3 V @ 150 mA, converts slight rocking motion of plate shaft into electrical signal; plate moves w/ respect to grid. Originally promoted for use as phono pickup. (RCA) {C798A, B, C} (10-19-48).

5736 [AKA WL-473]; xmtg. triode, external-anode, for radar modulators and RF use, T-T fil. 6.0 V @ 60 A, μ 22, 2.5 kW, 110 MHz (ML: 60 MHz), forced-air-cooled, intro. 1944 (ML, W̲) {WX322} (3-30-49). Two as drivers for 5891s in Gates BC-100B 100-kW AM xmtr. Also see 6623.

5737 Xmtg. triode (6-30-48).

5738 Beam-switching tube (electron-beam commutator) (10-6-48).

5739 Ionization gauge (7-20-48).

5740 [AKA FP-54]; rcvg. space-charge-grid (electrometer) tetrode, UX base w/ cap, fil. 2.5 V @ 90 mA, typical R_{in} 10^{16} ohms (GE) (7-15-48).

5741 [AKA FP-85A]; vac. rectifier, stepped tubular bulb w/ cap, UX base, fil. 10.0 V @ 5.0 A, 20 kV PIV, 20 mA (GE) (7-27-48).

5742 [AKA PJ-7]; "train control" triode, UX base, fil. 4.5 V @ 1.1 A, μ of 30, 7.5 W, 6 MHz (GE) (9-13-48). Introduced in early '20s for railroad signaling use, i. e., inductive transmission of block-signal indications into a locomotive cab.

5743 [AKA PJ-21]; triode sim. to PJ-8 but w/ μ of 3, for low-freq. amp and relay uses (GE) (7-27-48).

5744, WA, WB [AKA CK619CX]; T3 flat-press submin. UHF high-μ triode, heater 6.3 V @ 200 mA, μ 70 (GE, RA, SY, TS) (3-17-49) <5744>. Variant: 6151.

5745-48 X-ray sources (all 5-5-49).

5749 Min. pentode, 6BA6W computer-rated (GE, RCA, SY, TS) (3-20-50) <7BK>.

5750 Pentagrid converter, high-reliability 6BE6 (GE, RCA, SY, TS (3-6-50) <7CH>.

5751, WA, WB Duotriode, ruggedized computer-rated 12AX7; μ reduced to 70, heater raised to 175 mA per section (GE, RCA, SY, TS) {A4654} (10-26-49) <9A>.

5755 [AKA (WE) 420A]; min. 9-pin duotriode, high-stability for DC amps. etc., heater 12.6 VCT @ 180 mA, μ 70, g_m 0.5 mS (HY, RA, WE) (7-3-50). Variant: 6832.

5757 [AKA 592]; xmtg. triode, glass ring-seal construction, grid out ring, T-T fil. 5.0 V @ 110 A, μ 28, 5 kW, 100 MHz, water-cooled via external cooler (AX).

5758 [AKA 592R]; xmtg. triode, radiator variant of 5757, same ratings (AX).

5759 [AKA 501R]; RF-heating HF triode, glass ring-seal construction, flying ribbon leads from crown, fil. 7.5 V @ 24 A, 1 kW, 150 Mhz, forced-air-cooled (AX) (11-27-51).

5760 [AKA 502]; xmtg. triode, glass ring-seal construction, flying ribbon leads from crown, fil. 7.5 V @ 24 A (AX) (1-30-53).

5761 [AKA 502R]; xmtg. triode, radiator variant of 5760 (AX) (1-30-53).

5762, A [AKA TY6-5000D]; RF-heating triode, improved 7C24 w/ smaller high-efficiency radiator; claimed to need half the air flow (EEV, F, RCA) (11-25-48).

5763 [AKA QQV03-12]; min. VHF beam tube, heater 6.0 V @ 750 mA, 8 W (W: 12 W), 50 MHz (GE, MU, RCA, SY, W) {A4541H} (3-14-49) <9K>. Variant: 6417.

5764-68 Disc-seal "rocket" triodes, heaters 6.3 V @ 400 mA (all SY) (all 1-15-49).

5764 2C37 w/ modified electrode rings, pulsed osc. usable to 3.3 GHz, 200 W output {SB846B, D}.

5765 5764 format, CW osc., 0.9-3.0 GHz, no external feedback needed, 250 mW output.

5766 2C36 format but w/ grid on disc and plate on ring.

5767 [AKA 6481]; 2C36 format but with both discs folded into rings, CW osc.-amp.-multiplier to 3 GHz, 2 W out (GE, LI, SY) {SB846H}.

5768 CW amp.-multiplier usable to 3 GHz, plate at cap on top, grid on disc (LI, SY). Variant: RT434.

5769 Image orthicon, 5655 format, lower noise performance than 5655 (RCA) (12-17-48).

5770 RF-heating triode, modified 9C21, for G-G use, T-T fil. 11 V @ 285 A, μ 39, 50 kW, 20 MHz, water-cooled via external cooler (RCA) (12-13-48).

5771 [AKA ML-356]; RF-heating triode, improved 880, T-T fil. 7.5 V @ 170 A, 22.5 kW (AX, EI, F, ML, RCA, W) (12-13-48).

5777 Reflex klystron, 0.6-2.4 GHz, EC, CO, heater 6.3 V @ 985 mA, 400 V, 160 mW output (RA).

5778 Reflex klystron, 1.8-4.62 GHz, EC, CO, heater 6.3 V @ 985 mA, 300 V, 150 mW output (RA).

5779 Ignitron, glass bulb, air-cooled demonstrator (supersedes GL-427), 2-½" dia. by 8-¼" long, 350 V PIV, 10 A (GE) (12-1-48).

5780, A Magnetron, tunes 8.5-9.6 GHz, IM, WG, 14.5 V @ 3.4 A, 40 kV @ 36 A, 250 kW pulse out, forced-air cooled (AX, BL, LI, WE) [1538} (5-29-50). In WE radars: Mod. IV instrumentation set, Nike-Ajax precision tracker, Nike-Hercules monopulse tracker, and modified M33 autotrack radar.

5781-82 X-ray sources (both 7-15-50).

5783, WA, WB [AKA M8190, 85A3]; T3 flat-press submin. voltage reference, 85 V, 1.5-3.5 mA, submin. 5651 (MU, RA, RCA, SY, TS) (3-15-50) <5783>.

5784, WA, WB T3 flat-press submin. VHF dual-control pentode, 6AS6 char.; heater 6.3 V @ 200 mA (GE, SY, RA, TS) (3-15-50) <5784>. Variant: 6150.

5785 T2X3 submin. HV rectifier, fil. 1.25 V @ 15 mA, 3.5 kV PIV, 0.1 mA (GE, RA, SY) (3-15-50) <5785>.

5786 RF-heating triode, modified 6C24, T-T fil. 11.0 V @ 12.5 V, 160 MHz (RCA) (1-5-49).

5787, WA T3 flat-press submin. voltage regulator, submin. 0B2, 100 V, 5-25 mA (RA, SY) (3-15-50) <5783>.

5788 Ignitron, 5555 redesigned mechanically for better cooling and reliability (GE) (4-25-55).

5789, A Magnetron, fixed at 34,860 ± 348 MHz, 22-vane anode, IM, WG, pressurized output window, 6.0 V @ 2.4 A, 14 kV @ 20 A, 32 kW pulse output, "first commercially available US type for K_a band" (MA, SY) (5-27-49).

5790 TR tube, tunes 33.8-35.9 GHz, glass gas reservoir (SY) (5-27-49).

5791 [AKA X6007]; ATR tube used w/ 5790, 34.86 GHz (SY) (5-27-49).

5792 [AKA ATR345]; ATR tube, fixed at 2950 MHz, 750 kW (BL) (5-27-49).

5793 [AKA ATR346]; ATR tube, fixed at 3050 MHz, 750 kW (BL) (5-27-49).

5794 Packaged 1680-MHz radiosonde osc. w/ pencil tube, integral grid and plate resonators, heater 5.2-6.6 V @ 160 mA, 3.6 W, 600 mW out (RA, RCA, SY) {A2342} (3-4-49). In AN/AMT-4 radiosonde. Variants: 6562/5794A, 7533.

5795 Magnetron, tunes 3.1-3.5 GHz, IM, WG, 107 V @ 3.65 A, 50 kV @ 50 A, 1.0 MW pulse output (WE) (5-29-50). In acquisition radars for Nike-Ajax (OA-653/MS) and Nike-Hercules.

5796 [AKA WT-210-0043, WT-210-0141, XR1-1600]; triode gas thyratron, fil. 2.5 V @ 8.5 A, 1.5 kV PIV, 1.6 A (MU, W) (2-10-49).

5797 T3 submin. semi-remote-cutoff pentode, heater 26.5 V @ 45 mA, computer-rated (GE) {Z-1723} (9-17-51) <8CY>.

5798 T3 submin. med.-μ duo-triode, suited for osc.-mixer use up to 400 MHz or general use; heater 26.5 V @ 90 mA, μ 24, computer-rated (GE) {Z-1724} (9-17-51) <8CZ>.

5799 [AKA VX-21]; submin. VTVM or electrometer clipper diode, fil. 1.25 V @ 10 mA, 3 kV PIV, 0.6 mA (VN) (12-10-48).

5800 [AKA VX-41A, VX-124]; submin. electrometer tetrode, fil. 1.25 V @ 10 mA, μ 13, 50 V (VN) (12-10-48).

5801 [AKA VX-33A]; submin. triode, fil. 1.25 V @ 10 mA, g_m 0.065 mS (12-10-48).

5802 [AKA ME1401, VX-32B]; submin. electrometer triode (MU, VN) (12-10-48).

5803 [AKA VX-55]; submin. triode; improved 5802, for one-tube electrometers for radiation measurement, fil. 1.25 V @ 10 mA, μ 2.1; has higher g_m at the cost of higher grid current (VN) (3-25-49).

5804 Submin. RF pentode, 1AK4 w/ higher fil. current.

5806 [AKA P10]; submin. electrometer tube.

5811 [AKA G10]; submin. electrometer tube.

5812 Min. xmtg. beam power tube, premium 2E30, heater 6.0 V @ 650 mA (HY) (6-9-50).

5813 Camera tube, "misc." type (1-17-49).

5814, A, WA [AKA 12AU7W]; 12AU7, ruggedized w/ complete redesign; heater raised to 175 mA per half (GE, RCA, RA, SY, TS) {A4653} (10-26-49) <9A>. "A" is computer-rated.

5817 [AKA SRC12]; klystron (SP) (6-25-49).

5818 [AKA SRC13]; klystron (SP) (6-25-49).

5819 Photomultiplier, for scintillation counters, 10-stage, S-11 response, T16 head-on bulb 5-5/8" high w/ curved faceplate, diheptal base (RCA) (7-18-49).

5820, A, A/L, B Image orthicon, 5655 format, 3", for outdoors use; replaces 2P23 w/ 2.5 X sensitivity; replaces 5655 with better gray scale but more noise. Sensitive to visible light but not IR. Heater 6.3 V @ 600 mA, anode voltage 1650 (GE, RCA, W) (7-18-49). Used in early color cameras as well as monochrome; also in X-ray linkage systems. Variant: 8775.

5821 Ignitron (9-28-49).

5822, A [AKA WT-210-0075]; ignitron; 5552 format; in welder-control use, 220-600 V applied, 188 A; in rectifier use, 250-600 V applied, 355 A; water-cooled (AX, GE, RCA, W) (9-28-47). "A" has mounting for cooling thermostat. Variant: 6511.

5823 [AKA 1C21, q. v.]; min. cold-cathode triode relay or rectifier, 200 V PIV, 25 mA avg. (AX, GE, HY, MU, RCA, SY, W) {Z900T} (6-25-49) <4CK>.

5824 Power pentode, long-life/high-rel variant of 25B6G, interchangeable with it; introduced for use in power-line carrier-current systems (using 120-V station battery for B+?) (GE, RCA, SY) (7-27-49) <7AC>.

5825 Vacuum rectifier, ST16 bulb w/ cap, UX base, fil. 1.6 V @ 1.25 A, 60 kV PIV, 2 mA (GE, RCA) (4-22-49) <4P>.

5826 Image orthicon, for indoors use; replaces 5655 w/ 7 X sensitivity. Responds to visible light but not IR. Heater 6.3 V @ 600 mA, anode voltage 1500 (GE, RCA) (4-5-50).

5828 Submin. med.-μ DC-amp. triode, fil. 1.25 V @ 10 mA, μ 17.5 (VN) (4-22-49).

5829, WA T2X3 submin. VHF duodiode, submin. 6AL5, heater 6.3 V @ 150 mA, 330 v PIV, 5 mA per plate (GE, RA, TS) (4-23-51) <5829>.

5830 [AKA FG-41, WT-210-0054]; triode mercury thyratron for grid-controlled rectifiers; tubular/spherical bulb like early UV-203, w/ cap, 4-pin industrial base, heater 5.0 V @ 20 A, 10 kV PIV, 12.5 A, convection cooled; could oscillate [!] up to 37 MHz (GE, W) (intro. 1929, regis. 8-19-49).

5831 Xmtg. HF beam triode, bolted-seal construction, T-T fil. 6.0 V @ 2220 A, μ 25, 150 kW, water-cooled via integral grid and anode coolers (RCA) (3-16-50). Contains 48 independent electron-optical systems in parallel. "First commercial 1-MW CW beam power triode." Four in Navy Jim Creek 1-MW xmtr, ca. 1950; also in linear-accelerator drivers. Also promoted for RF heating.

5833 Xmtg. triode (8-1-50).

5834 [AKA NL-649]; mercury rectifier, 4-pin base, fil. 2.5 V @ 7 A (8-25-50).

5835 [AKA NL-653]; mercury rect, 4-pin base, fil. 2.5 V @ 10 A (8-25-50).

5836 Reflex klystron, 1.6-5.5 GHz depending on external cavity, pulse-or-CW variant of 6BL6, 75 mW output (RA, SY) (12-19-49).

5837 Reflex klystron, 0.55-3.8 GHz depending on external cavity, pulse-or-CW variant of 6BM6, 150 mW CW out (RA, SY) (12-19-49).

5838 [AKA TE-3]; full-wave rectifier, ruggedized 6X5GT, heater 12 V @ 600 mA (BX, GE) (10-16-50) <6S>.

5839 [AKA TE-2]; full-wave rectifier, ruggedized 6X5GT; heater 26.5 V @ 285 mA (BX, GE, TS) (10-16-50) <6S>.

5840, W T3 submin. ruggedized sharp-cutoff UHF pentode, heater 6.3 V @ 150 mA, computer-rated (GE, MU, RCA, ST, SY, TS) {A4669, SN973C} (3-16-50) <8DE>. Variants: 6205, 6225. Also see 5901.

5841 Submin. voltage regulator, corona type, 900 V @ 2-50 μA (VN) (9-15-49). Typical use, Geiger counters.

5842 [AKA (WE) 417A]; rcvg. "10,000-hr." frame-grid triode, heater 6.3 V @ 300 mA, μ 43, g_m 25 mS, 4.5 W (AX, GE, HY, RA, RCA, SY, WE) (6-20-50) <9V>. In Nems-Clarke 1302 surveillance rcvr.

5844 Duotriode, sharp-cutoff computer-rated variant of 6J6; μ reduced to 28, g_m reduced to 3.7 mS, heater reduced to 300 mA (GE, RCA) (5-31-51) <7BF>.

5845 Min. duodiode noise generator, fil. 2.15 V @ 435 mA (HY, SY) (9-21-50) <5CA>. Also see 5722.

5846 Reference cavity, 9280 ± 0.5 MHz (BL, W) (7-8-49). Also see 1Q23.

5847, A [AKA (WE) 404A]; min. 9-pin sharp-cutoff frame-grid pentode, heater 6.3 V @ 300 mA, g_m 12.5 mS (GE, HY, RA, RCA, SY, WE) (12-15-49) <9X>.

5851 T3 submin. RF beam power tube, for Class C use, fil. 2.5 VCT @ 55 mA, 1.5 W (RA, NU, SY) (8-15-50) <6CL>. Also see 6147.

5852 [AKA TE-5]; full-wave rectifier, ruggedized 6X5, heater curr. raised to 1.2 A (BX, GE, TS) (10-16-50) <6S>.

5853 TR tube, bandpass type fixed-tuned to 2.9-3.2 GHz, 750 kW, RG-48/U WG in and out (BL) (4-23-54).

5854 [AKA CK503A]; hearing-aid power pentode, CK501 format, fil. 1.25 V @ 30 mA, g_m 0.6 mS, 10 mW out (GE, RA) (11-28-52) <2E31>.

5855 [AKA XR1-12]; triode gas thyratron for motor control, quick-heating, tubular all-glass bulb, connections to lugs at ends, grid-out-side via ring seal, fil. 2.5 V @ 34 A, 1.5 kV PIV, 18 A (GE, HY, MU) (12-5-50) <5855>.

5857 Min. 9-pin secondary-emission VHF amp., heater 6.3 V @ 450 mA (9-11-50). Possibly same as EFP60.

5859-60 X-ray rectifier tubes (W) (1-24-50).

5861 [AKA EC55, TD03-10]; rcvg. disc-seal triode, fil. 6.3 V @ 400 mA, 10 W, oscillates to 3 GHz (AX, MU, NEC, PH, VO) (10-31-49).

5863 TR tube, bandpass type fixed-tuned to 8.5-9.6 GHz, 250 kW, RG-51/U WG in/out (MA, SY) (8-21-53).

5864 [AKA ATR321]; ATR tube, 9375 MHz, 500 kW, in RG-51/U WG package (BL) (3-19-56).

5865 [AKA TR361]; TR tube, bandpass type, fixed-tuned

for 5.4-5.9 GHz, 300 kW (BL, SY) (5-21-54). Variant: 6640.

5866-68 Xmtg./RF-heating VHF triodes, all-glass construction w/ plate pin, ringed graphite anode w/ heat-storage capability, forced-air-cooled (all 9-4-53)

5866, A [AKA AX9900, TY2-12S]; fil. 6.3 V @ 5.8 A, μ 25, 150 W, 150 MHz (AX, MU). "A" is AKA TY3-250, 7986, q. v.].

5867, A [AKA AX9901, TY3-250, TB3/750]; fil. 5.0 V @ 14.1 A, 250 W, 100 MHz, (AX, EEV, EI). EI version uses fluted rather than ringed anode. "A" is uprated to 350 W.

5868 [AKA AX9902, TY4-500]; fil. 10.0 V @ 10.0 A, 450 W, 100 MHz (AX, MU).

5869 [AKA AGR9950]; mercury triode thyratron, fil. 5.0 V @ 6.5 A, 13 kV PIV, 1 A (AX) (1-30-53). Also see 5870.

5870, A [AKA AGR9951]; mercury triode thyratron, long tubular body w/ cap and conical corona shield, large 4-pin base, fil. 5.0 V @ 14.0 A, 27 kV, 2.5 A (AX) (1-30-53). For grid-controlled rectifiers in xmtrs.

5871 Beam power tube, mobile-rated 6V6GT (SY) (5-4-50) <7S>.

5873 T3 submin. med.-μ duotriode, submin. 12AU7, heater 6.3 V @ 300 mA, μ 22, gm 2.9 mS, 1.6 W (GE, RA) (9-13-51) <5873>.

5874 Xmtg. audio triode, fil. 7.0 V @ 110 A, μ 6, 7.5 kW, water-cooled (F) (11-10-49).

5875 T2X3 submin. radiosonde beam power tube, fil. 1.25 V @ 100 mA, spray-shielded; 1AD4 uprated from 45 V to 90 V, g_m raised to 2.5 mS (TS) (4-9-51) <1AD4>.

5876, A Rcvg. high-μ glass pencil triode, CW osc.-amp.-multiplier for G-G use, 5675 format, heater 6.3 V @ 135 mA, μ 56, usable to 3 GHz (BL, GE, PH, RA, RCA, SY, TS, THF, VO) (6-9-50) <5675>.

5877 [AKA 22884]; triode gas thyratron, fil. 2.5 V @ 10.8 A, 1.5 kV PIV, 3.2 A (W) (4-15-57). Also see 5544.

5878 Triode gas thyratron, fil. 2.5 V @ 21 A, 1.5 kV PIV, 6.4 A, interchangeable w/ 5C21 (W) (1-6-58). Also see 5545.

5879 Min. 9-pin sharp-cutoff pentode; audio preamp, heater 6.3 V @ 150 mA, g_m 1.0 mS (GE, RCA, SY, TS) (3-13-50) <9AD>.

5881 [AKA 6L6WGB]; beam power tube; redesigned by TS w/ T11 bulb, button stem, noncontaminating cathode, carbonized gold-plated grid, Zr anode coating, and Micanol barrier base; uprated to 23 W (GE, RCA, SY, TS, W) (10-10-50) <7S>.

5883 ATR tube (11-8-54).

5884 "Multiple section" tube (7-1-57).

5885 T3 submin. duotetrode, fil. 1.25 V @ 20 mA, g_m 0.16 mS (GE) (8-13-62) <5885>.

5886 [AKA CK571AX, DF703]; T2X3 submin. electrometer pentode, fil. 1.25 V @ 10 mA, g_m 0.175 mS (RA, TF, TS) (12-17-51).

5887-88 Counter tubes (1-30-53).

5889 [AKA ME1403]; T3 flat-press electrometer pentode, 5886 w/ grid lead at top, w/ guard ring; fil. reduced to 7.5 mA (as low a fil. current as ever found!) (MU, RA, SY, TS) (3-30-50) <5889>.

5890 Rcvg. remote-cutoff HV beam power tube, for shunt HV regulators, T12 tubular bulb 6" high w/ cap, octal base, heater 6.3 V @ 600 mA, 10 W, 30 kV (GE) (4-24-50) <12J>.

5891 Xmtg. triode, radiator-type, for audio/sonar and RF use, two-section fil., 11 V @ 95 A each, μ 36, 25 kW, 10 MHz (W) (10-30-50). Two in Gates BC-50B/-100B 50-/100-kW xmtrs, modulated by two more.

5892 [AKA EL6B]; xenon rectifier, "industrial" metal base, fil. 2.5 V @ 21 A, 1250 V PIV, 6.4 A, replaceable by 6B (EL, HY) (5-4-50).

5893 Med.-μ glass pencil triode, heater 6.3 V @ 330 mA, μ 27, 6 W, 2 GHz, usable to 3.3 GHz (PH, RA, RCA, SY) (10-5-51). In AN/DPN-19 radar beacon, giving 50 W out at 2850-2950 MHz.

5894, A, B [AKA AX9903, QQV06-40A]; xmtg. duopentode; updated straight-sided variant of 829B, heater 12.6 VCT @ 900 mA, 40 W, uprated to 500 MHz (AX, MU, RCA) (8-14-51). "A" is smaller, redesigned w/ sintered-glass wall at anode end and internal bracing. "B" is AKA 8738, q. v. In Swan VHF-150 2-m FM amp., Motorola 80D high-band base station, etc. Besides usual communications uses, was pulser in AN/ULT-T5 Ku-band countermeasures trainer; was also deflection amplifier in high-frequency oscilloscopes (DuM 329A and 336A; Tektronix 536). Variants: 8509, 8643. Also see 6524.

5895 [AKA AX9905, QQZ04-15]; xmtg. dual VHF tetrode, tall tubular bulb on loktal base, quick-heating fil. 6.3 VCT @ 700 mA, ½ sec. heatup time, 12 W, radiation-cooled (AX, MU) (9-4-53).

5896 T3 submin. full-wave rectifier, heater 6.3 V @ 300 mA, 460 V PIV, 20 mA; also suited for UHF use (GE, RCA, SY, TS) (3-16-50) <8DJ>.

5897 [AKA 6K4]; T3 submin. power triode, repl. by 5718 (GE) (3-16-50) <8DK>.

5898 [AKA 6AD4]; T3 submin. high-μ triode sim. to 5637, repl. by 5719 (GE, MU) {SN980D} (3-16-50) <8DK>.

5899 T3 submin. UHF semi-remote-cutoff pentode, heater 6.3 V @ 150 mA, computer-rated (GE, RCA, ST, SY, TS) (3-16-50) <8DL>. Variant: 6206.

5900 T3 submin. semi-remote-cutoff RF pentode, repl. by 5899 (GE) (3-16-50) <8DL>.

5901 T3 submin. sharp-cutoff pentode, repl. by 5840 (GE) (1-25-50) <8DL>.

5902, WA T3 submin. AF beam power tube, heater 6.3 V @ 450 mA, 3.7 W (GE, MU, RCA, SY, TS) (4-12-50) <8DL>. Related to (ruggedized?) 5640.

5903 T3 submin. full-wave rectifier/HF diode, heater 26.5 V @ 75 mA, 460 V PIV, 10 mA per plate (GE, ST, SY) {SN989B} (4-12-50) <8DJ>.

5904 T3 submin. med.-μ triode, heater 26.5 V @ 45 mA, μ 20 (GE, SY) (5-18-50) <8DK>. Optimized for 26 V B+. Also see 6055.

5905 T3 submin. sharp-cutoff RF pentode, heater 26.5 V @ 45 mA, g_m 2.9 mS, optimized for 26.5-V B+ (GE, SY) (4-12-50) <8DL>.

5906 T3 submin. sharp-cutoff RF pentode, variant of 5905 for normal voltages, g_m 5.0 mS, 1.1 W (GE, SY) (5-18-50) <8DL>.

5907 T3 submin. remote-control RF pentode, heater 26.5 V @ 45 mA, g_m 3.0 mS, optimized for 26.5-V B+ (GE, SY) (5-18-50) <8DL>.

5908 T3 submin. dual-control RF pentode, heater 26.5 V @ 45 mA, g_m 2.2 mS, optimized for 26.5-V B+ (GE, SY) (5-18-50) <8DC>.

5910 Premium 1U4 (GE, SY) (1-25-50) <6AR>.

5911 [AKA DF67, 6008]; submin. tetrode, 0.625 V @ 13.3 mA, 6 mW out (AX).

5913 [AKA DL67, 6007]; submin. power tetrode, 1.25 V @ 13.3 mA, 6 mW out (AX).

5915, A [AKA EK90]; 6BE6, dual-control, sharp-cutoff, rated for computer use (GE, RCA, SY) (7-5-50) <7CH>.

5916 T3 submin. dual-control pentode equiv. to 6AS6, heater 26.5 V @ 45 mA, g_m 3.2 mS (GE, SY) {SN-986B} (5-18-50) <8DC>.

5917 Xmtg. triode, fil. 7.5 V @ 51 A, μ 20, 5 kW, 110 MHz, water-cooled (F) (7-15-50).

5918, A Xmtg. triode, all-glass except external anode, T-T fil. 11.0 V @ 286 A, μ 37, 70 kW, 22 MHz, water-cooled (EI, F, ML) (10-10-50). In pairs in three VOA 200-kW xmtrs at Mason, OH site.

5919 Xmtg. triode, air-cooled variant of 5918, derated to 35 kW (F) (12-22-52).

5920 [AKA E90CC]; min. rcvg. "10,000-hr." HF duotriode, 6J6 3/8" taller, w/ heater reduced to 400 mA, μ reduced to 27, g_m raised to 6.0 mS, diss. raised to 4.0 W total (AX, HY, MU) (7-15-51).

5921 ATR tube, fixed-tuned to 3200 MHz, 750 kW (BL) (11-8-54).

5922 ATR tube, 5921 for 3400 MHz (BL) (11-8-54).

5923 [AKA AX9904, TY6-5000W]; xmtg. VHF triode, glass ring-seal construction, thumbscrew clamps for fil. leads, grid out ring, T-T fil. 12.6 V @ 33 A, 6 kW, 220 MHz, water-cooled, for VHF TV and RF heating (AX, MU) (9-18-53).

5924, A [AKA AX9904R, TY6-5000A]; xmtg. VHF triode, radiator version of 5923, derated to 5 kW (AX, MU) (1-30-53). "A" is rerated to 6 kW, promoted for use in FM-TV xmtrs (AX). In Standard Electronics TL/TH653 low/high-band 5-kW TV xmtrs.

5925 TR tube, bandpass type, fixed-tuned for 5200-5530 MHz, 1 MW (BL).

5927 TR tube, bandpass type fixed-tuned to 3.1-3.5 GHz, RG-48/U WG in and out, 750 kW (BL) (11-8-54).

5928 Rcvg. pentode; see 6267 (ST).

5929 TWA, 4.0-5.0 GHz, EM, WG, octal base, 10 W output, gain 20 dB (F) (5-29-50).

5930 [AKA 2A3W]; power triode, ruggedized 2A3 in T12 bulb w/ wide base (GE) (7-5-50) <4D>.

5931 [AKA 5U4WG]; full-wave rectifier, premium 5U4G (GE, SY) (7-5-50) <5T>.

5932 [AKA 6L6WGA]; beam power tube, premium 6L6GA (GE, SY) (7-5-50) <7S>.

5933 [AKA 807W]; beam power tube, ruggedized 807, T12 bulb on wide base (SY, TS) (7-5-50) <5AW>.

5934 Rectifier, fil. 2.5 V @ 6 A, 20 kV PIV, 25 mA (W) (3-24-50).

5935 Rcvg. pentode (6-18-51).

5936 Xmtg. triode, multisection fil., 20 V @ 143 A each, μ 37, 70 kW, 15 MHz, water-cooled (EI, ML, W) (9-20-50). Also see 5986 and 6379.

5939 Pre-TR tube ("gas switching window"), fixed-tuned to 1250-1350 MHz, 10-500 kW pulse (W) (8-22-52).

5942 Counter tube (11-15-54).

5944 Counter tube (1-30-53).

5946 [AKA TT@]; xmtg. glass ring-seal triode, for G-G use, heater 6.3 V @ 3.4 A, μ 27, 250 W, 1.3 GHz, can give 14 kW pulse at 1250 MHz, forced-air cooled (RCA) (1-20-51).

5947 Rectifier, fil. 4.4 V @ 1.75 A (3-28-51).

5948, A [AKA 1754, A]; hydrogen thyratron, radar pulser, plate cap on glass stalk, flange-based, flying leads, fil. 6 V @ 30 A, 25 kV @ 1000 A, 1 A avg. (CH, GE, KL, TS) (1-31-55). H_2 reservoir draws 11 A @ 2.5-5.5 V.

5949, A [AKA 1907, XH25-500]; hydrogen thyratron, high-power radar pulser, plug-in base, fil. 6.3 V @ 22 A, 25 kV @ 500 A, 500 mA avg., 6.25-MW pulses (CH, KL, MU, TS) (5-21-54). In AN/MPQ-21.

5950 Submin. voltage reg., corona type, 700 V @ 2-50 μ A (VN) (5-26-50). Typical use, Geiger counters.

5956 [AKA E36]; hydrogen thyratron, compact all-glass, for airborne use, w/ hydrogen reservoir, fil. 6.3 V @ 6.7 A, 8 kV @ 90 A, 100 mA avg. (IT) (4-16-56).

5957 [AKA E37B]; hydrogen thyratron, 5956 w/ ceramic ring at base (IT) (3-12-56). Pulser in E-34 radar (see 6551).

5958 [AKA E40]; hydrogen thyratron, compact all-glass, for airborne use, plate pin at top on stalk, 4 pins at bottom, fil. 6.3 V @ 2.5 A (BL, KL).

5959 [AKA E41, SA10B]; hydrogen thyratron, fil. 6.3 V @ 2.5 A, 35 A pk., 45 mA avg. (BL, IT) (4-16-56).

5960 [AKA S340]; "misc." type (3-28-51).

5961 Pentagrid converter, premium 6SA7 (7-18-50).

5962 See 7160.

5963 Min. 9-pin duotriode, computer-rated 12AU7. Not intended for conventional amplifiers as to noise, microphonics, or plate-current balance (SM, GE, RCA, SY, TS) (7-5-50) <9A>.

5964 Min. duotriode, computer-rated 6J6 w/ increased g_m. See 5963 about non-use in analog circuits (GE, RCA, SY, TS) (7-5-50) <7BF>.

5965, A Min. 9-pin duotriode, computer-rated 12AV7 (GE,

RCA). "A" has controlled cutoff balance and 10,000-hr rated life; is derated to 2.2 W per section (GE, RCA, SY, TF) (6-2-52) <9A>. Also see 7062.

5966 Ionization gauge, fil. 6.0 V @ 2.5 A, 150 V, reads vacuum of 10^{-3} to 10^{-10} mm Hg, 2-1/16" dia. by 5-3/8" long (W) (7-21-52).

5967 T3 submin. med.-μ duotriode, fil. 1.25 V @ 120 mA, μ 17 (GE) (9-12-55) <8DQ>.

5968 T3 submin. high-μ VHF duotriode, fil. 1.25 V @ 120 mA, μ 50 (GE, SY) (7-1-57) <8DQ>.

5969 T3 submin. VHF duotetrode, fil. 1.25 V @ 200 mA, g_m 1.7 mS (GE, SY) (7-1-57) <8DR>.

5970 T3 submin. duopentode, fil. 1.25 V @ 160 mA (GE, SY) (7-1-57) <8DS>.

5971 T2X3 submin. med.-μ triode, fil. 1.25 V @ 80 mA, μ 23 (GE) (1-7-52).

5972 T2X3 submin. remote-cutoff RF pentode, fil. 1.25 V @ 60 mA (GE) (7-1-57) <1AD4>.

5973, A Vacuum rectifier, stepped tubular bulb w/ cap, bipin base; fil. 16 V @ 19.1 A, 1.25 A, 40 kV PIV. Promoted as radar charging diode and as rectifier in smoke precipitators and HV cable testers (GE) (1-5-51).

5974 Vacuum rectifier, fil. 10 V @ 50 A, 125 kV, 300 mA, oil-cooled, 3-3/16" dia. by 11-¼" long (W) (12-5-50).

5975 T3 submin. med.-μ triode, heater 6.3 V @ 175 mA, μ 16 (GE, RA) (4-9-51) <5975>.

5976 Reflex klystron, 6.20-7.425 GHz, 723A/B format, heater 6.3 V @ 440 mA, 330 V, 110 mW out (KK, RA) (9-7-50). In RA KTR-100 portable microwave TV relay system.

5977 T3 submin. med.-μ power triode, high-reliability 6K4, heater 6.3 V @ 150 mA, μ 16, uprated to 3.3 W (ST, SY, TS) (9-28-50) <8DK>.

5978 Counter tube (5-15-51).

5979 [AKA BS1]; submin. triode, fil. 1.25 V (9-25-51).

5980 [AKA BS2]; gamma counter (9-25-51).

5981 Reflex klystron, 1245-1460 MHz, IC, CO, heater 6.3 V @ 455 mA, 250 V, 175 mW output, replaces 5649 and 5650, 6L6 format w/ coax output line out base and tuning screw out top (RA, SFR, SP) (10-10-50).

5982 Magnetron, fixed at 9375 ± 30 MHz, 75 kW output (2-6-53).

5986 Xmtg. triode, apparent variant of 5936, for audio/-sonar and RF use, multisection fil., 12 V @ 103 A each, μ 37, 70 kW, 6 MHz, 150 kW output in Class C, water-cooled (W) (5-18-59).

5987 T3 submin. low-μ power control triode, heater 6.3 V @ 450 mA, μ 4.1, 4.0 W (SY) {SC968B} (1-15-51) <8DM>.

5991 [AKA 8422, q. v.]; "nixie" readout.

5992 [AKA TE-8]; beam power tube, premium 6V6GT, heater raised to 600 mA (BX, GE) (3-28-51) <7AC>.

5993 [AKA TE-10]; full-wave rectifier, premium 6X4, special pinout (BX, GE) (3-28-51) <5993>.

5994 Xmtg. VHF beam power tube, all-glass construction w/ cap, 50 W, 120 MHz, radiation-cooled; roughly sim. to EI 4-65 (HK) (1950).

5995 T3 submin. flat-press rectifier, heater 6.3 V @ 300 mA, 850 V PIV, 45 mA (GE, RA) (4-9-51) <5995>.

5996 Xmtg. triode, fil. 13.0 V @ 36 A, μ 25, 2.5 kW, 30 MHz, forced-air cooled (F) (9-7-50).

5997 Image-intensifier tube for converting X-rays to visible light; gives 200X increase in brightness in fluoroscopes (W) (8-29-55).

5998, A Dual power triode, 6AS7G w/ μ raised to 5.4, uprated to 15 W (GE, SY, TS) (10-3-51) <8BD>. "A" has T12 bulb. Variant: 7236.

6000, W Xmtg. beam power tube, for Class C use, T11 bulb, octal base, heater 26.5 V @ 280 mA, 25 W (GE, SY) (1-23-61) <6CK>.

6002 [AKA QK221]; magnetron, fixed at 9317 ± 87 MHz, IM, WG, 4.0 V @ 44 A, 30 kV @ 32 A, 250 kW pulse output (RA, SY) (2-6-53). In AN/CPS-9 weather radar; also in AN/TPS-10D.

6003 Magnetron, fixed freq. (2-6-53).

6004 Full-wave rectifier, 5Y3GT w/ plate caps, rated for shock and 90,000-foot altitude, derated to 1 kV PIV (GE) (10-27-50) <2AJ>.

6005 [AKA 6095, M8245]; beam power tube, premium 6AQ5, mobile- and computer-rated (GE, HY, MU, RCA, SY, TS) (11-21-50) <7BZ>.

6006 Rcvg. pentode, premium 6SG7 (1-20-51).

6007 Dual numbering of 5913, q. v. (AX) (1-30-53).

6008 Dual numbering of 5911, q. v. (AX) (1-30-53).

6009 Xmtg. triode, fil. 13.0 V @ 36 A, μ 25, 6 kW, 30 MHz, water-cooled (F) (10-16-50).

6011 [AKA 710, 7307A]; gas-mercury thyratron, for welder control, fast-heating, T13 tubular bulb w/ cap, jumbo 4-pin base, fil. 2.5 VCT @ 9 A, 1.5 kV PIV, 2.5 A (GE, EL, HY, RCA, W) (11-9-51). Variants: 7518, 7725, 7726.

6012 Xenon tetrode thyratron for motor control and inverters, T12 octal bulb, button stem, heater 6.3 V @ 2.6 A, 1.3 kV, 500 mA (HY, RCA, TS) (2-6-52) <6CO>.

6013 [AKA EL3B]; rectifier, fil. 2.5 V @ 9.0 A, 920 V PIV, 2.5 A (EL, HY) (1-30-53).

6014 [AKA C1K]; xenon triode thyratron, T12 bulb, 4-pin base, fil. 2.5 V @ 6.3 A, 1 kV PIV, 1 A (EL, HY, NL, RCA, W) (3-28-51) <4AX>. Also see 5683.

6015 [AKA EL6F]; rectifier, fil. 2.5 V @ 21 A, 920 V PIV, 6.4 A (EL) (3-28-51).

6017 Xmtg. ring-seal cermet VHF triode, for G-G TV amps., fil. 10.0 V @ 17 A, 1 kW, 400 MHz (6-20-52).

6018 Xmtg. triode, 2C36-style "rocket" CW amp.-multiplier usable to 3 GHz, fil. 6.3 V @ 400 mA (SY) (1-22-54).

6019 Xmtg. tetrode, fil. 6.3 V @ 24 A, 1 kW peak output, 900 MHz, promoted in 1954 as part of a full line for UHF TV including 6183, 6192, 6237-6242, and 6283 (GE) (6-15-51). Variant: 6183.

6021, WA T3 submin. med.-μ duotriode, ruggedized 6BF7 w/ g_m raised to 5.4 mS, 1/8" longer (GE, MU, RA, RCA, ST, SY, TS) (6-10-52) <8DG>.

6022 [AKA ATR332]; ATR tube, fixed-tuned to 5365 MHz, 1 MW (BL).

6024 [AKA ATR387]; ATR tube, fixed-tuned to 2800 MHz, 1 MW (BL) (3-17-52).

6026 T3 submin. UHF xmtg. osc. triode for 400-MHz radiosondes, heater 5.2-6.6 V @ 200 mA, 3 W, 1.25 W out @ 400 MHz (RCA, TS) (5-31-51).

6027, H [AKA 2J42A, q. v.]; magnetron, fixed at 9375 ± 30 MHz, IM, WG, heater 6.3 V @ 600 mA, 8 kV @ 8 A, 20 kW output (EEV, LI, MA, SY, TSU) (4-9-56). "H" is probably high-altitude variant like 2J42H.

6028 [AKA (WE) 408A]; 6AK5 w/ heater 20 V @ 50 mA (GE, SY, TS, WE) (6-28-51) <7BD>.

6029 [AKA CK573AX]; T2X3 submin. med.-μ rcvg. triode, fil. 1.25 V @ 200 mA, μ 8.5 (GE, RA) (9-11-53) <5676>.

6030 [AKA ECC91]; min. duotriode, variant of 6J6.

6032, A Image converter, for viewing via IR light, max. response at 800 nm, flanged body, 2-3/32" max. dia. by 4-15/32" high, photocathode window 1.6" dia, screen end 1-15/32" dia., grid voltage 20 kV (RCA) (10-31-55). "A" has tighter performance limits.

6033 ATR tube, 9300 MHz, 250 kW, 1B37A w/ mounting flange flat to ± 0.003 in. (BL) (9-19-51).

6034 ATR tube, 6033 for 8750 MHz (BL) (9-19-51).

6035 TR tube, 8.5-9.6 GHz, bandpass type, 200 kW, like 1B63A (BL) (9-19-51).

6037 [AKA QK243]; reflex klystron, 5.12-5.43 GHz, 30 mW (RA).

6038 ATR tube, 9.0-9.6 GHz, waveguide mount, 100 kW, 4 μsec. recovery (GE) (9-19-51).

6039 VHF xmtg. disc-seal triode, for TV and FM xmtrs, redesigned 9C24 (replaces it w/ smaller fil. xfmr.), T-T fil. 5.0 V @ 78 A, μ 21, 7 kW, 220 MHz, water-cooled (GE) (3-2-51).

6040 Reference cavity, 9308 MHz, loaded Q 2100, loaded Q 5 (BL, W) (7-6-51).

6041 Reference cavity, 6040 for 9312 MHz (W) (7-6-51).

6042 Duotriode, premium 6SN7GT w/ heater 25 V @ 150 mA (5-28-51).

6043 Reflex klystron, 2950-3275 MHz, IC, CO, heater 6.3 V @ 650 mA, 300 V, 150 mW output (RA) (1-4-51).

6044 Xenon triode thyratron, all-glass bulb mounted on straps from pins, 5528 ratings but half-size, fil. 2.5 V @ 18 A, 550 V PIV, 6.4 A (HY, T) (2-20-51).

6045 Duotriode, premium 6J6, heater curr. reduced to 350 mA (GE) (4-17-53) <7BF>.

6046 Beam power tube, computer-rated 25L6GT for industrial relay use (GE) (6-28-51) <7AC>.

6047 Beam-switching tube (3-20-51).

6049 T3 submin. semi-remote-cutoff pentode, heater 6.3 V @ 150 mA, 1.1 W (GE) (1-17-55) <8DL>.

6050 T2X3 submin. med.-μ UHF triode, fil. 1.25 V @ 120 mA, g_m 1.6 mS; UHF variant of 5676 w/ μ raised to 16 (GE, RA) (7-12-51) <5676>. Three in AN/URC-4 xcvr.

6051 T2X3 submin. pentode, fil. 1.25 V @ 120 mA, 370 mW (7-16-62) <6051>.

6052 Submin. full-wave rectifier, heater 6.3 V @ 300 mA (2-20-51).

6053 Submin. full-wave rectifier, 6052 w/ heater 26.5 V @ 45 mA (2-20-51).

6055 Submin. triode, dual-numbering of 5904 (2-20-51).

6056 Submin. pentode, dual-numbering of 5907 (2-20-51).

6057 [AKA M8137]; min. duotriode, premium 12AX7 (MU) (4-18-51).

6058 [AKA M8079]; min. duodiode, premium 6AL5, 3/8" taller, higher voltage rating (MU) (5-18-51).

6059 Min. pentode, premium 6BR7 (10-25-51).

6060 [AKA ECC801]; duotriode, premium 12AT7 (4-18-51).

6061 Min. pentode, premium 6BW6 (4-18-51).

6062 [AKA M8086]; min. beam power tube, premium 5763 (MU) (4-18-51).

6063 Min. full-wave rectifier, premium 6X4 (4-18-51).

6064 [AKA M8083]; min. pentode, premium 6AM6 (MU) (4-18-51).

6065 [AKA M8161]; min. pentode, premium 6BH6 w/ reduced gm and higher heater current (MU) (5-28-51).

6066 [AKA M8136]; min. duodiode-triode, premium 6AT6 (4-18-51).

6067 [AKA M8136]; min. duotriode, premium 12AU7 (5-18-51).

6072, A Min. duotriode, long-life and low-microphonics variant of 12AY7; heater curr. raised to 175 mA per section (GE, RCA, SY). "A" is computer-rated; has higher leakage resistance (GE) (3-5-52) <9A>.

6073 Min. voltage regulator, ruggedized/high-stability 0A2 (RCA) (9-4-51).

6074 Min. voltage reg., 0B2 as for 6073 (RCA) (9-4-51).

6075 [AKA AX9907, QY5-3000W]; xmtg. glass ring-seal VHF tetrode, fil. 6.3 V @ 33.5 A, 3 kW, 220 MHz, water-cooled, for SSB and FM-TV xmtrs; screen out ring, grid and fil. out posts at crown; two tubes gave 5 kW peak (AX, MU) (9-18-53).

6076, A [AKA AX9907R, QY5-3000A]; xmtg. tetrode, radiator version of 6075, same ratings (AX, MU) (9-18-53). "A" is cermet. Two in Gates FM-5B 5-kW FM xmtr.

6077 [AKA AX9906, TY12-50W]; xmtg. glass ring-seal HF triode, grid-out-ring, w/ corona ring over anode, T-T fil. 17.5 V @ 196 A, 50 kW, 30 MHz, water-cooled via external cooler, promoted for pulse applications (AX, MU) (9-18-53).

6078 [AKA TY12-50A, AX-9906R]; RF-heating/xmtg. triode, forced-air-cooled variant of 6077 derated to 45 kW, w/ fluted anode w/ cellular radiator; promoted for pulse applications (AX, MU) (9-18-53).

6079 [AKA AX9908, QY5-500]; xmtg. VHF tetrode, all-glass tubular bulb, graphite anode, fil. 10 V @ 9.7 A, 500 W, forced-air cooled, for AM and FM xmtrs (AX, MU) (9-4-53).

6080, WA, WB Power duotriode, ruggedized 6AS7G, T12 bulb, button stem, heater current raised to 2.5 A, derated to 13 W per section, for pass tube in voltage regs. and booster-scanning use in projection TV rcvrs. (MU, RCA, SY, TS) {A4535} (12-4-51) <8BD>. "WB" [AKA TE-46] is hard-glass type (BX).

6081 [AKA ATR407]; ATR tube, fixed-tuned to 5640 MHz, 300 kW (BL, SY) (5-21-54).

6082, A Power duotriode, 6080 w/ heater 26.5 V @ 600 mA (RCA, SY) (1-3-52) <8BD>. "A" is in shortened bulb; [AKA TE-46] is hard-glass type (BX).

6083 [AKA AX9909]; xmtg. VHF pentode, fil. 12.6 V @ 1.35 A, 45 W, radiation-cooled (AX) (9-4-53).

6084 [AKA E80F]; min. 9-pin audio sharp-cutoff "10,000-hr." pentode, heater 6.3 V @ 0.3 A, 1.3 W (AX, HY, MU) (10-16-51).

6085 [AKA E80CC]; "10,000-hr." duotriode, heater 12.6 VCT @ 300 mA, μ 27, g_m 2.7 mS, 4.0 W (AX, MU) (10-16-51). Intermediate in μ and g_m between 12AU7 and 12AZ7.

6086 Min. 9-pin pentode, heater 18 V @ 100 mA (HY) (10-16-51).

6087 [AKA 5Y3WGTB]; full-wave rectifier, indirectly heated, ruggedized 5Y3GT, same ratings (GE) (11-7-52) <5L>.

6088 [AKA CK522AX]; T2X3 submin. power pentode, fil. 1.25 V @ 20 mA, 1.2 mW output (GE, RA, TS) (8-27-51).

6089 [AKA 1Q23]; reference cavity (5-25-51).

6090 Beam-switching tube (2-20-56).

6091 [AKA 8423]; beam-switching tube (NL) (3-26-56).

6092 T2X3 submin. power pentode, fil. 1.25 V @ 50 mA (GE) (3-19-56) <2E31>.

6094 [AKA TE-18]; beam power tube, "10,000-hr." version of 6AQ5 w/ hard-glass bulb, tungsten pins, ceramic spacers, gold-plated pins, etc., heater 600 mA, uprated to 12.5 W (BX, GE, TS) (12-27-54) <9DH>.

6095 [AKA 6AQ5W, q. v.]; beam power tube, ruggedized 6AQ5 (GE, SY) (11-5-62) <7BZ>.

6096 [AKA 5654/6AK5W, q. v.]; min. pentode, ruggedized 6AK5 (GE, RCA, SY) (2-4-63) <7BZ>.

6097 Min. duodiode, premium 6AL5 (GE) (7-1-57) <6BT>.

6098 [AKA 6AR6WA]; beam power tube, premium 6AR6 (GE, SY, TS) (11-5-62) <6BQ>. Also see 6384.

6099 Min. duotriode, premium 6J6 w/ plate current balanced between sections (SY) (5-20-63) <7BF>.

6100 [AKA M8080]; min. triode, premium 6C4, 3/8" taller (GE, MU, SY) (11-5-62) <6BG>.

6101 [AKA 6J6WA, M8081]; min. duotriode, ruggedized 6J6, derated to 850 mW per section (GE, MU, RCA, SY) (9-18-53) <7BF>.

6102 Rectifier, 2-27/32" dia. by 2-3/16" long, fil. 5 V @ 7.2 A, 40 kV PIV, 150 mA, oil-cooled (W) (3-26-54).

6103 Rectifier, 6102 derated to 20 kV (W) (3-26-54).

6106 [AKA TE-22]; full-wave rectifier, premium 5Y3GT, cathode-type, curr. reduced to 1.7 A, uprated to 1.55 kV (BX, GE) (11-1-54) <5L>. Also see 6853.

6107 [AKA BS212].

6108 [AKA BS213.

6109 [AKA BS404].

6110 T3 submin. full-wave rectifier, heater 6.3 V @ 150 mA, 460 V PIV, 4.4 mA (GE, RA, SY, TS) (3-17-52) <8DJ>.

6111, WA T3 submin. med.-μ audio duotriode, heater 6.3 V @ 300 mA, μ 20, computer-rated (GE, RA, RCA, ST, SY, TS) (11-13-51) <8DG>.

6112 T3 submin. high-μ duotriode, general use, heater 6.3 V @ 300 mA, μ 70, computer-rated (GE, RA, RCA, SY, TS) (11-13-51) <8DG>.

6113 [AKA 6SL7WGT]; duotriode, ruggedized 6SL7GT (GE) (8-9-51) <7V>.

6115, A [AKA QK351]; reflex klystron, 5.1-5.9 GHz, IC, CO, heater 6.3 V @ 440 mA, 330 V, 100 mW out (KK, RA) (7-9-51).

6116 Reflex klystron, improved 2K45, 8.5-9.6 GHz, IC, CO, htr. 6.3 V @ 520 mA, 350 V, 20 mW out (BX, RA, THF) (2-6-53).

6117 TR switch, 1B58 w/ extra holes for pressurization (BL, MA) (4-24-52).

6118 Rcvg. duodiode-triode, rugged 6Q7 (SY) (11-1-54).

6119 Submin. voltage regulator, corona type, 2.1 kV @ 2-50 μA (VN) (10-15-52).

6120 Min. voltage regulator, corona type, 15.6 kV @ 5-200 μA (VN).

6121 T2X3 submin. med.-μ triode, fil. 1.25 V @ 120 mA, μ 15, 1.1 W; variant of 5676 (GE) (9-12-55) <5676>.

6122-27 "Classified type"(s) (all 5-7-54).

6128 ATR tube (12-5-51).

6130 Hydrogen thyratron, improved 3C45, q. v. (BL, GE, RCA) (7-31-53). BL version has longer stalk on cap than 3C45. GE version has corona-grading ring below cap.

6132 [AKA 7D10]; min. pentode, premium 6CH6 (2-29-52).

6133 Reflex klystron, 1.5-3.75 GHz, EC, CO, heater 6.3 V @ 650 mA, 300 V, 140 mW output (RA) (12-17-51).

6134 [AKA 6AC7WA]; rcvg. pentode, computer-rated 6AC7 (GE) (5-22-53) <8N>.

6135 Min. triode, ruggedized computer-rated 6C4WA, heater raised to 175 mA (GE, SY) (5-14-52) <6BG>.

6136 Min. pentode, ruggedized computer-rated 6AU6 (GE, RCA, SY, TS) (5-14-52) <7BK>.

6137 [AKA 6SK7WA]; rcvg. pentode, ruggedized computer-rated 6SK7 (GE) (2-18-52) <8N>.

6140 [AKA (WE) 423A]; min. 9-pin voltage reference, 100 V, 4-8 mA (HY, WE) (9-21-51).

6141 [AKA (WE) 427A]; min. 9-pin voltage reg. w/ starting electrode, 100 V, 5-40 mA (HY, WE) (9-21-51).

6142 Voltage regulator (7-10-53).

6143 Submin. voltage reg., corona type, 1.2 kV @ 5-100 μA (VN) (10-15-52).

6144 [AKA TT-1]; noise diode, analogous to 5722 but w/ coaxial construction; plate and fil. form elements of coax line (BX, RCA) {R6212A} (6-19-53).

6145 Loktal power/video pentode, 7AD7 uprated to 11 W (GE, SY) (11-13-51) <8V>.

6146 A, B, W [AKA QV06-20] xmtg. beam power tube, T12 bulb w/ cap, metal-ringed octal base, heater 6.3 V @ 1.25 A, 20 W, 60 MHz (AX, GE, MU, RCA, SY, W) (1-3-52) <7CK>. Variants: 6146B/8298A, 6159, 6293, 6883, 7212, 7607. "W" [AKA 7212, q. v.] is ruggedized. "A" is for mobile use, w/ "dark" heater; still rated at 20 W (GE, RCA, SY) (9-4-61). Variants: 4604, 4652/8042. "B" [AKA 8298A, S2001] is uprated to 27 W, w/ "dark" heater reduced to 1.125 A and "external-barrier" metal ring on base (GE, RCA, SY) (11-11-63). Variant: 6883B/8032A/8552 (see 6883B).

6147 T3 submin. RF pentode, fil. 2.5 VCT @ 125 mA, 1.5 W (GE, SY) (7-1-57) <6CL>. Appears to be 5851 w/ higher-curr. fil. Two in AN/URC-4.

6148 [AKA CK605CX, 5702WA]; submin. pentode, heater 6.3 V @ 200 mA (RA).

6149 [AKA 5703WA]; submin. triode, heater 6.3 V @ 200 mA (RA) (7-3-61).

6150 [AKA 5874WA]; submin. pentode (RA). [flat?]

6151 [AKA CK619CX, 5744, q. v.]; submin. triode.

6152 T3 submin. flat-press low-μ triode, heater 6.3 V @ 200 mA, μ 17.5, g_m 5.1 mS (GE, RA) (3-19-56) <5975>.

6153 Submin. sharp-cutoff dual-control pentode, heater 6.3 V (SY).

6154 See 6153.

6155 [AKA QY3-125]; xmtg. tetrode, improved 4D21/4-125A (interchanges w/ it), all-glass, graphite anode, same ratings in smaller size, for FM broadcast xmtrs and audio (AX, EI, MC, MU, RCA) (9-4-53).

6156 [AKA QY4-250]; xmtg. tetrode, improved 5D22/4-250A (interchanges w/ it), all-glass, same ratings in smaller size (AX, EI, ML, MU, RCA) (9-4-53).

6157 Min. 9-pin rectifier, heater 6.3 V @ 800 mA, 500 V in @ 75 mA; 350 V in @ 125 mA (HY) (2-29-52).

6158 Min. 9-pin medium-μ duotriode, heater 12.6 VCT @ 300 mA (HY) (2-29-52).

6159, A, B, W Beam power tube, 6146 w/ heater 26.5 V @ 300 mA (AX, RCA, SY, W) (1-3-52) <7CK>. "B" is uprated to 27 W. "W" is AKA 7357, q. v.

6161 Xmtg. ring-seal glass triode, for G-G FM-TV use, heater 6.3 V @ 3.4 A, μ 27, 250 W, 900 MHz, forced-air cooled (RCA) (1-21-52). In AN/ALT-7 jammer, giving 100 W at 24-350 MHz. Also see 5588.

6162 ATR tube (3-17-52).

6163 [AKA ATR428]; ATR tube, 8.5-9.6 GHz, used in tandem in special mount for RG-51/U WG (MA, SY) (8-14-53).

6164 [AKA TR429]; TR tube, fixed-tuned to 8.5-9.6 GHz, continuous phase and VSWR, 200 kW, RG-51/U WG in and out (MA, SY) (12-27-54).

6166, A [AKA 8F66R]; xmtg. beam tetrode, glass ring-seal coaxial construction but w/ pins for fil., T-T fil. 5.0 V @ 181 A, 10 kW, 30 MHz (W: 216 MHz), forced-air cooled (EEV, F, GE, NEC, RCA, W) (2-29-52). Five used in parallel in RCA TT-50AH TV xmtr. to give 50 kW peak on Ch. 7-13; two in TT-25DH for 25 kW; also in Collins 10-kW VHF-scatter xmtr. "A" [AKA 7007] is cermet variant, uprated to 12 kW and 220 MHz (RCA) (6-20-60).

6167 [AKA (WE) 439A]; beam-switching tube, 10-stage counter, T9 bulb w/ 16 pins in two rings; fires at 225 V, 10 mA, counts up to 1000 pps; can be used in decade chains, etc. (WE) (4-28-52).

6169 T3 submin. HF triode, heater 6.3 V @ 150 mA, 3.0 W (GE) (12-10-51) <8EE>.

6170 Beam-switching tube (11-21-51).

6173 Glass pencil diode for pulse-detection and peak-power measurement use, coaxial construction, 0.3" dia. by 2.3" long, heater 6.3 V @ 135 mA, 1 kV PIV, 1 mA, usable to 3.3 GHz (GE, RCA) (11-21-52).

6174 Min. cold-cathode rectifier, 2.8 kv PIV, 3 mA (RA) (3-19-56).

6177 [AKA QK-363]; CW magnetron, fixed at 4255-4370 MHz, electronically freq.-modulable ± 2%, IM, CO, heater 6.3 V @ 600 mA, 330 V @ 30 mA, 1.0 W output (RA, W) (3-28-55).

6178 Reflex klystron, 15.75-16.25 GHz, IC, WG, heater 6.3 V @ 600 mA, 330 V, 25 mW output (RA) (3-28-55).

6179 Xmtg. audio triode, fil. 27.2 V @ 179 A, μ 4.8, 50 kW, water-cooled (F) (8-1-52).

6180 Premium 6SN7GT (4-21-52).

6181 Xmtg. ring-seal coaxial beam tetrode, early cermet, for UHF TV, heater 120 V (yes, 120!) @ 1.6 A, 2 kW, 900 MHz, forced-air cooled, louvered fins on anode (RCA) (4-1-52).

6182 Xmtg. tetrode, cermet coaxial ring-seal, fil. 6.7 V @ 14 A, 5 kW peak output, 900 MHz, water-cooled via integral cooler (GE) (3-27-53). Also see 6019, 6183.

6183 Xmtg. tetrode, for UHF TV, fil. 6.3 V @ 24 A, 900 MHz, transverse-flow air-cooled; apparent variant of 6019, q. v. (GE) (3-19-52).

6184 T3 submin. UHF duodiode, heater 6.3 V @ 150 mA, 450 V PIV, 20 mA (GE) (3-17-52) <8EH>.

6185 Duotriode, premium 2C51.

6186, W Pentode, premium 6AG5WA (GE, RCA, SY) <7BD>.

6187 [AKA 6AS6W]; min. pentode, ruggedized 6AS6 (GE, SY) (4-11-60) <7CM>.

6188 Duotriode, ruggedized 6SU7WGT (GE, SY, TS) (1-30-61) <8BD>.

6189 [AKA M8136]; min. duotriode, ruggedized 12AU7, (GE, MU, RCA, SY, TS (11-5-62) <9A>.

6193 T3 submin. HF duotriode, heater 6.3 V @ 300 mA, μ 55, 2.0 W (GE) (7-24-52) <6193>.

6195 [AKA DL700]; T3 submin. beam power tube, fil. 2.5 VCT @ 110 mA, 2.5 W (GE) (6-25-52) <6CL>.

6196 Rcvg. tetrode (10-10-52).

6197 Min. pentode, computer-rated 6CL6 (GE, RCA, SY) (12-18-53) <9BV>.

6198, A Vidicon, 7697 format, "first commercial vidicon," for industrial use, resolution at center only 350 lines (Resitron Laboratories, RCA, W) (6-2-52). "A" (GC) has no sealoff tip out side. Also see 7351.

6199 Photomultiplier, 10-stage, for scintillation counters, S-11 response, head-on flat-face bulb 1-9/16" dia. by 3-7/8" seated height, duodecal base (RCA) (4-1-52). Used heavily in portable scintillation counters during the '50s "uranium rush." Variant: 2060.

6201 [AKA ECC801]; min. duotriode, ruggedized computer-rated 12AT7 (GE, RCA, SY, TF, TS) (5-14-52) <9A>.

6202 Min. full-wave rectifier, ruggedized 6X4, rated to 60,000 feet, derated to 50 mA (GE, RCA, TS) (5-18-53) <5BS>.

6203 Min. full-wave rectifier, ruggedized 6X4 on 9-pin base; heater current raised to 900 mA, rated to 60,000 feet (GE) (5-8-53) <9CD>.

6205 T3 submin. dual-control pentode, 5840 w/ separate suppressor lead (GE, ST, SY, TS) (6-10-52) <8DC>.

6206 T3 submin. dual-control pentode, 5899 w/ separate suppressor lead (GE, RCA, ST, SY, TS) (6-10-52) <8DC>.

6207-10 Counter tubes (all 1-30-53).

6211, A Min. duotriode, 9-pin computer-rated variant of 5844 w/ separate cathodes (AX, GE, HY, MU, RCA, SY, TF) (8-8-52) <9A>. "A" is a "10-000-hr" type.

6213, A [AKA VXR130]; submin. voltage regulator, 130 V @ 1-2.5 mA (RA, VN) (11-28-52).

6214 ATR tube, for 9375-MHz radar beacons, 1B35A w/ keep-alive electrode added for quick firing, allows single T-R ant. for beacons (BL, SY) (8-14-53).

6215 HV rectifier, premium 1B3GT, derated to 18 kV, uprated to 1 mA (GE) (8-1-52) <3C>.

6216 Min. 9-pin beam power tube, heater 6.3 V @ 1.2 A, 10 W, for audio, regulators, or VHF xmtg. (AX, GE, HY, SY, TS (3-13-53) <9CE>. Promoted for use as active 100-mA filter choke.

6217 Photomultiplier, 10-stage, S-10 response, 5819 format (RCA) (7-18-52). For color densitometers, spectrometers, color comparators, etc. Also see 6292.

6218 [AKA E80T]; beam-deflection tube for phase-discrim. use, ruggedized, heater 6.3 V @ 150 mA (AX, HY) (8-25-52).

6221 T3 submin. med.-μ triode, heater 6.3 V @ 175 mA, μ 27, 3.3 W (GE, ST) (12-12-55) <8HF>.

6222 T3 submin. high-μ audio triode, high-rel. type, heater 6.3 V @ 175 mA, μ 70 (GE, ST) (12-12-55) <8HF>.

6223 T3 submin. pentode, heater 6.3 V @ 175 mA, 1.1 W (GE, ST) (12-12-55) <8DL>. Variant: 7443.

6224 T3 submin. pentode, heater 6.3 V @ 450 mA, 5 W (GE, ST) (5-7-56) <8DL>.

6225 T3 submin. pentode, 5840 w/ different pinout, heater raised to 175 mA (GE, ST) (12-12-55) <8DL>.

6227 [AKA E80L]; rcvg. "10,000-hr." power pentode, heater 6.3 V @ 750 mA, 8 W (AX, MU) (7-16-56).

6228 [AKA GL-506]; ignitron, stainless steel case 9" dia. by 42" long, for frequency-changer use, or for capacitor discharge or crowbar ckts.; 20 kV PIV, 150 A, or 50 kW PIV and 30 kA peak; six tubes give 7.5 MW @ 17 kV (GE) (11-28-52).

6229 [AKA QK299]; magnetron, tunes 8.9-9.4 GHz, IM, WG, 5.0 V @ 500 mA, 4-5 kV @ 500 mA, 400 W pulse out (AX, MA, RA). In Nike-Ajax guidance section; also in AN/UPW-1 X-band data link.

6230 [AKA QK299B]; magnetron, 6229 uprated to 5 kV @ 1.0 A, 1 kW pulse output (AX, RA).

6231 [AKA ELC5F15]; thyratron, fil. 14.0 v @ 2.6 A, 500 V PIV, 5.0 A (EL) (1-30-53).

6232 [AKA TR437]; TR tube, 8.5-9.6 GHz, 250 kW, 1B63A for RG-51/U WG (BL, SY) (5-21-54).

6233 [AKA L-3023]; magnetron, fixed at 9312.5 ± 32.5 MHz, 7 kW out (LI) (12-5-52). In AN/APN-69A beacon.

6236 Reflex klystron, 3.8-7.6 GHz, EC, WG, heater 6.3 V @ 580 mA, 1.25 kV, 125 mW out (RA, VA) (12-12-55).

6237-42 Amplifier klystrons, for UHF TV xmtrs, three-cavity, EM, WG, heater 5.5 V @ 35 A, cathode bombardment voltage 3 kV, 18 kV @ 3 A, 27 dB gain, 15 kW out, water- and forced-air cooled via integral cooler. Developed by VA to GE specs. (all GE, VA) (all 11-14-52). Also see 6019.

6237	For 470-530 MHz.	6240	For 656-722 MHz.
6238	For 530-584 MHz.	6241	For 722-806 MHz.
6239	For 584-656 MHz.	6242	For 806-890 MHz.

6245 T3 flat-press submin. VHF pentode, sharp-cutoff variant of 5702, heater 6.3 V @ 200 mA, 1.85 W (GE, TS (9-12-55) <5702>.

6246 [AKA 508] xmtg. triode, fil. 30.0 V @ 80 A (AX) (1-30-53).

6247 [AKA CK628]; T3 submin. flat-press high-μ triode, low microphonics, heater 6.3 V @ 200 mA, μ 60, g_m 2.5 mS (GE, RA) (10-16-53) <8FO>.

6248 Magnetron, 8.7-8.9 GHz, 1.2 kW output. In AN/-APN-108 pulse-Doppler nav set on B-52E.

6249, A, B [AKA QK349]; magnetron, for airborne use, for motor-driven tuning, covers 8.5-9.6 GHz, IM, WG, fil. 9.0 V @ 14 A, 28 kV @ 25 A, 240 kW pulse out (RA, W) (11-20-61). "B" in AN/APG-51B.

6250 Receiving tetrode (10-10-52).

6251 Xmtg. VHF cermet tetrode, for G-G TV xmtrs, ring-seal coaxial construction, 14" high, T-T fil. 5.5 V @ 190 A, 25 kW, 220 MHz, water cooled via integral cooler (GE) (6-5-53).

6252 [AKA AX9910, QQV03-20A]; xmtg. dual tetrode, shortened 5894, dual fils., 6.3 V @ 650 mA each, derated to 20 W, forced-air-cooled, usable to 700 MHz (AX, MU) (11-13-53). In Sonar CD2 xmtr.

6253 Reflex klystron, 18.0-22.0 GHz, 40 mW.

6254 Reflex klystron, 22.0-25.0 GHz, 40 mW.

6256-58 RF-heating triodes, glass ring-seal, for 3-kW heaters, fil. 12.6 V @ 29 A, μ 21, 110 MHz (ML) (all 5-13-57).

6256 Water-cooled via external cooler, 5 kW.
6257 Water-cooled via integral cooler, 5 kW.
6258 Forced-air-cooled, 3 kW.

6259 Voltage regulator (11-14-52).

6260 Pre-TR or ATR switch, 1.2-1.4 GHz, 10-3000 kW (used optionally w/ 7166) (BL, W) (10-31-55).

6263, A Xmtg. glass power pencil triode, heater 6.0 V @ 280 mA, μ 27, 8 W, 500 MHz, optionally forced-air cooled, transverse air flow (PH, RA, RCA, SY, TS, VO) (11-20-53). Variant: 6264, A.

6264, A Xmtg. power triode, 6263(A) w/ μ raised to 40, for use in freq. multipliers (RCA, SY, TS) (11-20-53).

6265 Rcvg. pentode, ruggedized 6BH6; heater raised to 175 mA, derated to 2.0 W (GE) (3-26-54) <7CM>.

6267 [AKA EF86]; min. pentode, spiral heater, 6.3 V @ 0.2 A, g_m 2.0 mS, for audio preamps (AX, GE, MU, TF) (11-28-55) <9CQ>.

6268 [AKA AX-9911]; hydrogen thyratron, 4C35A w/ self-contained hydrogen source, rated 1000 hours min.; interchanges w/ 4C35 (AX) (11-13-53).

6269 Rectifier, heater 26.5 V @ 350 mA.

6270 Magnetron, fixed at 9550 ± 50 MHz, 8 kW output (11-6-53).

6271 Magnetron, fixed at 9453 ± 47 MHz (11-6-53).

6272-75 Magnetrons, equiv. to 2J42, fixed-freq., 8 kW out, within ± 50 MHz of nominal freq. (all MA) (all 11-6-53).
6272 For 9250 MHz. 6274 For 9050 MHz.
6273 For 9150 MHz. 6275 For 9273 MHz.

6276 ATR tube, 9280 MHz, flange-mounted version of 1B35 (MA) (1-1-53). Variant: 6284.

6277 [AKA 3B28A]; xenon rectifier (HY).

6278 [AKA (C) 5F14]; xenon triode thyratron, flying leads, heater 14 V @ 2.0 A, 500 V PIV, 5 A, similar to C3R14 (HY, T) (6-11-54).

6279 [AKA AX9912]; hydrogen thyratron, 5C22 w/ self-contained hydrogen source, rated 1000 hours min., interchanges w/ 5C22 (AX) (11-13-53.

6280 [AKA (WE) 416B, q. v.]; SHF triode (5-8-53). In low-noise preamps and in Government-owned TD-2 microwave systems as part of the "White Alice" Alaskan comm. system, 1961. In Defense Electronics Inc. G-175-H and G-166D surveillance rcvrs., 30-260 and 260-1000 MHz. Preamp in early REL 900-MHz troposcatter system.

6281 [AKA CK512AX]; T2X3 submin. sharp-cutoff AF pentode, fil. 0.625 V @ 20 mA (GE, RA) (3-19-56) <2E31>.

6282 [AKA BL-11]; TR tube, 24 GHz, bandpass type, fixed-tuned, 35 kW (BL) (4-25-55).

6283 Xmtg. tetrode, ring-seal cermet construction, for 250-W G-G UHF TV xmtrs, heater 6.3 V @ 3.6 A, 300 W, 900 MHz, forced-air cooled w/ transverse flow (GE) (3-27-53). Variant: 7399. Also see 6019.

6284 ATR tube, 6276 for 8750 MHz (MA).

6285 Magnetron, fixed at 1.3 GHz, EM, WG, 15.0 V @ 285 A, 70 kV @ 350 A, 10 MW pulse output, "highest power magnetron commercially available" in mid-'50s (W) (7-9-62).

6286 T2X3 submin. med.-μ triode, fil. 1.25 V @ 125 mA, μ 11.5, 1.25 W (GE, TS) (9-12-55) <5676>.

6287 Min. 9-pin audio pentode, heater 6.3 V @ 600 mA, 13.2 W (GE, HY) (6-12-53) <9CT>.

6288 Mercury rectifier, premium 816.

6291 Photomultiplier, 10-stage, S-11 response, head-on T12 bulb 4-¼" high, duodecal base (DuM, RA) (2-27-53). In OA-1007/FSQ SAGE video mapper console.

6292 Photomultiplier, version of 6217 w/ S-11 response (DuM) (2-27-53).

6293 Beam power tube, 6146 for pulse-modulator use, 10 W, 350 V, 3 A peak (RCA, RA) {A4423A} (7-3-53).

6297 Min. pentode, premium 6CL6; also see 6677.

6299 Planar UHF triode, cermet construction, ½" dia. by 1" high, metal surfaces gold-plated, for G-G preamps, heater 6.3 V @ 300 mA, μ 110, 2 W, NF 8.5 dB @ 1.2 GHz and 14 dB @ 3 GHz (GE) (9-18-53). Variants: 7644, 7784.

6301 Vacuum reference cavity, 9270 MHz ± 0.4 MHz, 5 kW, transmission-type w/ 6.5 dB loss, loaded Q 1250, low freq. drift w/ temperature (GE) (4-16-54). Variants: 1Q26A, 6452.

6303 [AKA X80]; radar clipper diode and rectifier, stepped tubular bulb w/ cap on glass stalk, 4-pin (industrial?) base, ringed graphite anode, T-T fil. 11.0 V @ 15.5 A, 40 kV PIV, 800 mA (UE) (intro. 1952, regis. 7-6-59). Variant: 561.

6304 [AKA BL-43]; ATR tube, 9.3 GHz, 250 kW, min. 1B35A, contact mount (BL) (6-17-57).

6305 [AKA R10]; min. half-wave rectifier, w/ cap, heater 4.0 V @ 500 mA, 12.5 kV PIV @ 5 mA (HY) (3-20-53).

6306 Gamma-ray counter, 1B85 w/ bismuth-coated screen (VN).

6307 Neutron counter, boron trifluoride, 2 kV, sensitivity 4.5 counts/neutron/cm^2/sec., aluminum body 1-1/32" dia. by 12-1/8" long, HN coaxial connector (W) (8-22-55). Also see 6938, 6998, 7087, 7243, 7317, and 7429.

6308 T3 submin. voltage reference, 87 V @ 3.5 mA (SY) (10-9-53) <8EX>.

6310-15 Reflex klystrons, for radar/beacon local oscs., 8.5-10.0 GHz, IC, WG, heaters 6.3 V @ 1.2 A, 350 V (all 7-11-55).
6310 [AKA V-260, V-262]; slotted shaft for tuning, 3-pin base, 125 mW out (BL, RA, VA, W).
6311 Specs as for 6310, 70 mW out (VA).
6312 [AKA V-270]; w/ flying leads potted for high-altitude use, 125 mW out (BL, MET, RA, VA, W).
6313 [AKA V-280]; locknut tuner, 140 mW output (VA).
6314 [AKA V-290]; locknut tuner and flying leads potted for high-altitude use, 120 mW out (MET, VA, W).
6315 [AKA V153]; w/ flying leads, 55 mW out (MET, VA, W). In Hughes AAI transponder, 9318 MHz.

6316 [AKA V-151]; reflex klystron, short-6L6 envelope,

coax lead out base, specs as for 6310, 70 mW output, thermally tuned (MET, RA, VA, W) (7-11-55).

6320 T3 submin. high-μ duotriode, heater 6.3 V @ 85 mA, μ 60 (GE) (12-27-55) <8DG>.

6321 T3 submin. low-μ duotriode, heater 6.3 V @ 85 mA, μ 16 (GE) (12-27-55) <8DG>.

6322 [AKA BL-25]; TR tube, 1285 MHz tunable, 450 kW (BL) (6-17-57).

6323 Photomultiplier, sim. to 6328 (8-7-53).

6324 Beam-switching tube (2-20-56).

6325 Full-wave rectifier, T9 bulb, octal base, heater 6.3 V @ 2.7 A, 2.2 kV PIV, 250 mA (GE) (6-5-53) <6325>.

6326, A Vidicon, mainly for B&W or color film chains, 7697 format (RCA, W) (12-18-53). "A" (GC) has no seal-off tip out side.

6327 Beam power tube, for radar sweep, sim. to two 6AR6s in parallel, T12 bulb w/ button stem, octal base w/ cap, heater 6.3 V @ 1.8 A, 35 W, rated to 60,000 feet (GE, TS) (6-6-55) <6327>.

6328 Photomultiplier, for control uses like car headlight dimmers, 9-stage, S-4 response, T9 bulb, 2.7" seated height, neosubmagnal 11-pin base, for use w/ AC voltages on electrodes (RCA) (11-20-53). Variants: 6472, 7117.

6330-31 "Misc." types (5-1-53).

6332 Voltage regulator (6-19-53).

6333 RF-heating HF triode, ruggedized 892 w/ fil. leads out crown, grid on ring, fil. 22 V @ 60 A, 10 kW (AX) (12-12-55). Variants: 6445, 6446.

6334 [AKA BL27]; TR tube, dual parallel 1B63As in same WG, 8.5-9.6 GHz, 250 kW, dual keep-alive electrodes, RG-52/U WG in/out (BL, SY) (5-21-54). Variants: 6642, 6643, 6648.

6336, A Duotriode, TT-16 stepped tubular hard-glass bulb, jumbo octal base, 30 W or 400 mA per plate, sim. to two 6080s in parallel, heater 6 V @ 4.75 A, μ 2.7, 60 W total (GE, RA, RCA, SY, TS) (9-11-53) <8BD>. "A" is raised to 5.0 A. Variants: 6394, 6528.

6337 Low-μ power triode, three triodes in parallel, hard-class tubular bulb w/ cap, button stem, large (7-pin giant?) base, heater 6.3 V @ 7.25 A, μ, gm 45 mS, for pass tube in volt. regulators (CH). Also see 7241.

6339 Vacuum clipper diode/rectifier, min. version of 3B29 for liquid-immersion cooling or air operation, fil. 6.3 V @ 1.5 A, 16 kV PIV and 65 mA if liquid-cooled (AX) (12-12-55).

6342, A, A/V1 Photomultiplier, 6217 w/ S-11 response, for scintillation counters (RCA) (11-20-53). Variant: 2061.

6343 Hughes thermo-gauge (2-5-54).

6344 Magnetron, tunes 5450-5825 MHz, IM, WG, fil. 11.0 V @ 12 A, 24 kV @ 30 A, 260 kW pulse out (LI, RA) (12-12-55).

6345 Ballast tube, 140-160 V @ 11.2 mA (7-9-62).

6346-48 Ignitrons, w/ thermostatic valves to stop water flow when not needed (all GE) (all 9-25-53).

6346 Variant of 5551.

6347 Variant of 5552.

6348 Variant of 5553.

6350 Min. 9-pin med.-μ duotriode, for computer use, heater 12.6 VCT @ 300 mA, μ 18, 7 W total (GE, HY, RCA, SY) (1-18-53) <9CZ>.

6351 [AKA Z319]; rcvg. pentode, heater 6.3 V @ 300 mA (10-30-53).

6352 T3 submin. temperature-limited (noise-generator) duodiode, fil. 3.0 V CT @ 360 mA, 250 V, 1.1 mA (GE, SY) (11-20-53) <8EY>.

6353 Voltage regulator, 20-kV corona type, for color-TV beam supplies, long metal tubular body w/ center pin like counter tube (VN) (2-5-54). Variant: 6392.

6354 [AKA 150B2]; voltage reference, 150 V, 10 mA (AX, MU, SY) (2-6-61) <7DU>.

6355 Min. twin indicator based roughly on 6AF6G, heater 6.3 V @ 140 mA (GE, HY) (12-11-53) <6355>.

6356-59 Gas noise sources (all 10-9-53).

6360, A [AKA QQE03-10]; min. 9-pin xmtg. VHF twin tetrode, heater 12.6 VCT @ 410 mA, 10 W, internally neutralized (AX, HY, MU, RCA) (6-20-55) <9PW>. In Gates BFE-10B 10-W FM xmtr. and AN/DMQ-2 beacon. In VHF xcvrs.: Drake ML-2, Heath HW-20, and Gonset Communicator II and IV-220.

6361 "Misc." type (94-53).

6362 Photomultiplier (5-11-59).

6363 Photomultiplier, 6292 w/ bulb flared to 3" dia. at faceplate (DuM) (3-19-54). Sim. to (RCA) 8054.

6364 Photomultiplier, 6292 w/ bulb flared to 5-¼"" dia. at faceplate (DuM) (3-19-54). Sim. to (RCA) 8055.

6365 Photomultiplier, min. 6-stage, S-11 response, head-on bulb 3/4" dia. by 2-3/16", 9-pin base w/ sealoff tip out base (DuM) (3-7-55).

6366 RF heating/xmtg. triode, flying ribbon leads, T-T fil. 11.0 V @ 29 A, μ 25, 3 kW, 30 MHz, forced-air cooled (F) (8-14-53).

6367 RF heating/xmtg. triode, flying ribbon leads, 5fil. 13.0 V @ 36 A, μ 25, 5 kW, 30 MHz, forced-air cooled (F) (8-28-53).

6370 [AKA E1T]; beam-deflection decade counter, fil. 6.3 V @ 300 mA, counts @ 100 kHz (AX, MU) (7-9-56).

6372 Photomultiplier, 10-stage, S-4 response, side-on, cathode 3" x 4-1/8", bulb 2-¼" dia. by 7-¼" high, 22-pin "twentyninar" base (RCA) (5-14-54).

6373 [AKA DL70]; rcvg. pentode, fil. 1.25 V @ 110 mA (MU) (8-14-53).

6374 [AKA EY84]; min. 9-pin half-wave rectifier, heater 6.3 V @ 1.0 A, 2 kV PIV, 150 mA (HY, MU) (8-14-53).

6375 [AKA DC70]; rcvg. triode, fil. 1.25 V @ 200 mA (MU) (8-14-53).

6376, A Fission counter, uranium oxide type, 300 V, sensitivity 0.7 count/neutron/cm^2/sec., aluminum body 2-3/32" dia. by 11-7/8" long, HN coaxial connector (W) (8-22-55). "A" is ion-chamber version rated for 800 V and high-temp. use. Also see 6971 and 7186-7188.

6377 Neutron counter, boron ion-chamber type, 300-800

V, magnesium alloy body 3-3/16" dia. by 24" long, three HN coaxial connectors (W) (8-22-55). Also see 6937, 6941, 7353.

6378 [AKA BL-62]; TR tube, min. 1B24A, 8.5-9.6 GHz, 100 kW, less gas reservoir; does not protrude beyond UG-39/U and UG-40/U WG flanges (BL) (1-17-55).

6379 Xmtg. triode, apparently an audio variant of 5936, fil. 11 V @ 285 A, μ 4.8, 70 kW (F, W) (3-28-55).

6381 Storage tube (8-27-62).

6383 Xmtg. triode, glass ring-seal construction, heater 6.3 V @ 3.4 A, μ 27, 600 W, 2 GHz, liquid-cooled via integral cooler (RCA) (6-28-54).

6384 [AKA TE-27]; premium 6AR6, hard-glass bulb, uprated to 30 W (BX, GE, HK) (4-9-56) <6BQ>. Also see 6098.

6385 [AKA TE-21]; min. duotriode, 5670 w/ shield added, heater 500 mA (BX, GE, HY) (2-7-55) <8CJ>.

6386 Min. duotriode, ruggedized computer-rated 2C51, heater 350 mA (GE, RCA, SY) (10-2-53) <8CJ>.

6388 [AKA (WE) 443A]; min. 9-pin cold-cathode volt. reg. and control tube, w/ starter electrode; 180 V peak; regulates 69 V @ 10 mA (HY, TS, WE) (5-14-54).

6389 Klystron (9-20-54).

6390 Reflex klystron, 6.87-10.75 GHz, IC, WG, heater 6.3 V @ 580 mA, 1.25 kV, 80 mW out (RA) (12-12-55).

6391 [AKA EF74]; submin. rcvg. pentode, heater 6.3 V @ 200 mA (MU) (9-11-53). Variant: 7434.

6392 Voltage regulator, 18-kV variant of 6352 (VN).

6393 [AKA BL-68]; ATR tube, 1B35A w/ special flange (BL) (1-17-55).

6394, A Duotriode, 6336A w/ heater 26.5 V @ 1.2 A (SY, TS) (10-29-56) <8BD>. "A" draws heater curr. of 1.3 A.

6396 [AKA BL650]; ATR tube, tunable 6393, in half-height WG 0.9" x 0.2" inside (BL) (12-27-54).

6397 T3 submin. power pentode, fil. 2.5 VCT @ 125 mA, 1.5 W (RA, TF) (3-19-56) <6CL>. Two in BX MRT-9 VHF pack VHF xcvr.

6398 Xmtg. triode, fil. 15.5 V @ 420 A, μ 21, 225 kW, 22 MHz, water-cooled (F) (4-10-61).

6399 Xmtg. triode, fil. 11 V @ 29 A, μ 25, 6 kW, 30 MHz, water-cooled (F) (5-7-54).

6400 Xmtg. triode, fil. 13 V @ 36 A, μ 25, 6 kW, 30 MHz, water-cooled (F) (5-7-54).

6401 Xmtg. pulse triode, fil. 4.25 V @ 400 A, μ 12, 1 kW, 600 MHz, water-cooled (F) (11-15-54).

6402 Magnetron, tunes 3.43-3.57 GHz, IM, WG, 8.3 V @ 32 A, 57 kV @ 55 A, 700 kW pulse output (RA).

6403 Submin. cold-cathode trigger tube, 1 W (SY).

6405 [AKA 1640]; gas phototube, S-1 response, low-microphonic, side-on T8 bulb on 4-pin base, 3-11/16" seated height, UX base, 135 μA/lumen (RCA) (5-14-54).

6406, A Magnetron, fixed at 2.85-2.91 GHz, IM, WG, 8.3 V @ 85 A, 56 kV @ 95 A, 2 MW pulse output (RA) (5-31-55). In AN/APS-20B, giving 2 MW at 2880 MHz.

6410, A [AKA QK338]; magnetron, fixed at 2805 ± 55 MHz, IM, WG, 8.3 V @ 85 A, 76 kV @ 135 A, 5 MW pulse output, water-cooled (GE, RA, TSU) (1-30-56).

6411 Storage tube (2-5-54).

6414 Min. 9-pin med.-μ sharp-cutoff duotriode, heater 12.6 V @ 225 mA, computer-rated (GE) (7-2-56) <9A>.

6416 "Misc." type (2-12-54).

6417 Beam power tube, 5763 w/ heater 12.6 V @ 375 mA (HY, RCA, W) (4-9-54).

6418 [AKA CK548DX]; T2 flat-press submin. pentode, fil. 1.25 V @ 10 mA (GE, RA, SY, TS) (7-1-57) <6418>.

6419 [AKA CK549DX]; pentode, 6418 w/ fil. 1.25 V @ 10 mA (GE, RA) (7-1-57) <6418>.

6420-26 Xmtg. triodes, glass coaxial ring-seal construction, T-T fils., rated to 30 MHz (all 11-9-56 exc. 6427).

6420 Fil. 7.0 V @ 85 A, μ 20, 12.5 kW, water-cooled via external cooler (ML, W). Variants: 6421, 6422.

6421, F Forced-air cooled 6420, derated to 10 kW (GE, ML).

6422 Fil. 7.0 V @ 85 A, μ 90, 20 kW, water-cooled (EI, ML, W).

6423, F Forced-air cooled 6422, derated to 12.5 kW (EI, GE, ML).

6424 Fil. 7.0 V @ 120 A, μ 20, 20 kW, water-cooled (EI, GE, ML).

6425, F Transverse-forced-air cooled variant of 6424, derated to 12.5 kW (EI, GE, ML).

6426 Fil. 8.0 V @ 200 A, μ 20, 20 kW, water-cooled via bayonet-mount water jacket (EI, ML, W).

6427 Forced-air cooled variant of 6426, aluminum radiator, derated to 12.5 kW (GE, ML) (2-11-57).

6435 Reference cavity (3-19-56).

6436 [AKA CK1036]; rectifier (9-12-55).

6437 [AKA CK1037]; flat submin. voltage regulator, 700 V @ 5-100 μ A (RA) (9-12-55).

6438 [AKA CK1039]; flat submin. voltage regulator, 1200 V @ 5-100 μA (RA) (9-12-55).

6440-41 "Misc." types (2-12-54).

6442, AL Planar cermet high-μ triode, for G-G use, heater 6.3 V @ 900 mA, 7.5 W, gives 2 kW pulses @ 3.5 GHz, oscillates to 5 GHz (GE) (2-28-55) <6442>.

6443 [AKA E484, M8091]; min. 9-pin diode w/ cap, heater 6.3 V @ 1.1 A, 1.8 kV PIV, 150 mA (HY, MU) (3-5-54) <9CZ>.

6444 [AKA ESM-48]; CW magnetron, 1.0 W, in 8.8-10.0 GHz area (MA). Can be pulsed w/ 450 V @ 60 mA, 5 W output. Variants are: F1, 9.8-10.0 GHz; F2, 9.6-9.8 GHz; F3, 9.4-9.6 GHz; F4, 9.2-9.4 GHz; F5, 9.0-9.2 GHz; F6, 8.8-9.0 GHz.

6445 Xmtg. triode, radiator version of 6333, derated to 5 kW (AX) (12-12-55).

6446 RF-heating triode, ruggedized, heavy-wall version of 892, fil. leads via crown, grid on ring, fil. 22 V @ 60 A, 20 kW, 7/16" thick anode wall for heat storage in RF-heating use (AX) (12-12-55). Variant: 6447.

6447 RF-heating triode, radiator version of 6446, derated to 10 kW (AX) (12-12-55).

6448 Xmtg. cermet beam tetrode, dual T-T fils., 1.35 V @ 1000 A each, 26 kW, 1 GHz, water-cooled via integral cooler and w/ water-cooled electrodes, in RCA TTU-12A UHF TV xmtr (RCA) (3-19-54).

6452 Reference cavity, 6301 for 9350 MHz (4-16-54).

6455 [AKA BL-61]; ATR tube, 5640 MHz, 300 kW, contact-mount (BL) (5-20-57).

6462 "Misc." type (5-9-55).

6463 [AKA ECC86]; rcvg. med.-μ "10,000-hr." computer duotriode, heater 12.6 VCT @ 300 mA, 4.4 W (AX, GE, MU, SY, TF) (7-23-54) <9CZ>.

6465 Klystron (9-27-54).

6466 Klystron (11-8-54).

6467, A Photomultiplier, 10-stage, S-11 response, head-on T10 bulb 4" high, diheptal base (DuM) (3-19-54).

6468-70 Reflex klystrons, IC, WG, heater 6.3 V @ 800 mA, 750 V, 1.5 W output (all SY) (all 9-20-54).

6468 [AKA K526]; for 6125-6425 MHz.

6469 [AKA K470]; for 6575-6875 MHz.

6470 [AKA K524]; for 7125-7425 MHz.

6472 Photomultiplier, variant of 6328, for car headlight dimmers, 11 wire leads for PC-board use, for use w/ AC or DC on electrodes (RCA) (3-7-55).

6474 [AKA 1854, M-561]; image orthicon, 5655 format, for color cameras (RCA) (4-30-54).

6476, A Beam-switching decade counter, all 10 cathodes pinned-out, counts at 4 kHz, puts out 25 V across cathode resistor (RA, SY) (9-20-54). Also see 6909. Variant: 6482.

6478 Xenon triode thyratron, hard-glass tubular bulb, flying plate lead out top, 4-pin base, fil. 2.5 V @ 7.5 A, 2.2 kV PIV, 1.5 A (HY, T) (9-30-63).

6481 [AKA RT434, 5767, q. v.]; planar triode (LI, SY) (11-8-54).

6482 Beam-switching tube, 6476 w/ single output cathode; puts out 15 V across cathode resistor (SY) (9-20-54).

6483 T3 submin. cold-cathode tetrode, 450 V (SY) (9-20-54) <6483>.

6484 [AKA 604, 4B24]; fullwave rectifier, jumbo base w/ cap, fil. 5 V @ 10 A, 15 kV PIV, 1.5 A (HY), like uprated vac. 8008.

6485 Min. pentode, computer-rated 6AH6 (GE, RCA, TS) (9-7-54) <7BK>.

6486, A [AKA TE-43]; min. pentode, premium 6AS6, hard-glass bulb, heater raised to 250 mA (BX, GE, SY) (4-9-56) <9DV>.

6487 [AKA EF70, M8125]; submin. pentode, heater 6.3 V @ 200 mA (MU) (5-14-54).

6488 [AKA EF73, M8122]; submin. rcvg. pentode, heater 6.3 V @ 200 mA (MU) (5-14-54). Variant: 7433.

6489 [AKA EA76]; rectifier, heater 6.3 V @ 150 mA (MU) (5-14-54).

6493 TWA, 2.0-4.0 GHz, EM, CO, heater 6.3 V @ 800 mA, 10 mW output, 35 dB gain (SY) (11-29-54). Input and output via BNC conns. on flying cables; octal plug on cable for power.

6494 TWA "dispersive amplifier," 2.0-4.0 GHz, 15-27 dB gain (SY).

6495 TWA, 2.0-4.0 GHz, 30 dB gain, 7 mW output, NF 11 dB (SY).

6496 BWO, voltage-tuned 2.0-4.0 GHz, EM, CO (BNC connector), heater 6.3 V @ 1.8 A, 2 kV, 200 mW out (SY) (12-27-54).

6498 Storage tube (3-7-55). Hughes 5" "memotron" storage CRT in 104D "memoscope."

6499 "Radechon" barrier-grid charge-storage tube for digital or analog signals, 3.4" dia. by 11-27/32" long, "twentyninar" base at reading end, diheptal base at writing end, heater 6.3 V @ 600 mA, 1.5 kV (RCA) (3-26-56).

6501 Dual TR tube, 8.5-9.6 GHz, 250 kW, large "X" WG, dual keep-alive electrodes (BL).

6503 Rcvg. "rocket" planar triode, 2C36 format, 25 mW out to 5750 MHz (LI, SY) (2-14-55).

6504 Ignitron, steel tank 9" dia. by 30-¼" long, 11 lb. each, for rectifier use in locomotives, three ignitors, coaxial current path, 4 kV PIV, 300 A; 12 tubes will supply 4000 hp. Has anti-splash baffles in the mercury pool (GE) (7-23-54). In "Virginian" 3300-hp locomotive, allowing use of ordinary diesel-electric traction motors from 11-kV 25-Hz supply. Also see 6509.

6507 Magnetron, 9-GHz, IM, WG (AX).

6508 Mercury rectifier, fil. 5.0 V @ 12.5 A, 21 kV PIV, 2.5 A (AX) (12-12-55).

6509 Ignitron, auxiliary rectifier in locomotives using 6504 or 6878, coaxial current path, 2.1 kV PIV, 150 A; built like 5555 but with splash baffles (GE) (7-23-54).

6510 Magnetron, fixed at 9375 ± 30 MHz, IM, WG, 12.6 V @ 2.3 A, 15 kV @ 15 A, 65 kW pulse output (LI). Presumably frequency-stable: promoted for use in MTI radars.

6511 Ignitron, 5822A w/ permanently adjusted thermostat built into the tube (12-6-54).

6512-15 Ignitrons, modifications of earlier types w/ integral overtemperature switch and thermostat to control water flow (all GE).

6512 Variant of 5554 (12-6-54).

6513 Variant of 5555 (12-6-54).

6514 Variant of 5788 (4-25-55).

6515 Variant of 5564 (12-6-54).

6516 [AKA M8082]; min. pentode, premium 6AM6 (MU, SY) (12-17-56) <6CH>.

6517 Magnetron, 1.25-1.35 GHz, 1.0 MW pulse output (11-6-61).

6518 Magnetron, "fixed at" 2.70-3.91 GHz, IM, CO, fil. 13.0 V @ 44 A, 48 kV @ 92 A, 1.5 MW pulse out (RA) (8-8-55).

6519 T1½X2 submin. power pentode, fil. 1.25 V @ 10 mA (GE) (3-19-56) <6418>.

6520 Duotriode, 6AS7G w/ additional voltage ratings (GE, SY, TS) (2-11-57) <8BD>.

6521, /R Magnetron, for weather and general radar use, fixed at 5400 ± 20 MHz, IM, WG, 10.0 V @ 3.2 A, 16 kV @ 16 A, 85 kW pulse output (RCA) (3-7-55). Designed for RCA AVQ-10, 1956 airline weather radar, operated at 16 kW; later used in AVQ-30X 1970 model w/ pulser, TR switch, and rcvr. modernized to solid-state. Sim. physically to 4J50.

6522 Thyratron.

6524 Xmtg. twin beam tetrode, like shorter variant of 5894, heater 6.3 V @ 1.25 A, derated to 20 W total and 100 MHz (RCA) (8-16-54). Final amp. in RCA E-line 450-MHz xmtr., 1961. Variant: 6850.

6525 Min. gas tetrode thyratron, for computer use, heater 6.3 V @ 150 mA, 500 V (GE) (3-26-56) <7BN>.

6526 T2X3 submin. beam power tube, fil. 1.25 V @ 125 mA (GE) (4-8-57) <6418>.

6527 Magnetron, fixed freq. (8-16-54).

6528 Duotriode, 6336A w/ higher g_m (37 mS) and μ (9) (GE, RA, SY, TS) (9-16-57) <8BD>.

6532 Rectifier (8-16-54).

6533 T3 submin. triode, nonmicrophonic 6247 for missile use, μ 53, g_m 1.75 mS (GE, RA, TS) (3-19-56) <8FY>.

6535 [AKA ECC91, M8081, T2M05]; duotriode, premium 6J6 (6-13-55).

6536 "Misc." type (6-13-55).

6538-39 "Misc." types (all 10-11-54).

6540 [AKA CK605CX]; T3 submin. sharp-cutoff pentode, variant of 5702, voltage ratings slightly different, g_m 5 mS, 1.1 W (GE, RA) (10-1-62) <5702>.

6541 Reflex klystron, 23.3-24.8 GHz, WG, heater 6.3 V @ 510 mA, 330 V, 10 mW output (BX) (4-9-56).

6542 [AKA CK1045]; T3 flat-press submin. VR tube, 150 V, 5-25 mA (RA, TS, VN) (3-19-56).

6543, A Magnetron, tunes 8.5-9.6 GHz, IM, WG, 12.6 V @ 2.3 A, 15 kV @ 15 A, 80 kW pulse output (LI) (10-15-62). Preumably frequency-stable: promoted for use in MTI radars.

6544 Xmtg. glass external-anode shielded-grid beam triode, for pulser use, fil. 6.0 V @ 60 A, μ 90, 25 kV @ 90 A, 1 kW, forced-air cooled (EI, ML) (12-28-59).

6545 TR tube, tunes 33.8-35.9 GHz, metal gas reservoir, ceramic window, 20 kW min. 100 kW max. (SY) (11-29-54).

6546 ATR tube, for 34.86 GHz; see 6544 for details (SY) (11-29-54).

6549, W Xmtg. beam pentode, PL-177A (q. v.) w/ conventional (non-vane) suppressor (PL) (12-27-54).

6550, A, B, C, V1 Beam power tube, ST-16 or long T16 button-stem bulb on wide octal Micanol base, heater 6.3 V @ 1.6 A, 35 W (RCA) or 42 W (TS) (GE, RCA, SV, SY, TS) (10-18-54) <7AC>. Offers carbonized, gold-plated grids; triple getter; and computer-rating. Screen has two dissipation ratings: 6 W steady-state and 10 W on speech or music peaks. "V1" is RCA matched pair. "A" has heater raised to 1.8 A. "B" and "C" are SV offerings. Twenty (!) in Audio Research Corp. D250 Mk. II 250-W stereo amp. 6550 is usually considered an audio tube, but LPB Inc. 50D 50-W AM broadcast xmtr. used two in parallel as final amp., modulated by two more.

6551 Magnetron, tunes 23.8-24.3 GHz, IM, WG, 5.0 V @ 3.2 A, 16 kV @ 18 A, 34 kW pulse output (BL, RA) (6-20-55). In E-23 radar.

6558 TWA, 1.7-4.0 GHz, EM, CO, heater 6.3 V @ 2.5 A, 1.55 kV, 1 W output, 35 dB gain, 30 db NF.

6559 TWA, 2.0-4.0 GHz, EM, CO (BNC conns.), heater 6.3 V @ 1.3 A, 1.0 kV, 1 W output, 37 dB gain, 2.0 W output (SY) (11-29-54).

6560 [AKA BL-35]; dual bandpass TR tube, 16 GHz, 100 kW (BL, GE) (4-22-57). In AN/MPQ-4A.

6562 [AKA 5794A]; pencil-triode oscillator, 5794 with leads for heater in place of pins (RCA, TS) (9-12-55).

6564 [AKA BL71]; dual TR tube, 8.5-9.6 GHz, 250 kW, four-element w/ dual keep-alive electrodes, for large "X" WG (BL) (1-31-55).

6565 TR tube (1-31-55).

6567 Xmtg. triode, for radar modulators and RF use, fil. 6.0 V @ 70 A, μ 33, 15 kW, 50 MHz, water-cooled (W) (5-25-59).

6568 TR tube (3-18-57).

6569 VHF power triode, 4-250A format, same fil., μ 45, 250 W, 60 MHz, for G-G and/or linear use (EI, PL) (3-28-55).

6570 Phototube, low-microphonics variant of 6405 w/ Micanol base, uprated to 500 V; sensitivity 30 μA/lumen (RCA) (3-7-55).

6571 Storage tube, for computer use, 3KP1-style CRT bulb 11-½" long, duodecal base w/ anode cap, heater 6.3 V @ 600 mA, 2.5 kV (RCA) (3-21-55).

6573 Reflex klystron, 15.5-17.0 GHz, 25 mW out (11-6-61).

6574 [AKA EN32]; thyratron rectifier, heater 6.3 V @ 950 mA (12-20-54).

6576 Xmtg. triode, fil. 7.5 V @ 170 A, μ 5.5, 22.5 kW, 25 MHz, water-cooled (ML) (11-9-56).

6577 Storage tube (12-20-54).

6578 "Misc." type (4-4-55).

6579 "Misc." type (4-25-55).

6580 Xmtg. triode, 400-W analog of 6569, w/ finned plate, other ratings same (EI, PL) (6-25-62).

6582, A [AKA TE-35]; min. 9-pin sharp-cutoff RF pentode, hard-glass bulb, heater 6.3 V @ 250 mA (BX, GE, SY) (2-11-63) <9EJ>.

6584 Reflex klystron, 5.1-5.9 GHz, CO, heater 6.3 V @ 500 mA, 330 V, 100 mW output (BX) (4-9-56).

6586 Reflex klystron, 7050-8125 MHz, IC, CO, heater 6.3 V @ 440 mA, 300 V, 140 mW output (RA).

6587 [AKA PL-174]; hydrogen thyratron, reformatted 5C22, glass ring-disc construction, external anode, grid on ring and base pin, w/ H_2 reservoir, fil. 6.3 V @ 11.6 A, 16 kV @ 325 A, 225 mA avg., 3.25 MW output (GE, KL, PL) (4-16-56).

6588 "Misc." type (3-4-57).

6589 Magnetron, tunes 3.35-3.50 GHz, IM, WG, 16.0 V @ 3.4 A, 30 kV @ 50 A, 500 kW pulse out (AX, PH).

6590 Thyratron.

6591 ATR tube, for 5400 MHz (SY) (5-2-55).

6592-95 TR tubes (all 3-18-57).

6596 TR tube (3-4-57).

6597 TR tube (3-18-57).

6599 TR tube (4-15-57).

6600 Reference cavity (4-15-57).

6601 TR tube (4-15-57).

6602-04 TR tubes (4-22-57).

6605 "Misc." type (4-15-57).

6611 [AKA DF61]; T2X3 submin. pentode, fil. 1.25 V @ 20 mA, g_m 1.0 mS (GE, MU, SY, TS) {QF394} (3-19-56) <6418>.

6612 T2X3 submin. pentode, fil. 1.25 V @ 80 mA, g_m 3.0 mS, 200 mW (GE, SY) (3-19-56) <6418>.

6615 TR tube (5-31-55).

6616 TR tube (6-27-55).

6617 [AKA TY12-25W]; RF-heating HF triode, fil. 8.0 V @ 98 A, 20 kW, water-cooled (AX, MU).

6618 [AKA TY12-25A]; xmtg. HF triode, appears to be a forced-air-cooled verson of 6617 derated to 15 kW, promoted for xmtr use (AX, MU).

6619 Gas-discharge tube: a 4-3/8" long glass tube sealed-off with gas; no electrodes, for use in ATR and pre-TR switches in high-power radars. Employed in groups of 10-20 in polarization-twist duplexers. Fires at 250 kW, 2.5 MW, 2.7-2.9 GHz; recovers in 25 µs (GE) (5-16-55).

6620 Gas-discharge tube, variant of 6619, 2-7/16" long, for use in pairs in 3.4-3.6 GHz branching-type duplexers; recovers in 40 µs; 1.5 MW (GE) (5-16-55).

6621 Gas-discharge tube, variant of 6619, 4-3/8" long, same use and freq.; recovers in 50 µs; 5 MW (GE) (5-16-55).

6623 Xmtg. triode, for audio/sonar use, fil. 6.0 V @ 60 A, µ 22, 2.5 kW, 30 MHz, forced-air cooled (ML, W) (11-14-55). Also see 5736.

6624 TR tube, broadband, 5370-5430 MHz (SY) (5-2-55).

6625 Amplifier klystron, three-cavity, electrostatically focused, 960-1215 MHz; CO (7/8" rigid line), forced-air cooled, heater 5.0 V @ 45 A, 20 kV @ 9.4 A, 22 kW pulse output (GE) (4-18-55).

6626 [AKA 0A2WA, q. v.]; voltage regulator, ruggedized/high-stability 0A2 (SY, RCA) (3-14-55) <5BO>.

6627 [AKA 0B2WA, q. v.]; voltage regulator, ruggedized/high-stability 0B2 (SY, TS) (3-14-55) <5BO>.

6628 Numeric readout.

6629 ATR tube (4-22-57).

6630-31 ATR tubes (all 4-15-57).

6632 TR tube (4-29-57).

6633 [AKA BL-37A]; TR tube, 1.2 GHz, 2 MW (BL) (4-15-57).

6634 [AKA BL90]; dual bandpass TR tube, 1215-1355 MHz, 2 MW, dual keep-alive electrodes (BL).

6635 TR tube (4-29-57).

6636 [AKA BL87]; dual TR tube, 2.6-3.0 GHz, 750 kW, dual keep-alive electrodes (BL) (4-29-57).

6637, 39 TR tubes (both 4-29-57).

6640 [AKA BL60]; TR tube, dual 5865 (BL) (5-6-57).

6641 TR tube (6-17-57).

6642 [AKA BL600]; dual TR tube, 6334 w/ large flange input and small flange out (BL) (5-6-57).

6643 [AKA BL84]; dual TR tube, 6334 w/ fast (1-µsec.) recovery time (BL).

6644 TR tube (4-29-57).

6645-48 Dual TR tubes, 8.6-9.6 GHz (all BL) (all 5-6-57)

6645 [AKA BL-95H]; 100 kW, w/ built-in heater.

6646 [AKA BL604]; dual 1B63A, 2-µsec. recovery time.

6647 [AKA BL604H]; 6646 w/ heater.

6648 [AKA BL615]; 6334 w/ saddle-type flange.

6649 TR tube (6-3-57).

6650 TR tube (5-20-57).

6651 TWA, 2.1-3.5 GHz, EM, CO, heater 6.3 V @ 2.1 A, 8.2 kV, 1 W output, 35 dB gain.

6655, A Photomultiplier, 6217 w/ flat face and S-11 response, for scintillation counters (RCA) (6-27-55). Variant: 2062.

6658 TWA, 1.7-4.0 GHz, heater 6.3 V @ 2.5 A, 2.0 W (10-31-55).

6659 [AKA CK1042]; rectifier (3-19-56).

6660-81 Series of tubes mobile-rated for heater supply of 6.3 ± 1.3 V (all GE, RCA, SY, TS). Also see 7054-7061.

6660 Equals 6BA6, ½" taller (5-9-55).

6661 Equals 6BH6 (5-9-55).

6662 Equals 6BJ6 (5-9-55).

6663 Equals 6AL5 (5-9-55).

6664 Equals 6AB4 (10-24-60).

(6668, see below.)

6669 Equals 6AQ5A, gives 4.1 W out w/ 5.0 V on heater (12-27-55).

(6670-71, see below.)

6676 Equals 6CB6A (10-24-60).

6677 Equals 6CL6 (12-27-55). Also see 6297.

6678 Equals 6U8A (8-6-56).

6679 Equals 12AT7 (11-14-55).

6680 Equals 12AU7 (11-14-55).

6681 Equals 12AX7 (11-14-55).

6668 TWA, 1.7-4.0 GHz, 10 W.

6670 Premium 12AU7.

6671 Premium 12AT7.

6682, A Min. 9-pin sharp-cutoff RF pentode, heater 6.3 V @ 250 mA (GE).

6684 Gas noise source (6-13-55).

6685 [AKA BL-616]; dual TR tube, 34.5 GHz, 20 kW (BL) (5-20-57).

6686 [AKA E81L]; rcvg. "10,000-hr." power pentode, heater 6.3 V @ 375 mA, 4.5 W (AX, HY, MU) (7-16-56).

6687 [AKA E91H]; rcvg. "10,000-hr." dual-control heptode for computer ckts., etc; repl. 5915A, heater 6.3 V @ 270 mA, 1.0 W (AX, MU) (4-16-56).

6688, A [AKA E180F]; 9-pin min. broadband pentode, ruggedized "10,000-hr." frame-grid type, heater 6.3 V @ 300 mA, 3.0 W, g_m 16.5 mS, promoted to repl.

5847 in new designs (AX, GE, MU, TS) (7-23-56) <9EQ>. Variant: 7737.

6689 [AKA E83F]; rcvg. "10,000-hr." wideband pentode, heater 6.3 V @ 300 mA, 2.1 W (AX, MU) (7-16-56).

6690 T3 submin. med.-μ video duotriode, heater 6.3 V @ 300 mA, μ 35, g_m 4.8 mS (GE, SY) (4-18-55) <8GQ>.

6691 Xmtg. triode, fil. 5.0 V @ 160 A, μ 37, 15 kW, 30 MHz, forced-air cooled (F) (3-28-55).

6692 Xmtg. triode, water-cooled variant of 6691, uprated to 30 kW (F) (11-14-60).

6693 [AKA RG4-3000]; mercury rectifier, stepped-tubulatr bulb w/ cap, 4-pin "industrial" base, rated between 575A and 869B, fil. 5.0 V @ 11.5 A, 15 kV PIV @ 3 A up to 35 °C, or 2.5 kV PIV @ 5 A up to 55 °C (AX, MU) (10-28-57).

6694, A Cadmium-sulfide photocell, S-12 response, rectangular body 0.35" wide, 0.2" deep, 0.3" high, two pins, 3 μA/fc @ 90 V (RCA) (5-9-55).

6695 [AKA QK253]; magnetron, 3.43-3.57 GHz, IM, CO, 16.0 V @ 3.1 A, 31 kV @ 60 A, 650 kW output (RA).

6696, A Xmtg. triode, variant of 6697A water-cooled via external cooler, uprated to 60 kW (EI, ML) (11-19-56). "A" has cermet construction.

6697, A RF-heating/xmtg. cermet medium-μ HF triode, fil. 13 V @ 205 A, grid 750 W, 35 kW, 30 MHz, anode radiator formed of stacked discs, for RF heating and broadcast xmtrs (EI, ML) (11-19-56). Two in 400-kW 100-kHz pulse source for C-Stellarator controlled-fusion system, 1961. Variants: 6696, 7480.

6698 TWA, 2.0-4.0 GHz, EM, CO, heater 6.3 V @ 2.2 A, 10 kV, 27 dB gain, 1 kW peak output (SY) (4-10-61).

6699 BWO, voltage-tuned 1.0-2.0 GHz, EM, CO, heater 6.3 V @ 3 A, 300 V, 10 mW out, CO (SY) (4-10-61).

6700 Min. beam-switching decade counter, 21-pin base, counts at 2 MHz (BR) (6-20-55).

6701 Decade counter, lower-voltage variant of 6700, counts at 1 MHz (BR) (2-20-56).

6702 Gas noise source (2-17-64).

6703-04 Beam-switching tubes (5-29-61).

6706 [AKA SFD-345]; magnetron, 30 kW at 16-17 GHz. In AN/APQ-118.

6710 Beam-switching tube (8-28-61).

6711-12 Beam-switching tubes (both 5-27-63).

6713 Gas noise source (2-17-64).

6714 [AKA BX4001]; "Beam-X" beam-switching counter (12-16-63).

6752 TWA, 1.0-2.0 GHz, EM, CO, heater 6.3 V @ 1.3 A, 2.0 kV, 2 W output, 37 dB gain (SY) (4-10-61).

6753 TWA, 1.0-2.0 GHz, EM, CO, heater 6.3 V @ 800 mA, 400 V, 10 mW output (SY) (4-10-61).

6754 [AKA (WE) 412A, TE-36]; min. 9-pin full-wave rectifier, heater 6.3 V @ 1 A, 1250 V PIV, 100 mA (BX, WE) (4-9-56) <9ET>. BX version is in hard glass.

6756 RF-heating HF triode, fil. 7.5 V @ 100 A, 20 kW, water-cooled (AX) (1-9-56).

6757 Xmtg. triode, glass ring-seal radiator style w/ strap flying leads, ca. 5 kW (AX) (1-9-56).

6758 RF-heating triode, fil. 12.6 V @ 33 A, 6 kW, water-cooled (AX) (11-25-63).

6759 RF-heating triode, fil. 12.6 V @ 33 A, 6 kW, forced-air-cooled (AX) (11-25-63).

6760 Min. 9-pin power pentode, heater 18.0 V @ 350 mA, for telephone-repeater use (HY) (6-13-55).

6761 Power pentode, 6760 w/ heater 6.3 V @ 1.0 A (HY) (6-13-55).

6762 "Misc." type (3-18-57).

6763 Min. cold-cathode rectifier, 2.8 kV, 100 mA (GE) (2-4-63) <6763>.

6771 Planar cermet high-μ triode, for G-G use, heater 6.3 V @ 575 mA, μ 90, 6.25 W, oscillates to 5 GHz (GE) (2-16-59).

6775 [AKA 4-400C]; xmtg. tetrode, ruggedized/long-life 8438, smaller plate structure, lower-temp. fil., one-piece plate cap, etc. (EI, PL) (9-3-62). Two in Wilkinson AM-1000B AM xmtr, modulated by two more.

6777 [AKA KU14]; hydrogen thyratron, fil. 6.3 V @ 2.7 A, 8 kV @ 35 A, 45 mA avg. (KL, RCA) (4-16-56).

6778 [AKA EC70, 6K4, 5718, q. v.]; submin. triode (MU) (6-13-55).

6779 [AKA Z803U]; min. cold-cathode trigger tube (AX, MU) (6-13-55).

6780 [AKA BL800]; reflex klystron, 8.5-10.0 GHz, locknut tuner, IC, UG-39/U WG, flying cable for voltages, heater 6.3 V @ 1.2 A, 350 V, 25 mW out (BL) (1-14-63).

6781 [AKA BL-803]; reflex klystron, 8.5-10.0 GHz, IC, WG, heater 6.3 V @ 1.2 A, 1 kV, 35 mW output (BL, MET, VA, W) (1-14-63).

6782 Gas noise source (4-29-57).

6783 Triode gas-mercury thyratron, fil. 2.5 V @ 8.5 A, 2.5 kV PIV, 1.6 A (W) (4-15-57).

6784 TWA, 3.1-3.5 GHz, IM, CO, heater 6.3 V @ 370 mA, 1.5 kV, 2 mW out, 18 dB gain, 6 dB NF (2-23-59)

6786 [AKA XGS-10]; mercury triode thyratron rectifier, fil. 5.0 V @ 15-20 A, 15 kV, 15 A (AX, MU) (12-9-57).

6787 Magnetron, for RF heating, fixed at 915 MHz, cermet construction, CO, EM, heater 12.5 V @ 61 A, 4.1 kV @ 1.5 A, 2.5 kW out, water-cooled (GE) (1-23-56).

6788 T3 submin. sharp-cutoff audio pentode, hard-glass "missile" type, heater 6.3 V @ 175 mA (GE, SY) (7-25-55) <8DL>.

6789 Voltage regulator (2-20-56).

6792 Beam tetrode, for shunt HV regulators and control ckts., tall tubular octal w. cap, heater 6.3 V @ 450 mA, 25 kV, 25 W (GE, HY) (11-7-55) <8GL>.

6793 [Type not listed] (2-4-57).

6794 Ionization gauge (8-29-55).

6795 TR tube, "bantam," for 8.5-9.6 GHz (SY) (11-28-55).

6796-97 TR tubes (11-28-55).

6799 Magnetron, fixed at 34,860 ± 348 MHz, IM, WG, heater 5.0 V @ 2.1 A, 20 kV @ 40 A, 120 kW pulse out (SY) (12-12-55).

6800 Xmtg. triode, fil. 7.5 V @ 100 A, μ 20, 20 kW, 22

MHz, water-cooled, for xmtrs and RF heating (AX, F) (2-13-56).

6801 Xmtg. triode, radiator version of 6800, derated to 10 kW (AX, F) (2-13-56).

6802 Beam-switching decade counter, 4 outputs, counts at 4 kHz (RA, SY) (11-28-55). also see 6910.

6803 Xmtg. triode, fil. 11.0 V @ 190 A (F) (1-3-56).

6804 Xmtg. triode, fil. 11.0 V @ 190 A (F) (2-13-56).

6805 TR tube (1-7-63).

6806 Xmtg. beam cermet tetrode, improved 6448, for UHF TV and general use, same fil., 35 kW, 1000 MHz, water-cooled via integral cooler (RCA) (3-26-56). In TTU-25B UHF TV xmtr.

6807 [AKA XR1-6400]; triode gas thyratron, reformatted into disc-seal construction, 4-pin industrial base for fil., grid on disc, plate on cap, fil. 2.5 V @ 21 A, 1.5 kV PIV, 6.4 A (GE, HY, MU, W) (10-10-55).

6808 Thyratron, 6807 w/ mounting strap and flying leads at filament end (GE) (5-7-56).

6809 Thyratron, 6807 w/ short rigid spade-tipped leads at filament end (GE) (10-10-55).

6810, A, B Photomultiplier, 14-stage, S-11 response, for scintillation counters, flat-face end-on bulb 2" dia. by 6.7" high, bidecal 20-pin Micanol base, w/ metal collar above base (RCA) (1-16-56). Designed to word directly into deflection plates of a CRT for low time delay.

6811 Reflex klystron, 2.5-5.0 GHz, 100 mW out (4-23-56).

6812 Reflex klystron, 2.5-5.0 GHz, 100 mW out (4-23-56).

6814 T3 submin. med.-μ triode, computer-rated, heater 6.3 V @ 150 mA, μ 29 (GE, RCA, SY) (11-21-55) <8DK>.

6816 Xmtg. UHF cermet beam tetrode, heater 6.3 V @ 2.1 A, 115 W, 1.2 GHz, usable to 2 GHz, forced-air-cooled, coaxial construction, aligned grids (AX, EI, RCA) (2-18-57). First RCA "cermolox" tube, dev. for use in xmtr. of AN/ARC-34. Variant: 6884.

6817-22 Magnetrons, fixed-freq., equiv. to 2J42 otherwise, 8 kW output. Freq. tolerance ± 50 MHz. Same freqs. as 7138-43 (all MA).

6817	For 9050 MHz.	6820	For 9350 MHz.
6818	For 9150 MHz.	6821	For 9450 MHz.
6819	For 9250 MHz.	6822	For 9550 MHz.

6823, 24 Magnetrons (AX).

6825 TWA, 2.0-4.0 GHz, EM, CO, heater 6.3 V @ 5 A, 9 kV, 2.0 W output (KL) (2-13-56).

6826, A TWA, 2.4-3.6 GHz, EM, CO, heater 6.3 V @ 5 A, 9 kV, 10 W output (750 W pulse), 30 dB gain (KL) (2-13-56). "A" has 40 dB gain.

6829 Duotriode, premium 12AV7 (GE) (3-26-56) <9A>.

6830 Min. VR tube, 0A2 w/ two flying leads (HY) (1-9-56).

6831 Min. VR tube, 0B2, w/ flying leads (HY) (1-9-56).

6832 T3 submin. med.-μ duotriode, for DC amps. and computer use, heater 6.3 V @ 400 mA, μ 26, g_m 1.0 ms, 100 mW (GE, RA, SY) (12-24-56) <8DG>.

6833 [AKA 1B85A, q. v.]; beta-gamma counter.

6835 Storage tube (2-13-56).

6836 Photomultiplier (2-6-61).

6839 Camera tube ("misc. type.") (5-28-62).

6840 Min. 9-pin med-μ duotriode, heater 12.6 VCT @ 400 mA, μ 20, 7.0 W total (GE, SY) (4-1-57) <9CZ>.

6841 [AKA QK400]; magnetron, fixed at 16,517.5 ± 107.5 MHz, IM, WG, 4.0 V @ 10 A, 19 kV @ 16 A, 50 kW output (RA) (11-6-61). In search radar in B-58s.

6842 Min. pentode HV regulator, shunt or series, 7 pins w/ cap, heater 6.3 V @ 150 mA, g_m 2.5 mS, 4 kV, 8 W (GE, HY, VN) (12-27-55) <7EQ>.

6843 Magnetron, 5450-5825 MHz, 250 kW output.

6844, A "Nixie" neon readout, digits 0-9, 170 V (BR) (1-7-57).

6845 Reflex klystron, 8.5-9.6 GHz, IC, CO, heater 6.3 V @ 520 mA, 350 V, 20 mW (BX, RA, TH) (11-20-61).

6846 Beam-switching tube (5-7-56).

6848 Xmtg. UHF cermet ring-seal tetrode, for G-G Class C amps, bombardment cathode, 6.7 V @ 14.5 A (13.5 A with bombardment), bombarding voltage 650 V, 2 kW, 800 MHz, forced-air cooled (GE) (3-27-61).

6849 Image orthicon, high sensitivity for observation of fluoroscopes, scenes under starlight, etc. (10 times as sensitive as 5820); blue-green sensitive, 5655 format, heater 6.3 V @ 600 mA, anode voltage 1350 (GE, RCA, W) (1-14-57).

6850 Xmtg. twin tetrode, 6524 with heater 12.6 V @ 625 mA (RCA) (4-9-56).

6851 [AKA TE-42]; duotriode, hard-glass ruggedized variant of 5751, heater 12.6 VCT @ 250 mA (BX, GE, SY) (3-18-57) <9A>.

6853 [AKA TE-45]; full-wave rectifier, hard-glass ruggedized 6106 (GE, SY) (2-11-63) <8HE>.

6854 [AKA TE-47]; duotriode, hard-glass ruggedized 6385, heater 6.3 V @ 500 mA. μ 35 (BX, GE, SY) (3-18-57) <8CJ>.

6855 [AKA NL716]; gas and mercury triode thyratron, T12 bulb, UX base, fil. 2.5 V @ 56.3 A, 500 V PIV, 1 A (NL, RCA) (4-2-56).

6856 [AKA 740, WT-210-0017]; triode gas-mercury thyratron, fil. 2.5 V @ 16 A, 1.5 kV PIV, 2.5 A (GE, W) (4-2-56) <6856>.

6857 [AKA 740P]; gas thyratron, fil. 2.5 V @ 16 A (GE) (4-2-56) <6807>.

6858 [AKA 760]; gas-mercury triode thyratron, bulb 2-9/16" dia. by 9-½" high w/ cap, super-jumbo 4-pin base, fil. 2.5 V @ 21 A, 1.5 kV PIV, 6.4 A (GE, RCA, W) (4-2-56) <6808>.

6859 [AKA 760P]; 6858 w/ mounting bracket (GE) (4-2-56).

6860 [AKA C6J/F]; triode gas thyratron, flying leads, fil. 2.5 V @ 21 A, 1250 V PIV, 6.4 A (EL, W) (4-2-56).

6861 TWA, low-noise amp., 2.7-3.5 GHz, EM, input/output via "N" fittings, 5.0 V @ 650 mA, 400 V, 250 mW out, 25 dB gain, 6 dB NF, 1 mW output (EEV, RCA) (6-4-56). Preamp for air-traffic-control radars. Said by RCA to be the "first commercial" TWT.

6864 [AKA ELC6J/KF]; thyratron, replaceable by 5878, fil. 2.5 V @ 21 A, 1 kV PIV, 6.4 A (EL) (6-18-62).

6865, A Magnetron, manually tunes 8.75-9.60 GHz, IM, WG, heater 13.8 V @ 3.2 A, 23 kV @ 28 A, 220 kW pulse output (RCA, W) (3-19-56). "Coupled-cavity" design: external variable cavity provides for tuning. Intended to replace 4J50.

6866 Storage tube, 4" direct-view type w/ 60-sec. persistence, aluminized, yellow phosphor, write and view guns, 5-1/8" dia. stepped to 3" dia., 15-½" long, "thirtyfivar" base and flying leads, heater 12.6 VCT @ 600 mA, 10 kV on viewing gun (RCA) (7-2-56).

6867 TWA, 8.0-9.6 GHz, EM, CO, heater 6.3 V @ 880 mA, 1.5 kV, 100 mW output, 30 dB gain (IT, RCA) (12-17-56).

6868 TWA, 1.7-4.0 GHz, EM, CO, heater 6.3 V @ 1.5 A, 1.75 kV, 10 W output, 40 dB gain, 30 dB NF (IT) (11-12-56).

6870 Min. 9-pin beam power tube, heater 12.6 VCT @ 300 mA (SY) (12-17-56) <9BF>.

6871 Thyratron, replaceable by 7299 (6-18-62).

6872 T3 flat-press submin. VHF pentode, heater 6.3 V @ 200 mA, 1.1 W (GE, SY, TS) (4-3-61) <5702>.

6873 "Misc." type (6-18-56).

6874 Magnetron, tunes 8.8-9.4 GHz, IM, WG, 3.8 V @ 3.4 A, 23 kV @ 30 A, 190 kW pulse out (SY) (2-11-57) <9GB>.

6877 [AKA TE-48]; 9-pin min. pass tube for voltage regulators, "½ 6080," hard-glass bulb, heater 6.3 V @ 800 mA, μ 3.75, 12 W (BX, GE, SY, TS) (3-18-57) <9GB>.

6878 Ignitron, double-ignitor, for rectifiers in locomotives, 13-1/8" dia. by 30-¼" long, 4 kV PIV, 675 A, includes anti-splash baffles in the mercury pool, coaxial current path (GE) (5-7-56). Also see 6509.

6879 [AKA CT-854]; beam-switching counter, 3 outputs, counts at 5 kHz (SY) (6-18-56).

6881 Gas noise source (10-28-57).

6882 Gas noise source (7-3-61).

6883 Xmtg. beam power tube, 6146 for 12-V use, heater
A, B 12.6 V @ 625 mA (AX, GE, RCA, SY, W) (8-27-56) <7CK>. "A" is mobile-rated (SY) (9-4-61). "B" [AKA 6883B/8032A/8552] has "dark" heater, 12.6 V @ 562 mA (GE, RCA) (3-16-64). In Clegg Thor 6 xcvr.

6884 Xmtg. beam tetrode, 6816 w/ heater 26.5 V @ 520 mA (AX, EI, RCA) (2-18-57).

6885-86 Xmtg. triodes (both 8-5-57).

6887 Rcvg. duodiode, computer-rated 6AL5, heater reduced to 200 mA (GE, RCA) (10-1-56) <6BT>.

6888 [AKA TE40]; dual-control pentode, for core drivers and gates, octal base, heater 6.3 V @ 800 mA, 8 W (GE, SY) (5-7-56) <8N>.

6889 [AKA TE-52]; beam power tube, ruggedized 6AR6, T11 hard-glass bulb w/ cap, octal base, heater 6.3 V @ 1.2 A, 30 W (BX, GE) (2-7-60) <8HG>.

6890 ATR tube (5-20-57).

6891 TWA, 8.5-9.6 GHz, IM, WG, 50 mW out (WE) (8-18-58).

6893 Xmtg. beam power tube, 2E26 w/ heater 12.6 V @ 400 mA (RCA, SY, W) (9-4-56) <7CK>.

6894 Mercury rectifier, T20 bulb w/ cap, jumbo 4-pin base, fil. 5 V @ 10 A, 20 kV PIV, 2.5 A (RCA) (2-25-57). Variant: 6895.

6895 Mercury rectifier, 6894 w/ super-jumbo base and different pinout (RCA) (2-25-57).

6896 [AKA 1855]; "graphechon" signal converter storage tube, electrostatic focus and 40° magnetic defl., for data-processing uses where signals are to to be translated from one time base to another, 2-1/32" dia. stepped to 1.1" dia. by 18-3/4" long, heater 12.6 VCT @ 600 mA, 13 kV (RCA) (2-24-58).

6897 Xmtg. triode, 2C39B w/ redesigned body and changes in minor parameters (e. g. μ 95 vs. 100 for 2C39A), for CW use (e. g. 20 W out at 1.9 GHz), "not generally interchangeable" (GE, RCA) (5-28-56).

6898 Rectifier, heater 5 V @ 8 A, 16 kV, 200 mA, forced-air-cooled (W) (3-25-57).

6900 [AKA TE-54]; duotriode, hard-glass ruggedized variant of 5687 (BX, GE, SY, TS) (3-18-57) <9H>.

6901 Gas-mercury thyratron (6-12-61).

6902 BWO, 17.5-27 GHz, EM, WG, 12.0 V @ 1.2 A, 2 kV, 100 mW (SY) (4-22-57).

6903 Photomultiplier, 10-stage, S-13 response, 5819 format lengthened to 5-5/8" high, flat face (RCA) (7-2-56).

6904 TR tube (6-17-57).

6905 [AKA BL-613]; TR tube, 5650 MHz, 5 kW (BL) (6-17-57).

6906 [AKA BL-643]; TR tube, as 6905 (BL) (5-20-57).

6907 Xmtg. dual tetrode, 5894 format, htr. 12.6 VCT @ 650 mA, 20 W, radiation-cooled, UHF mobile version of 832A, usable to 1 GHz (AX) (11-5-56).

6908 Rectifier, fil. 5.7 V @ 6.6 A, 150 kV PIV, 1 A peak, oil-cooled (ML) (8-1-60).

6909 Beam-switching decade counter, tubular bulb, 11-pin base w/ center pin, counts bidirectionally and displays, 4 outputs at "0," "5," "8," and "9," counts at 100 kHz (RA, SY) (11-12-56).

6910 Decade counter, 6909 w/ 10 outputs (RA) (11-12-56).

6911 Photomultiplier (3-11-57).

6912 Camera tube, "misc." type (9-17-56).

6913 Min. duotriode, prem. 12BH7 (GE) (8-13-56) <9A>.

6914, A Image converter, for IR-vision systems, S-1 response, glass body with metal end rings, 1.88" dia. by 2.9" high, photocathode at one end, viewing screen at other; supply voltage 16 kV (RCA) (3-30-59).

6915 Reflex klystron, 8.5-10.0 GHz, IC, WG, heater 6.3 V @ 1.2 A, 1 kV, 90 mW output (W).

6916 Counter tube (10-22-56).

6917 Magnetron, voltage-tunable, covers 2-4 GHz, cermet construction sim. to GE cermet triodes, 3/4" dia. by 5-1/8" high, EC, EM, ca. 50 W out (8-5-57).

6919 Min. duodiode, heater 6.3 V @ 200 mA, for gates and clamps in computers (GE, SY) (2-18-57) <6BT>.

6920 Xmtg. triode, for radar modulators, fil. 11 V, μ 41, forced-air cooled, 10 kW (F, W) (12-3-56).

6921 Xmtg. audio triode, fil. 5.0 V @ 260 A, μ 4.3, 30 kW, water-cooled (F) (9-24-62).

6922 [AKA E88CC]; "10,000-hr." frame-grid duotriode, premium 6DJ8, heater reduced to 300 mA (AX, GE, MU, RCA, SY) (5-26-58) <9AJ>. Variant: 7308.

6923 [AKA EA52]; disc-seal diode for UHF voltmeters, etc., 1 GHz, fil. 6.3 V @ 300 mA (AX, MU) (2-9-59).

6925 Xmtg. triode, fil. 13.0 V @ 36 A, μ 17, 3 kW, 50 MHz, forced-air cooled (F) (10-30-61).

6926, J Xmtg. triode, water-cooled variant of 6925, uprated to 6 kW (F) (10-30-61).

6927 Duotriode, premium 6J6, heater cut to 330 mA (10-15-56).

6928 Beam power tube, premium 6AQ5, heater reduced to 360 mA (10-15-56).

6929 Image converter, for IR vision, S-1 response, format like 6914 but only 1.35" dia. by 2.3" high; supply voltage 13 kV (RCA) (3-23-59). Variant: 7404.

6930 [AKA 635P]; rectifier, fil. 2.5 V @ 18 A (2-4-57).

6931 Voltage regulator (4-29-57).

6935 Photomultiplier (5-11-59).

6937, A Neutron counter, boron ion-chamber type, 300-800 V, aluminum body 3-1/16" dia. by 13-7/8" long, two HN coax connectors (W) (1-14-57). Also see 6377.

6938 Neutron counter, boron trifluoride, 2 kV, sensitivity 45 counts/neutron/cm^2/sec., aluminum body 7-15/16" dia. by 32-½" long, HN coax connector (W) (1-14-57). Also see 6307.

6939 [AKA QQV02-6]; min. xmtg. VHF frame-grid dual tetrode, internally neutralized, heater 12.6 VCT @ 300 mA, g_m 10.6 mS, 6 W total, 500 MHz (AX, MU, RA, RCA, SY) (3-16-59) <9HL>. Variants: 7645, 8577, 8595.

6940 Reflex klystron, 8.50-9.66 GHz, CO, heater 6.3 V @ 520 mA, 350 V, 20 mW output (BE, TH).

6941, A Neutron counter, uranium-oxide ion-chamber type, 300-800 V, aluminum body 2-3/32" dia. by 11-7/8" long, HN coax connector (W) (12-10-56). "A" is rated for high-temp. use. Also see 6377.

6942 Xmtg. UHF cermet tetrode, ring-seal construction, for TV transmitters and RF heaters, fil. 5.7 V @ 24 A, 1.5 kW, 900 MHz, sim. to 6848 but w/o bombardment cathode, forced-air cooled (GE) (1-7-57).

6943-48 Series of T3 submin. tubes, ruggedized for missile use, all w/ 6.3-V heaters (all GE, SY, TS) (all 5-27-57). Also see metal-cased analogs 8524-8530.

6943 Sharp-cutoff dual-control RF pentode, heater 175 mA, g_m 3.6 mS <8DC>.

6944 Semi-remote-cutoff dual-control pentode, heater 175 mA, g_m 3.2 mS <8DC>.

6945 Beam audio power tube, heater 350 mA, g_m 3.5 mS, 3.0 W <8DL>.

6946 Med.-μ triode, heater 175 mA, μ 16.5, g_m 3.8 mS <8DK>.

6947 Med.-μ duotriode, heater 350 mA, μ 35 <8DG>.

6948 High-μ duotriode, heater 350 mA, μ 70 <8DG>.

6949 Xmtg. cermet shielded-grid beam triode, T-T fil. 7.3 V @ 1040 A, μ 60, 400 kW, 75 MHz, water-cooled via integral cooler (RCA) {A2332C} (2-17-58). One in AN/FRT-32 300-kW VHF linear; two in RCA BTH-250A Ampliphase HF broadcast xmtr., the only one built, at border-blaster XERF, 1570 kHz, Ciudad Acuna, Mexico, ca. 1960; also widely used in particle accelerators like U. of Calif. Bevatron and HILAC.

6950 Xmtg. triode (9-26-60).

6952 Xmtg. cermet beam tetrode, for pulsed RF amp. use in long-range long-pulse search radars, etc.; coated fil. 1.0 V @ 495 A, 8 kW, 600 MHz, giving 2 MW peak @ 425 MHz, liquid-cooled via integral cooler. Claimed to have low fil. power for airborne use (RCA) (11-10-58). Variant: 8587.

6953 Phototube, S-1 response, 5819 format, 200 μA/-lumen, replaceable by CE-A30 (RCA) (1-7-57).

6954 Min. dual-control sharp-cutoff pentode, heater 6.3 V @ 300 mA (GE, SY) (2-25-57) <7CM>.

6955 Min. 9-pin med.-μ duotriode, heater 12.6 VCT @ 350 mA, μ 21 (GE) (1-14-57) <9A>.

6957, A Cadmium-sulfide photocell, S-15 response, short end-on bulb 1-11/16" high, octal base (RCA) (6-24-57).

6958 Ignitron, steel case, for rectifier use, dual ignitors, 4 kV PIV, 275 A, coaxial current path (GE) (11-11-57). Variant: 7042.

6959 Magnetron, 9.33-9.42 GHz, IM, WG, 6.0 V @ 30 A, 33 kV @ 67 A, 800 kW pulse output (RA) (11-6-61).

6960, A [AKA TY7-6000W]; RF-heating VHF triode, fil. 12.6 V @ 33 A, 6 kW, 30 MHz, glass ring-seal, water-cooled via integral cooler (AX, MU) (11-25-63).

6961 [AKA TY7-6000A]; xmtg. triode, forced-air-cooled version of 6960, same use (AX, MU) (6-9-58 - note discrepancy in registration date vice 6960).

6965, A Min. 9-pin duotriode, heater 12.6 V @ 225 mA (GE).

6967, A [AKA QK366]; magnetron, fixed at 9245 ± 40 MHz, 75 kW output (RA).

6968 Pentode, ruggedized, computer-rated 6AK5 (GE, SY, TS) (3-11-57) <7BD>.

6971, A Fission counter, uranium oxide type, 300 V, aluminum body 2-3/32" dia. by 11-7/8" long, HN coax connector (W) (3-25-57). "A" is rated for high-temp. use. Also see 6376.

6972 [AJA JP9-75]; magnetron, fixed at 9375 ± 30 MHz, IM, WG, 10.0 V @ 2.85 A, 20 kV @ 18 A, 75 kW pulse output (AX, MU, PH, RE, VO) (5-6-57).

6973 Min. beam power tube, 6CZ5 uprated to 12 W, w/ pinout changed slightly; heater 6.3 V @ 450 mA, intended for med.-power, high-quality hi-fi systems; has less power sensitivity than 6BQ5 but lower distortion (GE, RCA, SY, W) (8-12-57) <9EU>.

6974 [AKA K709]; reflex klystron, 4.6-5.0 GHz, IC, WG, heater 6.3 V @ 800 mA, 825 V, 1 W out (SY) (1-4-60).

6975 [AKA VA203B]; reflex klystron, tunable, covers 8.5-9.6 GHz, IC, WG, heater 6.3 V @ 450 mA, 350 V, 40 mW out (AX, MET, PH, VA, VO) (5-13-57).

6977 [AKA DM160]; T1 submin. computer indicator triode, blue-green glow, fil. 1.0 V @ 30 mA; shows no light at V_g -8, full glow at 0 V (AX, MU, RCA, SY, TS) (8-12-57).

6979 [AKA 4X250B]; xmtg. tetrode, glass-seal, fil. 6.0 V @ 2.6 A, 250 W, external anode, forced-air-cooled; superseded by 7034/4X250A (AX).

6981 Ionization gauge (12-2-57).

6982 [AKA EL3C/L]; full-wave rectifier, fil. 2.5 V @ 11.5 A, 725 V PIV, 2.5 A, lug base (EL) (6-19-61).

6983 [AKA EL6B/L]; rectifier, fil. 2.5 V @ 21 A, 920 V PIV, 6.4 A, lug base (EL) (7-17-61).

6984-86 Thyratrons, fil. 2.5 V @ 21 A, 900 V PIV, 4 A (all EL) (all 6-19-61).

6984 [AKA ELC4J]; 4-pin base.

6985 [AKA ELC4J/F]; bracket-mounted.

6986 [AKA ELC4J/L]; on lug base.

6987 [AKA ELC6C]; xenon thyratron, fil. 2.5 V @ 24 A, 4 kV PIV, 6.4 A (EL) (6-19-61).

6988 [AKA ELC6J/L]; thyratron, 5685 on lug base, replaceable by 7298 (EL) (6-19-61).

6989 [AKA C6J/KL]; thyratron, 5685 on lug base, replaceable by 7298 ((EL, GE) (6-19-61).

6990-92 Storage tubes (all 6-24-57).

6993 Counter tube (12-22-58).

6996 TWA, 8.0-9.6 GHz, EM, CO, heater 6.3 V @ 2.3 A, 3.5 kV, 10 W output, 30 dB gain (IT) (7-1-57).

6997 TWA, 2.7-2.9 GHz, 30 W pulse output (KL) (7-1-57)

6998 Neutron counter, boron trifluoride, 2 kV, sensitivity 13 counts/neutron/cm^2/sec., aluminum body 1-1/32" dia. by 30-½" long, HN coax connector (W) (3-25-57). Also see 6307.

6999 T2X3 submin. power pentode, fil. 2.64 V CT @ 50 mA, 750 mA (GE) (8-13-62) <6999>.

7000 Rcvg. pentode, sel. 6J7 (KR) (non-RMA no.) <7R>.

7001 [AKA M8167]; min. beam power tube, heater 6.3 V @ 450 mA, g_m 4.8 mS, 5.5 W (MU, SY) (6-23-58) <7EJ>.

7002 Xmtg. triode (12-12-60).

7003, A Xmtg. triode, glass ring-seal, for pulse modulators, heater 6.0 V @ 60 A, 3 kW, forced-air and liquid-cooled by immersion (EI, ML) (8-1-60). "A" is ruggedized.

7004 [AKA TD2-300A]; UHF xmtg. triode, glass or cermet construction per order, ring-seal coaxial, fil. 3.4 V @ 19 A, 300 W, 900 MHz (AX, MU).

7006 Magnetron, tunes 9.0-9.6 GHz, IM, WG, heater 13.8 V @ 3.2 A, 23 KV @ 30 A, 210 kW pulse out (LI, SY) (1-19-59).

7007 [AKA 6166A, q. v.].

7008 Magnetron, for servo-controlled motor-driven tuning, tunes 8.5-9.6 GHz, ruggedized, IM, WG, 13.8 V @ 3.2 A, 23 kV @ 28 A, 220 kW pulse output, replaces 4J50 (RCA, STC, W) (1-6-58). "Coupled-cavity" tuning design as with 6865.

7009 [AKA 2005]; numeric readout (11-10-58).

7012 Xmtg. triode, fil. 15.0 V @ 36 A, μ 25, 2.5 kW, 30 MHz, water-cooled (F) (2-13-61).

7014 [AKA 604]; full-wave gas-mercury rectifier, T16 bulb, super-jumbo 4-pin base, fil. 2.5 V @ 11.5 A, 900 V PIV, 2.5 A (HY, RCA, W) (7-29-57). Sim. to 4B24, replaceable by 6484.

7015 [AKA 604L]; rectifier, lug-based 7014 (W) (7-29-57).

7016 [AKA 606]; rectifier (7-29-57).

7017 [AKA 606L]; mercury rectifier, 6.4 A (RCA) (7-29-57).

7018 [AKA 615, CE-213A]; mercury rectifier, ST16 bulb w/ cap, UX base, fil. 2.5 V @ 7.0 A, 2 kV PIV, 2.5 A (HY, RCA, W) (7-29-57).

7019 [AKA 635]; gas-mercury rectifier, T16 bulb w/ cap, super-jumbo base, fil. 2.5 V @ 18 A, 1 kV PIV, 6.4 A (RCA) (7-29-57).

7020 [AKA 635L]; ignitron (7-29-57).

7021 [AKA 714]; gas and mercury triode thyratron sim. to FG-17; ST16 bulb w/ cap, UX base, fil. 2.5 V @ 5.0 A, 1250 V PIV, 1 A (RCA) (7-29-57).

7022 [AKA 740L]; gas thyratron (7-29-57).

7023 [AKA NL610, 760L]; gas thyratron (7-29-57).

7025, A Min. duotriode, low-hum/noise/microphonics version of 12AX7 (AX, GE, RCA, TS, W) (2-24-58) <9A>.

7027, A Beam power tube, T12 bulb, octal base, heater 6.3 V @ 900 mA, 25 W (GE, RCA) (3-31-58) <8HY>. Both of these are 6L6GB/GC variants w/ different bulb and button-stem construction, with two pins each for grid and screen for added heat dissipation through the base. Like 6L6GC, "A" uses thick plate material to get heat dissipation uprated to 35 W.

7028 [AKA JP9-2-5]; magnetron, fixed at 9410 ± 65 MHz, IM, WG, heater 6.3 V @ 500 mA, 3.5 kV @ 2.5 A, 3 kW pulse output (AX, MU, VO) (5-6-57).

7029 Photomultiplier, 10-stage, S-17 response, "dormer-window" side-on type with curved cathode, domed bulb 1-½" dia. by 3-¼" high, short duodecal base (RCA) (12-30-57).

7030 Ignitron, replaceable by 5552A (6-27-60).

7031 Triode gas thyratron, fil. 2.5 V @ 10.8 A, 1.5 kV PIV, 3.2 A (W) (6-10-57).

7032 Min. rcvg. heptode for gate use in computers, heater 6.3 V @ 300 mA (SY) (11-10-58) <7CH>.

7033 Storage tube (3-10-58).

7034 [AKA 7034/4X150A, 4F15R, QV1-150A]; xmtg. tetrode, glass body, quasi-octal base, heater 6.0 V @ 2.6 A, uprated from 150 to 250 W and from 1 to 2 kV, 150 MHz, forced-air cooled (AX, EI, GE, MU, NEC, RCA, SV) {X424D} (7-8-57). One each in Gates BT-100A 100-W VHF TV xmtr., and T-302/TRC of AN/-TRC-24. Interchanges physically w/ 4X250B. "Operationally equiv. to 4CX250B at freqs. below 150 MHz." Variants: 7035, 8172, 8296, 8297.

7035 [AKA 7035/4X150D]; xmtg. tetrode, 7034 w/ heater 26.5 V @ 580 mA (AX, EI, RCA) (7-8-57). Variant: 7609.

7036 Min. pentagrid converter, 5915A in larger envelope (2-3/8" tall) (GE) (6-10-57) <7CH>.

7037 Nixie readout, 250 V, 6-12 mA (11-25-57).

7038, **A, H, V, V4** Vidicon, 7697 format, sulfide-target, for live or film use in B&W or color; one-time "standard" for film chains; also for CCTV (GE, HI, RCA) (2-3-58). Variant: 8572.

7040 Magnetron, fixed at 5400 ± 20 MHz, IM, WG, heater 6.3 V @ 1.7 A, 14 kV @ 14 A, 85 kW pulse out (RCA).

7041 Duotriode, heater 12.6 V @ 450 mA.

7042 Ignitron, 6958 w/ integral overtemp. switch and thermostat for cooling-water control (GE) (2-17-58).

7043 [AKA CE881]; vacuum phototube, S-4 response, 929 format lengthened to 2-3/4" seated height, responds to light from any direction perpendicular to cathode, 45 μA/lumen (RCA) (8-23-57).

7044 Min. 9-pin med.-μ duotriode, computer-rated, for cathode followers; unique pinout, heater 12.6 VCT @ 450 mA, μ 21, g_m 10 mS, 4.5 W per side (GE, RCA, SY, TS) (7-1-57) <9H>.

7046 Photomultiplier, 15-stage, extended S-11 response (blue-UV sensitive), end-on bulb 5-¼" dia. stepped to 3" dia. by 9-3/4" long, 21-pin "thirtyfivar" base (RCA) (7-29-57).

7049 Reflex klystron, 1.5-6.0 GHz, EC, heater 6.3 V @ 680 mA, 350 V, 250 mW output (PO) (8-20-62).

7054-61 Series of tubes mobile-rated for heater supply of 13.5 ± 1.5 V (all GE, RCA, SY) (all 9-8-58). Analogous to 6675-6681 6-volt types. Series includes 7167 and 7258 (see below).

7054	Equals 12BY7.	7058	Equals 12AX7.
7055	Equals 12AL5.	7059	Equals 6U8.
7056	Equals 6CB6.	7060	Equals 6AU8.
7057	Equals 6BZ7.	7061	Equals 12AB5.

7062 [AKA E180CC]; "10,000-hr." duotriode, premium 12AV7, heater reduced to 200 mA (AX, MU) (8-3-59). To replace 5965; has longer bulb.

7064-65 Photomultipliers (8-5-57).

7066-68 TWAs, 8.0-12.0 GHz, EM, CO, heaters 6.3 V (all IT).

7066 Heater 850 mA, 500 V, 30 db gain, 50 mW out.

7067 Heater 2.3 A, 4.5 kV, 30 dB gain, 1 W out.

7068 Heater 5.2 A, 12 kV @ 2.5 A, 30 dB, 2 kW out.

7072 TWA, 2.0-4.0 GHz, IM, CO, heater 6.3 V @ 1.1 A, 975 V, 36 db gain, 1 W output (1-4-60).

7077 Rcvg. UHF planar cermet high-μ triode, for G-G amps, plate on top dome, grid on ring, heater on buttons at base; heater 6.3 V @ 240 mA, μ 90, 1.1 W, NF 5.5 dB @ 450 MHz (GE, SY, TSU) (3-31-58). Class B final amp. in Pioneer III space probe, 960 MHz. Variants: 7462, 7486.

7079 T3 submin. duotriode, heater 6.3 V @ 300 mA, μ 20 (GE) (10-23-61) <8DG>.

7083 T3 submin. RF pentode, sharp-cutoff variant of 5702, heater 6.3 V @ 200 mA, gm 5 mS, 1.1 W (GE) (10-23-61) <5702>.

7085 [AKA ML-356]; variant of 880 w/ T-T fil, for RF heaters and broadcast xmtrs; fil. 7.5 V @ 170 A, 22.5 kW, 25 MHz (EI, GE, ML) (12-16-57).

7086 Gas thyratron (10-21-57).

7087, A Neutron counter, boron trifluoride, 2 kV, sensitivity 40 count/neutron/cm^2/sec., aluminum body 5-7/8" dia. by 39-3/4" long, 12-foot coax cable attached (W) (10-21-57). "A" has 25-foot cable. Also see 6307.

7088 CW magnetron, tunes 5.4-5.9 MHz, IM, CO, 1.45 kV @ 800 mA, 100 W output (BL).

7090-91 CW magnetrons, for RF heating, fixed at 2450 ± 25 MHz, IM, CO (both 1-20-58).

7090 [AKA JP2-0-2]; heater 5.5 V @ 3.4 A, 1.7 kV @ 200 mA, 200 W out (AX, MU, RE, VO).

7091 [AKA JN2-2-5A]; heater 7.0 V @ 35 A, 4.5 kV @ 800 mA, 2 kW out, forced-air-cooled (AX, MU, PH, VO).

7092 [AKA TY6-800]; RF-heating VHF triode, fil. 6.3 V @ 32.5 A, 800 W (1300 W @ 50% duty cycle), radiation-cooled (AX, MU) (12-2-63).

7093 Magnetron, fixed at 34,860 ± 348 MHz, IM, WG, 4.0 V @ 4.3 A, 15 kV @ 16 A, 32 kW pulse output, pulse-width only 20 nsec., for high-resolution radars (AX, RE) (1-20-58).

7094 Xmtg. beam tetrode, all-glass construction w/ plate pin, heater 6.3 V @ 3.2 A, 100 W, 60 MHz, forced-air cooled at high power (RCA) (1-20-58). In Hallicrafters HT-41 G-G linear.

7096 BWO, voltage-tuned 2.0-4.0 GHz, heater 6.3 V @ 2.2 A, 1.4 kV, 100 mW pulse output, CO (SY).

7098 Magnetron, tunes 9.3-9.5 GHz, IM, WG, 5.0 V @ 650 mA, 2 kV @ 1.25 A, 60 W out (SY) (1-19-59).

7099 T2 flat-press submin. voltage regulator, 155 V @ 75-300 μA (TS) (7-13-59).

7100 Rectifier (5-21-62).

7102 Photomultiplier, 10-stage, S-1 response, 6199 format, Micanol base (RCA) (12-9-57).

7105 Dual power triode, 6080WA w/ heater 12.6-V @ 1.25 A (GE, TS) (3-24-58) <8BD>.

7110-12 Magnetrons, tune 8.5-9.6 GHz, IM, WG, 13.8 V @ 3.2 A, 23 kV @ 28 A, 230 kW pulse out. "Coupled-cavity" design as with 6865.

7110 Hand-tuned, replaces 4J50 (RCA, W) (1-6-58).

7111 Hand-tunable by push-to-tune mechanism and micrometer indicator (LI, RCA, SY, W) (11-17-58).

7112 Shaft for remote-controlled tuning (RCA) (11-17-58).

7116 TWA, 8.5-9.6 GHz, IM, CO, 3.5 V @ 860 mA, 1.6 kV, 43 dB gain, 20 dB NF, 200 mW output (IT) (3-16-59).

7117 Photomultiplier, 6328 for DC operation, same use, slightly different specs (RCA) (3-31-58).

7118 [AKA E181CC]; min. 9-pin duotriode, heater 6.3 V @ 400 mA (MU).

7119 [AKA E182CC]; rcvg. "10,000-hr." duotriode, heater 12.6 VCT @ 400 mA, 4.5 W (AX, MU) (2-9-59) <9H>.

7120-21 Xmtg. triodes (both 8-1-60).

7124-25 Xmtg. triodes (both 4-16-62).

7128 [Type not listed] (6-22-64).

7136 Mercury rectifier, current between 575A and 6693, elements like 6693 but w/ 575A base, fil. 5.0 V @ 11.5

A, 15 kV PIV @ 2.5 A (AX) (11-25-63).

7137 Min. triode, premium 6J4 (GE, SY) (11-25-57) <7BQ>.

7138-43 Magnetrons, fixed-freq., freq. tolerance ± 50 MHz, IM, WG, heater 6.3 V @ 520 mA, 7.4 kV @ 7.5 A, 18 kW pulse output (all CM) (all 3-31-58). Same freq. steps as for 6817-6822.

7138 For 9050 MHz.
7139 For 9150 MHz.
7140 For 9250 MHz.
7141 For 9350 MHz.
7142 For 9450 MHz.
7143 For 9550 MHz.

7147 Gas noise source (6-10-63).

7148 Gas noise source (7-3-61).

7150 Rcvg. VHF tetrode, T9 bulb, "special" (novar?) base, heater 6.3 V @ 450 mA, g_m 35.8 mS, 4 W, usable in G-G circuits, sim. to WE 436A but w/ different pinout (LME, SY) (2-24-58) <9JH>.

7151 Ignitron, 5551 format, for AC control use, 600 V, 900 A (GE) (1-20-58).

7153 Numeric readout (12-23-57).

7154 Gas noise source (1-13-58).

7155 Beam-switching counter (2-3-58).

7156 Magnetron, tunes 5450-5825 MHz, IM, WG, 5.0 V @ 5.4 A, 24 kV @ 24 A, 300 kW pulse output (LI, RA) (10-30-61). In C-band tracker section of M33 radar; also in airborne weather radars.

7158 Reflex klystron, 5925-6425 MHz, 120 mW output

7160-62 Min. "Corotron" corona voltage regulators, w/ anode cap; regulate at ±2 % of nominal voltage (all VN).

5962 [AKA BS101, CK1017]; regulates at 700 V @ 5-55 μA (AL, RA, VN) {QM380} (5-17-50).

7160 Regulates at 4.0 kV, 2-55 μA.

7161 Regulates at 3.5 kV 35-500 μA.

7162 Regulates at 2.5 kV, 25-500 μA.

7163 Cadmium-sulfide photocell, S-15 response, head-on case 1.24" dia. by 0.28" high w/ pins (RCA) (3-3-58).

7167 Min. pentode, mobile-rated 6CY5, heater 13.5 V @ 90 mA (GE, SY, W) (6-9-58) <7EW>.

7169-70 Ionization gauges (both 7-28-58).

7171 Ignitron, for capacitor-discharge use, stainless steel case 2-5/32" dia. by 8-3/4" long, 15 kV PIV, 35 kA peak (GE) (5-26-58).

7172-74 "Iatron" storage tubes, radar indicators (all IT) (all 1961). Also see 7423.

7172 2-½" dia., electrostatic focus and defl., coaxial gun.

7173 4" dia., magnetic focus and defl.

7174 4" dia., magnetic focus and defl.

7177 Camera tube, "misc." type (6-4-62).

7178 [AKA KU-53]; hydrogen diode, fil. 5.0 V @ 22 A (KL) (6-4-62).

7179-80 Ignitrons (both 3-10-58).

7182 Magnetron, 2.75-2.86 GHz, EM, EC, 12.0 V @ 14 A, 33 kV @ 185 A, 2.5 MW pulse output (CM, EEV, MC) (7-23-62). Does not look like a magnetron - has tubular metal body, glass end dome and neck, metal connecting ring.

7183, A Display storage tube, 4", electrostatic focus and 54° magnetic deflection, takes fixed yoke for TV or rotating yoke for radar use, P20 aluminized phosphor, 5.2" dia. body w/ writing and viewing gun, 11.6" overall, min. 7-pin base on viewing gun; 8-pin "neoditetrar" base on writing gun, three caps, heater 12.6 VCT @ 600 mA, 10 kV (RCA) (5-26-58). "A" has improved collimation system.

7184 [AKA VT-227]; rcvg. pentode (KR) sim. to 6V6; no CV no. or known use. (Non-RMA no.)

7185 Xmtg. triode, fil. 5.0 V @ 260 A (F) (11-14-60).

7186-88 Fission counters, uranium-oxide type, HN coax connector (all W) Also see 6971.

7186 250-400 V, sensitivity 0.001 count/neutron/cm^2/sec., 7/8" dia. by 53-9/16" long (3-3-58).

7187 300 V, sensitivity 0.52 count/neutron/cm^2/sec., 2-3/32" dia. by 11-7/8" long (3-3-58).

7188 300 V, sensitivity 0.07 count/neutron/cm^2/sec., 2-3/32" dia. by 11-7/8" long (7-7-58).

7189, A Rcvg. power pentode, sim. to 6BQ5 uprated to 400 V and 4W screen dissipation (AX, GE, RCA, SY, TS, W) (10-13-58) <9CV>. "A" has uprated screen voltage.

7190-92 Min. hydrogen thyratrons, radar pulsers, 7-pin min. hard-glass bulb, heater 6.3 V @ 1.8 A, 1.2 kV PIV, 20 A pk., 50 mA avg. (all TS) (all 3-31-58).

7190 [AKA CH-1062]; w/ regular min. base.

7191, H [AKA CH-1092]; w/ plate cap, cap offset from center axis. "H" [AKA 8653] presumably has added anode insulation for high-altitude use.

7192 [AKA CH-1055]; w/ stranded leads (TS) (3-31-58).

7193 [AKA 2C22, q. v.] (KR) (non-RMA no.) <4AM>. In BC-966A.

7198, A Image orthicon, ruggedized, 5655 format, glass target, heater 6.3 V @ 600 mA, S-10 response, 10 times as sensitive as 5820 (GE, RCA, W) (7-14-58).

7199 Min. 9-pin med.-μ triode - sharp-cutoff pentode, low-noise, for audio use, heater 6.3 V @ 450 mA, triode μ 17; pentode g_m 7 mS (GE, RCA, SY, W) (10-13-58) <9JT>.

7200 Photomultiplier, 9-stage, S-19 response, T9 tubular bulb, 5.7" seated height w/ 11-pin submagnal base (RCA) (4-14-58).

7203 [AKA 4CX250B]; xmtg. UHF cermet beam tetrode, quasi-octal base, heater 6.0 V @ 2.3-2.9 A, 250 W, 500 MHz, brazed external anode, for general use; will replace 4X250B (AX, EI, RCA, SV) (10-13-58). In Gates FM-250H3 and Wilkinson FM-250E 250-W FM xmtrs. Driver in Wilkinson FM-5000E, FM-7500E, FM-10,000E, and FM-20,000E FM xmtrs. Variant: 8560.

7204 [AKA 4CX250F]; xmtg. beam tetrode, 7203 w/ heater 26.5 V @ 580 mA (AX, RCA) (10-13-58). Six in each of 4 amps. in T-817/ALT-17. Variant: 8621.

7205 Xmtg. triode (4-14-58).

7206 Xmtg. triode (9-26-60).

7207 Xmtg. triode (11-14-60).

7208, A, B Magnetron, first WE coaxial design, tunes 15.5-17.5 GHz, IM, WG, heater 12.6 V @ 3.25 A, 20 kV @ 20

A, 125 kW pulse output, forced-air cooled (LI, WE) (8-29-60). "B" is ruggedized for airborne use (WE).

7209-10 Xmtg. triodes (8-1-60).

7211 Planar triode, 2C39 format, heater 6.3 V @ 1.1 A, 100 W, 2.5 GHz, pulse version rated for higher cathode curr. than 8250; longer body for higher insulation (EI, ML) (8-1-60). In AN/GRC-103.

7212 [AKA 6146W]; xmtg. tetrode, ruggedized 6146 (GE, RCA, SY) (7-1-58) <7CK>.

7213 Xmtg. cermet beam tetrode, ring-seal construction, aligned grids, heater 5.5 V @ 17.5 A, 1.5 kW, 1215 MHz, forced-air cooled (RCA) (10-13-58).

7214 Xmtg. tetrode, 7213 rated for pulsed amp. use, 10 kV, 65 kW pulse output (RCA) (7-6-59).

7215 Xmtg. triode, for radar modulators and RF use, fil. 6.0 V @ 60 A, μ 22, 9 kW, 30 MHz, water-cooled (W) (4-28-58).

7216 Thyratron, 5632 w/ industrial spade-lug base (EL, RCA) (2-5-62).

7217 Rectifier (3-30-59).

7222 Storage tube (6-15-59).

7223 Photocell, germanium P-N junction, S-14 response, metal body 0.08" dia., 0.52" long, lens at one end, two leads at other, for card readers, etc. (RCA) (7-7-58).

7224 Photocell, germanium P-N junction, S-14 response, glass bulb 0.3" dia., 0.46" long, for film-sound and computer uses (RCA) (8-18-58).

7225 Storage tube, 3" radechon barrier-grid type, for scan conversion, digital memories, etc.; electrostatic focusing and deflection, destructive-readout (W) (7-7-58). Also see 7566.

7226, A Vidicon, for broadcast/surveillance use, 1.14" dia. by 5-¼" long, heater 6.3 V @ 150 mA (GE) (10-27-58).

7227 Min. rcvg. power pentode, heater 27.5 V @ 175 mA, g_m 5.5 mS, 2.2 W (SY) (2-9-59) <9BA>.

7228 Storage tube (9-8-58).

7229 Storage tube (8-25-58).

7230-32 "Misc." types (all 8-25-58).

7233 Min. 9-pin low-μ power triode for series regulators, heater 6.3 V @ 1.0 A, 8.0 W (GE, SY) (8-15-60) <9FR>.

7234 HV pentode, control tube for series regulators, heater 6.3 V @ 150 mA, 8 kV, 10 W (GE, VN) (2-13-61) <9KD>.

7235 HV triode, shunt regulator or pass tube, T6-½ bulb w/ cap, heater 6.3 V @ 300 mA, μ 550, g_m 0.85 mS, 10 kV, 12 W; can substitute for 2C53 (GE, VN) (2-13-61) <9KE>.

7236 Dual power triode, computer-rated 5998A (GE, SY, TS) (2-15-58) <8BD>.

7237 Xmtg. triode, variant of 6961 w/ radiator design to interchange w/ 6366 and 6367; fil. 12.6 V @ 33 A, 6 kW, 55 MHz (AX) (6-16-58).

7239 Min. 9-pin beam pentode, w/ cap, for high-voltage pulse amps and shunt regulators, heater 6.3 V @ 300 mA, 4.0 W, 2.2 kV (GE, SY) (5-2-60) <9KH>.

7240 Min. hydrogen thyratron, 7190 w/ heater 28.0 V @ 420 mA {CH-1046} (TS) (4-27-59).

7241 Power triode, pass tube for regulators, in TT-18 stepped tubular bulb, on 7-pin "giant" base, 3 cathodes and grids w/ common plate, μ 2.7, 40 mS, 100 W, plate curr. rated at 1.2 A; repl. four 5998s in parallel (SY, TS) (11-24-58) <7241>. Variant: 7242.

7242 Power triode, 7241 w/ μ 9.0, g_m 110 mS, limited to 300 mA per cathode (SY, TS) (11-24-58) <7241>.

7243 Neutron counter, boron trifluoride, 2 kV, sensitivity 13 counts/neutron/cm^2/sec., 2-15/16" dia. by 15-3/8" long, HN coax connector (W) (8-25-58). Also see 6307.

7244, A Min. duotriode, premium 6J6; stacked ceramic-based frame-grid structure inside glass (GE, SY) (7-14-58) <7BF>.

7245, A Min. triode, premium 6J4; made as per 7244 (SY) (10-13-58) <7BQ>.

7246 T2X3 submin. UHF triode, variant of 5676, fil. 1.25 V @ 150 mA, μ 22, 700 mW, 500 MHz (GE, SY, TS) (4-3-61) <5676>.

7247 [AKA 12DW7]; min. 9-pin duotriode, equals ½ 7025 for preamp use plus ½ 12AU7 for phase inverter, etc. (GE, RCA) (5-4-59) <9A>.

7248 Xmtg. tetrode, glass bulb w/ cap, special pinned base, for pulse modulators, T-T fil. 6.3 V @ 11.7 A, 200 W, oil-cooled (EI, ML) (8-1-60).

7249 Power tetrode (8-1-60). Presumed variant of 7248.

7255 Xmtg. triode, for audio/sonar use, 15 kW (W) (6-20-60).

7256 Magnetron, tunable (7-7-58).

7257 Image intensifier ("light amplifier"), 5" photocathode, 1" viewing screen, all-glass body 8-11/16" dia. by 15-3/4" long, 30 kV (W) (10-20-58).

7258 Triode-pentode, mobile-rated 6AN8 (GE, RCA, SY, TS) (7-14-58) <9DA>.

7262, A Vidicon, sulfide target, broadcast variant of 7263, heater 6.3 V @ 80 mA (GE, HI, RCA) (10-13-58). "A" has sensitivity equal to 7735A. Variant: 8573.

7263, A Vidicon, 1" dia. by 5-¼" long, sulfide target, ruggedized 7262, low-power heater (6.3 V @ 90 mA), for small industrial live-scene solid-state cameras (RCA) (1-12-59). "A" has sensitivity equal to 7735A.

7264 Photomultiplier, 14-stage, S-11 response, 6810A format w/ rounded faceplate, fast response (RCA) (8-25-58). Variant: 7265.

7265 Photomultiplier, 7264 w/ S-20 response (RCA) (10-13-58).

7266 Rcvg. UHF planar button-base cermet diode, heater 6.3 V @ 215 mA, 600 V PIV, 2.2 mA (GE) (8-3-59). Variant: 7841.

7267 Backward-wave osc., heater 6.3 V @ 620 mA, 60 mW out @ 12.4-18 GHz (HU).

7268, A, B Display storage tube, for radar, etc., 4" ruggedized, electrostatic focus and deflection, P20 aluminized phosphor, 5-½" dia. body stepped to 3.4" dia. by 15-3/4" long w/ two writing guns and and viewing gun, 25-

pin two-row base w/ flying leads and three caps, heaters 6.3 V @ 600 mA, 11.5 kV (RCA, W) (7-20-59).

7269 Triode mercury thyratron, fil. 5 V @ 5 A, 21 kV PIV, 3.2 A (W) (4-6-59).

7270 Xmtg. beam tetrode, all-glass tubular construction w/ plate pin, for general use, heater 6.3 V @ 3.1 A, 60 W, 60 MHz, forced-air cooled at full power (RCA) (8-10-59). Variant: 7271.

7271 Xmtg. tetrode, mobile-rated 7270 w/ heater 13.5 V @ 1.25 A (RCA) (8-10-59).

7289 [AKA 3CX100A5]; planar pulse triode, cermet 2C39 format, usable to 3 GHz, unilateral replacement for 2C39A family (EI, GE, ML, RCA) (7-20-59).

7290 Vidicon, 7697 format, for slow-scan use (W) (9-8-58).

7291, A Vidicon (10-27-58).

7292 [AKA JN2-2-5W]; CW magnetron, liquid-cooled variant of 7091 (AX, MU) (6-13-60).

7293, A/L Image orthicon, for B&W broadcast studio and outdoors use, 5655 format, S-10 response (GE, RCA) (12-15-58). "A/L" has longer-life, non-burn-in glass target. Also see 8674.

7294 Image orthicon (12-15-58).

7295, A, B, B/L, C Image orthicon, 4-½", for B & W broadcast outdoors and studio use, S-10 response (RCA) (12-15-58). "C" has non-burn-in target. Variant: 8749.

7296 UHF planar cermet high-μ triode, stacked-and-lugged bolt-down construction, for osc. and power-amp use, heater 6.3 V @ 400 mA, μ 90, 5.5 W (GE) (8-3-59). Variant: 7588. Also see 8081.

7297 [AKA 23052]; triode gas thyratron, lug-base w/ plate cap, fil. 2.5 V @ 10.8 A, 1.5 kV PIV, 3.2 A (W) (11-6-61).

7298 Triode gas thyratron, lug-base, fil. 2.5 V @ 21 A, 1500 V PIV, 6.4 A (W) (1-5-59). Variant: 7299.

7299 Thyratron, 7298 in different bulb w/ plate cap, mounting strap, and flying leads (W) (4-6-59).

7306 [AKA WT-210-0106, C3J/A]; triode gas thyratron, lug-base w/ plate cap, fil. 2.5 V @ 9 A, 1250 V PIV, 2.5 A (W) (6-8-59). Variant: 7307.

7307 [AKA 6011, 710]; thyratron, gas-mercury variant of 7306, uprated to 1.5 kV, different bulb, UX base w/ cap (6-15-59).

7308 [AKA E188CC]; "10,000-hr." frame-grid UHF duotriode, premium 6922 w/ low microphonics, heater 6.3 V @ 335 mA, 2.0 W (AX, RCA, SY) (6-27-60.

7310 HV rectifier, 4-pin base w/ cap, fil. 5.0 V @ 6.5 A, 20 kV, 115 mA (GE) (5-2-66) <4P>.

7311-14 Small power tubes, *metal* min. envelope 7/8" dia. by 2.1" high, w/ cap, all w/ 6.3-V heaters (all GE) (all 2-4-63). Also see metal submins. 8524-8530.

7311 Beam power tube, heater 800 mA, 21 W <7311>.

7312 Low-μ triode, heater 1.25 A, μ 2, 20 W <7312>.

7313 Rectifier, heater 1.55 A, 2.8 kV PIV, 140 mA <7313>.

7314 Power pentode, heater 600 mA, 10 W <7314>.

7315 Display-storage tube, 3.8", for weather radar and spectrum analyzers, electrostatic focus and deflection, P20 aluminized phosphor, 5-½" dia. body stepped to 2" dia. by 13.6" long w/ writing and viewing guns, diheptal base w/ three caps, heater 12.6 VCT @ 600 mA, 11 kV (RCA) (12-15-58). Variant: 7448.

7316 Min. duotriode, premium 12AU7 for computer use (AX, SY) (8-31-59) <9A>.

7317 Neutron counter, boron trifluoride, 2 kV, sensitivity 13 counts/neutron/cm^2/sec., 1-1/32" dia. by 29-15/16" long, wire attached (W) (10-5-59). Also see 6307.

7318 Min. duotriode, premium 12AU7, heater curr. raised to 175 mA per section (GE, SY) (11-17-58) <9A>.

7319 [AKA ELC1K/B]; gas thyratron, fil. 2.5 V @ 6.3 A, 2 kV PIV, 1.0 A (EL) (6-18-62).

7320 [AKA E84L]; power pentode, mobile-rated 6BQ5 (2-9-59) <9CV>.

7321 [AKA C6J/L]; triode gas thyratron, lug-base, fil. 2.5 V @ 21 A, 1250 V PIV, 6.4 A (W) (12-29-58).

7322 Hydrogen thyratron, 25 kv PIV, 2 A (1-19-59).

7323 T2 flat-press submin. (neon?) thyratron, self-latching indicator, 1.25-V fil., to repl. 7979 (TS) (4-27-59).

7325 Vidicon, 7697 format, for industrial use, more sensitive than 6198A and replacing it (W) (1-12-59).

7326 Photomultiplier, 10-stage, head-on, flat-face type, S-20 response, T10 bulb, 5.8" seated height, metal collar above base, 14-pin diheptal base (RCA) (11-10-58).

7327 T3 submin. pulse-only med.-μ duotriode for blocking oscs, etc.; repl. 6111, 6021; heater 6.3 V @ 300 mA, 950 mW per section (GE, SY) (10-27-58) <8DG>. Also see 7550.

7328, A Xmtg. triode, fil. 7.0 V @ 270 A (F) (10-10-60).

7333 Xmtg. triode (12-12-60).

7334-35 Xmtg. triodes (both 2-26-62).

7336 Vidicon (1-19-59).

7351, A Vidicon, high-sensitivity unit for industrial low-light use, sim. to 6198 but 10X more sensitive; has long persistence (ML) (8-1-60).

7353 Neutron counter, boron ion-chamber type, 300-800 V, 3-5/32" dia. by 24-½" long, three HN coaxial connectors (W) (1-19-59). Also see 6377.

7355 Beam power tube, T9 octal, heater 6.3 V @ 0.8 A, g_m 7.6 mS, 18 W (GE, RCA, SY, TS) (11-16-59) <8KN>.

7356 Storage CRT, 5", for radar displays, two writing guns and readout gun, 30-sec storage time, electrostatic focus and defl.; one gun can be used for selective erasure (W) (ca. 1961).

7357 [AKA 6159W]; beam power tube, ruggedized 6159A, still rated at 20 W (GE, RCA) (4-24-61) <7CK>.

7358 Beam power tube, T12 bulb w/ cap, octal base, pulse modulator, heater 6.3 V @ 1.25 A, 3.5 kV, 10 W (GE, SY) (5-11-59) <7CK>.

7360 Min. 9-pin sheet-beam beam-deflection tube, gives carrier osc. and balanced modulator in one tube, two plates, heater 6.3 V @ 350 mA, 100 MHz (GE, RCA, SY) (9-14-59) <9KS>. Two in RCA Radiomarine SSB-5 HF transceiver, 1961. Also used in receivers (Squires-Sanders, Swan 500, etc.) as SSB demodulator. Note: color-TV tubes 6AR8, 6JH8 and 6ME8 are similar and are potential substitutes for 7360 w/ partial rewiring.

7363 Triode gas thyratron, fil. 2.5 V @ 8.5 A, 1.5 kV PIV, 1.6 A (W) (4-6-59).

7370 Min. duotriode, 5687 w/ heater 40 VCT @ 130 mA, 5 W per plate (SY, TS) (9-21-59) <9H>.

7371 Xmtg. pentode, for audio/sonar use, 250 W, 175 MHz (W) (12-7-59).

7377 [AKA QQE04/5]; xmtg. dual UHF tetrode, short all-glass cylindrical single-ended design, fil. 12.6 VCT @ 300 mA, 8 W (AX, VO) (4-6-59).

7378 [AKA QE08-200]; xmtg. HF beam tetrode, all-glass w/ cap, fil. 6.3 V @ 3.9 A, 100 W, 30 MHz, radiation-cooled, for linear amps. (AX) (12-2-63).

7381 Dual TR tube, 8.5-9.6 GHz, 150 kW (BL)

7382 Min. rcvg. single triode, equals ½ 12AX7, controlled for noise and hum; special pinout (SY) (8-31-59) <7382>.

7383 Storage tube, visual input, "permachon" sim. to vidicon, 30-min. storage time, mag. focus and defl., 800-line resolution, erased by flash of light (W) (10-5-59).

7384 Hydrogen thyratron, 25 kV, 1000 A pk., 1.25 A avg. (EEV) (7-23-62).

7386 Gas thyratron (5-18-59).

7389, A, B, C Image orthicon, 4-½" non-microphonic, 5820 format expanded, dev. for RCA TK-12 studio B & W camera; larger target gave improved resolution (EEV, RCA) (11-23-59). Also see 8748.

7390, A Hydrogen thyratron, cermet ring-seal construction, 6" dia. by 11' high, external anode, grid on ring, mounting flange, flying leads out base, for radar modulators, cathode 6.3 V @ 32 A, 33 kV PIV, 2000 A peak, 33 MW output, 4.0 A avg., convection cooled (GE, TS) (7-6-59). "A" has temperature indicator added at top.

7391 Planar UHF triode, cermet construction, for G-G oscs, heater 6.3 V @ 380 mA, μ 62, 2.25 W, same physical design as 6299; generates 65 mW @ 5.4 GHz (GE) (6-13-60) <6299>.

7393 TWA, packaged cermet construction, 4-8 GHZ, NF 10 dB max, heater 6.3 V @ 300 mA, 500 V, 5 mW out (GE) (10-9-61). Variant: 7394.

7394 TWA, variant of 7393 for 8-12 GHz (GE) (10-9-61).

7398 Magnetron, packaged voltage-tunable CW, 2200-3850 MHz, 2 W out, fil. 2.5 V @ 3.0 A, 2 kV (GE) (7-6-59).

7399 Xmtg. UHF tetrode, pulse version of 6283, ring-seal cermet construction, for L-band radars, heater 6.3 V @ 5.6 A, 300 W, usableto 1500 MHz, forced-air cooled (GE); for oscs. or G-G amps (5-23-60). In AN/UPX-14 IFF transponder, giving 10 kW at 1010-1050 MHz.

7400 Neon indicator thyratron, cold-cathode, sim. to WE 395A, T4 flat-ended body, for end-on viewing (TS) {CH-1116} (4-27-59). Variant: 7401.

7401 T3 flat-press submin. thyratron, 7400 less the indicator feature (TS) {CH1125} (4-27-59).

7402 Rectifier (11-26-62).

7403 [AKA 3D21WB]; pulse beam power tube, tubular variant of 3D21, T12 bulb, octal base w/ cap, heater 6.3 V @ 1.7 A, 40 W (GE) (7-18-60) <8JU>.

7404 Image converter tube, for "ultrascope" UV microscope or UV-illuminated night-vision systems, UV variant of 6929 (same format), S-21 response; contains a photocathode, focusing and accelerating grids, and a collector; runs at 12.5 kV (RCA) (3-23-59).

7408 [AKA OSW3106]; beam power tube, 6V6GT w/ controls on zero-bias characteristics (GE, RCA, SY, W) (6-22-59) <7S>.

7409 Image orthicon, ruggedized 7538 (GE) (5-29-61).

7410 Xenon thyratron, octal base, fil. 2.5 V @ 7.0 A (GE).

7412 Cadmium-sulfide photocell, S-15 response, bulb 0.29" dia. by 1.35" long with leads, end-on illumination (RCA) (3-9-59).

7413 Xmtg. triode, radar modulator, 10 kW (W) (12-7-59).

7414, 16 Rectifiers (both 3-23-59).

7423 "Iatron" storage CRT, radar indicator, 5", electrostatic focus and defl. (IT) (1961) (5-21-62). Also see 7172-74.

7427 Cadmium-sulfide photocell, min. 9-pin, for street-light controls, etc., S-15 response, large-area cathode ca. 0.6" by 1.5", 400 mW, operates relays directly (GE) (8-3-59) <9LN>.

7429, A Neutron counter, boron trifluoride, 2 kV, sensitivity 40 counts/neutron/cm^2/sec., 1-1/32" dia. by 12-1/8" long, cable attached (W) (4-13-59). Also see 6307.

7430 Rcvg. sharp-cutoff pentode, rectangular all-glass "pill-box" bulb 7/8" by 1.2" by ½", flying leads out all four sides, heater 6.3 V @ 200 mA, g_m 5 mS, 1.9 W (GE, SY) (11-9-59) <7430>. May be production version of W WX-4073.

7432 [AKA 6223]; submin. pentode (2-23-59).

7433 [AKA 6488]; submin. pentode (2-23-59).

7434 [AKA 6391]; submin. pentode (2-23-59).

7432-34 Receiving pentodes (both 2-23-59).

7435-36 Rectifiers (both 2-23-59).

7437 Rcvg. triode (2-23-59).

7438 Rcvg. pentode (2-23-59).

7439 "Krytron" cold-cathode trigger (AX) (3-30-59).

7440-41 "Misc." types (HY) (both 3-30-59).

7444 Magnetron, tunable (5-15-61).

7446 [AKA BL233].

7448 Display-storage tube, 7315 w/ 1000 times faster writing speed (300,000 in/sec. vs. 3000) (RCA) (6-1-59).

7449, A Magnetron, fixed freq., 45 GHz, IM, WG, 45 kW pulse output, forced-air cooled (9-14-59).

7452 [AKA QK324]; magnetron, fixed freq. (10-30-61). In AN/MPQ-4A, giving 70 kW at 16 GHz.

7454 Hydrogen triode thyratron, for radar clipper use instead of diode, glass bulb, anode at top, grid on ring, giant 4-pin base (TS) (12-18-60).

7455 Hydrogen thyratron, 7455 w/ flying leads instead of base (12-18-60).

7457 Xmtg. cermet beam tetrode, heater 6.3 V @ 3 A, 115 W, 1215 MHz, forced-air cooled (RCA) (11-2-59).

7459 Xmtg. VHF triode, glass ring-seal w/ strap flying leads, fil. 12.6 V @ 30 A, 4 kW, forced-air-cooled, for FM-TV xmtrs; replaces 5762/7C24 (AX) (5-25-59).

7460 Magnetron, tunes 5450-5825 MHz, 250 kW out (LI). In AN/SPS-5D.

7461 Magnetron, fixed freq. (10-12-59).

7462 Planar cermet triode, variant of 7077 w/ stacked, lugged construction (GE) (8-3-59). Subvariant: 7720.

7463 Xmtg. triode, for radar modulators and RF use, 4.5 kW, 75 MHz (W) (12-21-59).

7464 Xmtg. triode, for audio/sonar and RF use, 5.5 kW (W) (12-14-59).

7467 Photocell, germanium P-N junction, side-on type, S-14 response, glass bulb 0.28" dia. by 7/8" long (RCA) (5-18-59).

7475 [AKA LS620]; neon voltage regulator, on Br. 4-pin base (MU).

7479, A Xmtg. triode, fil. 8.0 V @ 200 A (ML) (10-29-62).

7480, A Xmtg. triode, vapor-cooled variant of 6697A, uprated to 80 kW (EI, ML) (2-25-63). Final amp. in Gates VP50 50-kW AM xmtr.

7482 RF-heating/xmtg. cermet triode, vapor-cooled variant of 8317, 200 kW (EI, ML) (2-26-62).

7484 Magnetron, tunable (11-23-59).

7486, AL Planar triode, osc./amp variant of 7077, 1.0 W, puts out 300 mW @ 450 MHz (GE) (7-6-59).

7489 Min. audio duotriode, premium 12AU7 (SY) (4-3-61) <9A>.

7490 Min. 9-pin audio/DC power duotriode, heater 12.6 VCT @ 300 mA, μ 32, 5.0 W (SY) (4-3-61) <9A>.

7492 Min. 9-pin RF/computer high-μ duotriode, heater 12.6 VCT @ 150 mA, μ 60, 28 W (SY) (4-3-61) <9A>.

7494 Min. 9-pin duotriode, premium 12AX7WA (SY) (3-13-61) <9A>.

7495 Min. 9-pin VHF beam power tube, heater 6.0 V @ 750 mA, g_m 7 mS, 12 W (SY) (3-13-61) <9K>.

7496 Min. rcvg. pentode, heater 6.3 V @ 300 mA, g_m 4.4 mS, 3.3 W (SY) (4-3-61) <7BK>.

7498 Min. rcvg. pentode, heater 6.3 V @ 300 mA, g_m 7.6 mS, 3 W (SY) (4-3-61) <7DB>.

7499 Min. pentode, premium 6CH6 (SY) (4-3-61) <9BA>.

7500 Min. rcvg. beam power tube, heater 6.3 V @ 450 mA, g_m 4.1 mS, 13 W (SY) (3-13-61) <9AH>.

7502 Min. heptode converter, heater 6.3 V @ 300 mA (SY) (3-13-61) <7CH>.

7503 Magnetron, tunes 9.3-9.6 GHz, IM, CO, 5.0 V @ 650 mA, 2 kV @ 850 mA, 200 W pulse output (LI, SY) (2-7-60). In AN/DPN-62 beacon.

7504 Amplifier klystron, 1.24-1.36 GHz, three int. cavities, WG, heater 16 V @ 8.25 A, 125 kV @ 87 A, 2 MW pulse output, WG (LI) (8-17-59).

7505 Reflex klystron, 0.7-3.0 GHz, EC, heater 6.3 V @ 680 mA, 350 V, 250 mW output (PO) (8-20-62). Variant: 7506.

7506 Klystron, 7505 for 0.5-3.0 GHz (8-20-62).

7509 Triode gas-mercury thyratron, lug-base, fil. 2.5 V @ 9 A, 1.5 kV PIV, 2.5 A (W) (12-14-59).

7513, B/L Image orthicon, 5655 format, for B&W and color cameras (RCA) (6-15-59). "B/L" has longer-life glass target. Also see 8673.

7518 [AKA 710L, 7509A]; thyratron, 6011 w/ lugged base; EL uprates to 1.7 kV PIV (EL, GE) (10-5-59).

7521 [AKA QK797]; magnetron, tunes 8.9-9.4 MHz, WG, heater 5.0 V @ 450 mA, 4.3 kV @ 950 mA, 950 W pulse out (RA) (10-30-61). In AN/PPS-4.

7522 Vidicon (1-9-61).

7524 TWA, 8.0-12.0 GHz, EM, CO, heater 6.3 V @ 1.7 A, 4 kV, 20 dB gain, 5 W output (IT). Variant: 7525.

7525 TWA, variant of 7524.

7526 TWA, 8.0-12.0 GHz, heater 6.3 V @ 850 mA, 1.6 kV, 30 db gain, 50 mW output.

7527 [AKA 4-400B]; xmtg. VHF tetrode, all-glass variant of 8438, same ratings (AX, EI) (11-27-61).

7528 Magnetron, 1.25-1.35 GHz, IM, WG, heater 3.7 V @ 90 A, 60 kV @ 90 A, 2.0 MW pulse out (RA) (11-23-59).

7529 Magnetron, tunes 2.70-2.85 GHz, ceramic insulation and output window, IM, WG, heater 8.2 V @ 78 A, 62 kV @ 115 A, 3 MW pulse out, liquid-cooled (RA) (2-7-60).

7532 Xmtg. triode (6-26-61).

7533 Pencil-triode osc., variant of 6562, tunes 1660-1700 MHz with dual adjustment screws in place of single, 600 mW output (RCA, SY, TS) (12-28-59).

7534 [AKA E130L]; "10,000-hr." output pentode, tubular bulb w/ cap, wide octal base, dual frame grids, heater 6.3 V @ 1.7 A, 6 kV peak, g_m 25 mS, 27.5 W; two give 60 W output in Class B (AX) (6-6-60).

7535 Magnetron, tunes 8.5-9.6 GHz, IM, WG, heater 20.0 V @ 4.0 A, 42 kV @ 45 A, 550 kW pulse out (WE) (6-6-60).

7536 Cadmium-sulfide photocell, side-on type, S-15 response, bulb 0.29" dia. by 1.35" long with leads (RCA) (8-10-59).

7537 TWA, 4.4-5.0 GHz, EM, WG, heater 6.3 V @ 800 mA, 1.5 kV, 34 dB gain, 30 dB NF, 6 W output, natural cooling (AX, PH, VO) (12-2-63).

7538 Image orthicon, MgO target, S-10 response, for low-light surveillance, etc. (GE) (4-3-61). Variant: 7409.

7539 [AKA 1855]; storage tube (4-11-60).

7540 Xmtg. triode, for audio/sonar use, 20 kW, water-cooled via integral cooler, flying leads (W) (4-25-60).

7541 Magnetron, tunes 8.5-9.6 GHz, IM, WG, heater 13.8 V @ 3.5 A, 23 kV @ 28 A, 250 kW pulse out (W) (9-11-61).

7543 [AKA EF94]; min. pentode, low-hum variant of 6AU6 (GE, RCA, W) (7-27-59) <7BK>.

7547 Magnetron, 406-450 MHz, IM, CO, heater 6.5 V @ 55 A, 55 kV @ 97 A, 2.0 MW output (RA) (4-3-61).

7548 Min. 9-pin secondary-emission hexode, VHF pulse

amp. w/ risetime < 4 nsec., heater 6.3 V @ 700 mA (GE, HY) (5-8-61) <9LJ>.

7550 T3 submin. med.-μ pulse duotriode, premium 7327, heater raised to 525 mA, uprated to 2 W per section (GE, SY) (8-31-59) <8DG>.

7551 Min. 9-pin beam power tube, mobile-rated for 12-15 V, 10 W, 175 MHz (RCA, SY) (10-12-59) <9LK>. Final in Hallicrafters SR-42 and Poly-Comm 62B.

7552 Cermet pencil high-μ rcvg. triode, heater 6.3 V @ 225 mA, μ 70, NF 16.5 dB @ 550 MHz, usable to 1.5 GHz (RCA) (10-19-59). Variant: 7553.

7553 Pencil triode, slightly quieter 7552 w/ more gain (RCA) (2-15-60).

7554 Xmtg. cermet high-μ pencil triode, heater 6.3 V @ 225 mA, μ 70, 2.5 W, 1 GHz, Class C amps, for G-G use up to 5 GHz (RCA) (10-19-59).

7556, A [AKA C3J/AL]; triode gas thyratron, lug-base, fil. 2.5 V @ 9 A, 1250 V PIV, 2.5 A (W) (2-7-60).

7558 Beam power tube, 7551 w/ heater 6.3 V ± 5% @ 800 mA (GE, RCA) (10-12-59) <9LK>.

7559 [AKA CH-1096?]; hydrogen thyratron, for crowbar use, tubular bulb w/ mounting flange, flying leads, 25 kV @ 1500 A peak (TS) (10-17-60).

7560, V RF-heating/xmtg. cermet triode, water-cooled variant of 8317, 175 kW (EI, ML) (2-25-63).

7561 Rcvg. audio beam power tube, T9 bulb, octal base, heater 25 V @ 300 mA, g_m 10 mS, 13 W (SY, TF) (1-11-60) <7561>.

7565 Xmtg. triode (8-28-61).

7566 Storage tube, radechon barrier-grid type, two writing guns, for digital memories; electrostatic focusing and defl., can be read out multiple times before refreshing reqd. (W) (12-21-59). Also see 7225.

7568 [AKA CH-1095?]; hydrogen thyratron, for crowbar use, tubular bulb w/ plate cap on stalk, 4-pin giant base, 25 kV @ 800 A peak (TS) (6-20-60).

7571 [AKA QK685]; storage tube (RA) (2-3-64). In MTI section of AN/FPS-20.

7575 Storage tube (7-30-62).

7576 T3 submin. high-μ VHF triode, for G-G use, heater 6.3 V @ 450 mA, μ 46, puts out 3 W @ 235 MHz (GE, SY) (8-22-60) <8KM>.

7577 Amplitron, 1.3-1.4 GHz, IM, WG, 3.8 V @ 60 A, 97 kV @ 98 A, 5 mW pulse output (RA) (11-6-61).

7578 Magnetron, tunes 5.4-5.9 GHz, for airborne beacons, ca. 4" dia. by 7" high, IM, CO, heater 5.0 V @ 1.75 A, 2.8 kV @ 2 A, 1 kW pulse out (RA) (10-30-61).

7579 Magnetron, fixed at 800 ± 25 MHz, 800 V @ 200 mA, 40 W *pulse* output.

7580, W [AKA 4CX250R]; xmtg. tetrode, ruggedized 7203 w/ higher perveance, same ratings (AX, EI, RCA, SV, SY) (11-30-59).

7581, A Beam power tube, 6L6GC w/ low-loss base; has controls on zero-bias characteristics (GE, RCA, SY, TS) (11-2-59) <7S>. Electrically sim. to KT66. "A" is uprated to 35 W.

7583 [AKA KU-82]; hydrogen thyratron (11-26-62).

7586 Nuvistor med.-μ triode, first one announced, general industrial and UHF use, heater 6.3 V @ 135 mA, μ 35, g_m 11.5 mS, 1 W (GE, RCA, SY) (2-15-60) <12AQ>. Could give 650 mW output at 160 MHz from B+ of 120 V; could oscillate to 1 GHz. Used as low-microphonics video preamp in RCA TK-12 TV camera. Variants: 7895, 8056, 8393.

7587 Nuvistor sharp-cutoff UHF tetrode, w/ plate cap, heater 6.3 V @ 150 mA, 2.2 W, g_m 10.6 mS (GE, RCA, SY) (3-20-61) <12AS>. Used in King aircraft xcvrs, radar-beacon IF strips, etc.

7588 Cermet triode, amplifier variant of 7296, usable to 500 MHz, μ 175, NF 3.0 dB @ 200 MHz (GE) (7-18-60).

7589 Magnetron, pin-tuned design, rapidly tunes 8.5-9.6 GHz, IM, WG, heater 20.0 V @ 4 A, 40 kV @ 40 A, 425 kW pulse output, forced-air cooled (WE) (12-4-61).

7590 Hydrogen thyratron, 30 kV (11-26-62).

7591, A Beam power tube, promoted for Ultra-Linear use, T9 bulb on octal base, heater 6.3 V @ 800 mA, 450 V, 19 W (GE, RCA, SY, TS, W) (2-1-60) <8KQ>. "A" has same ratings. Variants: 6GM5, 7868.

7592 TWA, 5.4-5.9 GHz, EM, CO, heater 6.3 V @ 3 A, 4.5 kV, 30 dB gain, 2 W output (SP) (12-10-62).

7595-99 "Misc." types (HY) (all 6-20-60).

7600-02 "Misc." types (HY) (all 6-20-60).

7603 Hydrogen thyratron (1-14-63).

7604 Xmtg. triode (3-13-61).

7605 [AKA CH-1097?]; hydrogen thyratron, for crowbar use, dual-stepped tubular body, mounting flange, flying leads, 30 kV @ 2500 A peak (TS) (10-17-60).

7606 Counter tube (12-7-59).

7607 Beam power tube, 6146 format, for RF use incl. linears, heater 6.3 V @ 1.6 A, 23 W (between 6146A and 6146B ratings) (GE, TS) (1-11-60) <7CK>.

7609 Xmtg. tetrode, glass body, heater 26.5 V @ 620 mA, 250 W, 150 MHz, sim. to 4X150D but w/ extra resistance to vibration (AX, EI) (9-19-60).

7610 [AKA SS502]; gas-discharge tube (KR).

7611 Image orthicon (11-23-59).

7619 Magnetron, fixed at 34,860 ± 348 MHz, 40 kW output (LI) (3-6-61).

7621 [AKA KU-70C, HY-2]; hydrogen thyratron, cermet construction, anode fitting at top, grid on ring, cathode on mounting flange, stud heater terminals at base, 8 kV, 100 mA, 350 kW pulse output, w/ H_2 reservoir, replaces 1258, 3C45, and 4C35 (EG&G) (5-21-62).

7623 Beam power tube, T16 bulb 4.5" high, octal base w/ cap, heater 6.3 V @ 1.6 A, 37.5 W (GE) (8-6-62) <6AM>. Variant: 7624.

7624 Beam power tube, 7623 w/ heater 12.6 v @ 800 mA (GE) (8-6-62) <6AM>.

7625 Rcvg. high-μ triode; stacked, lugged cermet construction like that of 7462; for audio preamps, heater

6.3 V @ 215 mA, μ 80 (GE) (2-15-60).

7626 T2X3 submin. xmtg. power pentode, fil. 1.25 V @ 125 mA, 1.1 W (GE) (8-13-62) <7626>.

7629, A Image orthicon, semiconductive target, S-10 response, for studio and remote B&W and color use (GE, RCA) (12-21-59).

7630 Magnetron, tunes 16.2-22.0 GHz, IM, WG, heater 4.8 V @ 12 A, 24 kV @ 52 A, 70 kW pulse out (RA) (10-30-61).

7631 Min. duodiode, premium 6AL5 (SY) (4-3-61) <6BT>.

7632-34 Phototubes (12-28-59).

7635 BWO, 7.0-11.5 GHz, WG, heater 6.3 V @ 1.7 A, 350 V, 180 mW pulse output (MU) (12-28-59).

7636 Backward-wave osc. (12-28-59).

7638 TWA, 7.0-11.5 GHz, WG, heater 6.3 V @ 500 mA, 1.35 kV, 20 dB gain, 25 dB NF, 6 mW out (MU) (12-28-59).

7639 TWA, 11.5-18 GHz, WG, heater 6.3 V @ 400 mA, 1.85 kV, 20 db gain, 28 dB NF, 3 mW out (12-28-59).

7640 TWA, 2.0-4.0 GHz, IM, CO, heater 6.3 V @ 3 A, 8 kV, 30 dB gain, 1 W output (IT) (5-21-62).

7641 TWA (11-12-62).

7642 TWA, 1.7-2.3 GHz, packaged, IM, CO via "N" fitting, heater 6.3 V @ 1.75 A via flying leads, 3 kV, 30 dB gain, 20 W output (RCA) (10-23-61). Used in RCA MM-600-2 600-channel microwave system.

7643 [AKA E80CF]; triode-pentode, premium 6BL8 w/ internal shield, heater cut to 330 mA (AX, SY) (5-9-60) <9AE>.

7644 Planar triode, 6299 for in radar receivers where grid current flows due to leakage from xmtr (GE) (3-14-60).

7645 Min. dual tetrode, 6939 w/ shortened bulb, 5.5 W (AX, GE) (12-23-63) <9HL>.

7649 Xmtg. cermet beam tetrode related to 4622, appears to be pulse-rated variant of 6816, heater 6.3 V @ 3.2 A, 115 W, 2 GHz, forced-air cooled (RCA) (8-1-60).

7650 Xmtg. UHF cermet tetrode, fil. 6.3 V @ 7.85 A, 700 W, 1215 MHz, coaxial construction, aligned grids, forced-air-cooled. For general use, incl. UHF TV in RCA Y1246 cavity. Ruggedized: promoted for missile use (AX, RCA) (3-14-60). Variant: 7651. Also see 4628.

7651 Xmtg. tetrode, pulse-rated 7650, 8 kV @ 9 A, giving 15 kW pulse at 1215 MHz (RCA) (3-14-60). Usable in Y1015 cavity for 950-1225 MHz.

7656 Rectifier (2-13-61).

7657 Counter tube (4-25-60).

7658 Rectifier (12-18-60).

7664 Photomultiplier (2-15-60).

7665 [AKA KU-72].

7667 [AKA KU-74]; hydrogen thyratron (11-26-62).

7668 HV triode, pulse modulator, 5-1/8" dia. by 11-5/8" long, heater 12.6 V @ 29 A, 150 kV, 6.5 A pk., 750 W (ML) (4-16-62).

7669-73 Ignitrons, stainless steel case, water-cooled, coaxial current path (all GE).

7669 1.5 kV PIV, 18 A (9-19-60).

7670 For welder-control use; 2.5 kV PIV, 10 A avg. out (9-26-60).

7671 Larger (5552 format), for AC control or rectifier use, 500 V PIV, 100 A (8-22-60).

7672 Larger (5552 format), for frequency-changer welder use, 1.5 kV PIV, 16 A avg. (5-1-61).

7673 5553 format, for rectifier use, 1.5 kV PIV, 84 A (5-22-61).

7681 Larger ("jumbo size C"), for frequency-changer welder use, etc., 1.5 kV PIV, 16 A avg. (9-5-60).

7675 Ionization gauge (8-28-61).

7676 Ionization gauge (7-18-60).

7677 Ionization gauge (4-24-61).

7680 CW magnetron, voltage-tuned 1.0-2.3 GHz, IM, heater 2.6 V @ 3.0 A, 2 kV @ 20 mA, 1 W out (GE) (12-31-62).

7681 See 7669.

7682 Storage tube (4-3-61).

7683 Min. 9-pin HV pentode, for shunt regulators or as control tube for series regs., heater 6.3 V @ 150 mA, g_m 5.6 mS, 1 kV, 15 W (GE, VN) (6-13-60) <9MN>.

7685 Xmtg. triode (11-28-60).

7686 Gas noise source (2-6-61).

7687 Min. 9-pin triode-pentode, low hum and microphonics, heater 6.3 V @ 500 mA, triode μ 18, pentode g_m 5.8 mS (GE, SY) (3-7-60) <9AE>.

7688-90 Compactron triple-triodes, T7½ bulbs, 12-pin bases, heaters 6.3 V @ 450 mA (GE, SY) (8-1-60) <12BA>.

7688 Med.-μ, μ 17; equals 1-½ 12AU7s. Variant of 6MJ8, different pinout.

7689 High-μ, μ 100; equals 1-½ 12AX7s. Variant of 6C10, different pinout.

7690 High-μ, μ 60; equals 1-½ 12AT7s.

7692, A [AKA M-4193B]; magnetron, hand- or servo-tuned, tunes 9.2-9.6 GHz, IM, WG, heater 13.8 V @ 3.2 A, 22 kV @ 21 A, 220 kW pulse output (LI, SY) (5-22-61). "A" tunes 8.5-9.6 GHz.

7693 [AKA E90F]; min. pentode, mobile-rated 6BH6 (AX, SY) (6-13-60) <7EN>.

7694 [AKA E99F]; rcvg. RF pentode, premium 6BJ6 (AX, SY) (6-13-60) <7EN>.

7695 Audio beam power tube, 9-T9 format, heater 50 V @ 150 mA, g_m 11.0 mS, 16 W, for low B+ voltages like 150 V (GE, RCA, SY, W) (3-29-60) <9PX>. Variant: 7754.

7696 Photomultiplier (9-19-60).

7697 Vidicon, for B&W industrial use, S-18 response, 1" dia. by 6-¼ long, all-glass, "ditetrar" 8-pin base (RCA) (6-13-60).

7698 Planar pulse triode, 2C39 format w/o radiator, w/ knurled knob on anode, conduction-cooled, heater 6.3 V @ 1.3 A, 10 W, 3 GHz; can give 2.5-kW pulses at 3 GHz (EI) (4-10-61).

7699 Rcvg. wideband dual pentode (?) for test gear, distributed amps., etc.; heater 6.3 V @ 600 mA, 6 W (AX) (3-7-60).

7700 Pentode, premium 6C6 (KR) (non-RMA no.) <6F>.

7701 Min. 9-pin xmtg. beam power tube, heater 13.6 V @ 160 mA, 9 W, for Class C amps (GE, SY, TS) (5-23-60) <9MS>.

7702 Storage tube (5-29-61).

7703 Ignitron, stainless steel case 2-¼" dia. by 7-5/8" long, for capacitor-discharge and crowbar use, 20 kV PIV, 100 kA peak (GE) (5-26-58).

7704 Xmtg. tetrode (1-7-63).

7709-11 Submin. cold-cathode trigger tubes (AX).

7709 [AKA Z70W]; for DC ckts. Variant: 7710.

7710 [AKA Z70U]; 7709 w/ higher starter current.

7711 [AKA Z71U]; w/ two starters for counting and switching in DC ckts.

7713 [AKA Z804U]; min. cold-cathode trigger (AX).

7714 [AKA Z805U]; min. cold-cathode relay tube for AC ckts. (AX).

7715 Xmtg. shielded-grid beam triode, for pulse modulators, glass external-anode construction, heater 6.0 V @ 60 A, 3 kW, forced-air- and liquid-cooled by immersion (EI, ML) (12-12-60). Ruggedized variant is ML-LPT-62.

7716 Rcvg. min. 9-pin triode-pentode, mobile-rated, heater 13.6 V @ 350 mA. Triode is high-μ (μ 102); pentode is sharp-cutoff, gm 10 mS (GE) (10-24-60) <9DX>.

7717 Min. pentode, 6CY5 mobile-rated at 6.3 V +1.2/-0.3 V @ 200 mA (GE, RCA) (5-23-60) <7EW>.

7718 [AKA QK624]; magnetron, fixed at 9375 MHz, IM, WG, heater 3.6 V @ 36 A, 39 kV @ 69 A, 1 MW pulse output (RA) (11-6-61).

7719 Min. 9-pin med.-μ rcvg. triode, heater 12.6 VCT @ 225 mA, μ 25 (GE) (8-22-60) <9MX>.

7720 Planar cermet triode, 7462 for UHF oscs., μ 90, 1.0 W, gives 100 mW @ 450 MHz (GE, SY) (12-12-60).

7721 [AKA D3a]; min. rcvg. pentode, heater 6.3 V @ 320 mA, g_m 35 mS, 4 W (HY) (6-13-60) <9EQ>. Variant: 7722.

7722 [AKA E280F]; min. pentode, 7721 w/ g_m reduced to 26 ms (GE, HY) (6-13-60) <9EQ>.

7723 [AKA 610]; gas rectifier, w/ cap, fil. 2.5 V @ 9.0 A (GE) (11-7-60).

7724 Min. duodiode-triode, 14GT8 mobile-rated at 12-15 V @ 150 mA; rated for 2000 on-off cycles (GE, RCA) (10-24-60) <9KR>.

7725 [AKA 710A]; thyratron, 6011 uprated to 3.5 kV PIV (GE) (3-21-60).

7726 [AKA 710AL]; thyratron, 7518 uprated to 3.5 kV PIV (GE) (3-21-60).

7728-34 Series of premium tubes; parameters being optimized not stated (GE) (7-24-60).

7728 Equals 12AT7.

7729 Equals 12AX7.

7730 Equals 12AU7.

7731 Equals 6U8.

7732 Equals 6CB6.

7733 Equals 12BY7.

7734 Equals 6GE8.

7735, A, B Vidicon, 1" dia., 7697 format, sulfide target, for B&W or color industrial TV cameras and film chains (GE, HI, RCA) (4-25-60). "A" has doubled sensitivity, low lag. 7735 and "A" were originally rated for broadcast remote-pickup use (B&W or color) or industrial applications; later only for "noncritical industrial and consumer product" use, in favor of "B." Variants: 4478, 4488, 8507.

7736 Ignitron, 5564, for high-voltage and phase-retard operation (GE) (8-22-60).

7737 [AKA E186F]; min. 9-pin "10,000-hr." frame-grid RF pentode, non-microphonic 6688, heater raised to 320 mA (AX, GE) (10-7-63) <9MZ>.

7738 Min. triode, premium 6AN4, originally for sonobuoys, silvered pins, 5.0 W (GE, SY) (5-23-60) <7DK>.

7739 Xmtg. planar triode (4-18-60).

7740 Ignitron (6-25-62).

7741 Counter tube (3-6-61).

7745 "Multiple section" tube (11-21-60).

7746 Photomultiplier, 10-stage, T16 bulb, spherical-faceplate end-on type, S-11 response, diheptal base (RCA) (5-23-60).

7747 BWO, voltage-tuned 26.0-40 GHz, IM, WG, heater 6.3 V @ 300 mA, 700 V, 50 mW out (STC) (2-6-61).

7748 Xmtg. triode (4-3-61).

7750 Xmtg. triode (12-18-60).

7751 [AKA E235L]; rcvg. beam power tube, T10 bulb 4.1" high, octal base, heater 6.3 V @ 1.2 A, 10 W (GE, HY) (6-13-60) <8KB>.

7752 [AKA 6AS6, q. v.]; min. pentode.

7753 RF-heating VHF triode, fil. 6.3 V @ 65 A, 2.1 kW, forced-air-cooled, external anode (AX) (2-15-60).

7754 Rcvg. beam power tube, variant of 7695 w/ heater 6.3 V @ 1.2 A (SY) (5-30-60).

7755 [AKA 6AJ5, q. v.]; min. pentode.

7756 [AKA 6AR6, q. v.]; power tetrode.

7757 Min. 9-pin beam power tube for pulse/sweep use, hard-glass bulb w/ cap, heater 6.3 V @ 600 mA, g_m 4.1 mS, 14 W (GE, TS) (5-15-61) <9NE>.

7758 Beam power tube, premium 6AR6.

7759-62 T3 submin. types, 26.5-V heaters (GE, SY) (8-1-60).

7759 Med.-μ duotriode, heater 90 mA, 1.1 W <8DG>.

7760 Duotriode, 7759 for 26.5-V B+ supply, μ 20 <8DG>.

7761 Semi-remote-cutoff pentode, heater 110 mA, g_m 9 ms, 4 W <8DL>.

7762 Power pentode, heater 110 mA, gm 4.2 mS, 4 W <8DL>.

7763 Min. 9-pin double-plate sheet-beam tube, for IF amp.-limiter use, heater 6.3 V @ 300 mA (GE) (5-29-61) <9NF>.

7764 Photomultiplier, 6-stage, T6 bulb, flat-faceplate head-on type, S-11 response, all-glass, 0.78" dia. by 2-3/4" long, "ninar" 9-pin base (RCA) (7-18-60).

7767 Photomultiplier, 10-stage, T6 bulb, flat-faceplate head-on construction, S-11 response, high sensitivity, 0.78" dia. by 4" long, 12 flying leads (RCA) (7-18-60).

7768 High-μ cermet planar triode, for broadband RF amps., heater 6.3 V @ 400 mA, μ 225, 5.5 W (GE, RCA) (1-1-62).

7779 [AKA XD56]; rectifier (11-28-60).

7782 [AKA KU-71]; hydrogen thyratron (11-26-62).

7783 Backward-wave oscillator (11-11-63).

7784 Planar triode, 6299 w/ isolated heater (GE) (2-13-61).

7785 BWO, 12.4-18.0 GHz, heater 6.3 V @ 840 mA, 1.9 kV, 10 mw output (HU) (8-8-60).

7786 [AKA NL-660L]; xenon rectifier, fil. 2.5 V @ 21 A (GE) (5-15-61).

7788 [AKA E810F]; min. 9-pin rcvg. frame-grid "10,000-hr." pentode, heater 6.3 V @ 340 mA, g_m 50 mS, 45 mA, 5 W (AX, GE) (5-8-61) <9NK>. Low-noise RF amp. in Mackay Marine HF rcvr.

7789-93 Hydrogen rectifiers, for radar charging diodes, finned anode cap, flange-mounted, flying leads (all TS).

7789 Fil. 2.5 V, 15 kV PIV, 400 mA (4-3-61).

7790 Fil. 5.0 V, 20 kV PIV, 1 A (6-11-62).

7791 Fil. 5.0 V, 25 kV PIV, 2 A (6-11-62).

7792 Fil. 11.0 V, 25 kV PIV, 2 A (4-3-61).

7793 Fil. 5.0 V, 25 kV PIV, 4 A (4-6-64).

7794 CW magnetron, for telemetry and comm. use, 4.2-4.4 GHz, IM, CO, heater 6.3 V @ 600 mA, 600 V @ 100 mA, 5 W output (W) (4-10-61).

7795 CW magnetron, fixed at 4.3 GHz, IM, CO, heater 6.3 V @ 600 mA, 425 V @ 30 mA, 10 W output (W) (3-13-61).

7796 CW magnetron, voltage-tuned 4.2-4.4 GHz, IM, CO, heater 6.3 V @ 600 mA, 350 V @ 30 mA, 1.5 W output (W) (3-13-61).

7800 Xmtg. HF triode, fil. 8 V @ 130 A, 15 kW, forced-air-cooled, for all services including hard-tube radar modulators (AX) (10-9-61).

7801 Xmtg. cermet beam tetrode, heater 12.6 V @ 500 mA, ca. 15 W, 400 MHz, conduction-cooled (RCA) (2-13-61). Variant: 7870.

7802 Low-μ power duotriode, octal base, heater 6.3 V @ 2.5 A, μ 8.5, 13 W (GE) (3-6-61) <8BD>.

7803 Min. 9-pin frame-grid duotriode, premium 6FW8 w/ heater cut to 365 mA, g_m 12.5 mS, 3.5 W, dev. for 200-MHz sonobuoys (GE, TS) (10-24-60) <9AJ>.

7804 RF-heating HF triode, fil. 6.3 V @ 130 A, 10 kW, forced-air-cooled, external anode (AX) (11-21-60) (11-21-60). Variant: 7805.

7805 Xmtg. triode, water-cooled variant of 7804, uprated to 15 kW (AX) (11-21-60) (11-21-60).

7806 RF-heating triode, external anode, fil. 8 V @ 120 A, 15 kW, forced-air-cooled (AX) (10-10-60). Variant: 7807.

7807 RF-heating triode, water-cooled version of 7806 (AX) (10-10-60). Variant: 8594.

7813 Indicator thyratron.

7815, AL, R [AKA 3CPN10A5]; planar triode, 2C39 format w/o radiator, heater 6.0 V @ 1.0 A, 10 W, 3 GHz, conduction-cooled (EI, GE, ML, SY) (4-10-61). For transponders and DME: can give 1.6-kW pulses at 3 GHz. In AN/APX-46 IFF transponder; gave 500 W at 1010-1030 MHz. Variant: 7815R. "R" [AKA 3CPX100A5, 8745, q. v.] is radiator version, uprated to 100 W. Also see 7698.

7817 Photomultiplier (9-19-60).

7818-19 Photomultipliers (both 10-24-60).

7820 Xmtg. triode (12-12-60).

7821, 23 TR tubes (both 10-15-62).

7824 TWA, 5.2-5.9 GHz, EM, CO, heater 6.3 V @ 770 mA, 250 VV, 50 mW output (10-3-60).

7828 Storage tube (4-10-61).

7829, A Klystron (5-31-66).

7831 Xmtg. triode (2-18-63).

7832 Xmtg. triode (11-25-63).

7835 Xmtg. cermet double-ended triode, fil. 3.1-4.2 V @ 6600 A, 300 kW, 300 MHz, can be pulsed to 10 MW output; also for use in particle accelerators; water-cooled via integral cooler (RCA) {A2346} (2-13-61). Also see 2054.

7837-39 Xmtg. triodes (9-25-61).

7841 Cermet planar diode, 7266 w/ lower PIV (350 V) and higher current rating (5.5 mA) (GE, SY) (1-2-61).

7842 Xmtg. cermet tetrode, for linear amps., heater 6.3 V @ 3A, 115 W, 1215 MHz, conduction-cooled (RCA) (11-28-60). Variants: 7843, 7844.

7843 Xmtg. tetrode, 7842 w/ heater 26.5 V @ 520 mA (EI, RCA) (11-28-60).

7844 Xmtg. UHF cermet tetrode, for general use incl. linear amps., same package as 7842 but heater 6.3 V @ 2.1 A, 115 W, 1215 MHz, conduction-cooled (RCA) (11-28-60).

7845 Xmtg. triode (2-26-62).

7846 Cadmium-selenide photocell, bulb 0.29" dia by 0.5", two leads (RCA) (11-7-60).

7847 TWA, 5.0-6.0 GHz, EM, CO, heater 6.3 V @ 2.2 A, 3 kV, 27 dB gain, 10 W output (IT) (11-6-61).

7848 TWA, 5.0-6.0 GHz, 2 kW output.

7850 Photomultiplier, 12-stage, T16 bulb, spherical head-on faceplace, S-11 response, bidecal Micanol base (RCA) (2-13-61).

7851 Min. electrometer tetrode, grid on cap, heater 2.5 V @ 200 mA, optimized for B+ of 12 V, g_m 0.04 mS, μ 5 (3-6-61) <7GE>.

7854 Xmtg. dual VHF beam tetrode, 5894 format, fil. 12.6 VCT @ 900 mA, 60 W, heat-sink or forced-air-cooled, self-neutralized (AX) (9-18-61). Variant: 8624.

7855, KAL Planar triode, 2C39 format, heater 6.0 V @ 1.0 A, 100 W, 3 GHz, forced-air cooled; for high freq. stability as CW osc. to 2.5 GHz, or may be pulsed to 3.0 GHz (EI, GE, ML) (8-7-61). "KAL" has knurled knob on anode in place of fins; is derated to 10 W. Variant: Y-503.

7857 TWA, 2.4-2.7 GHz, 20 dB gain, 250 mW output.

7859 See 8089.

7860 Photomultiplier (2-6-61).

7861 Min. duotriode, 5670 w/ heater 12.6 V @ 175 mA (GE) (3-20-61) <8CJ>.

7866 [AKA KU-274]; hydrogen thyratron (12-10-62).

7867 Beam power tube, for audio use, octal base w/ cap, heater 6.3 V @ 2.5 A, 24 W (GE) (1-16-61) <5BT>. Same format and pinout as 6EX6 sweep tube; rated 2 W higher.

7868 Beam power tube, novar variant of 7591A (GE, RCA) (6-19-61) <9RW>.

7869 Rectifier (7-3-61).

7870 Xmtg. tetrode, 7801 w/ heater 6.3 V @ 1 A (RCA) (2-13-61).

7872 Hydrogen thyratron, 5C22 w/ different base.

7880 Gas noise source (7-3-61).

7887 T3 submin. med.-μ duotriode, heater 26.5 V @ 90 mA, μ 20, 2.2 W total (GE, SY) (1-23-61) <8DG>. Variant: 7889.

7888 T3 submin. med.-μ triode, heater 26.5 V @ 45 mA, μ 27 (GE, SY) (1-23-61) <8DK>.

7889 T3 submin. duotriode, 7887 w/ μ 70 and 1.1 W total (GE, SY) (1-23-61) <8DG>.

7890 Hydrogen thyratron, cermet ring-seal construction, for radar pulsers, linear accelerators, etc., heater 6.3 V @ 33 A, 40 kV PIV, 4.0 A avg., cooled by forced air and/or water via integral cooler (GE) (2-13-61).

7892 Min. 9-pin pulse duotriode, heater 12.6 VCT @ 450 mA, pulse cathode current 5 A, 8.4 W total (TS) (3-13-61) <9H>.

7894 See 8089.

7895 Nuvistor high-μ triode, 7586 with μ raised from 35 to 64, g_m 9.4 mS (GE, RCA) <12AQ>.

7898 Min. duotriode, 12AZ7 mobile-rated at 12-15 V @ 150 mA (no heater tap) (GE, RCA) (2-19-62) <9EP>.

7899 Xmtg. triode, fil. 8 V @ 130 A, 8500 W, forced-air-cooled, for hard-tube modulators (AX) (10-9-61).

7900 Xmtg. VHF triode, fil. 12.6 V @ 32 A, 4 kW, forced-air-cooled, for VHF TV xmtrs (AX) (5-22-61).

7901-02 TR tubes (both 10-15-62).

7903 Ionization gauge (4-3-61).

7904 Ionization gauge (4-10-61).

7905 Min. 9-pin beam power tube, quick-heating fil., mobile-rated at 6.3 V ± 20% @ 650 mA, 10 W, 175 MHz (GE, RCA) (12-18-61) <9PB>.

7906-07 Rectifiers (both 2-26-62).

7908-09 Photomultipliers (both 2-13-61).

7910 Cermet planar triode, for pulsed UHF use, htr. 6.3 V @ 275 mA, μ 200, 1.5 W, gives 100 W pulses @ 5.9 GHz (GE).

7911 Cermet planar triode, like better insulated 7910, for pulsed UHF use, htr. 6.3 V @ 550 mA, μ 58, 1.5 W, gives 2.2 kW pulses @ 4.1 GHz (GE).(11-9-64).

7913 Cermet UHF high-μ planar triode sim. to 7911, htr. 6.3 V @ 400 mA, μ 100, 5.5 W, gives 4 W output @ 400 MHz (GE) (12-30-63).

7950 Magnetron, for long- or short-pulse ground radars, tunes 8.5-9.6 GHz, pin-tuned design, IM, heater 22 V @ 1.9 A, 40 kV, 30 A, 200 kW pulse output, forced-air cooled (LI, WE) (12-18-61).

7959 Magnetron, 2.70-2.85 GHz, 3.5 MW output.

7962 T3 submin. duotriode, heater 6.3 V @ 240 mA, μ 21 (GE) (5-29-61) <8DG>.

7963 T3 submin. duotriode, heater 6.3 V @ 350 mA, μ 40 (GE) (5-29-61) <8DG>.

7964, -1 Triggered spark gap, cermet construction w/ plastic dip coat, solder lugs, body 1.32" high, switches up to 18 joules at high voltage, operating voltage 2.0-2.6 kV, trigger voltage 5 kV, peak pulse current 4000 A (GE) (3-6-61). "-1" has lugs perpendicular to body.

7967 Image orthicon, 5655 format, semiconductive target, S-20 response, high sensitivity, for low-light B & W use (GE, RCA) (4-3-61).

7969 Image orthicon, MgO target, UV-sensitive, for near-UV biological research, etc. (GE) (4-3-61).

7971 Rcvg. pentode (2-13-61). Also reported to be T16-size "Nixie" numeric readout.

7972 Xmtg. tetrode (2-13-61).

7973 Rcvg. tetrode (2-13-61).

7974 Xmtg. triode (2-13-61).

7975 Rectifier (4-23-62).

7976 Magnetron, tunes 13.6-15.6 GHz, IM, WG, heater 12.6 V @ 3.25 A, 20 kV @ 5 A, 135 kW pulse output (WE) (12-18-61).

7977, A [AKA 4032]; digital readout (RCA) (5-29-61).

7978 Beam-switching decade counter (12-16-63).

7979 [AKA CK1050]; T2 submin. gas indicator triode, 4 leads, fil. 1.25 V @ 250 mA, 3 mA (GE) (7-16-62) <7979>. Also see 7323.

7980 [AKA 83A1]; voltage reference, 83 V, 4.5 mA (AX) (2-13-61).

7981 Gas thyratron (2-13-61).

7982 Xmtg. triode (2-13-61).

7983 Min. 9-pin xmtg. tetrode, quick-heating fil., 3.15 V @ 1.65 A, 7 W, self-neutralized, for hybrid mobile xmtrs (AX, GE) (10-9-61) <9PS>.

7984 Compactron beam power tube, for VHF mobile xmtrs, single-ended, duodecar base, heater 13.5 V @ 580 mA, 35 W intermittent, 175 MHz (GE) (3-19-62) <12EU>. In GE mobile radios; in LPB Inc. RC-25B 25-watt AM carrier-current xmtr. Variant: 8149.

7985 Xmtg. UHF tetrode, ring-seal cermet construction, for G-G use, heater 6.7 V @ 14.5 A, bombardment cathode taking 650 V, 3.5 kW, 800 MHz, water-cooled via integral cooler (GE) (2-20-61).

7986 [AKA 5866A] (AX) (1-21-63).

7988 Gas noise source (2-11-63).

7989-91 Gas noise sources (all 7-3-61).

7992-93 Gas noise sources (both 6-10-63).

7994 T3 submin. triode, heater 6.3 V @ 250 mA, μ 42 (GE, SY) (4-10-61) <8KM>.

7995 T3 submin. sharp-cutoff pentode, heater 6.3 V @ 250 mA, g_m 13 mS (GE, SY) (4-10-61) <8KZ>.

7996 Rectifier (11-5-62).

7998 Ignitron, "jumbo size C" steel case, 7681 ratings but a little smaller, coaxial current path (GE) (6-18-62).

7999 Gas noise source (8-20-62).

8000-26 (Non-RMA numbers.)

8000 Xmtg. triode, low-μ variant of 810 (μ 16.5) for general use, intro. 1-41 (GE, RCA, W) (8-10-45) <2N>.

8001 [AKA 4E27, q. v.]; xmtg. pentode, intro. 6-41 (GE) <7BM>.

8002, R RF-heating VHF triode, unusually small: 3-11/16" dia. by 5-3/8" long; fil. 16 V @ 38 A, μ 21.5, 1.2 kW, 150 MHz, water-cooled (GE). "R" is radiator variant, derated to 120 MHz (GE). Pair in GE 3-kW FM xmtr and 1-kW TV xmtr ca. 8-40.

8003 [AKA WL-471]; xmtg. triode, 211 w/ cap (GE, RCA, W) {A2126}.

8004 {R6124}.

8005 Xmtg. triode, 203A w/ cap, pure-tung. fil. 16.0 V @ 38 A, derated to 75 W, 60 MHz {A2130A} (GE, RCA, W) (9-24-43) <3G>; roughly sim. to 75TH.

8007 [AKA VT-219]; xmtg. beam power tube; cancelled, superseded by improved 807 {A4051H}.

8008 Mercury rectifier, 872A/872 in T18 bulb on long-pin super-jumbo "industrial" base {R6146} (AX, CE, EE, GE, GN, HY, ML, RCA, W) (9-24-43) <2P>.

8009 Xmtg. triode, slenderized 880, four grid pins instead of two, same ratings (GE).

8010-R, A-R Xmtg. UHF triode, disc-seal, like 2C33 except w/ radiators at both ends, bombardment cathode, μ 30, 50 W, intro. 1-42 (GE). "A-R" is improved, rated to 350 MHz {ZP449}. Two in GE prewar 260-300 MHz studio-transmitter link xmtr.

8011 [AKA VT90 (Br.), (WE) 710A, REL1, 10E/97, WL-538, VT-240]; disc-seal Br. VHF xmtg. radar osc. triode, forced-air-cooled, first of several "micropup" types in experimental British airborne-interception (AI) and air-to-surface-vessel (ASV) radars, operable up to 600 MHz {A2137}. NE and STC (Aust.) made 710As in bulk. Two in BC-702A xmtr. of SCR-521 (200-MHz airborne radar), and in the ASE Navy equivalent.

8012, A [AKA VT-228, CV662]; xmtg. VHF triode, superseding 1628, tubular all-glass construction w/ two side pins each for plate and grid; center-tapped fil. on flying leads; fil. 6.3 V @ 1.92 A total; μ 18, 40 W w/ forced-air cooling, 500 MHz (GE, RCA) {A2139, ZP-430} (3-28-45) "A" in T-39/APQ-9 radar jammer (5-15-45).

8013, A Xmtg. vac. rectifier, ST16 or stepped tubular hard-glass bulb w/ corona-shielded cap, UX base, usable in oil bath; fil. 2.5 V @ 5.0 A, 40 kV PIV, 20 mA (GE, HY, RCA, T) {R6137B} (9-24-43) <4P>. Three used as tripler in 50-kV supply of RCA electron microscope, 1948.

8014, A VHF pulse triode, smaller version of 6C24, fil. 15 V @ 14.5 A, 13.5 kV, μ 30, 400 W {A2124M}.

8016 Vacuum HV rectifier; prototype of 1B3GT; octal w/ cap, fil. 1.25 V @ 200 mA, 10 kV, 2 mA {R6158} (RCA) <4AC>.

8017 Xmtg. triode {A2140B}.

8018 [AKA VT60 (Br.), 807, q. v.].

8019 [AKA DET12, VT62 (Br.), NT58, 10E/11443]; 834 on Br. 4-pin base {A2009J}.

8020, W [AKA 100-R, RY12-100, WL-578, (GL, ZP) 451, 2-100A, VT-267]; vac. rectifier, T18 tubular bulb w/ cap, jumbo 4-pin base, T-T fil. 5.0 V @ 6.0 A, 40 kV, 100 mA (HY, LK, RCA, W) {R6174} <4P>. Used in radars and electrostatic precipitators; may also appear to be 100TH, less grid (AX, EE, EI, GE, GN, HY, MU, RCA, W) (11-13-42). Damper in SCR-584. Two in xmtr of SC-2; one in SCR-602-T6.

8021 [AKA VT-255]; xmtg. vacuum radar rectifier, fil. 5.0 VCT @ 5 A, 30 kV PIV, 150 mA; WE 705A uprated on current.

8022 [AKA 4B30, q. v.]; xmtg. rectifier.

8023 [AKA 4C30, q. v.]; xmtg. triode.

8024 [AKA 832A, q. v.]; xmtg. dual tetrode {A359B}.

8025, A Xmtg. triode, 8012 w/ UX base and grid/plate caps, not pins; same ratings (2-26-43) <4AQ>. "A" (RCA, ML) (5-15-45) has slightly lower C_{gp}. In T-3/CRN-2.

8026 [AKA 4C27, q. v.]; xmtg. triode.

8030, 31 Gas noise sources (8-20-62).

8032, A [AKA 8552, q. v.]; xmtg. beam power tube, "dark" heater 13.5 V @ 585 mA (GE, RCA) (4-24-61) <7CK>. For "A" [AKA 6883B/8032A/8552], see 6883B (3-16-64).

8033 Xmtg. triode (12-3-62).

8034 Vacuum diode, external-anode, for rectifier or clipper use, flying leads for fil., fil. 7.5 V, 25 kV PIV, 3 kW dissipation, forced-air cooled (DuM) (11-27-61).

8035 [AKA CT-4251]; beam-switching counter (8-27-62).

8036 Hydrogen thyratron, cermet 5949A, w/ mounting flange, flying leads, 25 kV, 500 mA, 6.5 MW output (TS) (3-23-64).

8037 [AKA 5031]; "misc." type (5-29-61).

8038 Xmtg. triode (2-26-62).

8040-41 Xmtg. triodes (2-26-62).

8042 Beam power tube, quick-heating variant of 6146A, for two-way radios, fil. 1.6 V @ 3.2 A (AX, GE, RCA) (7-24-61) <8LJ>. Two in Transcom SBT-3 xcvr. AX version uses "harp" parallel-multiwire fil.; claims 100-msec warmup. Related to 4652, which is a lower-cost version w/ longer heating time of 1.25 sec.

8043 Thyratron, 4-pin base w/ cap, fil. 5.0 V @ 6.0 A (GE) (4-24-61).

8044 Xmtg. triode (7-17-61).

8045 Xmtg. triode (11-25-63).

8049 Phototube, S-4 response (3-27-61).

8051 Vidicon, all-glass,, 1-½" dia. by 7-3/4" long for high-resolution B&W or color film chains, S-18 response, "super-ditetrar" 8-pin base (RCA) (7-17-61). Has magnetic focusing and internal collimating lens.

8052 Reflex klystron, 4.0-11.0 GHz, heater 6.3 V @ 680 mA, 1.25 kV, 240 mW output (LI, PO) (8-20-62).

8053 Photomultiplier, 10-stage, S-11 response, 5819 format w/ flat face, for scintillation counters (RCA) (10-23-61). Variants: 2063, 8054, 8055.

8054 Photomultiplier, 8053 w/ bulb flared to 3" dia. at faceplate (RCA) (5-22-61). Variant: 2064B.

8055 Photomultiplier, 8053 w/ bulb flared to 5" dia. at faceplate (RCA) (9-25-61). Variant: 2065.

8056 Nuvistor triode, 7586 optimized for 24-V plate supply, μ 11.5, derated to 450 mW, g_m 7.5 mS (GE, RCA) {A15247A} (2-19-62) <12AQ>.

8057 Ionization gauge (8-21-61).

8058 Nuvistor high-μ UHF triode, optimized for rcvg. G-G amps., grid on base shell, ceramic ring body, plate cap, heater 6.3 V @ 135 mA, μ 70, g_m 12.4 mS, 1.5 W, NF 12.2 dB @ 1.2 GHz (GE, RCA) (1-8-62) <12CT>. Performance sim. to some planar types; both gave 10-dB NF at 890 MHz. Three used in RCA FD3101 3-stage TRF amplifier for AN/APX-72 1030-MHz IFF transponder.

8059 Gas noise source (8-20-62).

8062 Photomultiplier (1-15-62).

8063 Gas thyratron (5-1-61).

8064 T3 submin. semi-remote-cutoff pentode, heater 26.5 V @ 45 mA, g_m 4.5 mS (GE) (6-26-61) <8DL>.

8065-66 Rectifiers (both 6-19-61).

8067 Storage tube (2-4-63).

8068 Beam pentode, pass tube in regulators, T12 bulb w/ cap, octal base, heater 6.3 V @ 900 mA, 35 W, 3.5 kV, 100 mA (GE, TS) (10-16-61) <8LC>.

8069 Voltage reg., corona type (VN) (5-29-61).

8070 T3 submin. high-μ triode, heater 6.3 V @ 125 mA, μ 55, 1 W (GE) (1-8-62) <8LD>. Variant: 8071.

8071 T3 submin. triode, 8070 w/ different pinout, uprated to 2 W (GE) (5-28-62) <8LE>.

8072, W Cermet UHF beam tetrode, heater 13.5 V @ 1.3 A, ca. 50 W, 500 MHz, mobile-rated and conduction-cooled variant of 8121, for general RF use (EI, RCA) (10-30-61). In Signal One CX-7 xcvr. Variant: 8462.

8073-75 Counter tubes (all 6-19-61).

8077 [AKA 8077/7054]; min. pentode, 7054 w/ bulb shortened (GE, RCA) (9-4-61) <9GK>.

8078 RF-heating VHF triode, fil. 5 V @ 32.5 A, 500 W, radiation-cooled (AX) (5-22-61).

8079 Magnetron, tunes 17.4-19.5 GHz, IM, WG, heater 13.9 V @ 3.5 A, 20 kV @ 5 A, 135 kW pulse output (WE) (1-22-62).

8080 Hydrogen thyratron (9-2-63).

8081 Cermet high-μ triode, stacked-and-lugged bolt-down construction similar to 7296 but smaller, for audio preamps, heater 6.3 V @ 215 mA, μ 80 (GE) (9-4-61). Variants: 8082, 8083.

8082 Planar triode, variant of 8081 for LF-UHF amps., heater 6.3 V @ 240 mA, μ 94, 1.1 W, NF 5.5 dB @ 200 MHz (GE) (9-4-61).

8083 Planar triode, variant of 8081 for UHF oscs., heater 6.3 V @ 240 mA, μ 90, 1.0 W, gives 100 mW @ 450 MHz (GE) (9-4-61).

8084 Min. sharp-cutoff RF pentode, heater 13.5 V @ 160 mA, g_m 10.5 mS (GE) (3-11-63) <7CM>.

8087 [AKA 10E]; scan-conversion tube, for radar use, 900-line resolution, replaces ML 403x-series, in FAA radar use (ML) (9-23-63).

8088 TR tube (10-15-62).

8089-91 T4 submin. two-lead voltage regulators, "Corotron" corona type, regulating at ±2% of nominal voltage (all VN).

- 7859 Regulates at 1.75 kV, 20-800 μA (1-9-61).
- 7894 Regulates at 3.0 kV, 30-850 μA (GE, VN) (7-10-61).
- 8089 Regulates at 1.6 kV, 20-800 μA (7-3-61).
- 8090 Regulates at 3.5 kV (7-3-61).
- 8091 Regulates at 4.0 kV (7-3-61).
- 8256 Regulates at 3.5 kV, 35-700 μA (4-2-62).
- 8257 Regulates at 1.2 kV, 15-600 μA (4-2-62).

8092, A Image orthicon, semiconductive target, for low-light B&W and color broadcast use, S-10 response, 5655 bulb, diheptal base (RCA) (9-11-61).

8093, A, A/L, B Image orthicon, 5655 format, TV use (GE, RCA) (9-11-61). "A/L" has non-burn-in target. Also see 8673.

8094 [AKA 199]; rectifier, 4-pin base, fil. 12.0 V @ 23.0 A (GE) (10-30-61).

8096 T3 submin. high-μ triode, heater 6.3 V @ 200 mA, μ 54 (GE) (7-16-62) <8FY>.

8097 Hydrogen thyratron (2-15-65).

8098 Storage tube (1-8-62).

8099 TR tube (10-15-62).

8100 Cadmium-sulfide photocell, ½" dia. by 0.55" high (GE) (7-24-61).

8102 Min. 9-pin triode-pentode, mobile-rated, heater 13.5 V @ 230 mA; triode has μ of 46; pentode has g_m of 7.5 mS (GE) (2-12-62) <9PJ>.

8103 T3 submin. med.-μ UHF duotriode, heater 26.5 V @ 75 mA, μ 20 (GE) <8DG>.

8104 Xmtg. triode (2-26-62).

8105 Neutron counter, ionization chamber, cylindrical metal bullet-ended housing w/ three "N" coax fittings (W) (11-6-61).

8106 Min. 9-pin beam pentode, for mobile use, heater 13.5 V @ 250 mA, 6 W, g_m 9 mS (GE, RCA) (7-16-62) <9PL>.

8108 [AKA EC157]; rcvg. disc-seal glass med.-μ triode, octal w/ plate caps and threaded grid ring, fil. 6.3 V @ 730 mA, μ 43, 10 W, puts out 500 mW @ 4 GHz, usable to 6 GHz, promoted for common-carrier microwave systems (AX, GE) (12-9-63).

8109-10 Rectifiers (both 12-3-62).

8111 TWA, 3.4-3.7 GHz, IM, CO, heater 6.3 V @ 2.8 A, 33 dB gain, 2 W output (5-21-62).

8113 Min. high-reliability sharp-cutoff RF tetrode, heater 6.3 V @ 200 mA, g_m 7 mS (GE) (8-19-63) <7EW>.

8116 Xmtg. dual tetrode, fil. 26.5 VCT @ 433 mA, 60 W, conduction-cooled, for SSB xmtrs (AX) (5-14-62). Variant: 8117.

8117, A Xmtg. dual tetrode, 6-V/12-V version of 8116, fil. 12.6 VCT @ 900 mA, large-diameter plate pins for heat removal, 60 MHz, promoted for use with both sections in parallel in SSB amps. (AX) (5-14-62). Variant: 8300.

8118 Xmtg. dual UHF tetrode, shortened 5894 format, quick-heating "harp" fil. 1.6 V @ 4.25 A, 20 W, 500 MHz, for hybrid mobile xmtrs (AX) (12-9-63).

8119 Xmtg. UHF coaxial cermet triode, fil. 3.4 V @ 19 A, 400 W, forced-air-cooled, usable to 900 MHz (AX).

8120 Xmtg. UHF triode, fil. 3.4 V @ 19 A, 500 W, forced-air-cooled, coaxial ceramic design, for xmtr use to 1000 MHz (AX).

8121, W Cermet UHF beam tetrode, "elevenar" base, heater 13.5 V @ 1.3 A, 150 W, 500 MHz, 11-disc radiator for transverse air flow, for general RF use (EI, RCA) (10-30-61). Variants: 8072, 8122, 8646.

8122, V1, W Xmtg. beam tetrode, 8121 for axial air flow, uprated to 400 W (EI, RCA) (10-30-61). Related to 8072, 8121, 8828. For VHF TV translator use in Y1222 cavity. "V1" is a matched pair. One in National NCX-1000 xcvr. Two in Hallicrafters SR-2000 and National NCL-2000 linears. Two as drivers in RCA BTA-5L/-10L 5-/10-kW AM xmtrs.

8123 Magnetron, variant of 7608, for airborne MTI applications, tunes 16.0-17.0 GHz, IM, WG, heater 12.6 V @ 3.2 A, 16 kV @ 14 A, 70 kW pulse output, forced-air cooled (WE) (9-10-62).

8124 Storage tube (5-14-62).

8128 TWA, 2.9-3.1 GHz, EM, WG, 8.0 V @ 8 A, 34 kV @ 12 A, 20 dB gain, 60 kW output (RA) (11-11-63).

8129 Amplitron, 2.9-3.1 GHz, IM, WG, 54 kV @ 66 A, 3 mW pulse output (RA) (11-11-63).

8130 Storage tube (5-13-63).

8134, V1 [AKA 8134/4811]; vidicon, sulfide target, 7697 format, electrostatically focused, for B&W or color transistor industrial cameras and X-ray linkage systems, also for B & W or color fil chains; low-power "dark" heater, 6.3 V @ 95 mA (GE, RCA) (4-16-62). "V1" is tested to closer requirements on shading, etc. Used in back-pack portable camera for NBC coverage of 1964 political conventions. Variant: 8567.

8136 Min 9-pin sharp-cutoff UHF pentode, heater 6.3 V @ 300 mA, 2.2 W, g_m 9.8 mS (GE, RCA) (1-1-62) <7CM>.

8137 Counter tube (12-4-61).

8138 Spark gap (9-10-62).

8139 Storage tube (9-23-63).

8142-44 Cadmium-sulfide photocells, 8100 format (GE, SY) (11-6-61).

8146-47 Xmtg. triodes (both 4-6-64).

8148 Xmtg. triode (12-17-62).

8149 Compactron VHF beam power tube, 7984 w/ CT heater, 12.6 VCT @ 600 mA, 35 W ICAS, 175 MHz (GE, SY, TS) (4-16-62) <12DT>. Variant: 8150.

8150 Xmtg. beam power tube, 8149 w/ plate cap (SY, TS) (4-16-62) <12DU>.

8151-52 Gas noise sources (both 8-20-62).

8153 Magnetron, fixed at 34860 ± 348 MHz, 80 kW (AX).

8154 TWA, 7.0-11.0 GHz, EM, CO, heater 6.3 V @ 300 mA, 1 kV, 25 dB gain, 10 dB NF, 5 mW output (GE) (1-22-62).

8156 Compactron xmtg. beam pentode, single-ended, for mobile Class C amps, heater 13.5 V @ 300 mA, 15 W, 175 MHz (GE, SY) (7-9-62) <12EU>. In Justin Mobiltrans 40 xmtr.-conv.; in Heathkit HW-17 2-m xcvr. and SB-500 transverter.

8158 [AKA 3CX10,000A1]; VHF cermet triode, coaxial base, fil. 7.5 V @ 94-104 A, μ 5, 12 kW, 140 MHz, intended for audio or RF linears w/ zero drive power (EI) (7-22-63). One in Wilkinson AM10000B 10-kW AM xmtr. Variants: 3CV30,000A1, 3CW20,000A1, 8159.

8159 [AKA 3CX10,000A3]; RF-heating triode, 8158 w/ μ raised to 20; grid dissipation raised from 100 W to 250 W; anode w/ high thermal inertia for short overloads; also promoted for 20-kW FM amps. (EI, SV) (1-29-62). In sound PAs in Gates TV-30L (6-kW aural) and TV-20L (4.5 kW aural) VHF TV xmtrs. Variants: 8160, 3CW20,000A3, 3CX10,000A7, 3CX-10,000H3.

8160 [AKA 3CX10,000A7]; xmtg. triode, variant of 8159, "first cermet tetrode"; μ 200, grid uprated to 500 W; for 20-kW grounded-grid linears, or push-pull zero-bias audio amps giving 45 kW (EI, SV) (11-5-62). 24 tubes used in parallel as cathode-follower driver (capable of 400 kW peak) for RCA A15030 triodes in C-Stellarator controlled-fusion experiment, 1961. One each in following xmtrs.: RCA SSB-T3 20-kW HF and BTF-5B 5-kW FM, Gates FM-5-H3 5-kW and FM-3H3 3-kW FM, Continental 816R-1A 10-kW FM, and Wilkinson FM-5000E and FM-7500E 5- and 7.5-kW FM. Variant: 3CW20,000A7.

8161 [AKA 3CX2500A3]; RF-heating/xmtg. VHF cermet ext.-anode triode, fil. 7.5 V @ 49-54 A, 2.5 kW, 75 MHz, forced-air cooled, coaxial base (EI, SV) (6-25-62). Two in Gates BC-10H 10-kW AM xmtr, modulated by two more. One, mod. by two, in BC-5H. One in Wilkinson AM5000B 5-kW AM xmtr. Also said to be common in FM-TV xmtrs. Variants: 8251, 8262.

8162 [AKA 3CX3000F7]; xmtg. triode, 3CX3000A7 w/ flying leads for fil. and grid; same ratings except 30 MHz max., for audio use (EI, SV) (11-5-62).

8163 [AKA 3-400Z]; xmtg. all-glass VHF triode, fil. 5.0 V @ 14.1 A, 400 W, μ 200, 400 MHz, for SSB linears; also see 3-500Z (AX, EI) (1-29-62). Two each in linears: Hallicrafters HT-45, Henry 2K (early), Hunter Bandit, and RCA SBA-1K 2-14 MHz.

8164 [AKA 3-1000Z]; mtg. VHF zero-bias triode, all-glass w/ cap, fil. 7.5 V @ 21.3 A, 1 kW, μ 200, 110 MHz, for grounded-grid and/or linear use (EI) (4-16-62). In BTI LK-2000D linear.

8165 [AKA 4-65A]; xmtg. tetrode, all-glass w/ cap, quick-heating T-T fil., 6 V @ 3.5 A, 65 W, 150 MHz, radiation cooled, promoted for mobile and SSB use (AX, EI, PL, RCA) (1-29-62). One in Kaar FM-179X 50-W mobile xmtr. Two in Gates CMG-1 250-W HF comm. xmtr. Also used in early 250-W FM broadcast xmtrs.

8166 [AKA 4-1000A]; xmtg. tetrode, for general use, T-T fil. 7.5 V @ 20.0-22.7 A, 1 kW, 110 MHz, radiation- and forced-air cooled (EI, PL, RCA) (11-12-62). Superseded 4-750A (1947). In Gates HF-15A 1-kW SW xmtr. Two as modulators in Wilkinson AM-5000B and AM10000B 5- and 10-kW AM xmtrs.

8167 [AKA 4CX300A]; cermet UHF tetrode, breech-block base, heater 6.0 V @ 2.6-3.1 A, 300 W, 500 MHz, forced-air cooled, for general use in severe environments (EI, RCA, RA) (1-29-62). Variants: 4CN-15A, 4CX125C, 4CX125F.

8168 [AKA 4CX1000A]; cermet VHF tetrode, breech-block base, heater 6.0 V @ 8.1-9.9 A, 1 kW, 110 MHz, forced-air cooled, for AB1 AF or RF amps. (EI, RCA) (6-25-62). In Gates FM-1H3 and FM-1K 1-kW FM xmtrs. Variant: 8244.

8169 [AKA 4CX3000A]; cermet VHF tetrode, fil. 9.0 V @ 43.5 A, 3 kW, 150 MHz, forced-air cooled, for linears and Class C amps. (EI, RCA) (6-25-62). Driver in Gates VP50 50-kW AM xmtr. and Continental 418D-2 100-kW SW xmtr.; final in Gates HFL-3000 SSB linear amp. Variant: 4CV8000A.

8170, W [AKA 4CX5000A; "W" AKA 4CX5000R]; cermet HF tetrode, coaxial base, fil. 7.5 V @ 75 A, 5 kW, 30 MHz, forced-air cooled, for linears and Class C amps. (EI, RCA) (1-29-62). Variants: 8171, 4CV20,000A, 4CW10,000A.

8171 [AKA 4CX10,000D]; 8170 uprated to 10 kW via larger radiator, 30 MHz (EI, RCA, SV) (1-29-62). In Gates ATL-10 10-kW HF ISB amplifier and FM-10K 10-kW FM xmtr.; in Wilkinson FM-10,000E 10-kW FM xmtr.

8172 [AKA 4X150G]; xmtg. tetrode, coaxial-based variant of 7034, glass body, fil. 2.5 V @ 6.25 A, 150 W, top freq. raised to 500 MHz (CW) or 1500 MHz (pulse) (EI) (6-25-62).

8173 Cermet VHF water-cooled tetrode, coaxial construction, bombardment cathode (fil. 10 V @ 25 A), 20 kW, 220 MHz, promoted as TV amp. (EI) (1-29-62).

8177 Xmtg. UHF tetrode, fil. 4 V @ 60 A, 1200 W, coaxial ceramic construction (AX) (5-14-62).

8179 Xmtg. HF tetrode, all-glass bulb w/ cap, graphite anode, large close-spaced pins, fil. 7.5 V @ 22.6 A, 800 W, radiation- or forced-air-cooled, for AM or SSB amps. (AX) (5-21-62).

8184 Xmtg. cermet beam tetrode, for pulse use, heater 22 V @ 12.6 A, 10 kW, 500 MHz, 2 MW max pulse input, forced-air cooled (RCA) (12-10-62).

8185 T3 submin. med.-μ triode, heater 6.3 V @ 300 mA, μ 42 (GE, SY) (4-9-62) <8KM>. Variant: 8186.

8186 T3 submin. triode, 8185 w/ heater 26.5 V @ 75 mA (GE, SY) (4-9-62) <8KM>.

8187 [AKA 4PR65A]; radial-beam tetrode for pulse modulators; all-glass, radiation- and convection-cooled; suited for long pulses; fil. 6.0 V @ 3.2-3.8 A, 65 W, 15 kV, 1 A. Appears to be a pulser version of 8165/4-65A (EI) (4-22-63).

8188 [AKA 4PR400A]; radial-beam tetrode for pulse modulators (pulser version of 8438/4-400), suited to long pulses; T-T fil. 5.0 V @ 14.7 A, 400 W, 20 kV, 4 A, radiation- and forced-air cooled (EI) (11-5-62).

8189, W [AKA 4PR1000A; "W" AKA 4PR1000B]; radial-beam tetrode for pulse modulators, suited to long pulses; fil. 7.5 V @ 20.0-22.7 A, 1 kW, 30 kV, 8 A, radiation- and forced-air cooled; appears to be pulser version of 8166/4-1000A (EI) (9-17-62).

8191 Hydrogen thyratron; see 8036; 10 kV, 50 mA, 135 kW (TS).

8192 Hydrogen thyratron; see 8036; 9 kV, 100 mA, 450 kW (TS).

8193 Power triode, for pass tube in series regs., 100 mA.

8198 TWA, 5.0-11.0 GHz, IM, CO, heater 6.3 V @ 650 mA, 3 kV, 60 dB gain, 34 dB NF, 1 W out (3-12-62).

8203 Nuvistor xmtg. triode, heater 6.3 V @ 160 mA, 1.5 W, μ 35, g_m 6 mS (GE, RCA) (5-4-64) <12AQ>.

8205 Ignitron, "size E" steel case 9-1/8" dia. by 23" long, coaxial current path, 600 V, 900 A (GE) (6-18-62).

8206 Voltage regulator, corona type, 12 kV (VN) (2-12-62).

8207 Rectifier (9-10-62).

8208 Rectifier (4-23-62).

8210 T3 submin. sharp-cutoff RF pentode, special T3 bulb 0.4" dia. by 1.8", heater 6.3 V @ 125 mA, g_m 9 mS (GE, SY) (5-14-62) <8LS>.

8211 T3 submin. video pentode, heater 6.3 V @ 360 mA, g_m 15.5 mS, 4 W (GE, SY) (4-9-62) <8DL>.

8212 Min. 9-pin rcvg. triode, heater 12.6 VCT @ 230 mA, μ 28 (GE, SY) (2-25-63) <9PY>.

8213 T3 submin. med-μ triode, heater 12.6 VCT @ 190 mA, μ 31, 5 W, for cathode followers (GE, SY) (2-25-63) <8LT>.

8214 Counter tube (7-23-62).

8216 Indicator thyratron, blue display.

8217 Cadmium-sulfide photocell, 7/8" dia. by 1-½" long, side-on illumination, for street-light controllers, etc. (GE, TS) (3-12-62).

8223 [AKA E288CC]; min. 9-pin med.-μ duotriode, heater 6.3 V @ 475 mA, μ 25 (GE, SY) (10-8-62) <9AJ>.

8224 Rectifier (12-16-63).

8225 TWA (7-13-64).

8226 Xmtg. beam tetrode, cermet, heater 6.3 V @ 3.2 A, 300 W, 400 MHz, forced-air-cooled (RCA) (4-29-63).

8227 Nuvistor xmtg. tetrode, heater 6.3 V @ 150 mA, 2.5 W, g_m 13 mS (4-29-63).

8228 [AKA ZZ1000]; T2 submin. voltage reg., 82 V, 3 mA (AX, GE) (11-25-63) <7894>.

8232 TWA (5-13-63).

8233 [AKA E55L]; frame-grid power pentode, T9 bulb, magnoval base, heater 6.3 V @ 600 mA, 10 W, for power/video use (AX. GE) (7-29-63) <9PZ>. Also see 8608.

8236 HV pentode, T9 octal 4-3/4" high w/ cap, graphite anode, heater 6.3 V @ 2.5 A, 1 kV, 50 W (GE) (4-1-63) <8JC>. Two each in Clegg Apollo Six linear and Hallicrafters SR-500 xcvr. Usable to up-power linears using 6DQ5s like Hammarlund HX-50.

8238 [AKA 3CX3000A1]; cermet triode, fil. 7.5 V @ 49-54 A, μ 5, 3 kW, coaxial base, forced-air cooled, intended for AB1 audio amps (EI) (7-29-63). Variants: 8239, 3CX3000A7, 3CX3000H7.

8239 [AKA 3CX3000F1]; audio power triode, 8238 w/ flying leads for fil. and grid; same use (EI, RCA) (6-25-62). Variant: 8241.

8240 Xmtg. triode (7-8-63).

8241 [AKA 3CW5000F1]; audio power triode, 8239 water-cooled via integral cooler, uprated to 5 kW (EI) (7-8-63).

8242 [AKA 3CW5000A3]; xmtg. triode, 8161 water-cooled via integral cooler, uprated to 5 kW (EI) (6-25-62).

8243 [AKA 3CW5000F3]; xmtg. triode, 8251 water-cooled via integral cooler, uprated to 5 kW and 75 MHz (EI) (6-25-62).

8244 [AKA 4CW2000A]; xmtg. tetrode, water-cooled version of 8168, uprated to 2 kW (EI).

8245 [AKA 4CX250K]; coaxial UHF beam tetrode, like cermet version of 8172, heater 6.0 V @ 2.3-3.0 V, 1200 MHz (1500 MHz w/ pulses) (EI) (6-25-62). Variants: 8246, 8590.

8246 [AKA 4CX250M]; xmtg. tetrode, 8245 w/ heater 26.5 V @ 530-680 mA (EI) (6-25-62).

8247 [AKA 4PR125A]; radial-beam tetrode for pulse modulators, suited to long pulses; T-T fil. 5.0 V @ 6.0-7.0 A, 125 W, 18 kV, 1.8 A; radiation- and forced-air cooled. Appears to be a pulser variant of 4D21/4-125A (EI) (11-5-62).

8248 [AKA 4PR250C]; tetrode for pulse modulators and switch use; fil. 5.0 V @ 13.5-14.7 A, 250 W, 50 kV, 4 A, radiation- and forced-air cooled. Appears to be a 5D22/4-250 w/ elongated bulb and large-diameter plate (EI) (1-6-64).

8249 [AKA 4W300B]; water-cooled version of 4X150A, integral cooler, 9-pin spl. base, heater 6.0 V @ 2.3-2.9 A, 300 W, 500 MHz (AX, EI) (6-25-62).

8250 [AKA 3CX100F5]; planar triode, 2C39 format, heater 26.5 V @ 200-240 mA, 100 W, 2.5 GHz max, forced-air cooled (EI) (8-6-62).

8251 [AKA 3CX2500F3]; RF-heating triode, 8161 w/ flying leads on fil. and grid; 30 MHz (EI, RCA) (9-17-62). Variants: 8243, 3CX2500H3.

8252, W [AKA 4PR60B; "W" AKA 4PR60C]; ruggedized radial-beam tetrode for pulse modulators, heater 26.0 V @ 1.95-2.35 A, 60 W, 20 kV, 18 A, radiation and rfroced-air cooled; supersedes 4PR60A; replaces 715C and 5D21 (EI) (5-7-62). In AN/CPN-6.

8254 [AKA EC1000]; T3 submin. triode for use in test-probe amps., w/ grid pin (Z_{in} = 450 Ω @ 250 MHz), heater 6.3 V @ 185 mA, g_m 14.5 mS (AX, GE, SY) (9-23-63) <8LW>.

8255 [AKA E88C]; min. 9-pin triode, premium 6DL4 (GE) (10-22-62) <9NY>.

8256 See 8089.

8257 See 8089.

8258 TWA (7-13-64).

8261 Xmtg. UHF cermet tetrode, heater 26.5 V @ 560 mA, 250 W, external anode, forced-air-cooled, for SSB use (AX) (11-28-66).

8262 Counter tube (12-10-62).

8264 [AKA KU-52]; hydrogen diode, fil. 5.0 V @ 11.5 A (KL).

8268 VHF RF-heating triode, fil. 12.6 V @ 33 A, 6 kW, forced-air cooled (AX) (12-9-63). Variant: 8593.

8269 VHF RF-heating triode, water-cooled version of 8268 (AX) (12-16-63).

8270 Mercury triode thyratron, fil. 5 V @ 13 A, 21 kV, 2.5 A (AX) (10-1-62).

8373 Hydrogen diode, cermet construction, 12 kV, 250 mA.

8374 Hydrogen diode, cermet construction, 15 kV, 500 mA.

8275 [AKA KU-93]; hydrogen rectifier, 20 kV @ 2 A.

8376 Hydrogen rectifier, cermet construction, 25 kV, 2 A.

8377 Hydrogen rectifier, cermet construction, 26 kV, 7 A.

8278 [AKA EL503]; audio beam power tube, T9 novar, heater 6.3 V @ 1.2 A, g_m 24 mS, 25 W (AX, GE, SY) (7-15-63) <9QB>.

8279-80 TWAs (both 7-13-64).

8281 [AKA 4CX15,000A]; cermet coaxial tetrode sim. to 8171 but uprated, fil. 6.3 V @ 152-168 A, 15 kW, 110 MHz, for FM and VHF TV (EI, RCA, SV) (2-18-63). In FM xmtrs: Gates FM-20H3 and FM-20K 10-kW, Continental 816R-3 25-kW, and Wilkinson FM-20,000E 20-kW. Two as *linear* amps. in Continental 316F 10-kW AM xmtr. Variants: 4CW25,000A, 4CX15,000J.

8283 [AKA 3CX1000A7]; VHF cermet ext.-anode zero-bias triode, breechblock base, fil. 5.0 V @ 34 A, μ 200, 1 kW, 220 MHz, forced-air cooled, for linears (EI) (7-29-63).

8286-87 Gas noise sources (both 2-11-63).

8288 Gas noise source (4-1-63).

8289 Gas noise source (2-4-63).

8290-91 Gas noise sources (both 2-11-63).

8292 Gas noise source (6-10-63).

8293 Gas noise source (2-11-63).

8294 Klystron (4-27-64).

8295, A [AKA PL-172]; beam pentode, glass-seal, external anode, special 7-pin base, heater 6.0 V @ 8.2 A, 1 kW, 30 MHz, for linears (EI, PL) (7-9-62). "A" has cermet construction; uses less cooling air. Variant: 8432.

8296 [AKA 4X150R]; xmtg. tetrode, ruggedized 7034, glass body; heater raised to 2.4-3.0 A (EI) (9-17-62).

8297 [AKA 4X150S]; xmtg. tetrode, ruggedized 7035, glass body; heater current raised to 560-680 mA (9-17-62).

8298, A [AKA 6146B, q. v.]; beam power tube (GE) (11-12-62) <7CK>.

8300 Xmtg. dual VHF beam tetrode, instant-heating variant of 8117, 5894 format, quick-heating "harp" fil. 2.0 V @ 3.4 A, 60 W, radiation-cooled, for tube-transistor hybrid xmtrs (AX). Also promoted for use w/ both sections in parallel as SSB linear.

8306 Ignitron (7-23-62).

8307 Variant of 6922 (?).

8317 Xmtg. triode, coaxial cermet construction, T-T fil. 14.5 V @ 450 A, 60 kW (ML) (4-1-63). Variants: 7482, 7560.

8318 Photoconductive cell, 8474 format (GE) (6-4-62).

8319 T3 submin. high-μ triode, heater 6.3 V @ 150 mA, μ 55 (GE, SY) (5-6-63) <8LD>.

8320 Storage tube (9-23-63).

8321 [AKA 4CX350A]; xmtg. cermet beam UHF tetrode, ruggedized and uprated 7203, spl. 9-pin base, fil. 6.0 V @ 3.2 A, 350 W, 500 MHz, forced-air-cooled, for AB_1 linears (AX, EI) (12-18-63). Driver in Continental 416D 10-kW SW xmtr. Variants: 8322, 4CS-250H.

8322 [AKA 4CX350F]; xmtg. tetrode, variant of 8321, heater 26.5 V @ 930 mA, 350 W (ruggedized and uprated 7304) (AX, EI) (2-18-63).

8326, A Hydrogen thyratron (11-12-62).

8327 Min. 9-pin rcvg. power pentode, heater 6.3 V @ 150 mA, 1.0 W (GE) (1-18-65) <9CV>.

8330-31 TR tubes (both 10-15-62).

8332 Klystron (11-30-64).

8333 Xmtg. triode (2-18-63).

8334 [AKA 1218A]; min. triode, premium 6AN4 (SY) (12-31-62) <7DK>.

8335-37 Reflex klystrons, heater 6.3 V @ 1.1 A, 500 V, 760 mW output (all PO) (all 8-20-62).

8335 For 1.7-5.0 GHz.

8336 For 1.0-2.8 GHz.

8337 For 1.0-4.0 GHz.

8345 Photoconductive cell, 8100 format (GE) (10-22-62).

8346-47 Photoconductive cells, 8100 format (GE) (9-24-62).

8348 Min. 9-pin xmtg. VHF tetrode, quick-heating "harp" fil. 1.6 V @ 2.5 A, 10 W, 175 MHz, for hybrid mobile xmtrs, internally neutralized (AX, GE) (12-23-63) <9QN>.

8349 [AKA 4CX35,000C]; cermet tetrode, graduated-ring base, fil. 10.0 V @ 300 A, 35 kW, 30 MHz (EI) (2-4-63). One as final, one as switch modulator, in Gates MW-50 50-kW pulse-duration-modulated AM xmtr.

8350 Xmtg. tetrode (3-4-63).

8351 [AKA 4CV100,000C]; cermet HF tetrode, graduated-ring base, fil. 10.0 V @ 300 A, 100 kW, 30 MHz, vapor-cooled, for general use (EI, SV) (8-26-63). One as final, one as switch modulator, in Gates VP-100 100-kW pulse-duration-modulated AM xmtr; likewise, in Continental 418D-2 100-kW and 419F 250-kW SW xmtrs. Variant: 4CV75,000A.

8352 [AKA 4CX1000K]; xmtg. tetrode, 8168 w/ solid-ring screen terminal for added stability in UHF linear amps (EI). In Wilkinson FM-1000E 1-kW FM xmtr.

8353 Beam-switching decade counter (8-27-62).

8356 Magnetron, fixed at 9375 MHz, IM, WG, heater 6.3 V @ 500 mA, 8 kV @ 8 A, 20 kW pulse out (EEV) (3-9-64).

8357, A, E Xmtg. triode, fil. 10 V @ 10 A (F) (8-3-64). "E" has bipin base and grid and plate fittings.

8358 Min. 9-pin twin pentode, quick-heating fil. 1.9 V @ 3.15 A, 7.5 W total (GE, SY) (10-7-63) <9QR>.

8360 Ignitron, for pulse welding, 5551 format, coaxial current path, 2.5 kV PIV, 10 A (GE) (10-1-62).

8370 Hydrogen thyratron (11-26-62).

8375 Hydrogen cermet diode, finned anode cap, mounting flange, flying leads (TS) (3-23-64).

8376 Rectifier; see 8375, enlarged (IT) (4-6-64).

8379 TWA, 2320-2680 MHz, for rcvr. preamps, EM, heater 5.0 V @ 650 mA, NF 4.5 dB, "N" fittings for input and output (RCA) (1-13-64).

8380 Nuvistor power tetrode, w/ plate cap, for sonobuoys and other expendable equipment, heater drawing 1.1 W "within ± 10% of any specified center heater voltage between 6.0 and 8.5" V, 1.2 W, 80 MHz (GE, RCA, SY) (10-14-63) <12AS>.

8382 Nuvistor power triode, single-ended, uses and heater as for 8380, 2 W, 175 MHz (GE, RCA, SY) (10-14-63) <12AQ>. Variants: 8441, 8456.

8386-88 Xmtg. triodes (all 5-25-64).

8393 Nuvistor triode, 7586 w/ heater 13.5 V @ 60 mA (GE, RCA) (5-4-64) <12AQ>.

8401-02 Gas noise sources (both 6-10-63).

8403 Planar high-μ UHF pulse triode, 2C39 format, heater 6.0 V @ 1.3 A, 100 W, 3 GHz, 1 kW pulse output (EI, GE) (11-18-63).

8408 Min. 9-pin dual UHF frame-grid xmtg. tetrode, quick-heating "harp" fil. 1.1 V @ 3.0 A, 8 W, for hybrid xmtrs (AX, GE) (12-23-63) <9QV>.

8412 Planar high-μ triode, heater 6.0 V @ 800 mA, 30 W (GE). Variant: 8413.

8413 Planar triode, 8412 repackaged, derated to 25 W (GE).

8414 T3 submin. sharp-cutoff RF pentode, heater 26.5 V @ 45 mA, g_m 5 mS, optimized for 26.5-V B+ (GE, SY) (1-21-63) <8DC>.

8415 Xmtg. triode (12-16-63).

8416 Rcvg. 9-pin duotriode.

8417 Beam power tube, T12 bulb, octal base, heater 6.3 V @ 1.6 A, 35 W (G, RCA) (7-1-63) <7S>. Nominally an audio type, but SBE SB-35 xcvr used two.

8420 Ignitron, for electronic-contactor use in locomotives, coaxial current path, 10-11/16" dia. by 22-15/16" long, 1 kV PIV @ 200 A (GE) (3-4-63).

8421 [AKA 5092]; "Nixie" readout (5-27-63).

8422 [AKA 5991]; "Nixie" readout (5-27-63).

8423 [AKA NL6091]; "Nixie" readout (5-27-63).

8424 Hydrogen thyratron (6-22-64).

8425, A Premium 6AU6 (SY) (7-1-63) <7BK>.

8426, A Premium 12AU6 (SY) (7-1-63) <7BK>.

8427 TWA, 2.7-3.2 GHz, IM, CO, 7.5 kV, 33 dB gain, 1 W output.

8428 "Misc." type (4-1-63).

8429 Xmtg. tetrode (12-16-63).

8431 Min. 9-pin med.-μ duotriode, heater 12.6 V @ 180 mA, μ 33 (GE, SY) <9AJ>.

8432 Cermet xmtg. tetrode, variant of 8295A, smaller diameter, same ratings (EI, PL) (8-18-63).

8433 Beam-switching tube (1-28-63).

8434-35 Rectifiers (both 9-2-63).

8436 [AKA EC158]; disc-seal triode, heater 6.3 V @ 900 mA, usable to 4.2 GHz (AX).

8437 Xmtg. cermet beam UHF tetrode, fil. 8.5 V @ 88 A, 10 kW, 400 MHz, forced-air cooled (RCA) (8-12-63).

8438, A [AKA 4-400A]; xmtg. VHF tetrode, T-T fil. 5.0 V @ 14.5 A, 400 W, 110 MHz, forced-air-cooled; 4-250A w/ radiating fins on plate (AX, EI, PL, RCA) (7-22-63). Two in Gates FM-1B 1-kW FM xmtr, and BT-1A 500-W VHF TV xmtr. Variants: 6775, 7527, 8188.

8441 Nuvistor power triode, 8382 w/ μ of 64 (GE, RCA, SY) (10-14-63) <12AQ>.

8443 Storage tube (9-23-63).

8444 T3 submin. RF pentode, heater 6.3 V @ 125 mA, g_m 9 mS (GE, SY) (5-6-63) <8DC>.

8445-48 Series of mobile-rated 9-pin min. types (GE, SY, TS) (7-22-63).

8445 Equals 6U8, heater 6.75 V @ 440 mA. Variant: 8446.

8446 Equals 6FV8, 8445 w/ pinout <9FA>.

8447 Equals 12BR7.

8448 Equals 12BY7A.

8453 [AKA Z550M]; counter tube (7-29-63).

8454 Storage tube (9-23-63).

8456 Nuvistor power triode, 8382 w/ μ of 11.5 (GE, RCA, SY) (10-14-63) <12AQ>.

8457 Min. 9-pin xmtg. dual VHF tetrode, 13.5-V 6360, fil. 13.5 V CT @ 380 mA, 10 W, 175 MHz, mobile-rated (AX, GE) (12-23-63) <9PW>. Variant: 8458.

8458 [AKA YL1240]; xmtg. dual VHF tetrode, novar version of 8457 (scaled-up 6360), uprated to 15 W (AX, GE) (12-23-63) <9PW>.

8460 Klystron (6-29-64).

8461 Xmtg. triode (1-27-64).

8462 Cermet beam tetrode, quick-heating variant of 8072, fil. 2.9 V @ 4.6 A, 100 W, 500 MHz, conduction-cooled (RCA) (11-18-63). Variant: 8828.

8463 Min. 9-pin xmtg. VHF pentode, quick-heating fil., 1.1 V @ 1.05 A, 5 W, for mobile use (GE) (12-23-63) <9QX>.

8465 Ignitron, for motor control use, w/ aux. anode and dual igniters, 5-3/4" dia. by 17-15/16" long, 2.1 kV PIV, 150 A (GE) (6-27-66). Variant: 8466.

8466 8465 w/ integral thermostat (GE) (6-27-66).

8467 Klystron (7-8-63).

8468 Magnetron, tunable (2-24-64).

8469 Voltage regulator (5-20-63).

8474-78 Series of photoconductive cells, glass, ¼" dia., 0.55" high (all GE). Resistances below are specified at illumination of 2 foot-candles, vs. dark resistance.

8474 1.5 kΩ vs. 150 kΩ (8-31-64).

8475, A 3 kΩ vs. 300 kΩ; 2 kΩ vs. 200 kΩ for "A" (8-31-64).

8476 6 kΩ vs. 600 kΩ (8-31-64).

8477, A 12 kΩ vs. 1.2 MΩ; 8 kΩ vs. 800 kΩ for "A" (8-31-64).

8478 24 kΩ vs. 2.4 MΩ (8-2-65).

8480 [AKA 4810]; vidicon, sulfide target, 1-½" dia. by 10-¼" long, all-glass, "super-ditetrar" 8-pin base, electrostatically focused, for color film chains and low-power solid-state cameras (RCA) (2-17-64).

8483 Vidicon, 1", mag. focus and deflection, heater curr. 90 mA (AX) (12-16-63).

8484, H Vidicon, 7226 format, same uses, heater 6.3 V @ 600 mA (GE) (3-15-65).

8488 Hydrogen thyratron (6-22-64).

8489 Min. 9-pin triode-pentode, heater 6.3 V @ 450 mA, triode μ 21, pentode g_m 7.0 mS (SY) <9DA>.

8493 [AKA SAL89]; klystron (6-7-65).

8489 Min. triode-pentode, premium low-leakage 6AN8 (GE, SY, TS) (7-22-63) <9DA>.

8500 Xmtg. planar cermet tetrode, heater 6.3 V @ 3.8 A, forced-air cooled (GE) (6-1-64).

8501 Xmtg. cermet beam tetrode, for UHF TV and general use, fil. 4.5 V @ 125 A, ca. 10 kW, 900 MHz, forced-air-cooled (RCA) (4-6-64). In RCA UHF TV xmtrs: one in TTU-2A 2-kW; two in TTU-10A 10-kW.

8502 "Misc." type (2-17-64).

8503 Hydrogen thyratron (3-9-64).

8505, A [AKA DX245]; xmtg. VHF beam tetrode, single-ended, half-height hard-glass Compactron style w/ large-dia. plate pin outboard from normal pin circle, fils. 13.5 VCT @ 660 mA, 25 W (AX). Variants: 8603, 8647.

8506 Planar triode, heater 6.3 V @ 400 mA, μ 110, 5 W, usable to 1 GHz (TSU) (2-24-64).

8507, A Vidicon, 7697 format, sulfide target, for general CCTV and broadcast TV use incl. color, heater 6.3 V @ 600 mA (GE, HI, RCA) (11-11-63). Magnetically focused counterpart of 7735A w/ internal collimating lens. "A" is tested to tighter limits. Used in 1970-vintage RCA single-tube color camera, and in PK-610 three-tube color film system. Variant: 4589.

8509 Xmtg. dual tetrode, 5894 w/ quick-heating "harp" fil., 2.0 V @ 3.25 A, 40 W, 250 MHz (AX).

8511 Vidicon (4-20-64).

8513 Xmtg. cermet tetrode, heater 7.0 V @ 13.5 A (GE) (1-18-65).

8514-15 Voltage regs. (both 11-4-63).

8517 T3 submin. xmtg. VHF pentode, heater 6.3 V @ 150 mA, 800 mW (GE, TS) (2-10-64) <8DC>.

8521 Vidicon, 8051 format, sulfide target, for film chains and high-resolution/low-light industrial use (RCA) (3-16-64). Has magnetic focusing and internal collimating lens.

8522 T3 submin. pentode, closer-controlled 5636 (GE, TS) (5-11-64) <8DC>.

8524-30 Submin. *metal* rcvg. tubes, shell 0.43" dia. by 1.08" long, 8 pins, 6.3-V heaters, rated at 165 V B+. Analogs of five of the "missile" submin. types 6943 and 6945-6948, w/ same pinouts (GE) (5-18-64). Also see 7311-14.

8524 Sharp-cutoff dual-control pentode, heater 150 mA, g_m 3.2 mS <8DC>. Close to 6943.

8525 Med.-μ duotriode, heater 300 mA, μ 35 <8DG>. Close to 6947.

8526 Med.-μ duotriode, heater 300 mA, μ 20 <8DG>. Close to 6948.

8527 Med.-μ triode, heater 150 mA, μ 27, g_m 6.5 mS <8DK>. Roughly comparable to 6946.

8528 Beam power tube, heater 450 mA, g_m 4.2 mS, 3.7 W <8DE>. Close to 6945.

8529 Rcvg. semi-remote-cutoff pentode, heater 150 mA, g_m 4.5 mS <8DE>.

8530 Rcvg. sharp-cutoff pentode, heater 150 mA, g_m 5.0 mS <8DE>.

8532, W [AKA 8532/6J4WA]; min. triode, ruggedized 6J4 (RCA, SY) (5-4-64) <7BQ>.

8533, W Planar pulse triode, 2C39 format, heater 6.0 V @ 1.3 A, 100 W, 3 GHz, elongated body for use up to 8 kV (EI, ML).

8534 Xmtg. planar cermet triode, heater 6.3 V @ 1.3 A (GE).

8535 Xmtg. triode, 8534 for forced-air cooling (GE).

8536 Xmtg. planar cermet triode, heater 6.0 V @ 1.0 A (GE).

8537 Xmtg. triode, 8536 for forced-air cooling (GE).

8538, A, B Xmtg. high-μ planar cermet triode, coaxial base, threaded stud on anode for connection to heat sink, heater 6.3 V @ 1.3 A, μ 120, 2.5 GHz (CW) or 3.0 GHz (pulsed) (GE, EI, ML). Typical dissipation w/ external radiator is 150 W. In RT-868/APX-76, giving 1.8 kW at 1090 MHz. "B" is AKA 8933, q. v.; differences from earlier versions not known.

8539 Xmtg. high-μ planar cermet triode, heater 6.3 V @ 1.3 A, forced-air cooled (GE).

8541, A, AX Vidicon, 7697 format, sulfide target, "dark" heater, 6.3 V @ 100 mA, for film chains and high-resolution low-light use incl. CCTV and X-ray linkage systems (GE, RCA) (4-20-64). Also see 4569.

8543 Magnetron, fixed freq. (7-6-64).

8545-46 Xmtg. tetrodes, fil. 12.0 V @ 400 A (ML).

8547 Xmtg. tetrode, fil. 14.5 V @ 450 A (ML).

8552 [AKA 6883B/8032A/8552]; xmtg. beam power tube; see 6883B (AX, RCA) (3-16-54) <7CK>.

8553 [AKA KU-72Z]; probably hydrogen thyratron.

8556 [AKA EC8010]; min. UHF triode, heater 6.3 V @ 260 mA (AX, TF).

8559 Hydrogen thyratron (6-1-64).

8560, A, AS Xmtg. UHF cermet tetrode, conduction-cooled variant of 7203, quasi-octal base, derated to 200 W, 500 MHz, general use (AX, EI, GE, SV) (4-5-65).

8561 [AKA 4CX300Y]; xmtg. cermet tetrode, 8167 uprated to 400 W, for general use, heater 6.0 V @ 3.5 A (EI, GE) (5-18-64).

8562 [AKA 8255, q. v.]; min. triode.

8566 Vidicon (12-21-64).

8567 Vidicon, 7697 format, ruggedized variant of 8134, for industrial use (RCA) (9-7-64). Related to the ruggedized 8134 used in the hand-held TV camera in the Apollo spacecraft command module.

8568 Pulsed five-cavity amplifier klystron, fixed at 2856 MHz, for linear accelerators, heater 15 V @ 14 A, input 260 kV peak @ 270 A; 68 MW peak input; typ. output 12-21 MW; waveguide output, oil-cooled electron gun and water-cooled internal structure; integral ion vacuum pump, weight 150 lb. alone, 1100 lb w/ electromagnet and lead X-ray shields, or 1625 lb. w/ permanent magnet and X-ray shields (RCA) {A1230D} (5-17-65). 240 tubes used as initial equipment in Stanford Univ. linear accelerator, 1964; also in several others including Argonne and Brookhaven. Variant: 8840.

8569 T4 submin. computer indicator triode, end-view, blue-green, fil. 0.7 V @ 35 mA, 65 V (TS) (11-16-64).

8571 Photomultiplier, 9 stages, ruggedized, S-4 response, side-on bulb 0.51" dia. by 1.37", 11 flying leads (RCA) (3-29-65).

8572, A, V, V4 Vidicon, 7697 format, sulfide target, for B&W or color film chains and CCTV (GE, RCA) (5-18-64). Magnetically focused counterpart of 7038 w/ internal collimating lens.

8573, A Vidicon, 7697 format shortened to 5.1", for industrial TV (RCA) (5-18-64). Magnetically focused counterpart of 7262A w/ internal collimating lens.

8574 TR tube (4-20-64).

8575, /V1, /V2 Photomultiplier, for scintillation counters, 12-stage, 2" dia., head-on T16 all-glass bulb w/ 21-pin base, coated w/ plastic light shield (RCA) (5-3-65).

8576 [AKA PL-264]; cermet beam pentode, heater 6.0 V @ 17 A, 3 kW, forced-air cooled, for linears (EI, PL) (7-6-64).

8577 Min. dual tetrode, 6939 w/ heater 13.5 V @ 280 mA (AX).

8578 TR tube (4-27-64).

8579 Xmtg. VHF beam tetrode, tubular all-glass bulb w/ cap, novar base, fil. 12.6 VCT @ 1.3 A, 75 W, for SSB use (AX).

8580 Xmtg. dual UHF tetrode, quick-heating fil., 1.1 V @ 3.8 A, 16 W, in hybrid mobile xmtrs (AX).

8582, A Photoconductive cell (GE) (9-28-64).

8587 Xmtg. beam tetrode, 6952 w/ different coolant connector and height increased from 8.6" to 9.2" (RCA) (3-29-65).

8590 [AKA 4CPX250K]; cermet xmtg. tetrode, pulse-rated variant of 8245; produces 10 kW pulses (EI). One in each of 5,184 final amps. in AN/FPS-85 phased-arrar radar, each driven by a pair of 8745s.

8591 RF-heating HF triode, fil. 6.3 V @ 130 A, 1.5 kW, water-cooled by copper tube soldered to anode (AX).

8592 Xmtg. VHF triode, fil. 12.6 V @ 33 A, 6 kW, water-cooled by copper tube soldered to anode, for RF heating and large AF amps. (AX).

8593 VHF RF-heating triode, helically-cooled version of 8268 (AX).

8594 RF-heating triode, helically-cooled version of 7807, uprated to 20 kW (AX).

8595 Xmtg. dual UHF tetrode, 6939 with added vibration resistance, derated to 4 W (AX).

8596 Xmtg. tetrode, cermet, forced-air-cooled, heater 6.3 V @ 3.2 A, ca. 20 W, 1215 MHz (EI, RCA) (1-4-65).

8597 Xmtg. triode (6-7-65).

8603 Xmtg. beam tetrode, quick-heating version of 8505, fil. 1.1 V @ 5.0 A (AX).

8604 Vidicon, 7697 format, for B&W film chains (GE) (11-16-64).

8605 [AKA 8605-1 (8605/V1), or 8605 (8605/V2)]; image
/V1, V2 intensifier, S-20 response, P20 (yellow-green) screen, 3" dia. bulb 3.7" long, fiber-optic photocathode at one end, fiber-optic screen image surface at other (RCA) (ca. 2-71). Variant: 8606. Also see 8857.

8606 Image intensifier, 8605 w/ 3 stages and integral voltage multiplier, 3.7" dia. by 12" long; requires 2.8 kV P-P AC rather than 16 kV DC as for 8605 (RCA) (ca. 2-71). Also see 8858.

8608 Power pentode, heater 6.3 V @ 600 mA, g_m 45 mS, for wideband video amps., appears to be Compactron version of 8233 (AX).

8609 TWA (1-25-65).

8612 Voltage regulator (12-21-64).

8615 Voltage regulator (1-25-65).

8618 Xmtg. cermet beam triode, for modulator and LF use, beam formed by integral magnet (ML) (1964).

8621 [AKA 4CX250FG]; xmtg. beam tetrode, 7204 w/ heater 26.5 V @ 540 mA, made for "extra stability in airborne linear amplifier service"; repl. 4CX250F (AX, EI, RCA) (11-28-66).

8622 TR tube (2-15-65).

8623 TR tube (2-15-65).

8624 Xmtg. dual beam tetrode, quick-heating version of 7854, fil. 2.0 V @ 3.4 A, derated to 40 W (AX).

8625-26 [Type not listed] (both 8-16-65).

8627, A Nuvistor xmtg. power triode, heater 6.3 V @ 150 mA, g_m 13 mS, 2.5 W, 1.2 GHz (GE, RCA) (4-26-65) <12CT>.

8628 Nuvistor triode, for audio preamps, Heater 6.3 V @ 100 mA, μ 127 (GE, RCA) (4-26-65) <12AQ>.

8632 Rcvg. high-μ triode, T9 bulb w/ cap on octal base, for HV shunt regulators and pulse amps., heater 6.3 V @ 300 mA, 18 kV, 100 mA peak (GE, SY).

8635 Magnetron, tunable (6-6-66).

8637 Min. 9-pin beam VHF tetrode, heater 13.5 V @ 800 mA, 45 W intermittent, radiation-cooled, w/ high-thermal-inertia anodes (AX). Claimed "the only twin tetrode ever designed and rated for push-to-talk service." For 1-in-5-min. use.

8639 Beam tetrode, heater 12.6 VCT @ 900 mA, 40 W, for use as pass tube in voltage regulators, replaces 3D21WB or 7403 w/ change in top connector (AX).

8643 Xmtg. dual VHF tetrode dev. from 5894, fil. 7.25 V @ 2.0 A, 60 W, with "wide range" cathode (AX). "Push-to-talk rated" - see 8637.

8644 Photomultiplier, 10-stage, S-20 response, head-on T6 bulb 3.8" long, 12 flying leads to detachable 12-pin molded base, for low-level light-detection systems extending to infrared (RCA) (11-8-65). Variant: 8645.

8645 Photomultiplier, 8644 w/ integral voltage-divider network potted in plastic, three flying leads, 0.9" dia. by 4.5" long (RCA) (11-8-65).

8646 UHF beam tetrode, 8121 w/ heater 26.5 V @ 640 mA (RCA) (10-11-65).

8647 Xmtg. VHF tetrode, quick-heating variant of 8505, fil. 1.1 V @ 3.9 A, 25 W, for mobile xmtrs. (AX).

8649 "Misc." type (3-14-66).

8650 "Misc." type (5-31-66).

8653 [AKA 7191H, q. v.].

8654 Xmtg. beam VHF tetrode, heater 6.3 V @ 19.5 A, 1500 W, ceramic construction, for SSB linears (AX).

8656 TWA (2-28-66).

8660 [AKA 4CX1500B]; xmtg. triode; heater-type, low-distortion version of 4CX1500A; heater 6.0 V @ 11 A, for linear amps. (EI) (4-25-66).

8661 [AKA 4CW10,000A]; xmtg. tetrode, 8170 water-cooled via integral cooler, uprated to 10 kW (EI, GE).

8664, Photomultiplier, 10-stage, 2.1" dia. by 4" long, rugged-
/V1 ized cermet construction, all leads coming out as stacked rings (RCA) (ca. 8-70). "V1" has integral scintillation-crystal holder and voltage-divider network of resistors attached to the dynode rings; only three leads.

8665 Storage tube (6-13-66).

8666-68 RF-heating VHF triodes, fil. 5.8 V @ 130 A, 10 kW (all AX).

8666 Forced-air-cooled.

8667 Water-cooled w/ separate water jacket.

8668 Water-cooled w/ integral cooler.

8669-70 [Type not listed] (AX).

8673-74 Image orthicons, 5655 format, 3", for high-performance color studio cameras (RCA) (10-17-66). "/S" is matched trio.

8673, /S Intended to replace 4513, 7513, 7513/L, 8093, 8093A, and 8093A/L.

8674, /S For remote cameras also; intended to replace 4415, 4416, 7293, 7293A, and 7293A/L.

8679-80 [Type not listed] (AX).

8683 [Type not listed] (AX).

8684 CW cermet magnetron, for RF heating, fil. 12.5 V @ 115 A, 14 kV @ 3 A, 30 kW out @ 915 MHz, WG, electromagnetic focusing, liquid-cooled (RCA) (2-13-67).

8685 Image orthicon, 2" dia. bulb stepped to 1.5" by 10" long, all-glass, 15-pin base, for general portable use (RCA) (11-7-66).

8727 Xmtg. high-μ cermet pencil triode, heater 6.3 V @ 225 mA, μ 70, 2.5 w, 2 GHz, usable to 5 GHz (GE) (7-31-67).

8731 [Type not listed] (AX).

8732-36 [Type not listed] (all 8-8-66).

8737 [AKA 5894B]; dual tetrode, "wide-range" cathode said to give only 5% variation in output from 9 to 14 V (AX) (8-8-66).

8738-40 [Type not listed] (all 8-8-66).

8741 [Type not listed] (ML).

8744 [Type not listed] (AX).

8745 [AKA 7815R, 3CPX100A5]; planar pulse triode, radiator version of 7815; same elongated body for insulation; for pulse and high-altitude use to 70,000 ft.; uprated to 100 W, derated to 2.5 GHz, 2 kW pulse output (EI, GE). In AN/FPS-85 phased-arrar radar: 10,368 tubes used as two-stage drivers, feeding 8590/-4CPX250Ks.

8748 Image orthicon, low-microphonics, for B&W kinescope pickup; interchanges w/ 7389 family; appears to be 7389 w/ non-burn-in target (RCA) (3-6-67).

8749 Image orthicon, 7295B w/ non-burn-in target; to interchange w/ 7295 family (RCA) (3-27-67).

8751 Xmtg. cermet planar triode, heater 6.3 V @ 1.05 A, conduction-cooled (GE).

8752 [Type not listed] (AX).

8753 Xmtg. beam tetrode, fil. 2.0 V @ 3.45 A (AX).

8754 [AKA NL840, 840, q. v.]; pentode (GE) (1-23-67).

8755-57 Planar cermet pulse triodes, threaded anode for heat dissipation (all EI).

8755 Heater 6.3 V @ 1.3 A, 150 W, 10 kV, 3 GHz, forced-air or conduction cooled.

8756 Heater 6.0 V @ 700 mA, 100 W, 2.5 kV, 2.5 GHz.

8757 8755 heater, 150 W, 2.5 GHz, slightly shorter, voltage derated to 2.5 kV.

8758 [Type not listed] (7-3-67).

8760 Cadmium-sulfide photocell, glass, ¼" dia., 2 leads (GE) (8-21-67).

8764 [Type not listed] (8-28-67).

8765 [AKA KU-71Z]; hydrogen thyratron.

8773 [Type not listed] (EI, ML).

8775 Image orthicon, 5655 format, 3", for general TV use, to replace 5820 family (RCA) (ca. 5-68).

8785-86 Xmtg. tetrodes, fil. 15.0 V @ 400 A (ML)

8791/V1 Xmtg. cermet beam tetrode, for VHF TV, ruggedized, heater 6.3 V @ 7.5 A, 1 kW, 400 MHz, forced-air-cooled (RCA) (ca. 9-68).

8792/V1 Xmtg. VHF beam tetrode, for VHF TV (scaled-up 8791?), same body design as 4651, fil. 5.3 V @ 17.3 A, 1.5 kW, forced-air-cooled, for use in Y1170 cavity, gives 1350 W peak-sync or 1250 W sound-channel out (RCA) (ca. 9-68). In Gates BT-5H 5-kW VHF TV xmtr.

8793 Xmtg. cermet beam tetrode, for HF SSB or general use, fil. 5.7 V @ 115 A, 5 kW, 400 MHz, forced-air-cooled (RCA) (ca. 3-68).

8794 Xmtg. cermet beam tetrode, fil. 5.7 V @ 125 A, 12.5 kW, 400 MHz, forced-air-cooled (RCA) (ca. 3-68). Variant: 8806.

8795 [Type not listed] (EI, ML)

8806 Xmtg. VHF beam tetrode, 8794 for VHF FM-TV use in Y1116 or Y1171 cavity (RCA) (ca. 9-68). In Gates BT-25 VHF TV xmtr, and in aural side of BT-35H. Also see 8807.

8807 Xmtg. cermet VHF beam tetrode, appears to be 8794 w/ bigger radiator and different fil. terminal, for VHF FM-TV use in Y1116 or Y1171 cavity, fil. 9.5 V @ 145 A, 15 kW, 400 MHz (RCA) (ca. 9-68). In Gates BT-25 and BT-35A VHF TV xmtrs, and in aural side of TV-50H.

8808 Nuvistor high-μ power triode, w/ plate cap, grid on base shell, for G-G amps, heater 6.3 V @ 340 mA, μ 100, g_m 18 mS, 6 W ICAS, usable to 1.2 GHz, able to be pulsed (RCA) (ca. 5-69).

8809 [AKA 4CX600J]; xmtg. tetrode, 4CX600B w/ heater 6.0 V @ 5.4 A, for linears (EI). Variants: 4CW800B, 4CW800F, 8921.

8811 Klystron amplifier, 4.4-5 GHz, four gang-tuned cavities, heater 6.5 V @ 7.6 A, 8.5 kV, 1.5 kW out, forced-air cooled, integral magnet, waveguide input-output (RCA) (ca. 10-71). 4658 uprated from 7.5 kV to 8.0; no other difference. Also see 4659 and 4660.

8824-26 CW klystron amplifiers, for UHF TV, heater 6.0 V @ 16.4 A, 20 kV, 12-31 kW output, electromagnet-focused; seals forced-air cooled, cavities water-cooled, collector vapor-cooled; "N" coax input, waveguide output (all RCA) (ca. 2-72).

8824 For 470-556 MHz

8825 For 566-698 MHz.

8826 For 698-890 MHz.

8828, W Xmtg. cermet beam tetrode, for SSB or general use, 8462 w/ beryllia side-arm heat sink, uprated, heater 26.5 V @ 680 mA, 200 W, 500 MHz, conduction-cooled (RCA) (11-18-63). Related to 8122.

8840 Klystron pulse amplifier, 2856 MHz, uprated 8568, heater 17 V @ 15.5 A, input uprated to 270 kV @ 298 A, pulse output 16-28 MW (RCA) (ca. 11-72).

8843 Numeric readout.

8844 Vidicon, 1" sulfide-target, for CCTV (RCA).

8847, A Xmtg. planar triode, 7815 format, heater 6.3 V @ 1.3 A, 4 kW pulse output, conduction-cooled (EI, GE). Used in airborne DME and CAS eqpt. near 1.1 GHz.

8850 Photomultiplier, 12-stage, GaP first dynode, flat-faced head-on bulb 2.1" dia. by 5.7" long, in plastic light shield, 21-pin base (RCA) (ca. 5-70). Variant: 8851.

8851 Photomultiplier, 8850 w/ spherical window (RCA).

8852-53 Photomultipliers, head-on bulb 2.1" dia. by 5.7" long (RCA).

8854 Photomultiplier, variant of 4522.

8857 /V1, V2 Image intensifier, sim. to 8605 scaled down to 1-½" dia. by 1.9" long, same input and output responses (RCA) (ca. 2-71).

8858 Image intensifier, 18 mm, 3 stages, w/ built-in voltage multiplier, related to 4550/4551 (RCA) (ca. 2-71).

8859 Xmtg. high-μ cermet pencil triode, 0.52" dia. by 1.9" long, heater 6.3 V @ 350 mA (GE) (6-19-67).

8866 Xmtg. planar tetrode, heater 6.3 V @ 3.8 A (GE) (4-10-67).

8867 [Type not listed] (AX).

8869 [Type not listed] (9-18-67).

8873 Xmtg. cermet high-μ triode, large elevenar 11-pin base, fil. 6.3 V @ 3.0 A, μ 240, 400 W, 500 MHz, conduction-cooled (EI). (Eimac advertising had fun with the number, as comprising "88" and "73".) Two each in Henry Radio 2K Ultra and Heathkit SB-230 linears.

8874 [AKA 3CX400A7]; xmtg. focused triode, axial-air-flow variant of 8873 (EI). Three in Alpha 374 linear.

8875 Xmtg. triode, transverse-air-flow variant of 8873, derated to 300 W, large 5-fin radiator (EI).

8877 [AKA 3CX1500A7]; xmtg. cermet focused VHF triode, for general use incl. linears, special 7-pin base, fil. 5.0 V @ 10.5 A, μ 200, 1.5 kW, 250 MHz, forced-air cooled, 7-pin spl. base (EI).

8890 Xmtg. cermet beam tetrode, for FM-TV use, fil. 5.7 V @ 115 A, 5 kW, 400 MHz, forced-air cooled (RCA) (ca. 9-70).

8891 Xmtg. cermet tetrode, for FM-TV or SSB use, fil. 9.5 V @ 153 A, ca. 7.5 kW, 400 MHz, forced-air cooled (RCA) (ca. 9-70). Also see 8916.

8892 Xmtg. planar triode, heater 6.3 V @ 650 mA (GE).

8893 Xmtg. planar triode, heater 6.3 V @ 1.3 A, conduction-cooled (GE).

8904 [AKA 4CX350FJ]; xmtg. tetrode, external anode, heater 26.5 V @ 650 mA, 8904 w/ reduced intermodulation (EI).

8906, AL, BAL Xmtg. planar cermet triode, heater 6.0 V @ 1.0 A, conduction-cooled (EI, GE).

8907, AL Xmtg. planar cermet triode, heater 6.0 V @ 1.0 A, forced-air cooled (EI, GE).

8908 Xmtg. pentode, octal w/ cap, heater 6.3 V @ 2.65 A (GE).

8909 [AKA 4CX5000J]; xmtg. tetrode, 8170, ruggedized and w/ low distortion, fil. raised to 103 A (EI, GE).

8910 [AKA 4CX15000J]; xmtg. tetrode, highly linear, ruggedized 8281, fil. 7.5 V @ 158 A, 15 kW (EI, GE, SV).

8916 Xmtg. cermet VHF beam tetrode, for SSB or for FM-TV use in Y1165 cavity, fil. 9.5 V @ 147 A, 22 kW, 400 MHz, forced-air cooled (RCA) (ca. 9-70). Same body format as 8891. In Gates TV-18H and TV-25H VHF TV xmtrs.

8917 Xmtg. planar cermet triode, heater 6.3 V @ 1.2 A, forced-air cooled (GE).

8921 [AKA 4CX600JA]; xmtg. cermet tetrode, 8809 w/ larger radiator for reduced air-pressure drop (EI).

8930 Xmtg. tetrode, 7580W w/ larger radiator requiring lower air flow (EI)

8931 [Type not listed] (W).

8933 [AKA 8538B, q. v.]; planar triode (EI, ML).

8938 [AKA 3CX1500C7]; xmtg. cermet high-μ triode, coaxial focused-beam construction, fil. 15.0 V @ 10.5 A, for UHF G-G amps. (EI).

8940 Xmtg. cermet planar triode, threaded stud on anode for connection to heat sink, coaxial base, heater 6.3 V @ 2.25 A, μ 65, typical dissipation 750 W, 2.5 GHz (CW) or 3.0 GHz (pulse) (EI). Variants: 8941, 8942.

8941 Xmtg. triode, 8940 w/ μ raised to 200 (EI).

8942 Xmtg. triode, 8940 w/ μ raised to 115 (EI).

8950 Xmtg. compactron pentode, w/ cap, heater 13.0 V @ 1.1 A (GE). In Swan 300B xcvr.

8954 (EI).

8957 [AKA 4CX250BC]; xmtg. cermet tetrode, long-life variant of 7203 (EI, GE, SV).

8959 Xmtg. cermet tetrode, for general use, coaxial base, fil. 15.5 V @ 215 A, 100 kW, 108 MHz, water- and forced-air cooled (EI)

8960 Xmtg. tetrode, for pulse modulators or switches, tubular bulb w/ cap sim. to expanded 8166/4-1000, fil. 7.0 V @ 36 A, 50 kV, 12 A pk., 1.2 kW (EI).

8961 [AKA 3CX400U7]; xmtg. cermet UHF focused-beam triode, coaxial base, heater 6.3 V @ 3.0 A, μ 240, 400 W, 1 GHz, promoted for 900-MHz land-mobile base-station use, offered w/ CV850 cavity assembly (EI) (ca. 1974). "Focused" construction uses multiple strip electron guns for low grid interception/high μ, w/ claimed gain of a tetrode and simplicity of a triode.

8962 [AKA 3CX1500U7]; xmtg. external-anode triode, heater 5.0 V @ 11.7 A (EI).

8963 Cermet coaxial triode, for TV xmtrs, 25 kW, w/ large rippled-fin radiator (EI) (1975).

8964 Xmtg. planar triode (EI).

8965 Xmtg. external-anode triode (EI).

8966 [AKA 5CX3000A]; xmtg. cermet VHF pentode, ring-and-breechblock base, fil. 9.0 V @ 41.5 A, 4 (not 3) kW, 150 MHz, for AF-RF linears (EI).

8971 Xmtg. cermet triode, fil. 16.3 V @ 640 A (EI). Also see 8972.

8972 Xmtg. cermet triode, coaxial base, two T-T fils. 16.3 V @ 320 A each, μ 23, 1.5 MW, 30 MHz, water-cooled via integral coolers on anode and fil. support (EI).

8973 Xmtg. cermet triode, coaxial base, T-T fil. 15.5 V @

8973 Xmtg. cermet triode, coaxial base, T-T fil. 15.5 V @ 215 A, 1 MW, 110 MHz, water- and forced-air cooled; fil. supports water-cooled, integral ion vacuum pump (EI). Variant: 8974.

8974 Xmtg. cermet triode, 8973 w/ expanded cooler, dual T-T fil. 16.3 V @ 600 A, 1.5 MW, 30 MHz (EI).

8984 Cermet VHF tetrode, forced-air cooled, for G-G linears. In Gates TV-35H and TV-50H VHF TV xmtrs.

8988 Cermet VHF tetrode, forced-air cooled, for G-G linears. Driver for 8984 in Gates TV-35H xmtr.

8989 [AKA 4CX12,000A]; xmtg. external-anode tetrode, coaxial base, fil. 7.5 V @ 120 A, 12 kW, 220 MHz, high-efficiency radiator (EI).

8990 [AKA 4CX20,000A]; xmtg. VHF cermet triode, coaxial base, fil. 10.0 V @ 140 A, 20 kW, 110 MHz, high-efficiency radiator (EI). In Gates FM-25K 25-kW FM.

9000 [AKA 4CM30,000G]; xmtg. cermet tetrode, graphite grids, coaxial base, fil. 15 V @ 500 A, 300 kW (*grid* dissipation 2 kW), 30 MHz, water- and forced-air cooled (EI).

9001-06 (Non-RMA numbers.)

9001 VHF rcvg. pentode, 954 in min. bulb (GE, RCA, TS) <7PM>.

9002 [AKA VT-202, CV664]; VHF triode, 955 in min. bulb (GE, RCA, TS) <7TM>.

9003 [AKA VT-203, CV665]; VHF rcvg. pentode, 956 in min. bulb (GE, RCA, TS) <7PM>.

9004 [AKA 653B]; acorn UHF diode, heater 6.3 V @ 150 mA, 117 V RMS input, 4 mA, sim. to CW10 except for pinout (GE, RCA, TS) <4BJ>. Suspected of being U. S. production of the "Australian" 953. Two as detectors in AN/APN-1.

9005 Acorn UHF diode; smaller (lower-cap.) elements than 9004, heater 3.6 V @ 165 mA, 117 VRMS input, 1 mA, internal resonant freq. 850 MHz (GE, RCA, TS) <5BG>.

9006 Min. rcvg. UHF diode, heater 6.3 V @ 150 mA, 750 V PIV, 5 mA, internal resonant freq. 1500 MHz (GE, RCA, TS) (1-15-43) <6BH>.

9007 VHF xmtg. cermet tetrode, for G-G linears, forced-air cooled. In Gates TV-20L 20-kW and TV-30L 30-kW VHF TV xmtrs.

9008 [AKA 4CPW300,000A]; xmtg. cermet tetrode, for pulse modulators or regulators, fil. 12.0 V @ 660 A, 100 kV 300 kW, water-cooled via integral cooler, w/ integral ion vacuum pump (EI). Variant: 9009.

9009 [AKA 4CPW1000KA]; xmtg. cermet tetrode, 9008 w/ expanded cooler, same uses, 150 kV, 1 MW (EI).

9011 (Non-RMA n.); VR tube, 90 V, 5-40 mA; advertized by RCA in 1942; possibly renamed VR-90/0B3.

9013 Xmtg. cermet tetrode, for pulse modulators or regulators, coaxial base, fil. 15.5 V @ 215 A, 100 kW, 110 MHz, water- and forced-air cooled via integral cooler (EI) {Y-676A}.

9015 [AKA 4CX20,000D]; xmtg. cermet tetrode, coaxial base, fil. 7.5 V @ 145 A, 20 kW, 110 MHz, high-efficiency radiator (EI).

9016 [AKA 4CPW10,000R]; xmtg. cermet tetrode, for switch and regulator use, coaxial base, fil. 7.5 V @ 75 A, 10 kW, water-cooled via integral cooler (EI).

9019 Xmtg. cermet tetrode, coaxial base, fil. 7.5 V @ 160 A, 18 kW, 110 MHz (EI, SV) {YC-130}.

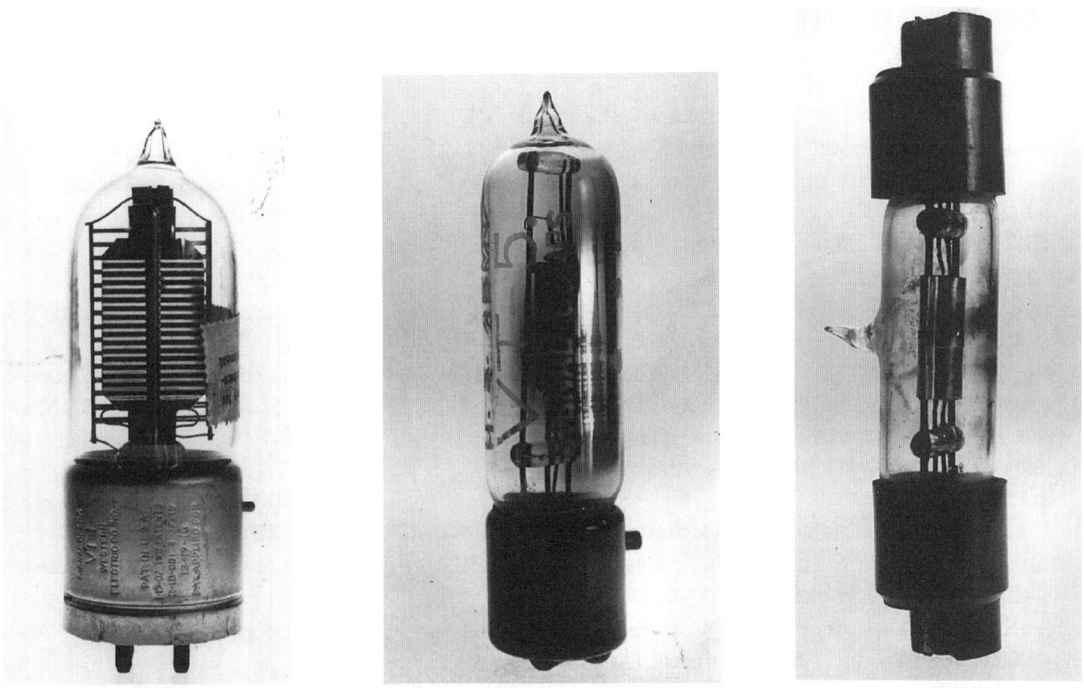

Three triodes liberally predating the 5500-series: Western Electric VT-1, WE VT-5, and Myers RAC.
Photos: Bob Dennison, W2HBE.

CRTs AND MONOSCOPES

This chapter lists early cathode-ray tubes and monoscopes whose specifications are often hard to find; many of them - radar indicators, for example - were obsoleted early. Most of these were introduced before 1950, but certain historically important later types are included. As a special case, RCA 1800- and 4000-series types are listed in the RCA chapter with 1600/2000/4000-series types. Likewise, Western Electric CRTs (224A, 302A, 325A, 326A, 330A, etc.) are in the WE section.

RMA-EIA registration of CRTs began in April, 1939 with the 5AP1. Variants of tubes with different phosphors were registered individually.

The phosphors in early Du Mont tubes were identified differently from industry "P" numbers. The conversion is Du Mont A5 = industry P1, B5 = P2, C5 = P5, and D5 = P4.

1AP5 CRT, 913 w/ P5 phosphor; not produced.

2-5/8P1 [AKA 3QP1, q. v.].

2AP1, A, W [AKA 1814P1] {C7483}; CRT (electrostatic focus and deflection), heater 6.3 V @ 600 mA, 1 kV, 6-7/8" long, 11-pin magnal base (GE, NU) (7-30-42). In A-scope of AN/MPG-1. "A" is xero-I_{b1} version. Also made in P5. "P11" is original DuM 2AP5; has longer persistence than later 2AP5.

2BP1, A CRT (electrostatic focus and deflection), heater 6.3 V @ 600 mA, 2.5 kV, 7-1/8" long, 12-pin duodecal base (GE) {C-73146} (5-10-46). Also made in P7, P11.

3AP1 [AKA 906P1, 2501A3, 34XH]; CRT (electrostatic focus and deflection), 10-7/8" long, 7-pin base, heater 2.5 V @ 2.1 A, 1.5 kV, one H plate tied to one V plate (GE, NU, RCA, SY) {C7350} (5-1-39). In monitor scope of SG. "A" is zero-I_{b1} version. Also made in P3 P5, P7. "P4" [AKA 906P4, 2501D3] is by (SY) (5-1-39). "P11" is original DuM 3AP5, longer persistence than later 3AP5.

3BP1, A, W CRT (electrostatic focus and deflection), heater 6.3 V @ 600 mA, 2 kV, 9-¼" long, 14-pin diheptal base (GE, NU) {C7488} (7-30-42). "A" is zero-I_{b1} version. "W" is ruggedized {C-7488B}. Also made in P4, P5, P7, and P11.

3BGP1 CRT, fil. 1.5 V @ 140 mA, 1-½" x 3" rectangular (SY) (4-18-60).

3BMP1 CRT, fil. 1.5 V @ 140 mA, 3" round (SY) (8-8-60).

3CP1, -S1 [AKA 1808P1, VT-248]; CRT, 3EP1 w/ radial-deflection electrode for circular-sweep "J-scope" use {C7510, C7332} (8-1-42). "-S1" has decalcomania inscription on faceplate (presumably the altitude scale for SCR-718 radar altimeter and/or AN/APN-84 SHORAN set). Also made in P4 and P7.

3DP1, A, -S2 CRT, 3BP1 w/ radial-deflection electrode for circular-sweep "J-scope" use (GE) {ZP526} (8-1-42). "A" is zero-I_{b1} version. "-S2," "A-S1" and "A-S2" have decal scales. "-S2" in SCR-718. Also made in P4, P5. "P11" is original DuM 3DP5; has longer persistence than later 3DP5.

3EP1 [AKA 1806P1, VT-226, CV817]; CRT (electrostatic focus and deflection), 9-3/8" long, 11-pin magnal base, heater 6.3 V @ 600 mA, 2.2 kV (11-6-42). Also made in P4.

3FP7, A CRT (electrostatic focus and deflection), accelerator type, 10" long, diheptal base, heater 6.3 V @ 600 mA, 4.4 kV (GE, NU) {ZP540} (8-1-42). "A" is zero-I_{b1} version.

3GP1, A [AKA 2537A3]; CRT (electrostatic focus and deflection) (9-24-42). "A" is zero-I_{b1} version. "P11" and "P11A" are original DuM 3GP5; have longer persistence than later 3GP5. "A" is zero-I_{b1} version. Also made in P4, P5.

3HP7 CRT (mag. focus and deflection), diheptal base, anode cap, heater 6.3 V, 5.5 kV {C7501, ZP-542} (1-20-43). Also made in P12, P14.

3JP1 K1012]; CRT (electrostatic focus and deflection), 9-¼" long, diheptal base, heater 6.3 V @ 600 mA, 4 kV {C7523B} (8-15-45). Also made in P2, P4, P5, P7 (2-10-47), P11 {C73146, C73147}, P12, and P14.

3KP1 CRT (electrostatic focus and deflection), heater 6.3 V @ 600 mA, 2.5 kV, 10-15/16" long, 11-pin magnal base (GE) {C7555} (4-25-46). Also made in P4 (8-3-48), as seen in Pilot TV-37 TV set; P7; and P11.

3MP1 CRT, short length (electrostatic focus and deflection), heater 6.3 V @ 600 mA, 1 kV, 7-½" long, 12-pin duodecal base (GE) (2-26-47).

3NP4 Projection CRT (magnetic focus and deflection), 10-½" long, slide base, 25 kV (N. Am. Philips) (3-20-48).

3QP1 [AKA 2-5/8P1]; CRT (electrostatic focus and deflection) (4-25-47).

3RP1, A CRT (electrostatic focus and deflection), 8-5/8" deep, duodecal base, heater 6.3 V @ 600 mA, 2.5 kV (3-4-49). In Waterman S14A; "A" in Hycon 617-T.

3SP1 See 3RP1 (4-6-50) Also made in P2, P4, P5, P7, and P11.

3U2P1 CRT, two guns (electrostatic focus and deflection). Possibly from Electronic Tube Corp.

3U3P1 CRT, three guns; see 3U2P1. Also made in P11.

3W2P1A, 7 CRT, two guns; see 3U2P1.

3WP1 Flat-face CRT, 11" long, duodecal base, heater

6.3 V @ 600 mA, 2.5 kV (10-31-52). In DuM 331 and Hycon 627R. Also made in P2 and P11.

4AP1 CRT (magnetic focus and deflection), octal base, anode cap, heater 6.6 V, 9.9 kV {ZP556, C7516B}. Also made in P10.

55P( ) Dual-gun CRT, 5", stepped cylindrical bulb, diheptal base, caps for deflection plates, ratings sim. to those for 5CP1, 1945 (DuM).

5AP1, A [AKA 1805P1]; CRT (electrostatic focus and deflection), 12-7/16" long, 11-pin magnal base, heater 6.3 V @ 600 mA, 2 kV (NU, SY). Also made in P4 (RCA, SY) (4-17-39).

5BP1, A [AKA 2525A5, 1802P1, (5HP1)]; CRT (electrostatic focus and deflection, 16-3/4" long, 11-pin magnal base, heater 6.3 V @ 600 mA, 2.2 kV (GE, NU) {ZP480} (5-1-39). In range indicator of SG; later in Heath OM-1, etc. "A" is zero-I_{b1} version. Also made in P2, P4, P5, P7, and P11. P4 [AKA 2525B5, 5HP4, VT-111, CV836, 1802P4] {ZP453} (5-1-39) was indicator in SC-1. P7A" is Can. type (5-15-47). "P11" original DuM 5BP5; longer persistence than 5BP5.

5CP1, A CRT (electrostatic focus and deflection), accelerator type, 16" long, diheptal base, heater 6.3 V @ 600 mA, 4 kV on accelerator, 2 kV on anode, 14-pin diheptal base (GE, NU) {ZP517, C7520} (8-18-42). In A-scope of SC-2. "A" is zero-I_{b1} version. Also made in P2 and P4 {ZP518, C7491} (8-18-42); P5, P7 {C7519} (8-18-42); P11, and P12 (7-20-52). "A" is zero-I_{b1} version. "P11" is original DuM 5CP5; longer persistence than later 5CP5. "A" is zero-I_{b1} version.

5DP1 CRT (electrostatic focus and deflection).

5EP1 CRT (electrostatic focus and deflection). Also made in P7.

5FP1A CRT and view-finder kinescope, magnetic focus and 53° deflection, 11-½" long, octal base, anode cap, heater 6.3 V @ 600 mA, 8 kV (GE, NU, RCA, W). Also made in P4, P5, and P7 [AKA 1812P7, GL-475] {C7474D, ZP475} (8-1-42). In AN/APS-3. 5FP7B is uprated to 12 kV (1-28-57). Also made in P11 (GE, W) (7-29-57), P14 (GE, W) (6-30-45) P15, and P25 (GE) (5-9-55).

5GP1, A CRT, 5BP1 format, replaceable by 5BP1; same electricals except lower deflection sensitivity (SY) (5-7-42). Also made in P4.

5HP1, A CRT, 5BP1A w/ Micanol base for high-altitude use (RCA) {C7508} (7-13-42). "A" is zero-I_{b1} version. Also made in P4 {C7512} (7-13-42).

5JP1, A [AKA 2529A5]; CRT (electrostatic focus and deflection), magnal 11-pin base; heater 6.3 V; 4.4 kV (5-7-42). Also made in P2 [AKA 2529B5] (9-25-47), P4 [AKA 2529D5] (5-7-42), P7 [AKA 2529C5] (9-25-47), and P11. "P11" is original DuM 5JP5; has longer persistence than later 5JP5 (9-25-47).

5LP1, A [AKA 2511A5]; CRT (electrostatic focus and deflection), magnal 11-pin base, heater 6.3 V, 4.4 kV (9-24-42). Also made in P2 [AKA 2511-B5] and P4 [AKA 2511D5], P5 [AKA 2511C5], P7, and P11. "P11" is original DuM 5LP5; has longer persistence than later 5LP5.

5LRP1 CRT, "potentiometer" (unverified).

5MP1 [AKA 2505A5]; CRT (electrostatic focus and deflection), 7-pin base, heater 2.5 V (9-24-42). Also made in P4 [AKA 2505D5], P5 [AKA 2505-C5], and P11. "P11" is original DuM 5MP5; has longer persistence than later 5MP5.

5NP1 Original 5BP1 w/ short gun (SY) (11-8-43).

5NP4 CRT, replaceable by 5NP1 (11-8-43).

5QP4-A CRT (mag. focus & 53° deflection), aluminized screen, heater 6.3 V @ 600 mA, 12 kV on anode, octal base w/ anode cap (GE) (2-12-52).

5RP1, A CRT, diheptal base, heater 6.3 V, 26 kV (4-30-48). Also made in P2, P4, P5, P7, P10 (4-30-48), and P11.

5SP1 CRT (electrostatic focus and deflection), diheptal base, heater 6.3 V, 4.4 kV (DuM) (8-25-45). Also made in P2, P4, P5, P7, and P11.

5TP4 CRT, mirror-backed projection (electrostatic focus and 50° mag. deflection), 11-¼" long, duodecal base, anode cap, heater 6.3 V @ 600 mA, 27 kV (RCA) (4-15-46). Variant: 4820.

5UP1 CRT (electrostatic focus and deflection), 14-¼" long, 12-pin duodecal base, heater 6.3 V @ 600 mA, 2.5 kV on accelerator, 1 kV on anode (GE, W) (1-3-47). In RCA WO-88A, Jackson CRO-2, Weston 983, etc. Also made in P7, P11, and P31.

5VP7 CRT, electrostatic focus and deflection, sim. to 5BP1-5GP1-5HP1, with clear port in bell for photographing the display from the rear (Radio Valve Co. of Canada, Ltd.) (4-16-48).

5WP11 CRT, metal-backed, transcriber type (electrostatic focus., 50° mag. deflection), 10-15/16" long, duodecal base, heater 6.3 V @ 600 mA, 27 kV on accelerator, 6 kV on first anode (RCA, W) (1-10-49). Also made in P15 for flying-spot scanners for slide transparencies.

5X2P1 CRT, dual gun (electrostatic focus and deflection). In Tektronix 513 and 517. Also made in P2, P4, and P5.

5X3P1 CRT, triple gun (electrostatic focus and deflection). Also made in P5 and P11.

5XP1 CRT, HV, high sensitivity (electrostatic focus and deflection) (12-9-48). Also made in P2, P7, and P11.

5Y3P1 CRT, triple gun. Also made in P3, P4, P5, and P11.

5Z2P1, 2, 4 CRT, dual gun (electrostatic focus and deflection). Also made in P5, P7, and P11.

5Z3P1, 2, 4 CRT, triple gun. Also made in P5, and P11.

5Z4P1 CRT, quadruple gun. Also made in P2, P4, P5, and P11.

7AP4 CRT (electrostatic focus, 45° magnetic deflection), 12-7/8" long, 5-pin base, heater 2.5 V @ 2.1 A, 3.5 kV (RCA) (10-19-39).

7BP1, Y CRT (magnetic focus and 53° deflection), 13-¼" long, octal base, anode cap, heater 6.3 V @ 600 mA, 8 kV (6-30-45). Also made in P4, P7, qnd P12. "P12" was B-scope in AN/MPG-1. "Y" has Micanol base. 7BP7, A, B [AKA 1813P7]; (GE, NU, RCA, W) {ZP468, C7475A} (8-1-42) was PPI scope in SJ and AN/MPG-1; also B-scope in AN/MPG-1. "B" has higher-resolution electron gun.

7CP1 [AKA 1811P1]; CRT (electrostatic focus, 55° magnetic deflection), 12-7/8" long, octal base, anode cap, heater 6.3 V @ 600 mA, 7.7 kV (NU, RCA) {C7466A} (11-6-42). Also made in P4 (6-25-46).

7DP1 CRT (electrostatic focus and 50° magnetic deflection), 13-9/16" long, duodecal base w/ anode cap, heater 6.3 V @ 600 mA, 8 kV on accelerator, 2.4 kV on first anode (RCA) (4-15-46). Also made in P4.

7EP4, A CRT (electrostatic focus and deflection), magnal 11-pin base, heater 6.3 V, 3.3 kV (11-30-45). "A" is zero-I_{b1} version.

7GP1 CRT (electrostatic focus and deflection), 13-3/4" long, diheptal base, heater 6.3 V @ 600 mA, 4 kV on accelerator, 1.5 kV on first anode (RCA) (4-15-46). Also made in P4.

7HP1 CRT, magnetic focus and deflection; introduced use of external Aquadag coating (North Am. Philips) (4-10-46). Also made in P4.

7JP1 CRT (electrostatic focus and deflection), 13-3/4" long, diheptal base, heater 6.3 V @ 600 mA, 6 kV on accelerator, 2.8 kV on first anode (RCA) (9-20-50). In Eico 470 and Precise 300B. Also made in P4.

7KA1P4 See 7HP4.

7MP7 CRT, magnetic focusing and 50° deflection, 12" long, duodecal base w/ anode cap, heater 6.3 V @ 600 mA, 8 kV (RCA, W) (4-17-50). Also made in P14.

7NP4 Projection kinescope, electrostatic focus and 35° mag. deflection, 19-½" long, diheptal base w/ anode cap, forced-aid cooled, bulb ridged for added HV insulation, heater 6.6 V @ 620 mA, 80 kV (RCA) (9-12-50). Also see 4486 and 4583-4584.

7UP7 CRT (2-20-52). In AN/APS-20B. No other information.

7Z4P11 Multiple-gun CRT (Electronic Tube Corp.?).

7Z5P2 CRT, five-gun (electrostatic focus and deflection). Also made in P7.

8AP4 TV pic tube, metal envelope, 14" deep, "triode" gun (GE) (3-14-49).

8BP4 TV pic tube, like expanded 7JP4 (NU) (8-15-49). In some versions of Motorola VT-71.

9AP1 [AKA 1804P1, 4]; CRT (electrostatic focus, 40° magnetic deflection), 11-1/8" long, 6-pin base, heater 2.5 V @ 2.1 A, 7 kV (RCA) (5-1-39). Also made in P4.

9BP4 CRT (electrostatic focus, magnetic deflection).

9CP4 TV pic tube (magnetic focus and defl.), heater 2.5 V @ 2.1 A, second anode 7 kV (8-20-40).

9DP1 [AKA 2514A9]; CRT (electrostatic focus and deflection). Also made in P2 [AKA 2514B9], P3 [AKA 2514C9], P4 [AKA 2514D9], P5 [AKA 2514E9], and P11. "P11" is original DuM 9DP5; longer persistence than later 9DP5.

9EP1 CRT (electrostatic focus and deflection). Also made in P4 and P7.

9FP1 CRT (electrostatic focus and deflection).

9GP7, A [AKA 9MP7, VCR140]; CRT (magnetic focus and deflection), octal base, anode cap, heater 6.3 V, 7.7 kV {ZP549, C7487} (6-15-44).

9HP7 CRT (magnetic focus and deflection) (2-15-44).

9JP1 [AKA 1809P1]; CRT (electrostatic focus, electrostatic deflection on one axis and 55° magnetic deflection on the other), 15-1/8" long, octal base w/ anode cap, heater 2.5 V @ 2.1 A, 5 kV (RCA) {C7431} (5-9-42).

9LP7 [AKA (C, GL, ZP) 455]; CRT (magnetic focus and deflection), octal base, anode cap, heater 6.3 V, 7.7 kV (6-15-44). PPI indicator in SG. Also made in P1.

9MP7 [AKA 9GP7, REL43, VCR140]; CRT (magnetic focus and deflection), replaceable by 9GP7 {ZP549, C7487} (6-6-43). In BC-1083 indicator of SCR-602A. Also made in P1, P8, P9.

9NP1 CRT (electrostatic focus and deflection) (9-25-43). Also made in P7.

9Z8P7 CRT, 8-gun (Electronic Tube Corp.?).

10BP4, A TV pic tube (magnetic focus and 50° deflection), 17-1/8" long, duodecal base w/ anode cap, heater 6.3 V @ 600 mA, 10 kV (RCA) (4-15-46). First mass-market pic tube: in RCA Victor 630 TV set and dozens of similar models. "A" is uprated to 12 kV.

10DP4 CRT (electrostatic focus, mag. defl.) (9-5-47).

10FP4, A TV pic tube {R6025} (magnetic focus and 50° deflection), aluminized screen, 17-1/8" long, duodecal base w/ anode cap, heater 6.3 V @ 600 mA, 12 kV (RCA) (5-27-47).

10GP4 CRT (electrostatic focus and defl.) (8-15-47).

10KP7, A, B CRT (mag. focus & 50° deflection), 17-1/8" long, aluminized screen, duodecal base, heater 6.3 V @ 600 mA, 10 kV (GE, W) {C73111} (4-10-48). "A" has high-resolution electron gun.

12AP4 [AKA 1803P4]; CRT (electrostatic focus, 38° magnetic deflection), 24-3/8" long, 6-pin base w/ anode cap, heater 2.5 V @ 2.1 A, 7 kV (RCA) (5-1-39). Also made in P7 {C7442}.

12CP4 TV pic tube, magnetic deflection, heater 2.5 V @ 2.1 A, 7 kV (8-28-40).

12DP7, A, C [AKA VCR516]; CRT (mag. focus and 50° deflection), 19-5/8" long, octal base w/ anode cap, heater 6.3 V @ 600 mA, 10 kV (GE, RCA) {ZP-477, C7485B} (8-18-42). "A" has beam-limiting aperture for sharper focus. "C" has aluminized screen. PPI scope in SC-2; also in BC-1239A PPI of SCR-270D.

12DP8 [AKA REL25]; Can. CRT (magnetic focus and deflection); replaceable by 12DP7 {ZP-477}; 1942-vintage tube developed by RCA. (Canadian GE registered the British-developed P8 and P9 radar-tube phosphors with the RMA.) In BC-1135 indicator of SCR-597.

12EP8 "Canadian CRT (obsol.)."

12FP7 [AKA REL45]. CRT (electrostatic focus and deflection) (REL) {ZP548} (6-6-43). Also made in P1.

12GP7 [AKA GL-488]; CRT (electrostatic focus and deflection), diheptal base, heater 6.3 V, 6.6 kV (6-15-44).

12HP1 [AKA REL33A]; CRT (electrostatic focus and deflection) (REL) (2-15-44). Also made in P3 [AKA REL33] and P7. "P7" [AKA REL37] (REL) (10-1-43) was in BC-1111 indicator of SCR-588 CHL/GCI radar, and of SCR-636.

12JP4 CRT (magnetic deflection), duodecal base, anode cap, heater 6.3 V, 13.2 kV.

12KP4, A TV pic tube (magnetic focus and 54° deflection), 17-5/16" long, duodecal base, heater 6.3 V @ 600 mA, 12 kV (RCA) (4-30-48).

12LP4, A TV pic tube (magnetic focus and 57° deflection), 18-¼" long, duodecal base, heater 6.3 V @ 600 mA, 12 kV (RCA) (5-25-48).

12SP7, B, D CRT (mag. focus and 54° deflection), 18-3/4" long, duodecal base w/ anode cap, heater 6.3 V @ 600 mA, 10 kV (GE, W). "B" has aluminized screen. "D" has tightened tolerances on dimensions and high-resolution electron gun (7-8-49).

12Z2P7 CRT, dual gun (electrostatic focus and deflec.).

12Z10P7 CRT, ten guns (Electronic Tube Co.).

14AP1 [AKA 2531A14]; CRT (electrostatic focus and deflection) (9-24-42). Also made in P2 [AKA 2531B14], P4 [AKA 2531D14], and P5 [AKA 2531C14].

14CP4, B TV pic tube (mag. focus and 70° deflection), rectangular, 16-3/4" long, duodecal base w/ anode cap, heater 6.3 V @ 600 mA, 14 kV (GE, RCA) (4-21-50). "B" is uprated to 15.5 kV and has aluminized screen.

15GP22 First commercial color pic tube (electrostatic focus and 45° magnetic deflection), 26-1/8" long, bidecal base w/ anode cap, heaters 6.3 V @ 1.8 A, 20 kV (RCA) (2-19-54).

16AP4, A TV pic tube, first metal-cone design (magnetic focus and 53° deflection), filter-glass faceplate, 22-5/16" long, duodecal base, heater 6.3 V @ 600 mA, 14 kV (RCA) (12-10-48).

20AP1 [AKA 2532A20]; CRT (electrostatic focus and deflection), 12-contact Br. slide base, heater 2.5 V, 8 kV; superseded by K-1019P1 et al. (DuM) (9-24-42). Also made in P2 [AKA 2532B20], P4 [AKA 2532C20], P5 [AKA 2532D20], and P11. "P11" is original DuM 20AP5; longer persistence than later 20AP5. "P4" in DuM prewar big-screen TV set.

22AP4, A CRT (magnetic focus and deflection) (4-21-50).

24XH CRT, 2", green short-persistence phosphor, electrostatic deflection, octal base, heater 6.3 V @ 600 mA, 600 V (DuM).

30BP4 CRT, 30", round w/ gray face, 30-1/8" dia. by 23-9/16" deep, magnetic focus and 95° deflection, duodecal base w/ cap, 30 kV (DuM) (7-16-51). In "Royal Sovereign" TV set. Also used as air-traffic-control radar indicator.

34XH CRT sim. to 906, octal base (DuM)

54H CRT (DuM).

54-8-H [AKA 2507]; CRT (DuM).

54-9-T CRT, 5", first intensifier-type (DuM) (1938).

54-11-T [AKA 5LP1, 2, 4 & 5; 2511]; CRT (DuM).

94-8-H [AKA 2512]; CRT (DuM).

212 "Television pickup tube (Remington-Rand)."

353A CRT (unconfirmed) (not Western Electric).

902, A [AKA 902P1]; 2" CRT (electrostatic focus and deflection), octal base, heater 6.3 V @ 600 mA, w/ "triode" gun (GE, NU) (2-21-45). Pulse indicator in SC-2. "A" [AKA 902P1A] is zero-I_{b1} version of 902P1 w/ "tetrode" gun (8-15-45), 7-7/16" long, 660 V (RCA).

903 CRT, 9", magnetic deflection, 6-pin base w/ cap, heater 2.5 V @ 2.1 A, 7 kV {C735} (NU, 1935).

904 CRT, 5", green (electrostatic focus, electrostatic and 39° mag. deflection), 15-5/8" long, 6-pin base w/ 2 caps for deflection plates, heater 2.5 V @ 2.1 A, 4.6 kV (GE, RCA) {C737}.

905, A CRT, 5", P1 phosphor, electrostatic focus and 22° deflection, 15-7/8" long, 5-pin base w/ 4 caps for deflection plates, heater 2.5 V @ 2.1 A, 2.2 kV (GE, RCA) {C743} (2-21-45). "A" is zero-I_{b1} version (8-15-45).

906P1 [AKA 2501A3, 3AP1, q. v.] {C744} (NU). In train indicator of FC-FD. Also made in P3 [AKA 3AP3, q. v.] (SY) and P4 [AKA 3AP4, q. v.]. Intro. by RCA but second-sourced (1938) by NU and SY. Further made in P5 and P7.

907, A 905, A w/ P5 blue phosphor) (GE, NU, RCA)

{C753} (1935).

908, A — 906P1/3AP1 w/ P5 phosphor (GE, W) {C756} (3-21-45).

909 — 905 w/ P2 bluish-white long-delay phosphor {C757} (RCA).

910 — 906/3AP1 w/ P2 bluish-white long-delay phosphor {C755} (RCA).

911 — 906/3AP1 w/ nonmagnetic gun {C7310}.

912 — CRT,5" (electrostatic focus and deflection), P1 phosphor, 17-15/16" long, 5-pin base w/ 4 caps for deflection plates, heater 2.5 V @ 2.1 A, 15 kV {C751} (RCA).

913 — CRT, metal, 1" dia., P1 phosphor (electrostatic focus and deflection), in 6L6-style metal octal envelope, heater 6.3 V @ 600 mA, 500 V, one H plate tied to one V plate (GE, RCA) (1936).

914, A — CRT, 9", P1 phosphor, (electrostatic focus and deflection), 19-7/16" long, 6-pin base, heater 2.5 V @ 2.1 A, 7 kV (GE, RCA) (8-14-50).

2001 — CRT, 1", apparently NU version of 913.

2002 — CRT, 2", P1 phosphor, electrostatic deflection, heater 6.3 V @ 600 mA, 600 V (NU).

2005 — CRT, 5", "fine trace for 441 line definition," electrostatic deflection, heater 2.5 V @ 2.1 A, 2 kV (NU "videotron", 1938).

2030 — Monoscope (NU "Monotron"), 1938, "spot small enough for 441 line definition." 5" CRT-style format, 14-3/16" long w/ electrode out faceplate, 6-pin base, electrostatic focus and magnetic deflection, heater 2.5 V @ 2.1 A, 1.7 kV. Original target had horizontal resolution wedges, images of man and woman superposed on "T" antenna.

2031 — NU monotron.

2109 — CRT, 9" "videotron," magnetic deflection, "fine trace for 441 line definition" (NU). See *Radio-Craft*, 12-38.

2112B — [AKA F16-T-72112-40]; CRT (NU, R).

2203 — NU monotron, 7-pin base, heater 2.5 V @ 2.1 A, 1 kV.

2501A3 — [AKA 906P1, 3AP1, q. v.].

2501C3 — [AKA 3AP5, q. v.].

2501D3 — [AKA 906P4, 3AP4, q. v.].

2505A5 — [AKA 5MP1, q. v.].

2505C5 — [AKA 5MP5, q. v.].

2505D5 — [AKA 5MP4, q. v.].

2507 — See 54-8-H.

2511A5 — [AKA 5LP1, q. v.].

2511B5 — [AKA 5LP2, q. v.].

2511C5 — [AKA 5LP5, q. v.].

2511D5 — [AKA 5LP4, q. v.].

2512 — See 94-8-H.

2514A9 — [AKA 9DP1, q. v.].

2514B9 — [AKA 9DP2, q. v.].

2514C9 — [AKA 9DP3, q. v.].

2514D9 — [AKA 9DP4, q. v.].

2525A5 — [AKA 1802P1, 5BP1, q. v.].

2525B5 — [AKA 1802P4, 5BP2, q. v.].

2525C5 — [AKA 5BP5, q. v.].

2525D5 — [AKA 5BP4, q. v.].

2529A5 — [AKA 5JP1, q. v.]. Intro. 3-41 (DuM).

2529B5 — [AKA 5JP2, q. v.].

2529C5 — [AKA 5JP5, q. v.].

2529D5 — [AKA 5JP4, q. v.].

2531A14 — [AKA 14AP1, q. v.].

2531B14 — [AKA 14AP2, q. v.].

2531C14 — [AKA 14AP5, q. v.].

2531D14 — [AKA 14AP4, q. v.].

2532A20 — [AKA 20AP1, q. v.].

2532B20 — [AKA 20AP2, q. v.].

2532C20 — [AKA 20AP5, q. v.].

2532D20 — [AKA 20AP4, q. v.].

2537A3 — [AKA 3GP1, q. v.].

2537C3 — [AKA 3GP5, q. v.].

2537D3 — [AKA 3GP4, q. v.].

ET3000 — CRT (electrostatic focus and deflection).

FP-53 — CRT, 5", 1930 (GE); not made commercially for patent reasons.

K1002P1 — CRT (DuM).

K1003P1 — CRT (DuM) (electrostatic focus and deflection). Also made in P2, P4, P7, and P11.

K1012 — [AKA 3JP1, q. v.].

K1017P1 — CRT, HV (electrostatic focus and deflection), diheptal base, heater 6.3 V, 37.5 kV (DuM). Also made in P11.

K1019P1 — CRT, 20 inch (electrostatic focus and deflection) (DuM). Also made in P7.

K1230P7 — 14" CRT in OA-1007/FSQ Sage video mapper console.

K1725P16 — CRT, 5" high-resolution, 1-mil spot size (DuM).

KR-3B — Microwave spectrum-analyzer display tube, down 3 dB at 2 GHz, usable to 3 GHz, w/ traveling-wave deflection section, sim. to SY "wamoscope." In EG&G 2236A spectrum analyzer.

MW22-2 — CRT.

Phasmajector #1 — DuMont monoscope, Lincoln-head pattern, electrostatic deflection, cylindrical bulb ca. 5" dia. stepped to CRT neck, electrode out front, five (?) pin base (4-38).

RD6025 — CRT (magnetic focus and deflection).

SC-3016 — CRT, fil. 1.5 V @ 140 mA, 1-1/8" round (SY)

VCR139A (Br.) [AKA 10E/466, 1810P1, q. v.].

Z1304 CRT (GE).

SPECIAL-PURPOSE TYPES: NONREGISTERED

FNH [AKA 6A3 Special, q. v.]; power triode.
GL [AKA 1B32/532A, q. v.]; TR tube.
LVX [AKA 2J42, q. v.]; magnetron.
Navy Type A [AKA ELC1A, q. v.]; thyratron.
Navy Type B [AKA ELC5A, q. v.]; thyratron.
Navy Type C [AKA ELC6A, q. v.]; thyratron.
Resnatron Tetrode UHF oscillator, 340-625 MHz, internal cavities, demountable (the U. S.' only production demountable tube ever?), pumped in operation, 50-kW noise-modulated CW output, coax output into WG, water-cooled via integral cooler, 13 kV @ 6.2 A, p/o AN/MRT-2 radar jammer. Type number either not assigned or not declassified.
SA, SA, SC Gas switching tubes.
TVTA [AKA 892, q. v.]; RF-heating triode (TL).
XP [AKA CV74, q. v.].
XXB [AKA 3C6, q. v.]; loktal triode.
XXD [AKA 14AF7, q. v.]; duotriode.
XXFM [AKA 7X7, q. v.]; loktal duodiode-triode.
XXL [AKA 7A4, q. v.] loktal triode.
K-0 series [AKA 2J23-29, q. v.]; magnetron.
BL1 TR tube.
BS1 [AKA 5979, 1B80, q. v.].
(CE, NU) **1** ( ) Family of gas phototubes, general purpose, 868 format; S-1 response. Lettered grades A/B, C, and D are progressively less sensitive. "D" is equiv. to 918; "E" is equiv. to 868. "RBL" and "RBS" [AKA 59AP, G1] are rebased, as is "S" [AKA G13, CE59AR, 4GSM]. "V/VC" are vacuum variants equiv. to WL-734.
FE1 Spark-gap radar modulator, improved 1B42.
G1 Gas phototube.
GC1-7 Geiger-Muller counter tube.
Mk 1 Beta-gamma counter. Mod. 10A is indoors type; Mod. 10D is outdoors type.
J1 [AKA 2J22, q. v.]; magnetron.
J1RF [AKA 2J22, q. v.].
NRC001V [AKA REL53, q. v.]; klystron (?).
TGC1 [AKA 1B83, q. v.].
TL1 Mercury rectifier, 866 format w/ ceramic-ringed cap, fil. 2.5 V @ 5 A, 160 V PIV, 2.5 A (TL).
TT1 [AKA 6414, q. v.]; noise diode.
V1 Vacuum phototube.
VG1 Ionization vac. gauge, range 10^{-3} to 10^{-9} mm Hg, 200 μA/micron (Distillation Products Co.).
1E [AKA 1B69, q. v.]; counter.
1M [AKA 1B70, q. v.]; counter.
1N Gamma counter tube.
A2 "Diffraction tube, copper target" (X-ray?).
B2A [AKA 2050, q. v.]; thyratron.
BL2 TR tube.
BS2 [AKA 5980, q. v.]; counter.
(CE, NU) **2** ( ) [AKA G6, 71A, WL-737]; family of gas phototubes, for sound repro., 4-pin dwarf (UX-199) base, short tubular bulb, side-on cathode, S-1 response; grades A/B, C, and D are progressively less sensitive. "C" is equivalent to 71A3 and 1P30. "V" is a vacuum variant on a med. 4-pin base. "RB" is rebased.
D2, D2RN1 [AKA WL-442, 2J21, q. v.]; magnetron.
EM2GB Electrostatic storage tube.
G2 [AKA C2]; hearing-aid pentode (ST).
GY-2 [AKA D-161831, VT-279, 1278-GY2]; thyratron (WE); no known use or no.; not seen in surplus; never made?
H2-10 HV rectifier, ca. 1938, repl. by 879.
KRN2 [AKA CV87, q. v.]; klystron.
KRN2A [AKA CV129, q. v.]; klystron.
LD2 [AKA 2J42, q. v.]; magnetron.
NGT2 [AKA VGT128, q. v.]; gas tube.
TGC2 [AKA 1B84, q. v.]; counter.
TT2 [AKA 5946, q. v.]; xmtg. triode.
Y2 [AKA 2J21, q. v.]; magnetron.
ZA2 [AKA 954, q. v.]; acorn triode.
2RA3 [AKA CE213, q. v.]; rectifier.
2RA5 [AKA CE205, q. v.]; rectifier.
2RA6 [AKA CE206, q. v.]; rectifier.
2RA15 [AKA CE215, q. v.]; rectifier.
2A15 Diode rectifier.
2AC15 Diode rectifier.
EE2BT Vacuum rectifier for projection TV receivers, stepped-tubular bulb w/ cap on 4-pin base, 40 kV PIV (EE).
2FV20 "Osc., medical and surgical."
2KB72 Klystron.
BL3 TR tube.
C3A14 Xenon thyratron.
(CE, NU) **3** ( ) [AKA WL-738, 79A, 803A, 4G3A, G10, (WE) 3A]; family of gas phototubes for movie sound systems; patterned after WE 3A, spherical bulb, flying leads, S-1 response. Grades A/B, C, D, and E are progressively less sensitive. "D" is equiv. to G10, 79A4, WE 3A, and 1P31.
EM3GA Electrostatic storage tube.
KRN3/1 Reflex klystron, 9.6-9.9 GHz, external electromagnet, EC, WG, 4.0 V @ 1.3 A, 1.5 kV, 25 mW output (EMI).
P3 Hearing-aid tube.
PM3 Loktal power-measurement lamp (SY); dual filament; user compares brightness of RF-lighted fil. vs. DC-lighted fil.; 1-10 W range. Variants: PM4-PM7.
TR3SW [AKA REL30, q. v.].
3B/501A [AKA 4C25, q. v.]; xmtg. triode.
3BHP [AKA REL3BX, q. v.]; magnetron.
3DHP [AKA CV56, q. v.]; magnetron.
BL4 TR tube.

(CE) **4** ( ) [AKA 58A]; family of gas phototubes, for sound repro., dwarf 4-pin (UX-199) base, tubular bulb, side-on cathode, S-1 response. Grades A/B, C, and D are progressively less sensitive; also see 1P36. "C" is equiv. to 1P36, 58A3. "RB" is equiv. to 58AL and G12. "RBS" [AKA 58A, 4GW] is "RB" rebased. "V" is vacuum variant. "WB" is equiv. to 53AWB, 4GW, and G5.

G4 [AKA CE-5]; gas phototube.

GA4 [AKA (WL) 532, q. v.]; TR tube.

GE-N4 [AKA 2D29, q. v.]; tetrode.

PM4 Power-meas. lamp (SY), 0.8-5.5 W; see PM3.

SN4 [AKA 1D21]; flash tube.

4E [AKA 1B71, q. v.]; counter.

4M [AKA 1B72, q. v.]; counter.

AU5 [AKA VU111, q. v.]; rectifier.

(CE, NU) **5** ( ) Family of gas phototubes, 927 format, S-1 response. "AB" is AKA 53AB. "BB" is AKA G16B, and 53ABB. "WB" is AKA 53AWB, WL-739, 4GTP, and G4.

(CE) **CR5** Mercury thyratron, fil. 2.0 V @ 12 A, 175 V PIV, 5 A, external metal-mesh grid; mogul base w/ cap (CE, HY); also see CR10.

E5A [AKA 3J31, q. v.]; magnetron.

EM5GC Electrostatic storage tube.

G5 [AKA CE-4]; gas phototube.

GA5, A [AKA 1B32, q. v.]; TR tube.

PM5 Power-meas. lamp (SY), 0.7-3.5 W; see PM3.

5DC-5, 5M "Kalotron" temperature-limited diode, T9 bulb on octal base, heater 5.6 V, 500 V, cathode current 450-575 μA (Thermosen, Inc.). 5DC-5M is min. version.

Tungar **5A** [AKA 4B36, q. v.]; rectifier.

V5 [AKA CE-4V]; vacuum phototube.

B6 [AKA 1B75, q. v.]; counter.

CA6 X-ray tube, crystal-analyzing.

CE6 ( ) Family of phototubes w/ grades A/B, C, and D.

FR6 Neon voltage regulator.

G6 [AKA CE-2]; gas phototube.

PM6 Power-meas. lamp (SY), 0.3-1.7 W; see PM3.

QF-6 [AKA RK-61, q. v.]; rcvg. gas triode.

R6A [AKA 4B26, q. v.]; rectifier.

TGC-6 Geiger-Muller counter tube.

V6 [AKA CE-2V]; vacuum phototube.

6PEB Photomultiplier.

B-7 Geiger-Muller beta-gamma counter tube.

BC-7 Geiger-Muller counter tube.

(CE, NU) **7** ( ) Family of phototubes, 4-pin base; grades A/B, C, D.

G7 Gas phototube.

HK7 [AKA 4J31-35, q. v.]; magnetrons.

HK7-T [AKA 5586, q. v.]; magnetron.

K7, RF4 [AKA WL-441, 2J30-34, q. v.]; magnetrons. "RF4" is 2J32.

K7T1-4 [AKA 2J66-2J69, q. v.]; magnetrons.

PM7 Power meas. lamp, 120-950 mW; see PM3.

V7 Vacuum phototube.

(CE) **8** ( ) Family of phototubes; grades A/B, C, D.

G8 Gas phototube.

L8 Planar tetrode.

Navy VG-8 [AKA 991, q. v.]; voltage regulator.

PM8 Power-meas. lamp (SY), 5-100 mW; see PM3.

RC8 Klystron.

V8 Vacuum phototube.

8DC15 Diode rectifier.

G9 [AKA 868, PJ-23, CE-1]; gas phototube.

V9 [AKA CE-1V, WL-734, q. v.]; phototube.

A10 RF-heating HF triode, sim. to 111H.

(CE, NU) **10** ( ) Family of phototubes; grades A/B, C, D.

CR10 Thyratron - "external grid control tube," 10 A (CE); also see CR5.

CW10 [AKA 655B, REL10, "6048" (R6048?)]; acorn UHF diode, sim. to 9004 except for pinout (Can.); probably same as Can. "Westinghouse No. 10." As "6048," used in BC-701B and I-143 indicator of IE-48 test set for 200-MHz radars and IFF sets. Also see 953.

G10 [AKA VG4, CE3D, (WE) 3A, 79A4, 1P31, q. v.]; photocell.

GLC10 Geiger-Muller tube.

P10 [AKA 5806, q. v.]; electrometer.

VG10A [AKA 1B67, q. v.]; counter.

(UX-2)**10** [AKA PT-10-A, PJ-13, 210, VT-25, CV603]; power triode, 12 W, fil. 7.5 V @ 1.25 A, μ 8, 12 W (RA, RCA). Variants: RK-10, 10Y, 210T, 841, 842, 1602.

10 Spl [AKA VT-25A, 2C25, q. v.]; xmtg. triode.

10 Spl (RCA) Original designation for 1602, q. v. Was "not listed for distributor sales" by RCA.

10Y [AKA 2C23, VT-25A]; 10 w/ Micanol base, uprated to 15 W, 8 MHz (HY, RCA).

10-54448 Cold-cathode discharge tube.

10E [AKA 1B73, q. v.]; counter.

10M [AKA 1B74, q. v.]; counter.

10PEK Photomultiplier.

10S Power triode, 10 w/ plate cap (Triad).

10X Power triode, 10 w/ ceramic base.

A11 RF-heating HF triode.

BW-11 [AKA 834, q. v.]; VHF triode (UE).

CE11 ( ) Family of phototubes, for relay and photometer use, 917 format, S-1 response; see CE11V. Grades include A/B, D. "V" is vacuum variant equiv. to 917.

CV-11 RF-heating triode (UE), sim. to UX-CV11, 211H.

G11 Gas phototube.

K11 [AKA 1B68, q. v.]; counter.

NU11V [AKA 917, q. v.]; phototube.

UX-CV11 RF-heating triode, CV11 w/ UX base (sim. to HY-51-A, 332, T841W).

V11 Vacuum phototube.

(CE, NU) **12** ( ) Family of gas phototubes, same as (CE, NU) 1 but bulb is short tubular and base is med. 4-pin. Grades include A/B, C, and D.
G12 [AKA CE4]; gas phototube.
HV12 [AKA 4C24, q. v.]; xmtg. triode (UE).
T12(V) Xmtg. triode.
V12 [AKA CE-4V]; vacuum phototube.
X-12 Klystron.
12H [AKA 1630, q. v.]; secondary-emission amplifier.
(CE, NU) **13** ( ) Family of phototubes, CE2 w/ anode at top. Grades include A/B, C, and D. "V" is vacuum variant.
G13 Gas phototube.
L-B13 [AKA "Special PT-210," q. v.]; power triode.
V13 Vacuum phototube.
VG13 [AKA 1B68, q. v.]; counter.
X-13 Klystron.
(CE, NU) **14** Phototube (sim. to (CE, NU) 13 w/ grid surrounding anode). Grades include A/B, C, D.
G14 [AKA 920, q. v.]; gas phototube.
G14T [AKA CE-21]; gas phototube.
HK14 Xmtg. triode.
KU14 [AKA 6777, q. v.]; hydrogen thyratron.
V14T Vacuum phototube.
(CE, NU) **15** ( ) Family of phototubes, rectangular anode. Grades include A/B, C, and D.
G15 Gas phototube.
G15F [AKA CE-25, 927]; gas phototube.
H15S [AKA 4J36-41, q. v.]; klystrons.
HW15 [AKA AC203, CE203, q. v.]; rectifier.
LM15 [AKA 2P21, q. v.]; TV pickup tube.
Mk. 15 Beta-gamma counter tube (Mk. 15 Mod. 1F)..
R15A [AKA 5B21, q. v.]; rectifier.
TL15 Mercury rectifier, 875 format, fil. 5 V @ 20 A, 160 V PIV, 35 A (TL).
V15F [AKA CE-25V]; vacuum phototube.
VG15 [AKA BS-7]; Geiger-Muller tube.
G16 Gas phototube.
G16B [AKA CE-25]; gas phototube.
GE **16X897** [AKA 4B27, q. v.]; rectifier.
NE16 [AKA 991, q. v.]; neon lamp/regulator.
NU16 Phototube (sim. to NU1 w/ cathode in independent halves).
(GE) **12X825** [AKA 4B35; q. v.]; mercury rectifier (supersedes 277465).
P-0016A X-ray tube (GE).
P-0016B [AKA GL-152, q. v.].
V16 Vacuum phototube.
V16B [AKA CE-25V]; vacuum phototube.
(CE, NU) **16** ( ) [AKA 920]; family of phototubes, 4-pin base, twin cathodes; grades include A/B, C, and D.
(CE, NU) **17** ( ) Family of phototubes (sim. to NU2 w/ flat cathode). Grades include A/B, C, and D.
(FG) **17** [AKA 38217, WT-272, (UE) 967, SE-4261, CE-309, NL715, O517, DEO517, 4261, (WL) 5557, q. v.]; gas-mercury thyratron rectifier (CH, EE, GE, T, W) (1929). Also see 5594.
G17 [AKA CE-22]; gas phototube.
G17F Gas phototube.
TD17 [AKA 5643, q. v.]; submin. thyratron.
V17 [AKA CE-22V]; vacuum phototube.
V17F Vacuum phototube.
VC17 [AKA 6D4, q. v.]; thyratron.
(CE, NU) **18** ( ) Family of gas phototubes; grades are A/B, C, and D.
G18T Gas phototube.
HV-18 RF-heating triode, T23 bulb w/ cap, grid-outside, graphite anode, T-T fil. 10 V @ 3.85 A, μ 18, 200 W, 50 MHz, interchangeable w/ HF-200 but rated higher (UE).
M-18A Geiger-Muller counter tube.
V18T Vacuum phototube.
DET19 [AKA 2C34, q. v.]; duotriode.
G19 [AKA CE-26]; gas phototube.
(CE, NU) **20** [AKA 920]; gas phototube, 3-pin min. base, superseded by CE25. Grades include A/B, C, D.
(UE) **FV20** Triode vacuum control tube.
G20 [AKA CE-23]; gas phototube.
NU20, C [AKA 927, q. v.].
U20 [AKA CV31, q. v.].
20X672 [AKA 4B36, q. v.]; gas rectifier.
A21 RF-heating HF triode.
AW21 Triode vacuum control tube sim. to 203A.
(CE, NU) **21** ( ) [AKA G14, 60A]; family of gas phototubes, for sound repro., 868 format, S-1 response. Grades are A/B [AKA G14, 60A], C [AKA 60A3, 920], and D [AKA G14, NU21A/B, 60A].
G21 Gas phototube.
KY-21, A Mercury triode thyratron rectifier, grid-controlled version of RX21, same ratings (1938) (EI, HY).
RX21, A Xmtg. mercury rectifier, fil. 2.5 V @ 10 A, 11 kV PIV, 750 mA (1938) (EI, HY). KY-21 is thyratron version.
RA21 Xmtg. triode.
X-21 Klystron.
(CE, NU) **22** ( ) [AKA 1P41]; family of gas phototubes, general-purpose, 1P41 format, S-1 response, 100 μA/-lumen. Grades include A/B, C, and D [AKA 924]. "WB-AB" and "WB-S1" are unbased.
CE-B22, V Gas phototube, CE-22 w/ flat-ended tubular bulb w/ larger dia.. Grades A/B, C, and D are progressively less sensitive. "V" is vacuum variant.
D22, S Geiger-Muller counter tube.
G22 [AKA CE-36]; gas phototube.
(CE, NU) **23** ( ) [AKA 923]; family of phototubes, 923 format, sound repro and relay use, S-1 response. Grades A/B, C, and D are progressively less sensitive. "D" is replaceable by 923.
G23 [AKA CE-1]; gas phototube.
KU-23 [AKA NU-300, q. v.]; triode (UE).

V23 [AKA CE-1V]; vacuum phototube.

CE24C Phototube.

24G [AKA 3C24, q. v.]; xmtg. triode (GN, HK) (10-20-43). In BC-640.

CE25 ( ) Family of gas phototubes, for sound repro., 927 format, S-1 response. Grades A/B, C, and D are progressively less sensitive. "D" [AKA 1P32, 51A3, 927] is a common projector phototube. "V" is vacuum variant. "CE-B25" is another variant.

KU25 Hydrogen thyratron, 5C22 (q. v.) adapted for high pulse rate (KL).

NU25C [AKA 927, q. v.]; phototube.

S25A [AKA CV16, q. v.]; UHF triode.

WT25 Triode vacuum control tube.

25TG [AKA HK-24G, 3C24, q. v.].

(CE, NU) **26** ( ) [AKA (WE) 6A, R78A]; family of gas phototubes, sound repro. (CE3 on med. 4-pin base). Grades A/B, C, and D are progressively less sensitive. "D" is equiv. to 1P33. "CE-A26" is variant w/ short anode.

HK-27 Rcvg. VHF beam pentode.

HV-27 [AKA 822]; xmtg. triode, T21 bulb w/ cap, grid-out-side, 4-pin jumbo base, graphite anode, T-T fil. 10 V @ 4 A, μ 26, 200 W, 30 MHz; high-μ variant of HV12 (UE).

CE28 ( ) Family of gas phototubes. Grades include A/B, C, and D. "D" is equiv. to 928.

CW28 RF-heating triode.

(CE) **29** ( ) [AKA R61BV]; family of vacuum phototubes, 929 format, S-4 response. Grades include A/B, C, D, Q, and R.

(CE, NU) **30** ( ) [AKA R61A, 930]; family of phototubes, gas versions of CE-29 (sim. to CE15), for sound and relay control; S-1 response. Grades A/B, C, and D are progressively less sensitive. "V" [AKA R61AV] is vacuum version.

CE-A30 Variant w/ "shadowless" anode, 160 μA/lumen.

L30C [AKA 2C38, q. v.]; planar triode.

L-30-P [AKA 2C41, q. v.]; planar triode.

RW30 Rectifier sim. to UV-218 and CG-1861.

30 Spl [AKA VT-67]; see HY-24.

(CE) **31** ( ) Family of vacuum phototubes, small 4-pin base. Grades A/B, C, and D are progressively less sensitive. "V" [AKA 919] is vacuum version (CE11V w/ cathode at top) for photometers, S-1 response. Other variants are "Q" and "R." "VD" is equiv. to 919.

(CE, NU) **32** ( ) Family of vacuum phototubes; grades include A/B, C, and D.

32C406G5( ) [AKA 1N49-52, q. v.]; Ge diodes.

32C406G5E [AKA 1N63, q. v.]; Ge diode.

32E [AKA 932, q. v.]; electrometer, 1229 replaces.

(CE, NU) **33** Phototube (CE13 w/ cathode at top).

(FG, NFG) **33** [AKA 5720, q. v.]; thyratron.

CE-34 [AKA R51BV]; vacuum phototube, 927 format, S-4 response; "Q" variant is sensitive than "R."

34XH [AKA 3AP1, q. v.]; CRT (DuM).

C35 [AKA NC35, RX219]; special-purpose (small).

CE36 ( ) Family of gas phototube, 927 format, S-1 response; common in projectors. Grades A/B,C, and D are progressively less sensitive. "C" [AKA 19210] is sim. to 927, but shorter in base. "V" and "CE-B36" are vacuum versions.

E36 Hydrogen thyratron.

CE37 Gas phototube.

RL37 [AKA CV66]; UHF amp.

CE38-39 Gas phototubes.

CE40 Phototube.

CE41, C Gas phototube. "C" is sim. to 921.

CE42, C Vacuum phototube. "C" is sim. to 922.

M-42A Geiger-Muller counter tube.

H45 [AKA ZG547, 3C45, q. v.]; hydrogen thyratron.

45 Spl [AKA 2C45, q. v.]; power triode. In BC-180, -230, -430.

(GE) **45X674** [AKA 1674, 5B24, q. v.]; rectifier.

CE46, 47 Phototubes.

CE48 [AKA 1P34, q. v.]; phototube.

CE49 Phototube.

UE49 [AKA 849, q. v.].

(CE, NU) **50** Phototube.

HZ-50 [AKA 12Z3, q. v.]; rectifier.

EA50 [AKA 6923, q. v.]; submin. diode.

R50 [AKA CE-22]; phototube.

R50AV [AKA CE-22V]; vacuum phototube.

KP51 Submin. rcvg. HV rectifier.

KU51 Hydrogen diode, fil. 5.0 V @ 6.2 A (KU).

(R)51A [AKA CE-25]; phototube.

R51AV [AKA CE-25V]; phototube.

R51B [AKA CE-64]; phototube.

R51BV [AKA CE-34]; phototube.

51A3 [AKA 1P32, CE25C, 927, q. v.]; phototube.

51AV, SA Phototubes.

CE52, V Phototube, med. 5-pin base, short tubular bulb, side-on cathode, S-1 response, 26 μA/lumen. "V" is vacuum version.

KU-52 [AKA 8264, q. v.]; hydrogen diode.

52A3 [AKA 1P35, q. v.]; phototube.

52N Beta-gamma counter tube, long (ca. 8") thin-walled metal shell, self-quenching (AX).

(CE) **53** ( ) Family of gas phototubes. Variants include "AB," "ABB," "AWB" [AKA CE5, 927] and "D."

EA53 Planar test-probe diode, heater 6.3 V (AX). Also see (EI) 2-01C.

KU-53 [AKA 7178, q. v.]; hydrogen diode.

CE54 Phototube.

M54 Submin. power tetrode, fil. 0.625 V @ 40 mA, g_m 0.2 mS. Also see M64, M74. Seems to be same as 3403, q. v.

54B [AKA 4C26, q. v.]; xmtg. triode.

CE55 Phototube.

E55L [AKA 8233, q. v.]; power pentode.

EC55 [AKA 5861, q. v.]; disc-seal triode.
CE56, 56 Phototube.
HK-57 Rcvg. HF beam pentode.
CE58 ( ) Family of gas phototubes, red-sensitive, tubular bulb on 4-pin base, side-on cathode (CE, RL). For industrial controls and 16-mm movie sound. Variants include "A" [AKA CE4; see 1P36], "AL," "AWB" [AKA CE-4], "AV" [AKA CE-4V], vacuum version, "AVWB" [AKA CE-4V], "D," "R."
R58A [AKA CE-4]; gas phototube.
R58AV [AKA CE-4V]; vacuum phototube.
58A3 [AKA 1P36, q. v.].
CE59 ( ) [AKA R61B]; family of gas phototubes, GT bulb, side-on cathode, octal base, S-4 response, 125 μA/lumen. (RL R-59 series is in 868 format.) Variants include "A" [AKA CE-1; also see WL-734], "A4" [AKA 868], "AP," "AR,", and "R." Vacuum variants are "AV" [AKA CE-1V] (see PJ-22), "DV1/2" [AKA CE61A/B, GL-441], and "TAV3" [AKA 917].
R59 ( ) Family of phototubes. Gas variants are "A" [AKA CE-1], "B" [AKA CE-91], and "D" [AKA 928]. Vacuum variants are "AV" [AKA CE-1V], "AV3" [AKA PJ-22], "BV" [AKA CE-61], and "TAV" [AKA CE-11V].
CE60 [AKA 917, q. v.]; phototube.
F60 [AKA 4B26, q. v.]; industrial "Allen" rectifier.
EFP60 Secondary-emission pentode, w/ pair of target plates and added anode (AX). In high-freq. sweep gen. in EG&G 2236A microwave spectrum analyzer. May be same as 5857.
(HF, NU) **60** [AKA H330]; RF-heating/xmtg. triode, UX base w/ grid and plate caps, fil. 10 V @ 2.5 A, μ 20, 60 W (AX, NU).
(R) **60A** [AKA CE-21]; phototube, UV-868 format, dual anodes and cathodes, for 2-track movie sound, etc. (RL).
60A3 [AKA 920, q. v.]; phototube.
CE61A/B [AKA R-59BV, GL-441, q. v.]; phototube.
R61 ( ) Family of phototubes. Gas variants are "A" [AKA CE-30] and "B" [AKA CE-59]. Vacuum variants are "AV" [AKA CE-30V] and "BV" [AKA CE-29].
61A3 [AKA 930, q. v.]; phototube.
61DV3 [AKA 929, q. v.]; phototube.
ML-LPT-62 Vacuum switch tube, rugged 7715, q. v. (EI, ML).
CE64 [AKA 5583, q. v.]; phototube.
M64 Submin. tetrode, fil. 0.625 V @ 20 mA, g_m 0.11 mS. Also see M54, M74. Seems to be same as 3401 or 3402, q. v.
(CE) **64** ( ) [AKA R51B]; family of gas phototubes, 927 format, S-4 response. "Q" variant more sensitive than "R." "AV3" [AKA 925] is a vacuum variant.
RT66 [AKA REL8A, q. v.]; klystron (?).
CE68 Gas phototube.
(DT, TS) **70** [AKA 2E22, q. v.]; xmtg. pentode.
KU-70C [AKA 7621, q. v.]; hydrogen thyratron.
V70, A Xmtg. triode, 4-pin jumbo base w/ cap, fil. 10 V @ 3.25 A, μ 14, 70 W. "A" has μ raised to 25.
V70B, C, D Xmtg. triode, V70 w/ UX base. "C" has μ of 25. "D" has μ of 20.
70-D Xmtg. triode, hard glass, graphite anode, fil. 7.5 V @ 2.5 A, μ 20, 70 W (UE), upgrade from T-40, T-55, or 812. Differs from V70D.
KU-71 [AKA 7782, q. v.]; hydrogen thyratron.
KU-71Z [AKA 8765, q. v.]; hydrogen thyratron.
R71 ( ) Family of phototubes, tubular bulb w/ cap on 4-pin base (RL). Gas version "A" is AKA CE-2. Vacuum variants are "AV" [AKA CE-2V] and "TAV" [AKA CE-13V].
(CE) **71** ( ) Family of phototubes. Variants are "A3" [AKA CE2C, 1P30, q. v.] and "TA." Vacuum variants are "AV" and "D."
KU-72 [AKA 7665, q. v.].
KU-72Z [AKA 8553, q. v.].
(CE) **73A**, D Phototubes.
(CE) **73WB**-Q Phototube.
(CE) **73WB**-S4 Phototube.
KU-73 Hydrogen thyratron, fil. 6.3 V @ 22.0 A (KL).
KU-74 [AKA 7667, q. v.]; hydrogen thyratron.
M74 Submin. tetrode, fil. 0.625 V @ 20 mA, g_m 0.13 mS. Also see M54, M64. Seems to be same as 3401 or 3402, q. v.
75N Counter tube, long metal body w/ center pin (AX).
CE-75 ( ) Family of phototubes, small 4-pin base, silicone-treated, tubular bulb w/ side-on cathode, S-1 response, 20 μA/lumen. Gas variants are "A" and "D." Vacuum variants are "AV" and "V."
HF75 Xmtg. triode, UX base, grid and plate caps, fil. 10 V @ 3.25 A, μ 12.5, 75 W.
NU-75H [AKA HY75, q. v.].
TW75 Xmtg. triode, UX base, grid and plate caps, fil. 7.5 V @ 4.15 A, μ 20, 75 W.
VR-75/30 [AKA 0A3, q. v.]; voltage regulator.
(CE) **76** Phototube, CE36 w/ S-4 response.
76X13 Gas rectifier.
CE-77 Gas phototube, small 4-pin base, 10" slim tubular bulb, side-on cathode, S-4 response, 138 μA/lumen.
R78A [AKA CE-26]; phototube.
R79A [AKA CE3]; phototube.
(CE) **79** ( ) Family of gas phototubes Grades are "A" [AKA CE3A/B, q. v.], "A4" [AKA CE3D, (WE) 3A, G10; also see 1P31], and "D."
DC80 Min. triode.
EC80 Min. grounded-grid triode (PH, VO).
PG80 Phototube, smoke detector.
X80 [AKA 6303, q. v.]; radar clipper diode.
(CE) **81** Gas phototube.
EC81 Min. triode (AX, PH, VO).

KU-82 [AKA 7583, q. v.]; hydrogen thyratron.

R82A Phototube.

82AV4 [AKA 919, q. v.]; phototube.

(CE) **83** Phototube, blue-sensitive CE3.

83A1 [AKA 7980, q. v.]; voltage reference.

CE-84, A3 Vacuum phototube, single-contact bayonet base, flat-ended tubular bulb, side-on cathode, S-4 response, 20 μA/lumen. "A3" is AKA 923, q. v.

R85A, AV [AKA CE23A/B]; phototube, nondirectional, tubular flat-topped bulb on UV-199 base, cylindrical cathode, spiral wire anode (RL) "AV" is vacuum variant.

85A4 [AKA 928, q. v.]; phototube.

CE-86A/B Phototube, high-sensitivity CE-84, S-4 response, 80 μA/lumen.

VR-90/30 [AKA 0B3, q. v.]; voltage regulator.

CE-91 [AKA R59B]; gas phototube, for sound from dye-image track w/ incandescent light, 868 format, S-4 response, 120 μA/lumen.

PG94 CE1 sel. for Bell & Howell; also see WL-734.

A95 Phototube.

CE97 Phototube, blue-sensitive CE26.

RTS-98 [AKA 1R6, q. v.].

KU-99 Hydrogen thyratron, fil. 6.3 V @ 2.7 A (KL).

99X44, 45, 46 Gas rectifiers.

B-100 Bolometer element for UHF power measurement, 200 ohms (TS) (1945).

C-100-A Small gridless osc. tube, long T9 bulb on 6-pin base, two mesh electrostatic control elements outside bulb; six parallel wires as cathode; twin tubular anodes. Xtal osc. in 1935-37 Collins xmtrs.; made by AX. Some are stamped "Amperex" instead of "Collins."

C100D "Negative resistance" osc. tube, large tubular bulb on 6-pin base. Xtal osc. in 1937-38 Collins xmtrs. "Obsolete, convert eqpt. to 2A5." Also see C100A.

C100E Xmtg. triode.

HF-100 [AKA 4C22, q. v.]; RF-heating triode. HF-*100* is rated at 75 W. HF-100, -200, and -300 were promoted for HF applications as having a high ratio of g_m to capacitance.

RW100 [AKA 214, q. v.]; xmtg. rectifier.

TF100 [AKA HF-100, 4C22, q. v.]; RF-heating triode.

100A [AKA 1B75, q. v.].

100 IG Ionization gauge, fil. 3.5-7.5 V, body like VT-127 w/ sensing tube out bottom, range 10^{-3} to 10^{-8} mm of mercury (EI).

100N [AKA 1B76, q. v.]; counter tube (AX).

Boonton **101A**, C [AKA Boonton 105A]; duodiode-triode, 2A6 for Boonton Q-meter selected for low grid current, replaceable by 1659. Variants: Boonton 105A, 535B.

BS101 [AKA 5962, q. v.]; voltage regulator.

C101 [AKA HF-100, 4C22, q. v.]; Collins xmtg. triode.

Boonton **102A** Selected 45 for Boonton Q-meter. Variant: Boonton 536B.

X102B [AKA 6G4, q. v.]; UHF triode.

X102G [AKA 2C29, q. v.]; disc-seal triode.

X102L [AKA 2C31, q. v.]; VHF triode.

X103 [AKA 4D21]; xmtg. UHF tetrode.

Boonton **105A** [AKA Boonton 105A, q. v.]; duodiode-triode.

VR-105/30 [AKA 0C3, q. v.]; voltage regulator.

106C Beta-gamma counter tube.

108C Beta-gamma counter tube.

X111AD [AKA 8C23, q. v.]; xmtg. triode.

X111B [AKA 8C22, q. v.]; xmtg. triode.

111H RF-heating triode, UX base w/ cap, fil. 10 V @ 2.25 A, μ 23, 75 W.

A116 [AKA 2J42, q. v.]; magnetron.

X117 [AKA 5D22, q. v.]; xmtg. tetrode.

C120 [AKA HF-120, q. v.]; Collins.

HF-120 Xmtg. triode, 4-pin base, fil. 10 V @ 3.25 A, μ 12, 100 W (sim. to 311CT). Also see HF-140.

RX120, A Argon-mercury Tungar rectifier, mogul base, cathode 2.5 V @ 30 A, 120 PIV, 20 A.

ZB120 [AKA 3C21, q. v.]; xmtg. triode.

120C [AKA 1B77, q. v.]; counter tube (AX).

120N Neon self-quenched beta counter tube (AX).

A121 [AKA 2J41, q. v.]; magnetron.

A124 [AKA 2J64, q. v.]; magnetron.

FT125 Strobotron.

HF-125 [AKA 203H]; RF-heating triode, 4-pin jumbo base, plate at top, fil. 10 V @ 3.25 A, μ 25, 100 W.

X126H [AKA 6C23, q. v.]; xmtg. triode.

128A [AKA 2522N1, q. v.]; thyratron.

X129D [AKA 3C26, q. v.]; xmtg. triode.

L-129R RF-heating triode, 6" dia. by 14" high, grid-out-side, 25 kW, forced-air cooled.

HF-130 [AKA 211C, q,. v.]; xmtg. triode.

PE130A [AKA 3C24, q. v.]; xmtg. triode.

PE130B [AKA 3C28, q. v.]; xmtg. triode.

PE130C [AKA 3C34, q. v.]; xmtg. triode.

PG132 [AKA CE203, q. v.]; xmtg. triode.

QL134 [AKA 2C48, q. v.]; power duotriode.

A136M Magnetron.

HF-140 [AKA 211D]; xmtg. triode, HF-120 parameters except higher C_{gp} - single-ended HF-120?

VCR140 [AKA 9GP7, q. v.]; CRT.

143D [AKA 2X2/879]; HV rect. repl. by 879 ca. 1938.

146 [AKA 846, q. v.].

147SM [AKA 5J26, q. v.].

EP150 [AKA CE1VC, PJ-22]; phototube.

M150T RF-heating HF triode.

HF-150 RF-heating triode, 4-pin jumbo base, fil. 10 V @ 3.25 A, μ 12.5, 125 W, nearest equiv. 4C22 (AX).

TW150 Xmtg. triode, 4-pin jumbo base w/ cap., grid-out-side, fil. 10 V @ 4.1 A, μ 35. 150 W.

VR-150/30 [AKA 0D3, q. v.]; voltage reg. Intro. 1938.

150C [AKA 1B78, q. v.]; counter.

150M [AKA 1B79, q. v.]; counter.

150N, NA [AKA 1B80, q. v.]; counter.

150P Xmtg. pentode.

151N, NA [AKA 5980]; gamma counter tube.

DT153 [AKA 1U4, q. v.]; rcvg. pentode.

X155 Min. 9-pin VHF duotriode, for TV tuners, heater 6.3 V @ 400 mA, replaceable by 6BC8, 6BQ7, or 6BS8 (Philco) (1955) <9AJ>.

EC157 [AKA 8108, q. v.]; lighthouse triode.

QMG159 [AKA CK1005, q. v.]; cold-cathode rectifier.

(FG, FP) **162** "Special GE (obsol.)."

T-160-L [AKA 572, q. v.] (Cetron).

PL172 [AKA 8295A, q. v.]; xmtg. pentode.

HF-175 Xmtg. triode, 4-pin jumbo base, fil. 10 V @ 4 A, μ 18, 125 W.

PL-175A Beam pentode, 4-400 format, forced-air cooled, fil. 5.0 V @ 14.5 A, 400 W, vane suppressor brought out separately, for linears or Class C use (EI, PL). Said to replace 4-400 and give more output from higher efficiency.

PL-177, WA Beam pentode, 4-125 format but all-glass, fil. 6.0 V @ 3.2 A, 75 W, forced-air cooled, for linears; suppressor as on PL-175A (EI, PL). Variant: 6549.

O178 [AKA FG-178, q. v.].

QM190 Min rcvg. type.

PL-195 Xmtg. beam pentode, all-glass, external anode, fil. 6.0 V @ 17 A4 kW, forced-air cooled; 5 kW out as linear (PL).

TS195/1S5 Selected 1S5 for Bureau of Ordnance.

TS196/1L4 Selected 1L4 for Bureau of Ordnance.

QF-197 [AKA VT-284, 2B24, q. v.].

C200 [AKA HF-200, q. v.]; Collins.

EE200 Industrial vac. control triode (EE); see HF-200.

HF-200 Xmtg. and RF-heating VHF triode, tubular bulb w/ cap, grid-out-side, fil. 10.5 V @ 3.4 A (T: 6 A), μ 18, 150 W (T: 200 W), 45 MHz (AX, C, GN, ML, NU, T) (1936). Replaceable by HV-18.

QF-200A,B Submin. type.

QF-200C [AKA VT-285, 2C27, q. v.]; submin. triode.

CE-200A Full-wave mercury rectifier, med. 4-pin base, fil. 2.5 V @ 7 ("5 to 8") A, 900 PIV, 2 A (CE, CR, HY). Sim. to NL602.

200C [AKA 1B81, q. v.]; counter (AX).

200N Neon self-quenched alpha-beta-gamma counter (AX).

CE201 [AKA 16X897, 4B27, q. v.]; rectifier.

RM201 Thyratron.

CE-201A CE-200A on industrial base; can replace 4B27.

HF201A [AKA WL-468]; xmtg. triode, same ratings as HF-200 (AX).

201N Alpha-beta-gamma counter tube.

C202 [AKA 211H, q. v.]; xmtg. triode, Collins.

CE202B [AKA NL625]; mercury rectifier, mogul base, fil. 2.5 V @ 20A, 900 V PIV, 15 A (HY).

QF-202 Submin. type.

(UV) **202** [AKA 202, PR-2-A]; xmtg. triode, anncd. 3-21, superseded by UV-202A.

(UV) **202A** [AKA PR-2-B]; xmtg. triode.

(CE) **203** Mercury rectifier, mogul base, fil. 2.5 V @ 20 A, 500 V PIV @ 15 A (HY).

203 [AKA PG132, PR-3-A, HW-15, NL623, CG-1144A]; xmtg. triode, T18 bulb, jumbo 4-pin base, anncd. 3-21, pure-tung. fil., superseded by 203A. Four in QCB 26-kHz sonar.

203A [AKA 303A, (WE) 295A, 503A, 38103, PR-3-B]; xmtg. triode, T-T fil. version of 203, fil. 10 V @ 3.25 A, μ 25, 100 W, 15 MHz, radiation-cooled (higher-μ 211), anncd. mid-'23 (AX, C, GE, RCA, T, W).

VA-102B [AKA 6975, q. v.]; klystron.

HD203C RF-heating triode sim. to 303-U.

203H [AKA HF-125, q. v.]; RF-heating triode, 203A w/ cap, plate out top, grid-out-side, sim. to 303U (UE).

203W, WA Xmtg. triode, ruggedized 203A.

(UV) **204** [AKA VT-8, PR-4-A]; xmtg. triode, spherical-tubular bulb w/ cap, blade-and-pins base, pure-tung. fil. 11.0 V @ 14.75 A, 250 W, anncd. 3-21, superseded by 204A.

204A [AKA UN304A, 504A, 38104, VT-22, PR-4-B, CG-1860A, CV2563]; xmtg. triode, T-T fil. version of 204, stepped-tubular bulb w/ cap, blade-and-pins base, anncd. mid-'23, fil. 11 V @ 3.85 A, μ 23, 250 W, 3 MHz, radiation-cooled (C, F, GE, RCA, T, W). "Purchased outside and merchandized by RCA Mfg. Co." as of 1-20-38.

UV-205 Variant of UV-204; "never active."

CE-205 [AKA 2RA5]; mercury rectifier, mogul base, fil. 2.0 V @ 13 A, 1 kV PIV, 5 A (CE, HY).

CE206 [AKA 2RA6]; mercury rectifier, mogul base, fil. 2.0 V @ 13 A, 500 V PIV, 6 A (CE, HY).

QF-206 [AKA VT-283, 2E27, q. v.]; submin. pentode.

(UV) **206** [AKA 38106, CG-2933]; xmtg. triode; number assigned 6-21, fil. 11 V @ 14.75 A, 1 kW; grid-out-side (GE, RCA). "Purchased outside and merchandized by RCA Mfg. Co." as of 1-20-38.

RX-207 [AKA 45X674, EL6C, 4B25, q. v.].

(UV) **207** [AKA CG-1971, VT-34, F-307A, 38107]; xmtg. triode, gen'l use 1923, fil. 22 V @ 52 A, μ 20, 10 kW, 1.6 MHz, water-cooled via ext. cooler (F, GE, ML, RCA, W) (5-31-45). "Purchased outside and merchandized by RCA Mfg. Co." as of 1-20-38.

207M RF-heating variant of 207 (EP) (1947). Same source as 891M and 892M.

RM-208 Mercury triode thyratron "permatron" rectifier, 866A w/ collector electrode; 866A ratings; controllable by external magnetic field (RA). See *Electronics*, 4-39.

(UV) **208** [AKA CG-1353]; triode, 5 kW, number assigned 12-21; grid-out-side.

RM-209 Mercury triode thyratron "permatron" rectifier,

872 w/ collector electrode. Also see RM-208.

CE-210 Mercury rectifier, fil. 2.5 V @ 5 to 8 A, 900 V PIV, 2 A (HY).

PL-210 Xmtg. beam pentode, glass external-anode, T-T fil. 6.0 V @ 85 A, 5 kW, 10 kW out as linear amp. (PL).

210T Ceramic-based xmtg. 10 (SY).

(UV) **211**, W [AKA 38211, VT-4-B, PR-11-A]; 203A w/ μ reduced to 12.5, sim. to 242A, anncd. 1923 (RCA, W). Note: 1950 RCA data sheet for 211 shows lower capacitances than 1936 issue (C_{gp} 14.5 $\rightarrow$ 14 pF; C_{gf} 6.0 $\rightarrow$ 5.4 pF; C_{pf} 5.5 $\rightarrow$ 4.8 pF). This may be related to existence of 211 Spl/211C/-4C21 (see below) w/ lower capacitances and electode currents. W promoted 211 for RF heating, claiming 60 W output @ 50 MHz (AX, C, GE, NU, RCA, T, UE, W).

211 Spl [AKA 4C21, VT-4C]; 211 w/ "low grid-current" limits.

211C [AKA FP-285, 38111A, HF-130]; RF-heating HF triode, 100 W (T), 120 W (AX), 150 W (UE). 211 w/ lower capacitances (T).

211D RF-heating triode, replaceable by 311T (UE).

211H [AKA HF150, 311CH]; 211C w/ cap (AX), 10 V @ 3.25 A, 1500 V B+, 100 W (AX). Replaceable by 311CH (UE).

CE212 [AKA 289415]; gas rectifier.

RX212 [AKA NL627]; mercury rectifier, heater 2.5 V @ 30 A, 1 kV PIV, mogul base.

CE-213 [AKA 2RA3]; rectifier; see CE-213A.

CE-213A [AKA 7018, q. v.]; mercury rectifier.

J213AAA [AKA 4B26, q. v.]; "Allen" industrial rectifier.

QF-213, 214 Submin. types.

RM214 Thyratron.

(UV) **214** [AKA CG-2938, 38114, RW100]; xmtg. rectifier, gridless UV-207, same fil., 50 kV PIV, 2.4 A, number assigned 8-22 (C, GE, F, RCA, W) "Purchased outside and merchandized by RCA Mfg. Co." as of 1-20-38.

(UV) **214E** [AKA 217A]; xmtg. vacuum rectifier.

CE215 [AKA 2RA15]; mercury rectifier, fil. 2 V @ 18 A, 500 V PIV, 15 A (CE).

QF-215 [AKA 2C35, q. v.]; xmtg. triode.

RX215 Full-wave mercury rectifier, cathode 2.5 V @ 30 A, 500 V PIV, 30 A.

QF-217 Rcvg. UHF osc.

(UV) **217** A, B, C Rectifier, anncd. 10-21; superseded by "A." "A" [AKA CG-2936] is gridless 203A, number assigned 11-24, T18 bulb, jumbo 4-pin base, T-T fil. 10 V @ 3.25 A, 190 mA, 3.5 kV PIV (GE, RCA, W). "B" had number assigned 12-21; little known. "C" [AKA 38117, HK-235, (UE) 317C, CG-2936A] was 217A w/ plate cap and uprated to 7.5 kV PIV, number assigned mid-'26 (GE, W). In plate supply of QCB sonar.

(UV) **218** [AKA CG-1861, 38118, F218A, 318A]; xmtg. rectifier, gridless UV-204, 50 kV PIV, 750 mA pk., number assigned 1921 (F, GE, RCA, W). "Purchased outside and merchandized by RCA Mfg. Co." as of 1-20-38.

218S Xmtg. rectifier.

RX219 [AKA C35, q. v.].

(UV) **219** [AKA 38119, CG-2937, R319]; xmtg. vacuum rectifier, T48 stepped tubular bulb w/ cap, blade-and-pins base; appears to be a gridless triode, fil. 22 V @ 24.5 A, 50 kV PIV, 800 mA, number assigned 6-22 (RCA, W). "Purchased outside and merchandized by RCA Mfg. Co." as of 1-20-38.

CE-220 [AKA RKR72, 72, q. v.]; also see 8020.

CEP220 [AKA 38220, 3C21, 120, q. v.].

CE221 [AKA 4B25, q. v.]; rectifier.

CE222 [AKA EL1C, 3B22, q. v.]; rectifier.

CE224 [AKA 4B24, q. v.]; rectifier.

CE225 [AKA 4B28, WL-289414, GE189048]; gas rectifier; can repl. 4B28.

Z-225 Mercury rectifier, 866A in short T14 bulb (UE).

CE-226 [AKA 4B26, q. v.]; rectifier.

RX226A Diode rectifier (obsol.).

A227A AX water-cooled.

TS-229 [AKA 5687, q. v.]; min. duotriode.

CE230 [AKA 3B24, q. v.]; rectifier.

WT230 [AKA 873, q. v.]; thyratron.

RX232 [AKA 72]; rectifier, superseded by 72.

RX233, A [AKA 2C33, q. v.]; thyratron. In BC-704A.

233 RF-heating triode, external anode, dual grid-out-side leads, flying fil. leads, ca. 10 kw, 30 MHz, water-cooled (AX). Supersedes 232CH.

234A-R Xmtg. triode, external anode, six radial flying leads, 1 kW, 110 MHz, forced-air cooled (AX).

CE-235, A [AKA 5B21, q. v.]. "A" has flexible, not stiff, anode lead.

RX235 Diode rectifier (sim. to 72).

235R Radiator-type xmtg. tube, for RF heating, fil. 14.5-15 V @ 39 A, μ 14 (AX), 1945.

XP-248 VHF triode, water-cooled, dish base, exp., ca. 1933.

H250 [AKA 1223, q. v.]; rcvg. pentode.

HF250 Xmtg. triode, 4-pin jumbo base w/ cap, grid-out-side, fil. 10.5 V @ 4.0 A, μ 18, 150 W.

AT-257-C 4E27/AT-257 w/ Zr-coated plate, uprated from 75 W to 125 W (LE) (1946). Bulb differs from HK-257.

264 [AKA 864, q. v.]; nonmicrophonic rcvg. triode.

PL264 [AKA 8576, q. v.]; xmtg. beam pentode.

WT-272 [AKA 5557, q. v.]; thyratron.

KU-274 [AKA 7866, q. v.]; hydrogen thyratron.

DT275 [AKA 5A6, q. v.]; xmtg. beam pentode.

PL284 Tetrode for pulse modulators, suited to long pulses, fil. 7.5 V @ 20-22.7 A, 1 kW, 45 kV, 8 A; appears to be improved 8189 (EI).

PL-290, A Cermet beam pentode, heater 6.0 V @ 17 A, 5

kW, max. freq. not stated, forced-air cooled, for linears (EI, PL).

WT294 [AKA 0D3, q. v.]; voltage regulator.

QM296 [AKA 5517, q. v.]; cold-cathode rectifier.

(CE, CR) **300** HV discharge diode for electric-fence chargers, ST-14 bulb, 4-pin base. Also see Taylor 208.

(HF-) **300** [AKA KU-23]; Xmtg./RF-heating VHF triode, T24 tubular bulb w/ cap, grid-out-side, fil. 11 V @ 4 A, μ 23, 200 W, 45 MHz (AX, EE, GN, ML, NU, T) (1936); nearest equiv. WL-463.

V301 [AKA 1N40, q. v.]; Ge diode quad.

CE302 [AKA (WE) 287A, q. v.]; thyratron (HY).

GB-302 Geiger-Muller beta counter, 4-pin base, metal-foil window w/ slide-on protective cap (SY).

CE303 [AKA 3C31, q. v.]; thyratron.

(F, UE) **303A** [AKA 203A, q. v.]; xmtg. triode.

303C RF-heating triode, 150 W (T), sim. to 303U.

303-U RF heating triode (UE) (nearest equiv. 303C).

CE304 Mercury triode thyratron, conical tapered bulb w/ cap, industrial 4-pin base, fil. 2.5 V @ 22.5 A, 1 kV PIV, 12.5 A (HY).

304A [AKA 204A, q. v.]; xmtg. triode (F, UE, UN).

GB-304 Geiger-Muller gamma counter tube, tubular bulb ca. 8" long, 4-pin base (SY).

CE305 Thyratron.

305D UE version of WE 205D.

CE-306 [AKA C6J, q. v.] Intro. 6-44.

CE308 Thyratron rectifier (selected CE306).

CE309 [AKA FG-17, q. v.]; thyratron, ST-bulbed version (CE).

SA309 Strobotron (SY).

NU-300 [AKA KU-23]; xmtg. triode, 200 W, ca. 3-42.

CE310 Thyratron (UE) (1936).

GM-310A Geiger-Muller counter tube.

(UE) **310** [AKA 801A, q. v.]; xmtg. triode (UE variant).

UK310P [AKA NT98, q. v.]; magnetron.

CE311 Thyratron, 3C23 derated to 1 kV PIV and raised to 1.5 A (HY).

(UE) **311** [AKA 211, q. v.]; xmtg. triode..

(UE) **311C** RF-heating HF triode (UE); see 38111A (sim. to 835, 211).

(UE) **311CH** RF-heating HF triode, plate out top (nearest equiv. 211H).

(UE) **311CT** [AKA FP-285, T211C, 835]; RF-heating HF triode; see 38111A.

(UE) **311T** RF-heating HF triode.

312D [AKA 212D, q. v.]; xmtg. triode (1936)

312E [AKA 212E, q. v.]; xmtg. triode, mech. redesigned (F, UE)

BM313 [AKA REL32, q. v.]; magnetron.

CE315 [AKA C5B, q. v.]; rectifier.

315B [AKA CV38, q. v.]; magnetron.

315D [AKA NT98, q. v.]; magnetron.

TR317 TR tube, sim. to X7047.

(UE) **317C** [AKA UV-217C, 38117, q. v.]; vacuum rectifier.

R319 [AKA 219, q. v.]; vacuum rectifier.

CE320A Mercury-gas triode thyratron, fil. 2.5 VCT @ 9 A; 1.5 kV PIV, 2.5 A (HY).

QM322 [AKA 5656, q. v.]; rcvg. duotetrode.

CE326 [AKA C6L, q. v.]; thyratron.

QM328 [AKA 5686, q. v.]; beam pentode.

CE329 Thyratron, T16 tubular bulb w/ cap, bakelite base w/ four flying leads, fil. 14 V @ 2.4 A, 500 V PIV, 3 A (CE) (1949). Also see CE330.

CE330 Thyratron, CE329 uprated to 5 A, all-glass, eight flying leads.

H330 [AKA HF60]; RF-heating tube (AX).

330C [AKA 805, q. v.]; xmtg. triode.

A331 RF-heating triode sim. to CV-11 (AX).

A333 RF-heating triode (AX).

(AT, PE) **340** [AKA 4D21, q. v.]; xmtg. tetrode (LE).

(UE) **342B** [AKA 242B, q. v.]; xmtg. triode.

342C [AKA (GL) 242C, q. v.]; xmtg. triode.

ATR345 [AKA 5792, q. v.]; ATR tube.

ATR346 [AKA 5793, q. v.]; ATR tube.

C-350 Xmtg. tetrode (Collins/AX); slightly enlarged C-300; fil. 11.5 V @ 4.0 A, μ 24, 250 W (1938).

H350 [AKA 4C35, q. v.]; hydrogen thyratron.

350D [AKA 801A, q. v.]; xmtg. triode.

F-352A [AKA 852, q. v.]; xmtg. triode.

361A [AKA (WE) 261A, q. v.]; xmtg. triode, graphite-anode version; has insulating shoulder on plate pin; 30 MHz (UE) (1936).

(C, F) **375A** [AKA 575A, q. v.]; mercury rectifier.

(UE) **376A** [AKA (WE) 276A, q. v.]; xmtg. triode, graphite-anode version (UE) (1936).

QM380 Min. corona voltage regulator, sim. to BS101.

TS-383 Min. vacuum rectifier, fil. 6.3 V, used as 1500-V rectifier in DuM 331 scope.

384D [AKA (WE) 284D, q. v.]; xmtg. triode, graphite-anode version (UE) (1936).

(FJ) **401** [AKA 1P29, q. v.]; phototube.

404 Xmtg. vacuum rectifier

QF-408 [AKA 1AD4, q. v.]; submin. pentode.

QF-409 [AKA 1AE5, q. v.]; submin. mixer.

410-1L [AKA 3K21, q. v.]; klystron.

410R Klystron two-cavity amp. or osc., 3 GHz, IC, octal base, cavity inputs/outputs via four threaded coax fittings; prototype of 3K30. Produces 15 W pulsed output w/ 2 kV (SP).

412 [AKA 2K32, q. v.]; klystron.

S413 Strobotron (SY).

414 [AKA 4K34, q. v.]; klystron.

HT-415 [AKA 5C22, q. v.]; hydrogen thyratron.

416 [AKA 2K36, q. v.]; klystron.

418 [AKA 2K38, q. v.]; klystron.

420R [AKA 5K40, q. v.]; klystron.

421C [AKA 2K21, q. v.]; klystron.

(K, X) **424D** [AKA 4X150A, q. v.]; xmtg. tetrode.

RT434 Disc-seal "rocket" triode, CW osc. for butterfly tuned circuits, variant of 5768 (SY).

X434 UHF measurement diode.

C455 RF-heating triode, replaceable by 830B.
X461 [AKA 2-240]; xmtg. vacuum half-wave rectifier
466S [AKA 466A].
QF-477 Submin dual electrometer.
ZG489 [AKA VT-282]; thyratron, no known use or CV-no.
QF-491 Submin. dual electrometer.
S491R Xmtg. loktal high-μ UHF triode, heater 6.3 V @ 175 mA, μ 58, 7 W, oscillates to 400 MHz (SY).
492 [AKA 5757, q. v.]; xmtg. triode.
492R [AKA 5758, q. v.]; xmtg. triode.
498 Xmtg. triode, 508 char.
SS501 [AKA R7061]; argon U-shaped control tube.
501R [AKA 5759, q. v.]; xmtg. triode.
QM502 [AKA 5757, q. v.]; thyratron.
SS502 [AKA K7610]; gas discharge tube.
502 [AKA 5760, q. v.]; xmtg. triode.
502R [AKA 5761, q. v.]; xmtg. triode.
(DRH) **505** [AKA 2J45-2J46-2J47, q. v.]; magnetron.
VI507 [AKA REL21, q, v.]; spark-gap pulser tube.
508 [AKA 6246, q. v.]; xmtg. triode.
P509 Gas-glow tube.
VCR516 [AKA 12DP7, q. v.]; CRT.
(DEO) **517** [AKA DGO517, FG-17, et al., q. v.].
520B, M See 520B, M in De Forest chapter. "Purchased outside and merchandized by RCA Mfg. Co." as of 1-30-38.
Boonton **535B** [AKA 1659, Boonton 101A, q. v.]; selected 2A6.
Boonton **536B** [AKA Boonton 102A, q. v.]; selected 45.
554 Radar clipper diode, ext.-anode, flying fil. leads, fil. 6.3 V @ 3.5 A, 18 kV PIV @ 150 mA (UE).
558 Mercury rectifier, spl. 2-pin base, 2.5-V fil.; as of 4-1-34, RCA showed as "not listed for distributor sales." Suspected of being a DeF version of WE 258B, later transferred to RCA.
561 Clipper diode, low-impedance variant of X80, derated to 30 kV PIV (UE).
ZJ566 [AKA 7516B]; phototube.
WT-568 Rectifier, replaceable by CE-213A (HY).
WT-571 Rectifier, replaceable by CE-201A (HY).
572, B [AKA T-160-L]; zero-bias xmtg. triode, graphite-plate variant of 811A, uprated to 160 W ICAS (CE, SV, UE). Two each in Heathkit HW-14, Heathkit SB-200, Hammarlund HXL-1, and Yaesu FL2000 linears. Four in Gonset GSB-201 Mk. III and IV linears.
575 Mercury rectifier, original 575 w/ spherical-tubular bulb (DeF). See 575A.
575A, B [AKA (UE) 975A, F375A, GL-512A, VT-187, WT-210-0044]; mercury rectifier, tubular bulb w/ cap, super-jumbo (A4-29) base, T-T fil. 5 V @ 10 A, 15 kV PIV, 1.5 A (Arpin, AM, AX, C, EE, HY, GE, GN, ML, RCA, W) (3-15-43). Promoted for 5-kW xmtrs. otherwise requiring 869. Variant: 673.
576, A Radar clipper diode, tubular bulb w/ cap on glass stalk, 4-pin giant base, finned graphite anode, fil. 5.0 V @ 14 A, 25 kV PIV, 500 mA (UE).
577, W Radar clipper/rectifier diode, T16 bulb w/ ceramic-ringed cap, 4-pin metal base, ringed graphite anode, fil. 5.0 V @ 10.25 A, 25 kV, 1.5 A pk., 300 mA avg. (UE, CE) (1950).
578 Radar clipper/rectifier diode, T17 bulb w/ ceramic-ringed cap, 4-pin bakelite base, graphite anode, fil. 5.0 V @ 6.0 A, 40 kV, 750 mA pk., 100 mA avg. (UE) (1950).
579, B [AKA REL16, q. v.]; HV rect., fil. 2.5 V @ 6.0 A.
582 Vacuum rectifier, ruggedized, T12 bulb w/ cap, UX base, finned graphite anode, fil. 2.5 V @ 1.5 A, 9 kv PIV, 200 mA (UE).
583 Radar clipper-rectifier, T11 or T12 bulb w/ cap, octal base, finned graphite anode, fil. 2.5 V @ 5.0 A, 17 kV PIV, 65 mA (TS, UE).
588 Radar clipper diode, external-anode, glass bulb, flying leads for fil., oil- or air-cooled, heater 6.3 V @ 1.65 A, 16 kv PIV, 65 mA (UE).
589 Radar clipper diode - rectifier, glass external-anode, short flying leads for fil., < 2" long, 0.9" dia. by 1.9" long, fil. 6.3 V @ 1.9 A, 10 kV PIV (16 kV oil-immersed), 2 A pk., 65 mA avg., sim. to (possibly same as) 6339 (TS).
(EI, GL) **592** [AKA 3-200A3]; VHF RF-heating/xmtg. triode, all-glass w/ cap, dual grid terminals out side, radiation and forced-air cooled, for general use, center-tapped fil. 10.0 V @ 5.0 A, μ 25, 200 W, 150 MHz (EI, GE) (1-25-50). EI version is redesigned internally.
596 Full-wave rectifier, high-reliability 1641, T18 bulb, plates out angled pins at steps on bulb, fil. 5.0 V @ 3 A, 2.3 kV PIV, 275 mA (UE).
NL600 Gas rectifier sim. to EL10.
NL602 Gas rectifier sim. to CE200; replaceable by CE-211A (HY).
604 [AKA 6484, q. v.]; rectifier.
604L [AKA 7015, q. v.]; rectifier.
(NL) **604** [AKA 7014, q. v.]; rectifier.
606 [AKA 7016, q. v.]; rectifier.
(NL) **610** [AKA 5811, 7723, q. v.]; rectifier.
ZG610 [AKA 1B44, q. v.]; spark-gap modulator.
(NL) **615** [AKA 7018, q. v.]; rectifier.
NL-616 Gas-mercury rect., tubular bulb w/ cap, 4-pin base, fil. 2.5 V @ 9.0 A, 1.25 kV PIV, 2.5 A (NL).
(NL, WL) **617** Gas rectifier, replaceable by CE-205 (HY).
NL-618, P Rectifier, replaceable by CE-6B (HY).
NL-619 Mercury rectifier, tubular bulb w/ anode pin, medium screw base, fil. 2.0 V @ 12 A, 300 V PIV, 6 A (NL).
NL623 [AKA CE203, q. v.]; rectifier sim. to RX120A.
NL625 [AKA CE202B]; gas rectifier
NL627 [AKA RX212]; gas rectifier.
631A [AKA 805, q. v.]; xmtg. triode.

631P1 [AKA 1D21, q. v.]; flash tube.
NL-633 Mercury rectifier, tubular bulb w/ anode pin, bracket-mounted, flying leads at bottom for fil., fil. 2.5 V @ 50 A, 900 V PIV, 6 A (NL).
635 [AKA 7019, q. v.]; rectifier.
635L [AKA 7020, q. v.]; rectifier.
635P [AKA 6930, q. v.]; rectifier.
WT-644 Rectifier, replaceable by CE-200A (HY).
Z647 Magnetron.
648P1 Cold-cathode neon discharge tube.
NL-649 [AKA 5834, q. v.]; rectifier.
NL-653 [AKA 5835, q. v.]; rectifier.
653B [AKA 9004, q. v.]; acorn diode.
655B [AKA REL10, q. v.]; acorn diode.
NL-660L [AKA 7786, q. v.]; rectifier.
Z-668 Reflex klystron, 21.9-26.1 GHz, heater 6.3 V @ 650 mA, 1.7 kV, 20 mW out, forced-air cooled.
672, A [AKA WT-210-0100]; mercury tetrode thyratron, T18 stepped bulb w/ cap, 4-pin industrial base, heater 5.0 V @ 6 A, 1.5 kV, 2.5 A (RCA, W) {WX-3172} (12-20-46). "A" draws only 5 A heater current; is uprated to 2.5 kV and 3.2 A.
673 Mercury rectifier, 575A w/ super-jumbo 4-pin base w/ cap, stepped-tubular bulb, fil. 5.0 V @ 10.0 A, 10 kV PIV @ 1.75 A (25-55°C) or 15 kV @ 2.5 A (30-50°C) (GE, HY, RCA) (5-10-46). 12 in Navy's Jim Creek megawatt station, 1950.
(F, L, Z) **675** CW magnetron, supersedes 5J29.
G683 [AKA 5B21, q. v.]; rectifier.
Z694 Duodiode-triode, octal 7K7 (KR).
Z696 Duodiode-triode, octal 7X7 (KR).
Z698 [AKA X-6044]; TR tube.
V700 Xmtg. triode, nongettered hard glass in 805 format on tall metal base, Zr-impregnated graphite anode, fil. 7.5 V, 85 W, to replace soft-glass tubes (UE) (intro. 6-46).
(CE) **701** Lead-sulfide photoconductor.
(CE) **702** Lead-sulfide photoconductor.
(CE) **703** Submin. 2-pin lead-sulfide photoconductor (CE) (1950).
E-703A [AKA 203A, q. v.]; xmtg. triode (AS) (1932).
(CE) **705** Lead-sulfide photoconductor, glass bulb ½" dia. by 15/16" high incl. pins.
NL710 Thyratron rectifier (sim. to C3J); replaceable by 320A.
710 [AKA 6011, q. v.]; thyratron.
710A [AKA 7725, q. v.]; thyratron.
710AL [AKA 7726, q. v.]; thyratron.
710L [AKA 7509A, 7518, q. v.]; thyratron.
E-711 [AKA 211, q. v.]; xmtg. triode, 1932 (AS).
(J) **714** [AKA 2J22, q. v.]; magnetron.
(NL) **714** [AKA 7021, q. v.]; thyratron.
NL715 [AKA 5557, q. v.]; thyratron.
716 [AKA 6855, q. v.]; thyratron.
730 Gas-mercury thyratron, for motor control, tubular bulb w/ cap, 4-pin base, fil. 2.5 V @ 12 A, 1.5 kV PIV, 3.2 A (NL).
740 [AKA 6856, q. v.]; thyratron.
740L [AKA 7022, q. v.]; thyratron.
740P [AKA 6857, q. v.]; thyratron.
E-745 [AKA 845, q. v.]; xmtg. triode; 1932 (AS).
760 [AKA 6858, q. v.]; thyratron.
760L [AKA 7023, q. v.]; thyratron.
760P [AKA 6859, q. v.]; thyratron.
E-766 [AKA 866, q. v.]; rectifier (AS) (1932).
E-772 [AKA 872, q. v.]; rectifier (AS) (1932).
S829 TR tube; "no replacement required; break glass, leave air gap."
S836 [AKA 1960, q. v.]; gas switching diode.
S843B [AKA 1961, q. v.]; gas switching diode.
S853 [AKA 1B25, q. v.]; gas switching tube.
S856 [AKA 0A2, q. v.]; voltage regulator.
S859 [AKA 0C2, q. v.]; voltage regulator (SY).
VC861 HV rectifier, 15 kV PIV, 50-60 mA (SY).
KY866 Mercury triode thyratron, 866A w/ grid.
OS880 Bureau of Ordinance spec. for GL-299.
S901C [AKA 5651, q. v.]; voltage reference.
901 Crater glow-discharge tube for fax machines, small 4-pin base (GE, RCA) {C760}.
(UE) **905**, A Xmtg. triode, graphite-anode 805, q. v.
911CH Xmtg. triode, 150 W output (UE) (1936).
912NP Beta-gamma counter, 7" ribbed stainless steel shell, 900 V (AX).
917 [AKA CE-11V, 59TAV3]; vacuum phototube, for photometers and relays, low leakage, 868 format w/ cap, S-1 response, 20 μA/lumen (GE, HY, RCA, W) {C710} (6-16-47).
918 [AKA CE-1C, 59A3, VT-246, CV2692]; gas phototube, common in projectors, 868 format, S-1 response, high sensitivity, 150 μA/lumen (GE, HY, RCA, W) {C718} (6-16-47). Possibly same as NL-5560.
919 [AKA (CE, NU) 31V, 82AV4]; 917 w/ cathode connected to cap, for photometers (GE, RCA, W) {C719} (6-16-47).
920 [AKA (CE, NU)-16, -21C, G14, 60A3]; dual gas phototube for dual-track film sound, 868 format, S-1 response, 100 μA/lumen (GE, RCA, W) {C703} (6-16-47).
921 Gas phototube, for relay control and measurements, cartridge mount, short tubular bulb 0.9" dia. by 1-21/32" long, contact caps at ends,, S-1 response, 135 μA/lumen (CE, GE, RCA) {C7009} (6-16-47).
922 Phototube, vacuum version of 921 for relay control, S-1 response, 20 μA/lumen (CE, GE, RCA, W) {C7007} (6-16-47).
923 [AKA 84A3, WL-741, (CE, NU) 23C, VT-252]; gas phototube for sound on film, 868 format shortened to 2-15/16" above socket, S-1 response, high sensitivity, 135 μA/lumen. Sim. to 918 but shorter (GE, RCA, W) {C7005} (6-16-

47).

924 [AKA CE-22D]; 1P41 on screw base (GE, RCA) (6-16-47).

925 [AKA 64AV3]; vacuum phototube, for relay use. S-1 response, short side-on octal bulb, 2-1/16" above socket, 15 μA/lumen (GE, RCA, W) (6-16-47).

926 Vacuum phototube, S-3 response, 921 format, 6.5 μA/lumen (GE, RCA) (6-16-47).

927 [AKA 1P32, (CE, NU) 20C, G15F, 51A3, (CE, NU) 25C]; gas phototube, for 16-mm sound repro, relay control, and photometers, "peewee" 3-pin min. base, flat-ended tubular bulb 0.67" dia. by 1-15/16" above socket, side-on cathode, S-1 response, 125 μA/lumen; (sim. to CE36C but longer base) (GE, RCA, W) (6-16-47).

928 [AKA CE28D, 85A4]; gas phototube, nondirectional (red-, IR-sensitive) (GE, W) (6-16-47).

929 [AKA 59D, 71D, CE29, 61DV3]; vacuum phototube, octal base and GT bulb, side-on cathode, for relay control and photometers, S-4 response, high sensitivity, 45 μA/lumen (GE, RCA, W) (6-16-47). Variant: 1P39.

930 [AKA CE30C, 61A3]; gas version of 929, sound repro and relay control, S-1 response, high sensitivity, 135 μA/lumen (GE) (6-16-47). Variant: 1P40.

(UE) **930**, B Graphite-anode versions of 830, B (q. v.); 40 W and 60 W respectively (UE) (1936).

931, A, B, VA Photomultiplier, intro. 6-41 (first commercial photomultiplier), for scintillation clunters, relay use, and photometers; 9-stage, S-4 response, GT-style side-on bulb, submagnal 11-pin base, 20 A/lumen (GE, RCA, W). "A" (10-30-43) has doubled sensitivity. "B," ca. 1973, has lower dark current and greater sensitivity {C7075K}. Variants: 4471, 4472. Also see 1P21. "A" in AN/-APT-5 jammer, illuminated by small lamp, as broadband noise generator.

932 [AKA 32E]; electrometer tetrode, 1229 aged for grid stability.

933 Vacuum phototube.

934 [AKA CE-34]; vacuum phototube, sound repro. and facsimile use, 927 format, S-4 response, 30 μA/lumen, intro. 11-42 (RCA) (6-16-47).

935 Vacuum phototube, S-5 response, for photometers, tall side-on T9 bulb w/ cap, octal base, 30 μA/lumen, intro. 11-42 (GE, RCA) (6-16-47).

936 [AKA 1P24, q. v.]; phototube (RCA).

(UE) **938** Xmtg. triode, graphite-anode 838 (q. v.) (UE).

DE941 [AKA 341].

(UE) **941** Xmtg. triode, graphite-anode version of 841, q. v.; redesigned w/ ceramic base, etc., uprated to 20 W and 6 MHz (UE) (1936).

(UE) **942** Xmtg. triode, graphite-anode version of 842, q. v.; redesigned w/ ceramic base, etc., uprated to 15 W (UE) (1936).

(UE) **945** Xmtg. triode, graphite-anode version of 845, q. v. (UE) (1936).

(UE) **949**, A Xmtg. triode, graphite-anode version of 849, A, q. v.; redesigned internally, rated at 400 and 500 W respectively (UE) (1936).

(UE) **949H** RF-heating triode, graphite-anode grid-out-side version of (A, C) 849H, q. v., rated to only 20 MHz (UE).

950 Rcvg. power pentode (NU), sim. to 1J5G and 33, replaceable by 1B4.

951 [AKA 1B4P, q. v.] (NU).

(UE) **951** Graphite-anode version of 851, q. v. (UE).

(UE) **952** Graphite-anode version of 852, q. v., rated to 30 MHz (UE).

953 "Aust. acorn type." Acorn diode, 6.3 V heater, under dev. as of 1-20-38 {R6048}.

954 [AKA 38954, VT-120; (Br.) ZA2, VR95, CV1095, 10E/95]; rcvg. pentode, 5-pin acorn bulb w/ plate pin, heater 6.3 V @ 150 mA, g_m 1.4 mS (RCA, TS) {A449} (anncd. 3-35). In rcvr. of SC-1 radar, etc. Variant: 9001.

955 [AKA 38955, VT-121; (Br.) HA2, VR59, CV1059, 10E/11452]; xmtg-rcvg. triode; 5-pin acorn bulb, heater 6.3 V @ 150 mA, μ 25, 1.6 W, originally sold as "RCA-De Forest" (GE, HY, RCA, TS) {A179} (anncd. 3-35). Local osc. in rcvr. of SC-1, etc. Variants: 1650, 9002.

956 [AKA 38956, VT-238, CV649]; rcvg. remote-cutoff pentode, 5-pin acorn bulb w/ grid and plate pins; heater 6.3 V @ 150 mA, g_m 1.8 mS (GE, RCA) {A488} (anncd. 9-36). In rcvr. of SC-1 radar, etc. Variant: 9003.

957 [AKA VT-237, CV2700]; rcvg. triode, 5-pin acorn bulb, fil. 1.25 V @ 50 mA, μ 13.5 (RCA) (anncd. 12-38).

958, A [AKA VT-212, CV650]; triode, 5-pin acorn bulb, fil. 1.25 V @ 100 mA, μ 12, 600 mW, usable to 350 MHz; intended for xmtr use, anncd. 12-38. "A" is AKA 38958 (GE, RCA).

959 [AKA 38959]; xmtg-rcvg. remote-cutoff pentode, 5-pin acorn bulb, grid and plate pins, fil. 1.25 V, 50 mA, g_m 0.6 mS (GE, RCA) (anncd. 12-38).

V960 [AKA VO133].

(UE) **966**, A [AKA 866, A, q. v.]; rectifier.

(NU, UE) **967** [AKA 5557, q. v.]; thyratron.

(UE) **971** [AKA 871, q. v.]; rectifier.

(UE) **972**, A [AKA 872, A, q. v.]; rectifier.

(UX) **973** [AKA 873, q. v.]; thyratron (UE).

(UE) **975A** [AKA 575A, q. v.]; rectifier, tubular bulb stepped from 3" dia. at top to ca. 2" on bottom (UE).

991 [AKA NE16, CD2005, Navy VG-3]; neon volt. reg. or indicator (specially labeled NE16), T4-½ bulb on double-contact bayonet base, 48-68 V, 0.4-2 mA (GE, RCA) (7-6-43).

BX-1000 "Beam-X" decade counter/switch, 10 outputs, counts at 10 MHz, tubular bulb, internal rod magnets; no external shield or magnets (BR).

FM-1000 Loktal locked-oscillator FM det., heater 6.3 V @ 300 mA (SY-Philco)(1950).

H1000 [AKA 5C22, q. v.]; hydrogen thyratron.

V-1000 [AKA 1N33, q. v.]; germanium diode.

1000 Spec [AKA 6C21, ZP527, q. v.]; xmtg. triode.

NL1001 Ignitron, for resistance welders, glass bulb, anode and cathode connected to mounting straps at top and bottom, 600 V applied, 9 A, convection or forced-air cooled (NL).

R1001 [AKA 3C36, q. v.]; xmtg. triode.

SR1002 Submin. triode sim. to 6K4.

CH1008 Min. gas-triode pulse generator.

A1013 Magnetron.

VC1017 Min. 2X2 (NU).

NU-1018 [AKA 7D4]; never made? (NU).

CH1027 Series of submin. "curpistor" minute-current regulators, tubular bulb 0.24" dia. by 3/4" long (1-¼" long for CH1027-12), nitrogen-filled, w/ Ra-226 radioactive source. Operate with applied voltage from 25 to 500. Current passed is directly proportional to amount of Ra (TS).

CH1027-9 For 10^{-9} A (contains 18.75 μC Ra).

CH1027-10 For 10^{-10} A (contains 1.875 μC Ra).

CH1027-11 For 10^{-11} A (contains 0.1875 μC Ra).

CH1027-12 For 10^{-12} A (contains 0.01875 μC Ra).

1036 [AKA 6436, q. v.]; rectifier.

1037 [AKA 6437, q. v.]; voltage regulator.

(NU) **R1038** Ionization gauge, fil. 3.0 V @ 1.8 A, 200 V; ten times the ion current in A equals pressure in mm of mercury.

1038 [AKA 6438, q. v.]; voltage regulator.

1038AV Phototube.

NU-**M1039** [AKA 7AF7, q. v.]; duotriode.

R-1045-1 Min. HV rectifier, heater 6.3 V @ 300 mA, 10 kV output, 2 mA (NU).

R-1045-4 HV rectifier, R-1045-1 w/ 150-mA heater, 2 kV output, 20 mA (NU).

CH-1046 Min. hydrogen thyratron, heater 28 V (TS).

NL1051A [AKA 5551, q. v.]; ignitron.

NL1053A Ignitron, intended to be rebuilt, includes thermostat mounting plate; "Size D" (5553B, q. v.) ratings (NL).

1054 Xmtg. triode, 3 kW.

CH-1055 [AKA 7192, q. v.]; hydrogen thyratron.

M1060 [AKA 2C53, q. v.]; HV triode (NU).

CH-1062 [AKA 7190, q. v.]; hydrogen thyratron.

CH-1068 Power triple-triode, octal base, for pass tube in voltage regs., heaters 6.3 V. Has common grid and plate but separate cathodes; like scaled-down 7241. In ±250-V supplies of DuM 343 (17") and 345 (21") scopes.

CH-1092 [AKA 7191, q. v.]; hydrogen thyratron.

CH-1095-97 Hydrogen crowbar thyratrons (TS).

CH-1095 [AKA 7568 (?), q. v.]; rated 800 A.

CH-1096 [AKA 7559 (?), q. v.]; rated 1500 A.

CH-1097 [AKA 7605 (?), q. v.]; rated 2500 A.

R1100 Manometer tube (SY), 4-pin w/ gas port at top, thermocouple for measuring low gas pressures. Heater 3 V @ 125 mA.

1103D Submin. semi-remote cutoff pentode.

1103S Submin. semi-remote cutoff pentode.

1106 [AKA 6AZ6]; submin. duodiode, heater 6.3 V, video detector (NU).

R1111 Pirani gauge for measuring low pressures (10^{-1} to 10^{-4} mm Hg); operating current 100 mA; resistance 6.6 Ω (in air) to 17 Ω (in vacuum).

CH-1116 [AKA 7400, q. v.]; indicator thyratron.

1125A Submin. triode.

R1130, A, B [AKA 1B59, q. v.]; cold-cath. crater-glow lamp.

R1131, A Crater lamp, R1130B w/ 0.093" crater (SY).

1132B, D Submin. RF pentode, sharp-cutoff, two-ended.

1145 Neon lamp regulator (SY).

E1148 [AKA CV6, VT-232]; Brit. VHF xmtg. triode, HY-made, same as HY615 except for cutoff char.; export-only? (HY). In VHF section of Mk. 19 tank radio.

R1159 Min. voltage regulator (SY).

R1160A [AKA 1B46, q. v.]; voltage regulator.

EM-1271 Pair of spl. aged 12AX7s, for 1735D scope (IT).

(CE, CR) **1250** Mercury rect., fil. 2 V @ 7 A, 7.5 kV PIV, 1.5 A.

K1209 Photomultiplier, 12-stage, S-11 response, 6364 format except for 25-pin base (DuM).

K1213 Photomultiplier, 12-stage, S-11 response, end-on bulb 3" dial by 6-3/8", 25-pin base (DuM).

VC-1257 Hydrogen thyratron, radar pulser, tubular bulb w/ cap on glass stalk, flange-mounted, w/ H_2 reservoir, flying leads at base, 38 kV @ 2000 A, 2.5 A avg., 35 MW (CH, KL, TS).

VC-1258 Min. hydrogen thyratron, hard glass, for small radars up to 10 kW - claimed to be "industry standard" for small pulsers (BL, CH, KL, TS).

DG1290 [AKA FG-190, q. v.].

K1292 Photomultiplier, IR-sensitive 6292 (DuM).

K1295 Photomultiplier, 12-stage, S-11 response, 6292 format lengthened to 6-3/4" (DuM).

K1303 Photomultiplier, ruggedized 6365 (DuM).

K1306 Photomultiplier, UV-sensitive 6292 in fused-silica bulb (DuM).

K1328 Photomultiplier, 12-stage, S-11 response, 15-7/8" dia. by 22-¼", in same bulb as 16AP4 metal CRT (DuM).

Z1450 [AKA 5710, q. v.]; xmtg. triode.

1466 [AKA 5649, q. v.]; klystron.

A1480 TWT, 11-18 GHz, integral magnets, 20 W out, heater 6.3 V @ 350 mA, collector voltage 1700, SMA coax connectors for input and output.

E1468 [AKA CV172.]

Z1497 Duotriode, ruggedized 2C51.

Z1498 Duotriode, ruggedized 12AU7.

Z1499 Beam tetrode, ruggedized 6AQ5.

1530P1 Strobotron.

W1535 [AKA 6F6G, et al., q. v.]; power pentode.

1538 [AKA 5780, q. v.]; magnetron.

(CEL) **1552** [AKA 5C30, C5B, q. v.]; thyratron.

VC1568 [AKA 1R6, q. v.]; reactance modulator.

1596 [AKA 5650, q. v.]; klystron.

DG1672 [AKA FG-172]; thyratron.

(AX) **1701** [AKA 5557, q. v.]; thyratron.

Z-1723 [AKA 5797, q. v.]; submin. pentode.

Z1724 [AKA 5798, q. v.]; submin. duotriode.

1754, A [AKA 5948, q. v.]; hydrogen thyratron.

V1901 [AKA VU504, q. v.]; HV rectifier.

1902 "Oscillator," T48 bulb, double-ended, fil. 11 V; "not listed for distributor sales" by RCA 4-1-34. "Purchased outside but merchandized by RCA Mfg. Co." as of 1-20-38.

1903 "Oscillator," T32 bulb, fil. 22 V, water-cooled; "not listed for distributor sales" by RCA 4-1-34. "Purchased outside but merchandized by RCA Mfg. Co." as of 1-20-38.

(GL) **1904** Triode thyratron, heater 5 V @ 4.85 A, 1 kV PIV, 2.5 A avg.; variant of 91 for welder control {R636}.

1907, A [AKA 5949, A, q. v.]; hydrogen thyratron..

V1920 [AKA CV121, q. v.]; HV rectifier.

1923 [AKA REL38, q. v.]; cold-cathode radar pulser.

1923F [AKA CV85, q. v.]; cold-cathode radar pulser.

1924 [AKA CV100, q. v.]; cold-cathode radar pulser.

1945 Vacuum gauge, triode ionization type, hydrogen-sensitive, metal octal shell w/ vacuum port sealed into top, heater 6.3 V @ 750 mA, for leak-detection systems (RCA) (ca. 6-47).

1946 Vacuum gauge, thermocouple type, covers 0.0001 to 1.0 torr; T12 hard-glass bulb 6-¼" long necked to vacuum port at top, flying leads at base; heater ca. 1 V @ 70 mA; thermocouple ca. 5 Ω (RCA) {C781} (ca. 6-47).

1947 Vacuum gauge, Pirani type, covers 0.01 to 1.5 torr, T9 soft-glass bulb 7-9/16" long necked to vacuum port at top, 4-pin base; Pt-Ir fil., ca. 10 V @ 70-100 mA (RCA) {C778A} (ca. 6-47).

1948 (RCA) {C787A}.

1949 Vacuum gauge, triode ionization type, covers < 0.0001 torr, T16 stepped-tubular hard-glass bulb 11-½" long, w/ plate pin, vacuum port at angle off top, flying leads at base, fil. 5 V @ 3.5 A, w/ spare fil. (RCA) {C785} (ca. 3-54).

1950 Vacuum gauge, 1949 in ST19 soft-glass bulb; fil. current 3.5 A (RCA) {C791} (CA. 3-54).

1951 Vacuum capacitor, T10 bulb, no base, 31 pF, 9 kV, 6 A @ 30 MHz (RCA) {I1300} (1936).

1960 [AKA S836]; gas switching diode (11-3-44).

1961 [AKA S843B]; gas switching diode (3-44).

CD2000 Special neon diode.

DR2000 Min. 9-pin 7-segment "Numitron" incandescent digital readout tube, numerals 0-9, IC-compatible segments use 4.5 V @ 24 mA each, solderable pins (RCA) (ca. 2-71).

S-2001 [AKA 6146, q. v.]; xmtg. beam tetrode.

DR2010 Readout, DR2000 w/ decimal pt. added (RCA).

DR2020 Readout, DR2000 w/ only +/- signs and numeral 1 (RCA).

2027 [AKA CV125, q. v.]; cold-cathode radar pulser.

DR2030 Readout, DR2000 w/ only +/- signs.

J2041V3 Packaged pencil-tube pulsed amp. chain, 1030 MHz, 200 W pulse, 5.3" x 2.8" x 1.25" (RCA) (ca. 6-75).

2050, A, W [AKA WT-210-0004, B2A, VT-245, CV2565, CV276]; xenon octal tetrode thyratron, 6.3 V @ 600 mA, 1.3 kv PIV, 100 mA as rectifier (CH, HY, GE, RCA, TS, W) (5-19-44). "A" has GT bulb. Variants: 502A, 1657.

2050 Spl "Special" 2050 {R6237}.

2051 [AKA VT-109, CV1798]; thyratron, slightly derated 2050, replaceable by it, "for renewal use only" as of 1944 (GE, RCA) (5-19-44).

R2057E Secondary-emission amp. "sim. to 1630" (this may be the developmental no. for 1630).

2083 Gas rectifier.

DR2100/V1 "Numitron" readout (see DR2000), ½" dia, for 10-pin TO-5 style socket or soldering. "V1" has formed leads (RCA).

2179A [AKA 2C40, q. v.]; lighthouse triode.

2183 [AKA G683, 5B21, q. v.]; rectifier.

DR2200, V1 "Numitron" readout (see DR2000), ½" dia., segments 2.5 V @ 14 mA each. "V" has formed leads for soldering.

FD-2200 Packaged TRF 3-stage pencil-triode amp. for 1090 MHz, tubular case w/ coax input-output fittings, including filters, heater 6.3 V @ 405 mA, 150 V B+, NF 11.5 dB (RCA) (ca. 9-65).

DG2252 [AKA FG-166, q. v.]; rectifier.

SR-2495 Pentode, rugged stacked-in-glass 6AK5 (SY).

SR-2498 Duotrilde, rugged stacked-in-glass 6111 (SY).

W2510 [AKA NT98, q. v.]; magnetron.

2522N1 [AKA 128A]; sweep-gen. thyratron, heater 2.5 V @ 1.75 A, 1 kV, 3 mA (DuM).

2523N1 [AKA 128AS]; thyratron.

SR-2601 Beam tetrode, rugged stacked-in-glass 6AQ5 (SY).

2779A [AKA 1B23, q. v.]; TR cell.

K2900 [AKA T-300, q. v.]; xmtg. triode.

2911 [AKA SAL89A]; klystron; also see 8493.

X2935 Submin. magnetron, dynatron mode, voltage-tuned, EM, heater 6.3 V @ 150 mA, B+ 25 V, tunes 30-440 *Hz*, also tunable by varying magnetic field (GE) (1963).

HF-3000 Xmtg. triode, stepped tubular bulb, flying lead for plate, dual side pins for grid, fil. 21-22 V @ 40.5 A, μ 16 (AX).

L3002 Magnetron.

3064 [AKA 4J70-77, q. v.]; series of magnetrons.
3067 Xmtg. triode, grid-out-side, w/ corona grading rings, μ 44, 160 kW, water-cooled, for G-G SW broadcast use (F).
UK3108 [AKA CV38, q. v.]; magnetron.
L-3110 CW magnetron, 2.4-3.6 GHz, 150 W out. In AN/ALT-6B jammer.
3150 [AKA NT98, q. v.]; magnetron.
ZB3200 Xmtg./RF-heating triode, stepped tubular bulb, cragle at top for plate, grid-out-side, flying leads for fil. (AX).
3401 Submin. tetrode, nonmicrophonic for use as det. or first stage, 0.385" dia. by 1-1/8" long, fil. 0.625 V @ 20 mA (Microtubes, Inc.). Also see M64, M74.
3402 Submin. tetrode, 3401 less the control on microphonics. Also see M64, M74.
3403 Submin. power tetrode, 0.385" dia. by 1-1/8" long, fil. 0.625 V @ 40 mA (Microtubes, Inc.). Also see M54.
F3652 [AKA 820B, q. v.]; xmtg. triode.
4021 Digital readout.
4031 Digital readout.
4077A [AKA CV33, q. v.]; rectifier.
R4100 [AKA 1D21, q. v.]; flash tube.
R4101 [AKA 1D21, q. v.]; flash tube.
R4200 [AKA 648P1, q. v.]; flash tube.
4260 [AKA FG-17, q. v.]; thyratron.
4261 [AKA FG-17, q. v.]; thyratron.
4270A [AKA CV30, q. v.]; xmtg. triode.
4313C [AKA CV75, q. v.]; cold-cathode thyratron.
R4330 Flash tube, blue-white, w/ trigger electrode (SY, 1946).
R4340 Cold-cathode flash tube; coiled gas tube inside tubular bulb on six-pin base; produces 48M lumens or 4X the output of R4330 (SY) (1946).
R4350 Cold-cathode flash tube; see R4340 (SY).
R4410 [AKA ES-10816]; flash tube.
4604 Xmtg. pentode, 30 W.
H5015 [AKA 3D21, q. v.]; pulse tetrode.
5031 [AKA 8037, q. v.].
5022 Xmtg. tetrode, sim. to 4-400A.
B5032 [AKA B5018].
5092 [AKA 8421, q. v.].
5137 [AKA 3C45, q. v.]; hydrogen thyratron.
5311 [AKA 931A]; photomultiplier.
B5440 Digital readout.
A5586 [AKA VT-123]; secondary-emission amp., superseded by A5588A (1630/VT-128).
X6000 [AKA 3J30, q. v.]; magnetron.
CE6001 Full-wave mercury rectifier, like modernized, self-starting Cooper-Hewitt but w/o mercury pool, water-cooled via integral tubing, 250-600 V in, 25 A (Clark Electronic Labs.) (1947).
X6001 [AKA 2J42, q. v.]; magnetron.
X6002 [AKA 3J31, q. v.]; magnetron.
X6003 [AKA 3J31, q. v.]; magnetron.
X6007 [AKA 5791, q. v.]; ATR tube.
R6010 Reflex klystron, 4.4-4.8 GHz, IC, CO, heater 6.3 V @ 900 mA, 800 V, 37 W output (EMI).
X6010 [AKA 1B26, q. v.]; TR tube.
X6011 [AKA 1B24, q. v.]; TR tube.
X6012 [AKA 1B28, q. v.]; gas switching tube.
X6013 [AKA 1B27, q. v.]; TR cell.
X6015 [AKA 1B36, q. v.]; ATR tube.
X6020 [AKA 1N24, q. v.]; Si diode.
X6021 [AKA 1N26, q. v.]; Si diode.
R6025 [AKA 10FP4, q. v.]; CRT.
X6030 Diode rectifier.
X6031 [AKA 1B35, q. v.]; TR tube.
X6032 [AKA 1B50, q. v.]; TR tube.
X6033 [AKA 1B51, q. v.]; ATR tube.
X6034 [AKA 1B55, q. v.]; TR tube.
X6035 [AKA 1B52, q. v.]; ATR tube.
X6036 [AKA 1B54, q. v.]; Pre-TR tube.
X6044 [AKA 1B63A, q. v.]; TR tube.
(R?)6048 [AKA CW10, q. v.]; acorn diode; also see 953.
R6180-A Xmtg. triode thyratron, w/ cap, heater 2.5 V, modulator in SCR-602-T6.
R6212A [AKA 6144]; coaxial noise diode.
R6137B [AKA 8013A, q. v.]; HV rectifier (T).
R6146 [AKA 8008, q. v.]; rectifier (T).
R6271 Submin. lighthouse diode, same dimensions as 1N21 diode, heater 6.3 V @ 300 ma, 200 V PIV, 20 mA (RCA) Also see 5647.
R6285 Lighthouse diode, selected 2B22.
R6288 Thyratron, modified 2D21.
UX-6653 TR tube in SG radar (ca. 1941).
R7011 [AKA 1N21, q. v.]; Si diode.
R7013 [AKA 1N33; q. v.]; Si diode.
VRX7203A Varian klystron replacing WE 459A, q. v.
VRX7204A Varian klystron replacing WE 457A, q. v.
VRX7204B Varian klystron replacing WE 445A, q. v.
X7047 Gas switching tube.
X7103 [AKA 1N53, q. v.]; Si diode.
X7107 [AKA 5789, q. v.]; magnetron.
X7108 [AKA 5790, q. v.]; TR tube.
X7109 TR tube.
R7601 [AKA SS501, q. v.]; argon control tube.
R7850 Photocell, same format and ratings as WE 6A (NE).
XG7893 [AKA 2K43, q. v.]; klystron.
EC8010 [AKA 8556, q. v.]; triode.
XG8071 [AKA 2K42, q. v.]; klystron.
XG8072 [AKA 2K44, q. v.]; klystron.
XZA8073 [AKA 3K22, q. v.]; klystron.
ES-10816 [AKA R4410, q. v.]; flash tube.
19210 [AKA CE36C, q. v.]; phototube.
24154 [AKA GL-34 long]; vacuum switch, long operating arm.
24163 [AKA GL-34 short]; vacuum switch, short operating arm, superseded by 1S21.

GA-53798 TWA, packaged, 8.5-9.0 GHz, for receiver pre-amps in missile-borne guidance use, ruggedized, WG input and output, heater 3.5 V @ 900 mA, 1125 V, NF 23 dB, 50 mW out (WE) (ca. 4-64). Variant: GA-53979.

GA-53979 TWA, GA-53798 w/ different power connector.

62005 [AKA 1B69, q. v.]; gamma counter tube.

62006 [AKA 1B70, q. v.]; gamma counter tube.

62007 [AKA 1B71, q. v.]; gamma counter tube.

62008 [AKA 1B72, q. v.]; gamma counter tube.

62009 [AKA 1B73, q. v.]; counter.

62010 [AKA 1B74, q. v.]; counter.

62011 [AKA 1B75, q. v.]; counter.

62012 [AKA 1B76, q. v.]; counter.

62013 [AKA 1B77, q. v.]; counter.

62014 [AKA 1B78, q. v.]; counter.

62015 [AKA 1B79, q. v.]; counter.

62016 [AKA 1B80, q. v.]; counter.

62017 [AKA 1B81, q. v.]; counter.

75615 Phototube.

189048 [AKA 4B28]; gas rect.; CE-225 replaces (GE).

189049 [AKA 4B26, q. v.], gas rectifier, replaceable by CE-226 (GE, HY).

189049 [AKA 4B26, q. v.]; rectifier (GE).

A15030 Super-power pulse amp. (RCA), for C-Stellarator controlled-fusion project, 1961; six tubes produce pulse of 50,000 A for 1 ms (1959). Peak pulse rating is 20 MW, 50 kV, 660 A peak cathode current at 1% duty cycle.

A15236 to A15236 (even nos.) Packaged pencil-tube pulsed osc., 0.9-3.4 GHz, 80 to 100 W peak, 1" dia., 6" to 2" long (RCA) (ca. 6-75).

A15599A Packaged pencil-tube osc., 1090 MHz, 250 W pulse, 0.88" dia. by 3" long (RCA) (ca. 6-75).

A15610 Packaged pencil-tube osc., 1090 MHz, 325 W pulse, 0.88" dia. by 3" long (RCA) (ca. 6-75).

A15612 Packaged pencil-tube pulsed osc., 4.3 GHz, 10 W pulse, 1" dia. by 2.2" long (RCA) (ca. 6-75).

A15622 Packaged pencil-tube pulsed osc., 3.0 GHz, 100 W pulse, 0.88" dia. by 2.2" long (RCA) (ca. 6-75).

195528 Rectifier, superseded by 206501.

199698 Full-wave gas rectifier.

206501 [AKA WL-289415, q. v.]; rectifier, supersedes 195528.

217283 [AKA 5B21, q. v.]; rect.; CE-235 replaces (GE).

277465 [AKA 12X825]; rectifier, supersed. by 12X825.

289414 [AKA 4B28, q. v.]; rectifier, replaceable by CE-225 (HY, W).

289415 [AKA 206501, CE212, q. v.]; mercury rectifier, fil. 2.0 V @ 12 A, 275 V PIV, 2 A (W).

289416D [AKA 4B26, q. v.]; rectifier, replaceable by CE-226 (HY, W).

289881 Gas rectifier.

(WL) **766776** [AKA 5B21, q. v.]; rectifier.

859483 Gas rectifier, format and ratings of 289415.

XB767527-1 Temperature-limited (noise-generator?) diode.

966626 [AKA 4B35, q. v.]; rectifier.

R5660201 [AKA 3K23, q. v.]; klystron.

AMPEREX TYPES

AX9900 [AKA 5866, q. v.]; xmtg. triode.

AX9901 [AKA 5867, q. v.]; xmtg. triode.

AX9902 [AKA 5868, q. v.]; xmtg. triode.

AX9903 [AKA 5894], q. v.]; xmtg. duopentode.

AX9904 [AKA 5923], q. v.]; xmtg. triode.

AX9904R [AKA 5924], q. v.]; xmtg. triode.

AX9905 [AKA 5895, q. v.]; xmtg. duotetrode.

AX9906 [AKA 6077, q. v.]; xmtg. triode.

AX9906R [AKA 6078, q. v.]; xmtg. triode.

AX9907 [AKA 6075], q. v.]; xmtg. tetrode.

AX9907R [AKA 6076], q. v.]; xmtg. tetrode.

AX9908 [AKA 6079, q. v.]; xmtg. tetrode.

AX9909 [AKA 6083, q. v.]; xmtg. pentode.

AX9910 [AKA 6252, q. v.]; xmtg. duotetrode.

AX9911 [AKA 6268, q. v.]; hydrogen thyratron.

AX9912 [AKA 6279, q. v.]; hydrogen thyratron.

AGR9950 [AKA 5869, q. v.]; thyratron.

AGR9951 [AKA 5870, q. v.]; thyratron.

BENDIX - RED BANK PREMIUM TYPES

Part of "HY-G-300" line of ruggedized high-reliability types, some w/ hard-glass bulbs; octal types have melamine bases.

TE-2 [AKA 5839, q. v.]; full-wave rectifier.

TE-3 [AKA 5838, q. v.]; full-wave rectifier.

TE-5 [AKA 5852, q. v.]; full-wave rectifier.

TE-8 [AKA 5992, q. v.]; beam power tube.

TE-10 [AKA 5993, q. v.]; full-wave rectifier.

TE-18 [AKA 6094, q. v.]; beam power tube.

TE-21 [AKA 6385, q. v.]; min. duotriode.

TE-22 [AKA 6106, q. v.]; full-wave rectifier.

TE-27 [AKA 6384, q. v.]; beam power tube.

TE-36 [AKA 6754, q. v.]; full-wave rectifier.

TE-42 [AKA 6851, q. v.]; min. duotriode.

TE-43 [AKA 6486A, q. v.]; min. pentode.

TE-44 [AKA 6582A, q. v.]; min. pentode.

TE-45 [AKA 6853, q. v.]; full-wave rectifier.

TE-46 [AKA 6080WB, q. v.]; duotriode pass tube.

TE-47 [AKA 6854, q. v.]; min. duotriode.

TE-48 [AKA 6877, q. v.]; min. triode pass tube.

TE-52 [AKA 6889, q. v.]; beam power tube.

TE-54 [AKA 6900, q. v.]; min. duotriode.

TE-55 [AKA 6082A, q. v.]; duotriode pass tube.

ELECTRONS, INC.

EL1C [AKA 3B22, q. v.]; full-wave xenon rectifier. One in BC-460.

ELC1A [AKA Navy Type A]; thyratron {WX3169} (CE, EL, GE).

ELC1B [AKA 3C31, q. v.]; thyratron.

ELC1B/A [AKA 5664, q. v.]; thyratron.

ELC1J Thyratron rectifier.

ELC1J/A [AKA 5683, q. v.]; thyratron.

ELC1K [AKA 6014, q. v.]; thyratron.

ELC1K/B [AKA 7319, q. v.]; thyratron.

EL3B [AKA 6013, q. v.]; rectifier (HY).
EL3C [AKA 4B24, q. v.]; rectifier, replaceable w/ 6484. In BC-1241-A.
EL3C/L [AKA 6982, q. v.]; rectifier.
ELC3J [AKA 5632, 7306, q. v.]; thyratron. Also see 5684 (EL, T).
ELC3J/A [AKA 5684, q. v.]; thyratron (EL, W).
ELC3J/AL [AKA 7556A, q. v]; thyratron.
ELC3J/L [AKA 7216, q. v.]; thyratron.
ELC3P14 Xenon triode thyratron, for inverter motor control, "industrial" 4-pin base, heater 14 V @ 2.6 A, 1250 V PIV, 3 A (EL, HY, T).
ELC3R14 Xenon triode thyratron, ELC3P14 derated to 500 V PIV and w/ narrower temp. limits, same use (EL, HY, T). Also see 6278.
ELC4J [AKA 6984, q. v.]; thyratron.
ELC4J/F [AKA 6985, q. v.]; thyratron.
ELC4J/L [AKA 6986, q. v.]; thyratron.
EL5.0 [AKA ELC6A, q. v.]; thyratron.
EL5B, B2.5 [AKA 4B22, q. v.]; rectifier.
EL5BHD [AKA 4B23, 38403, q. v.]; rectifier.
ELC5A [AKA Navy Type B]; thyratron, superseded by C6A.
ELC5B [AKA 1552, 5C30, CE315]; thyratron. In MC-447-A.
ELC5F14 [AKA 6278, q. v.]; thyratron (EL, T).
ELC5F15 [AKA 6231, q. v.]; thyratron.
ELC6A [AKA 1661, Navy Type C]; xenon triode thyratron rectifier, fil. 2.5 V @ 18 A, 600 V PIV, 6.4 A, mogul base w/ cap; supersedes ELC5A (HY).
ELC6C Thyratron.
ELC6J [AKA 5C21, q. v.]; thyratron (EL, T).
ELC6J60 Xenon thyratron.
ELC6J-A [AKA 5685, q. v.]; thyratron (EL, T).
ELC6J-F [AKA 6860]; thyratron; C6J-A with flying leads (EL, HY, T).
ELC6J/K Thyratron, fast-deionizing version of C6J/A.
ELC6J/KL [AKA 6989, q. v.]; thyratron.
ELC6J/L [AKA 7321, q. v.]; thyratron.
ELC6L [AKA 5528, CE326, q. v.]; thyratron (EL, T).
ELC6M Thyratron, fast-deionizing C6J (EL, T).
ELC6C [AKA 6987, q. v.]; thyratron.
ELC6J/L [AKA 6988, q. v.]; thyratron.
ELC6J/KF [AKA 6864, q. v.]; thyratron.
ELC6J/KL [AKA 6989, q. v.]; thyratron.
EL6F [AKA 6015, q. v.]; rectifier.
EL6B [AKA 5892, q. v.]; xenon rectifier (EL, HY).
EL6B/L [AKA 6983, q. v.]; rectifier.
EL6CF [AKA 38674A]; rectifier, EL6C on special base, 1.54" dia; also see 4B25.
ELC16J [AKA 5665, q. v.]; thyratron.
EL16B Gas rectifier.
EL16F Gas rectifier, strap-mounted w/ flying leads, fil. 2.5 V @ 36 A, 620 V PIV, 16 A (EE).
EL302.5 [AKA 3B21, q. v.]; rectifier.
EL60B Rectifier, metal body w/ cooling fins, strap mount, flying leads at top and base, heater 115 V @ 1.3 A, 1.25 kV PIV, 50 A.

FEDERAL TELEGRAPH/IT&T TYPES

F-11X Xmtg. triode, water-cooled (20 kW).
F-100A Xmtg. triode, rated at 350 W output for broadcast use, fil. 11 V @ 25 A, μ 14, 500 W, 100 MHz.
F-101B Xmtg. triode sim. to WE 232A.
F-102A Xmtg. type rated at 100 W output in BC use.
F-103A Vacuum rectifier, 4-1/16" by 21", fil. 28 V @ 21 A, 50 kV PIV, 9 A peak (F).
F-104A Vacuum rectifier, 4-1/16" by 16-5/8", fil. 11.0 V @ 25 A, 30 kV PIV, 1.25 A peak (F).
F-108A Xmtg. type, general-purpose, rated at 100 W output for broadcast use; fil. 10 V @ 11 A, μ 12, 175 W, 30 MHz.
F-110X Xmtg. RF triode, supersedes F110A, fil. 28 V, μ 40, 25 kW, 3 MHz, water-cooled (F).
F-116A Xmtg. triode.
F-118B Xmtg. triode.
F-119A "Obsol.; superseded by F-127A which is not interchangeable."
F-121A Xmtg. triode.
F-122A Xmtg. triode.
F-123A, C Xmtg. triode, fil. 10.0 V @ 4.0 A, μ 14.5, 125 W.
F-124A, R Xmtg. triode, six-section fil., 27.2 V @ 68.7 A each, μ 42, 40 kW, 20 MHz, water cooled (F). "R" is radiator variant, derated to 20 kW.
F-125A, R Xmtg. audio triode, 3" dia. by 26" long, fil. 13.6 V @ 204 A, μ 4.8, 40 kW, water-cooled (F) (ca. 1938). "R" is radiator variant.
F-126A Xmtg. triode, water-cooled, sim. to F343A.
F-127A Xmtg. triode, fil. 10 V @ 4.0 A, μ 38, 200 W, 30 MHz (F).
F-128A Xmtg. triode, fil. 11 V, μ 36, 600 W, 30 MHz, forced-air cooled. In BC-467A.
F-129B, R Xmtg. triode, fil. 18 V, μ 26, 7.5 kW, 30 MHz, water-cooled (F). In BC-340. "R" [AKA L129R] is radiator variant, derated to 5 kW.
F-131A, R Xmtg. triode, water-cooled. "R" is rad. variant.
F-132A Xmtg. triode, for Class A or B audio, fil. 11.0 V, μ 10, 600 W, air-cooled (F).
F-134 Xmtg. triode, fil. 25 v, μ 21, 150 kW, 22 MHz, water-cooled (F).
F-307A [AKA 207, q. v.].
F-311A [AKA 860, q. v.]; xmtg. triode.
F-314A [AKA UV-214A, q. v.]; rectifier.
F-318A [AKA UV-218, q. v.]; vacuum rectifier.
F-319 [AKA UV-219, q. v.]; vacuum rectifier.
F-320A, B, C [AKA (WE) 220B, q. v.]; xmtg. triode. "C" [AKA (WE) 220C] was developmental, never applied.
F-322A [AKA 222A, q. v.].
F-328A, B [AKA 520M, (WE) 228A, q. v.]; xmtg. triode.
F-331A [AKA 861, q. v.]; xmtg. tetrode.
F-332A [AKA (WE) 232A]; xmtg. triode, water-cooled.

F-332C [AKA (WE) 232B]; xmtg. triode superseded by F342A.

F-336A [AKA (WE) 236A]; xmtg. triode, water-cooled.

F-337A [AKA (WE) 237A]; xmtg. rectifier, water-cooled.

F-342R [AKA 342A]; xmtg. triode, radiator version, 10 kW, 4 MHz, intro. 12-40.

F-346A [AKA 846, q. v.]; xmtg. triode.

F-348A [AKA 848]; xmtg. triode modulator.

F-349A [AKA 849, q. v.].

F-351A [AKA 851, q. v.]; xmtg. triode.

F-353A [AKA 872A, q. v.]; mercury rectifier. In BC-340A.

F-357A [AKA 857B, q. v.].

F-358A [AKA 858, q. v.]; xmtg. triode; actual manufacture in doubt.

F-363A [AKA 863, q. v.]; xmtg. triode.

F-366A [AKA 866A, q. v.]; mercury rectifier.

F-367A Xmtg. rectifier, (WE) 315A char.

F-369A, B [AKA 869B, q. v.]; mercury rectifier.

(F) **600N** [AKA 6C22, q. v.]; xmtg. triode.

F-606 Vacuum rectifier sim. to 5625.

F-631 Xmtg. triode, fil. 28 V, μ 40, 25 kW, 3 MHz, water-cooled (F).

F-661 Xmtg. triode sim. to 893.

L600, N [AKA 6C22, q. v.]; xmtg. triode.

F-5302 Xmtg. triode, water-cooled variant of F-5303.

F-5303 [AKA 7C25, q. v.]; xmtg. triode.

HEINTZ & KAUFMAN TYPES

VG-2 Ionization gauge. Also see VG-54.

HK-24 [AKA 3C34, q. v.]; xmtg. triode. In T-29/GRT.

VG-24G Ionization gauge. Also see VG-54.

HK-54 [AKA 4C25, q. v.] (intro. 1938) (10-20-43).

HK-54S HK-54 w/ fil. current reduced to 3.35 A.

VG-54 Ionization gauge, HK-54 w/ glass spigot out top side.

HK-55 Gridless "Gammatron," fil. 6.0 V @ 3 A, μ 3.5, 75 W; grid is replaced by a tantalum "gamma plate" which is operated w/ negative bias and provides grid action. Usable for Class A and B audio and Class B RF amps.

HK-152L Xmtg. triode, fil. 5 V @ 13 A or 10 V @ 6.5 A, 150 W.

HK-153 Vacuum rectifier, 5 kV, repl. by HK-253.

HK-154 Xmtg. triode, T16 bulb, metal UX base, grid and plate out sides, tantalum plate fil. 5 V @ 6.5 A, μ 6.7, 50 W.

HK-155 Gridless "Gammatron," fil. 5 V @ 10 A, μ 2, 150 W. Also see HK55.

HK-158 Xmtg. triode, UX base w/ cap, grid-out-side, fil. 12.6 V @ 2.5 A, μ 25, 50 W.

HK-252L [AKA 152TL, q. v.]; xmtg. triode.

HK-253 [AKA 217C]; xmtg. vacuum rectifier, T20 bulb w/ cap, jumbo 4-pin base, T-T fil. 5 V @ 10 A, 15 kV PIV, 350 mA (HK, EI).

HK-253A Xmtg. vacuum rectifier, fil. 5 V @ 10 A, 10 kV PIV, large 4-pin base.

HK-253B HK-253A on industrial 4-pin base.

HK-253C Xmtg. vacuum rectifier.

HK-254 Xmtg. triode, hard-glass bulb w/ plate pin, grid-out-side, fil. 5 V @ 7.5 A, μ 25, 100 W (sim. to 100TH w/ slightly less fil. curr. and different bias; also sim. to RK-38 and 4C22/HF-100). Intro. 5-38. LK version, 1952, had Zr-coated plate, uprated to 125 W. Made by PL as PL-254W.

HK-255 Gridless "Gammatron," fil. 14 V @ 30 A, μ 3, 500 W. Also see HK55.

HK-257 [AKA 4E27, q. v.]; xmtg. pentode, intro. 4-41.

HK-257B [AKA 4E27A, q. v.]; xmtg. pentode.

304H [AKA 304TH]; xmtg. triode; orig. tall version and shorter, ruggedized variant. Had plate pin, not cap, and ring around bulb to hold grid cap.

HK-304L [AKA HK304TL, q. v.].

HK-353 Xmtg. mercury rectifier, fil. 1.25 V @ 13 A, 1 kV PIV, 2.5 A.

HK-354 ( ) Family of xmtg. triodes, T24 tubular bulb 8-¼" high w/ cap, 4-pin jumbo base, grid out base, fil. 5 V @ 10 A, 150 W, 50 MHz; FCC-listed for 250-W AM xmtrs. HK-354 has μ 14. "A" has μ 9.5. "C" is HF version of HK-354, same electricals except lower capacitance. "D" has μ 22. "E" has μ 35 (very sim. to 250TH). "F" has μ 50.

HK-454A, C Xmtg. triode, 250 W.

HK-454H, L [AKA 250TH, TL, q. v.]; xmtg. triodes.

HK-645 Xmtg. triode.

HK-653A Xmtg. mercury rectifier, fil. 5 V @ 6.5 A, 10 kV PIV, output 1.75 A, giant 4-pin base.

HK-653B HK-653A on industrial 4-pin base.

HK-654 Xmtg. triode, 4-pin jumbo base w/ cap., grid-out-side, tantalum grid, fil. 7.5 V @ 15 A, μ 22, 300 W, 50 MHz, intro. 1938.

HK-753A, B Xmtg. mercury rectifier, 5 V @ 6.5 A, 10 kV PIV, 2.5 A. "A" is on 4-pin giant base; "B" is on 4-pin industrial base.

HK-854H, L [AKA 450TH, TL, q. v.]; xmtg triodes, 125 MHz.

HK-953B [AKA 4B33, q. v.]; rectifier.

HK-953D [AKA 5B33, q. v.]; rectifier.

HK-953E [AKA 7B33, q. v.]; rectifier.

HK-1054L [AKA 750TL, q. v.]; xmtg. triode.

HK-1554 Xmtg. triode, T72 stepped tubular bulb w/ cap, 18" high, grid-out-side, fil. 11 V @ 17.5 A, μ 14.5, 750 W (1500 W w/ forced air), 30 MHz; FCC-listed for 1-kW AM xmtrs (10-20-43).

HK-2054A [AKA (WE) 279A]; xmtg. triode, fil. 10 V @ 22 A, μ 10, 1.2 kW, 20 MHz.

HK-3054 Xmtg. triode, T72 stepped tubular bulb w/ cap, 31" high, fil. 14 V @ 45 A, μ 20, 1.5 kW (3 kW w/ forced air), 30 MHz, intro. 1936; FCC-listed for 2.5-kW AM xmtrs (10-20-43).

HYTRON TYPES

{HD10} [AKA 2C26, q. v.]; xmtg. triode.

{HD25} [AKA 2E25, q. v.]; xmtg. tetrode.

{HD27} [AKA 6AL5, q. v.]; duotriode.

{HD28} [AKA 6AQ6, q. v.]; duopdiode-triode.

{HD30} [AKA 3B4, q. v.].

{HD34} Min. beam tetrode, pulse modulator like min. 3D21, heater 6.3 V @ 540 mA, 1 kV, 700-mA pulses (dev. type).

{HD50} [AKA 3C30, q. v.]; xmtg. triode.

{HD51} [AKA 0A2, q. v.]; voltage regulator.

{HD52} [AKA 0B2, q. v.]; voltage regulator.

{HD57} [AKA 5516, q. v.]; beam tetrode.

{HD59} [AKA 2E30, q. v.]; beam pentode.

{HD69} [AKA 5514, q. v.]; power triode.

HY24 [AKA RK-24, VT-67, 30 Spl]; triode, VHF xmtg. version of 30, w/ ceramic UX base, fil. 2 V @ 120 mA (double current), μ 8, 1.5 W. On the market in 1934, about the time the BC-222/223 "walkie-talkie" was being developed; this tube's greatest user.

HY25 Xmtg. triode, ST bulb w/ cap, UX base, fil. 7.5 V @ 2.25 A, μ 55, 25 W.

(HY, NU) **30Z** Xmtg. triode, T-T fil. 6.3 V @ 2.25 A, μ 87, 30 W, for zero-bias use, sim. (not identical) to 809.

(HY, NU) **31Z** Xmtg. duo-triode, ceramic UX base and dual plate caps, quick-heating T-T fil., 6.3 V @ 2.6 A, μ 45, 30 W, for zero-bias use; roughly similar to 2C34. Variant: HY1231Z.

(HY, NU) **40**, Z [AKA NU40T]; xmtg. triode, ST bulb w/ cap, ceramic UX base, fil. 7.5 V @ 2.25 A, 40 W. "Z" is high-μ variant for zero-bias use, μ 80 (sim. to NU40TZ).

(HY, NU) **51A**, B, X, Z [AKA RK-51]; xmtg. triode, ST bulb w/ cap, ceramic UX base, T-T fil. 7.5 V @ 3.75 A, μ 25, 65 W. "B" is HY-51A w/ fil. 10.0 V @ 3.25 A. "Z" is HY-51A for zero-bias use, μ raised to 85.

HY57 Xmtg. triode, ST bulb w/ cap, ceramic UX base, graphite anode, fil. 6.3 V @ 2.25 A, μ 50, 40 W.

HY60 Beam tetrode, RK-56 uprated to 15 W.

HY61 [AKA 807, q. v.]; beam tetrode, ceramic base.

HY63 Xmtg. tetrode, octal base w/ cap, fil. 3.0 V CT @ 112.5 mA, 3 W.

HY64 Xmtg. beam tetrode, octal w/ cap, fil. 6.3 V @ 850 mA, 15 W.,

HY65 Xmtg. tetrode, quick-heating tubular octal w/ cap, T-T fil. 6.0 V @ 800 mA, 15 W, to repl. 6V6 and 6L6, repl. by 2E25, intro. 8-41. Also see 5516.

HY67 Xmtg. beam tetrode, 811 format w/ cap, graphite anode, two T-T fils., 6.0 V @ 2.15 A each, 65 W, intro. 8-41.

(HY, NU) **69** [AKA NU69, RK-66, q. v.]; beam tetrode. Variant: HY1269.

HY75 [AKA NU-75H]; xmtg. VHF triode, ceramic octal base and dual caps, T-T fil. 6.3 V @ 2.5 A, μ 10, 15 W.

HY113 [AKA HY123, q. v.]; rcvg. min. ("bantam junior") VHF triode, min. style tipped bulb w/ small 5-pin base, fil. 1.4 V @ 70 mA, μ 6.3.

HY114A, B [AKA 2C24, VT-234, CV3505]; VHF triode (HY, NU). In CPR-60AAC.

HY115 [AKA VT-281]; audio pentode, HY113 format (q. v.), fil. 1.4 V @ 70 mA, B+ 90 V max.; VT-no. cancelled. Also see HY145.

HY123 [AKA HY113, q. v.]; min. triode.

HY125 [AKA HY155, q. v.]; audio power pentode, HY113 format, fil. 1.4 V @ 70 mA, g_m 0.31 mS, max. B+ 90 V.

HY145 [AKA HY115, q. v.]; audio pentode.

HY150 [AKA 211H]; RF-heating triode (sim. to 311CH).

HY145ZT [AKA 2E28, q. v.]; min. pentode.

HY155 [AKA HY125, q. v.]; power pentode.

HY165 Hearing-aid power pentode, fil. 1.4 V @ 50 mA, g_m 0.45 mS (1940).

HY205 Hearing-aid pentode.

HY245 Hearing-aid pentode, T4 flat-press bulb, fil. 1.25 V @ 28 mA, g_m 0.375 mS, w/ flying leads but shipped w/ octal plug for testing (1940).

HY255, X Hearing-aid power pentode, HY245 format, same fil., 10-18 mW out.

(HY, NU) **615** [AKA VT-235, CV3506]; xmtg-rcvg. VHF triode, variant of HY114/2C24 w/ heater 6.3 V @ 150 mA and uprated to 3.5 W; same as E1148 except for cutoff char. (sim. to Br. VR135). In BC-647A and CPR-60AAK.

HY1231Z Xmtg. duotriode, HY31Z w/ T-T fil., 12.6 V CT @ 1.6 A.

HY1269 Xmtg. beam tetrode, HY69 w/ fil. 12.6 V CT @ 1.7 A.

MACHLETT TYPES

ML-1E Rectifier, "medical and surgical."

ML-1K X-ray tube.

ML-3W Rectifier, "medical and surgical."

ML-20 Vacuum rectifier for X-ray use, ca. 1932.

ML-100 [AKA 5575, q. v.].

ML-101 [AKA GL-605, q. v.].

ML-102A Vacuum rectifier, fil. 20 V @ 19 A, 75 kV PIV, 750 mA peak, convection-cooled.

ML-103 Vacuum rectifier, fil. 10 V @ 11.5 A, 125 kV PIV, 78 mA peak, oil-immersed.

ML-108 Vacuum rectifier, fil. 13 V @ 12.5 A, 140 kV PIV, 200 mA peak, oil-immersed.

ML-110 Vacuum rectifier, fil. 10 V @ 11.5 A, 140 kV PIV, 28 mA peak, convection-cooled.

ML-115 Vacuum rectifier, fil. 10 V @ 11.5 A, 125 kV PIV, 100 mA peak, convection-cooled.

ML-120 Vacuum rectifier, fil. 13 V @ 12.5 A, 140 kV PIV, 200 mA peak, convection-cooled.

ML-121 Vacuum rectifier, fil. 10 V @ 11.5 A, 140 kV PIV, 100 mA peak, oil-immersed.

ML-126 Vacuum rectifier, fil. 13 V @ 12.5 A, 150 kV PIV, 200 mA peak, convection-cooled.
ML-141 Vacuum rectifier, fil. 5.5 V @ 6.5 A, 125 kV, 750 mA peak, oil-immersed.
ML-142 Vacuum rectifier, fil. 3.8 V @ 6.6 A, 100 kV, 250 mA peak, oil-immersed.
ML-148 Vacuum rectifier, fil. 5.7 V @ 6.6 A, 150 kV, 1 A peak, oil-immersed.
ML-170 Vacuum rectifier, fil. 13 V @ 12.5 A, 200 kV, 200 mA peak.
ML-180 Vacuum rectifier, fil. 13 V @ 12.5 A, 200 kV, 200 mA peak, oil-immersed.
ML-199 Vacuum rectifier, fil. 12 V @ 23 A, 110 kV, 10 A peak, oil-immersed.
ML-200 [AKA 5576, q. v.].
ML-308, R [AKA 5668, 5669, q. v.].
ML-309 [AKA 5666, q. v.].
ML-309R [AKA 5667, q. v.].
ML-322 Ceramic-metal VHF power diode, appears to be gridless 2C39, used in TR switches and VOR antenna systems; heater 6.3 V @ 0.9-1.0 A, 800 V PIV, 125 mA, 100 W diss. (EI, ML)
ML-354 Xmtg. triode, glass ring-seal construction, for G-G amps. and RF heaters, T-T fil. 12.5 V @ 220 A, μ 25, 50 kW, 110 MHz (75 kW @ 50 MHz), water-cooled via external cooler (1949).
ML-356 [AKA 5771, 7085, q. v.]
ML-381 Variant of 7289, rated for pulse use, glass or ceramic (EI, ML).
ML-394 See 5745 and 5747.
ML-560 HF xmtg. triode, stepped tuubular bulb, electrodes outside via stems, 500 W, 50 MHz.
ML-728 [AKA 5557, q. v.].
ML-23165 No further information.
ML-23185 No further information.
ML-23931 No further information.
ML-23646 No further information.

SPERRY KLYSTRONS

SRC2 [AKA 2K33, q. v.]; klystron.
SRL7C Xmtg. klystron, tunes 1850-2160 MHz, for radio-relay use, octal base, CO, 1 kV, 10 W out.
SAC9G Xmtg. klystron, tunes 4790-5034 MHz, octal base, 9 W out, 1 kV, CO.
SMC11 Klystron freq. multiplier, CO, 5 GHz output from 832 MHz input, 1 kV, 400 mW out, to feed SMX-32, 1949.
SRC12 [AKA 5817, q. v.]; klystron.
SRC13 [AKA 5818, q. v.]; klystron.
SRX-16 Reflex klystron.
SRL17 Reflex klystron, tunes 750-990 MHz, 1 kV, 5 W out.
SRX18 Reflex klystron.
SAC19 Xmtg. klystron, "synchrodyne" heterodyne mixer for 6.0-6.3 GHz, two internal cavities, octal base, WG in for local osc. and RF out, IF signal applied to electrodes, 625 V, 1-1.7 W out, forced-air cooled, for radio-relay use, dev. for Philco TLR-2 system.
SAX22 Klystron.
SAK23 Klystron.
SRC24 Klystron.
SOS25 Klystron (diathermy).
SAC26 Klystron.
SMS27 Klystron.
SAS28 Xmtg. klystron, three internal cavities, tunes 2.6-2.8 GHz, flying leads, WG, 4 kV, 225 W out, 30 dB gain
SAC29 Klystron.
SAX30 Klystron.
SMX32 Klystron freq. doubler, 5 GHz in from SMC-11, octal base, coaxial input, WG output, 1 kV, 10 GHz CW output @ 1 W, 1949.
SAC33 Xmtg. klystron, three cavities, tunes 4.8-5.3 GHz, IM, WG, flying leads, 5.2 kV, 500 W out, forced-air cooled, beam focused by permanent magnet.
SMS34 Klystron osc.
SAC35 Klystron osc.
SRC36 Klystron.
SAL39 Xmtg. pulsed klystron, tunes 960-1215 MHz, CO, heater on pins, 20 kV, 30 kW pulse out, forced-air cooled.
SAC41 Xmtg. klystron, tunes 3.7-4.2 GHz, dual-cavity, octal base, WG, 750 V, 10 W out, forced-air cooled. In 25 stations of Western Union New York - Pittsburgh-Washington - New York microwave route, ca. 1952.
SRU55 Reflex klystron, 14-17.5 MHz, octal base, heater 6.3 V @ 600 mA, 60 mW output, 300 V, UG-419/U WG out.
(SP) **411** [AKA 2K31, q. v.].

TAYLOR TYPES

T-20 Xmtg. triode, general-purpose, "S" bulb w/ cap, Isolantite 4-pin base w/ cap, fil. 7.5 V @ 1.75 A, μ 20, 20 W, 56 MHz. Variant: TZ-20.
TUF-20 Xmtg. VHF triode, tall GT octal w/ grid and plate caps, T-T fil. 6.3 V @ 2.75 A, μ 10, 20 W, 250 MHz (1945).
TZ-20 Xmtg. triode, variant of T20 w/ μ 62, primarily for Class B modulators.
T-21 [AKA RK-49]; beam power tube; 6L6G bulb on 6-pin ceramic base, diss. uprated to 21 W.
TB-35 [AKA 3D23]; xmtg. beam tetrode.
T-40 [AKA NU40T]; xmtg. triode, tubular bulb w/ cap, UX base, fil. 7.5 V @ 2.5 A, μ 25, 40 W (1938); variant: TZ-40.
TR-40M Vacuum rectifier, GT27 hard-class bulb 9-7/8" high, 4-pin jumbo base, T-T fil. 5.0 V @ 10.5 A, 60 kV, 250 mA; replaceable by 250-R.

TZ-40 [AKA NU40TZ]; xmtg. triode, "zero-bias" version of T-40 w/ μ 62 (nearest equiv. Brit. VT76 and DA41).

T-55, Z Xmtg. triode sim. to V70D; hard-glass bulb, "S" bulb w/ cap, UX base, graphite anode, fil. 7.5 V @ 3 A, μ 20, 55 W. "Z" is high-μ zero-bias variant.

T-60 RF-heating type.

R-110 RF-heating type.

T-124 HF RF-heating triode.

T-125 RF-heating/xmtg. triode, hard-glass bulb, 4-pin jumbo base w/ cap, grid-out-side, graphite anode, fil. 10 V @ 4.5 A, μ 25, 125 W, 30 MHz (sim. to FV20), replacement for 203A.

T-150 Xmtg. triode.

T-155 Xmtg. triode, tubular bulb w/ cap, grid-out-side, large 4-pin base, fil. 10 V @ 4 A, μ 20, 155 W, 150 MHz.

T-200 Xmtg. and RF-heating VHF triode, tubular hard-glass bulb w/ cap, large 4-pin base, grid-out-side, fil. 10 V @ 5.75 A, μ 16, 200 W. Replaceable by HV-18; sim. to HF-300.

HD-203A, C Xmtg. triode, 203A w/ cap, grid-out-side, fil. 4.0 A, μ 25, uprated to 150 W. "C" is RF-heating variant, μ 20.

203B Class B mod., 10 V @ 3.25 A, superseded by 203D and F.

T-203Z Xmtg. audio triode, "zero-bias" variant of 203, hard-glass bulb w/ cap, fil. 10 V @ 3.25 A, μ 85.

207 Vacuum rectifier for electric-fence controllers; ST-16 clear bulb on 4-pin base; fil. 2.5 V @ 2.5 A, 1250 V in, 125 mA (1944). Also see 208.

208 Glow-discharge tube for electric-fence controllers; ST-16 clear bulb on 4-pin base; drops 875-950 v @ 8 mA; works w/ 207 (1944).

HD-211C RF-heating triode, 211C uprated to 150 W, plate out top. Replaceable by 311CH (UE).

T-282 [AKA WE 282A, q. v.]; xmtg. tetrode as made by Taylor.

(T, WL) **756** Xmtg. triode, "amateur type," general-purpose, Isolantite UX base, fil. 7.5 V @ 2 A, μ 25, 40 W.

VICTOREEN TYPES

VW-21 Submin. diode, fil. 1.25 V @ 15 mA.

VW-32 Submin. low-μ triode.

VX-10 Submin. electrometer switch (1952).

VXG-11 [AKA 1B87, q. v.]; counter.

VX-21 [AKA 5799, q. v.]; submin. diode.

VX-32, -B [AKA 5802, q. v.]; submin. triode.

VX-33, A [AKA 5801, q. v.]; submin. triode.

VX-41, A, C [AKA 5800, q. v.], submin. electrometer tetrode. "C" is "A" in plastic sleeve.

VX-53 Submin. pentode.

VX-55 [AKA 5803, q. v.]; electrometer triode.

VX-68 High-voltage rectifier, 27 kV PIV.

VX-124 [AKA 5800, q. v.]; electrometer tetrode.

VXR-130 [AKA 6213, q. v.]; voltage regulator.

VXR-1000 Voltage regulator, 1 kV.

VXR-1500 Voltage regulator, 1.5 kV.

WELTRONIC TYPES

WT-210-**0001** [AKA 2D21, q. v.]; min. thyratron.

WT-210-**0004** [AKA 2050, q. v.]; thyratron.

WT-210-**0008** [AKA 866A, q. v.]; rectifier.

WT-210-**0015** [AKA 5557, q. v.]; thyratron.

WT-210-**0016** [AKA 0B3, q. v.]; voltage regulator.

WT-210-**0017** [AKA 6856, q. v.]; thyratron.

WT-210-**0018** [AKA 0D3, q. v.]; voltage regulator.

WT-210-**0027** [AKA 872A, q. v.]; rectifier.

WT-210-**0038** [AKA FG-172, q. v.]; thyratron.

WT-210-**0043** [AKA 5796, q. v.]; thyratron.

WT-210-**0044** [AKA 575A, q. v.]; rectifier.

WT-210-**0045** [AKA 892, q. v.]; xmtg. triode.

WT-210-**0047** [AKA 3C23, q. v.]; thyratron.

WT-210-**0051** [AKA 0A2, q. v.]; voltage regulator.

WT-210-**0054** [AKA 5830, q. v.]; thyratron.

WT-210-**0056** [AKA 5559, q. v.]; thyratron.

WT-210-**0057** [AKA WL-632B, q. v.]; thyratron.

WT-210-**0058** [AKA KU-676, q. v.]; thyratron.

WT-210-**0059** [AKA WL-677, q. v.]; thyratron.

WT-210-**0062** [AKA 5557, q. v.]; thyratron.

WT-210-**0063** [AKA 5685, q. v.]; thyratron.

WT-210-**0066** [AKA 5665, q. v.]; thyratron.

WT-210-**0069** [AKA 5557, q. v.]; thyratron.

WT-210-**0070** [AKA 5550, q. v.]; ignitron.

WT-210-**0071** [AKA 5551A, q. v.]; ignitron.

WT-210-**0072** [AKA 5552A, q. v.]; ignitron.

WT-210-**0073** [AKA 5553B, q. v.]; ignitron.

WT-210-**0074** [AKA FG-105, q. v.]; thyratron.

WT-210-**0075** [AKA 5822A, q. v.]; ignitron.

WT-210-**0077** [AKA 5727, q. v.]; thyratron.

WT-210-**0078** [AKA FG-172, q. v.]; thyratron.

WT-210-**0079** [AKA FG-105, q. v.]; thyratron.

WT-210-**0086** [AKA 833A, q. v.]; xmtg. triode.

WT-210-**0093** [AKA 5668, q. v.]; xmtg. triode.

WT-210-**0100** [AKA 672A, q. v.]; rectifier.

WT-210-**0106** [AKA 7306, q. v.]; thyratron.

WT-210-**0116** [AKA WL-632B, q. v.]; thyratron.

WT-210-**0120** [AKA KU-610, q. v.]; thyratron.

WT-210-**0134** [AKA WL-678, q. v.]; thyratron.

WT-210-**0137** [AKA 502A, q. v.]; thyratron.

WT-210-**0141** [AKA 5796, q. v.]; thyratron.

WT-210-**0149** [AKA 5551A, q. v.]; ignitron.

WT-210-**0156** [AKA 5551A, q. v.]; ignitron.

WT-210-**0157** [AKA 5552A, q. v.]; ignitron.

WT-210-**0179** [AKA C5H-1, q. v.]; thyratron.

WT-210-**0181** [AKA C6H-3, q. v.]; thyratron.

MILITARY TYPES

1. ARMY VT-NUMBERS

The U. S. Army assigned "VT" numbers to its radio tubes from 1916/17 to 1942. Conversion tables to list the civilian equivalents have been around, one place or another, for generations. Most collector-restorers are content to see that the VT-86 required by a freshly acquired BC-348 receiver, for example, "is" a 6K7. However, the usual list talks guardedly of types like "Special Tube," "K-7," "441," or "C7063," without giving an idea of what these really were. This is not surprising, considering that such tubes were "classified" at the time. This section attempts to de-mystify these odd types, adding a few types to the usual lists (VT-15, -208A, -219), correcting a couple of listings (VT-67, -251), and so on.

There is insight to be gained from looking at the growth of the VT-list. From a fast start during WW I, the growth of new types lagged badly during the '20s and early '30s, a reflection of the starvation budget that the Signal Corps faced until the late '30s. Then, in the 1940-42 period, the list mushroomed almost out of control. The cure was the Joint Army-Navy (JAN) designation system, which basically used RMA type numbers. That was the end of new VT- and Navy numbers.

Most of the VT-tubes were off-the-shelf commercial types. However, the first ones and many later types were unique developments for such uses as radar, where civilian technology was unavailable.

As an index of military cooperation during WW II, the British assigned CV (Common Valve) numbers to the VT-4C, -17, -19, -22, and almost all the other VT-numbered tubes then active. This corresponds to the furnishing of large amounts of U. S. gear to Britain under the Lend-Lease program.

Coonversely, a few VTs were supplied primarily to Britain: the VT-232/E1148 (Hytron) and VT-248/EF50 (Sylvania) were British designs having little use in U. S. equipment, but made in large numbers for export. The VT-240/710A was a British design adapted by Bell Labs for quantity production in the U. S. and Canada, and used in REL Canadian-built radars supplied to the U. S. and Commonwealth nations.

One early tube didn't make the VT-list: while the Army was giving out nomenclature like Battery BA-3, Hammer HM-1, Reel RL-2, etc., it coded a Tube TB-1. This was a brass-based, tipped vacuum diode, used as an electronic voltage regulator in Generator GN-1-A, the wind-driven power source for the SCR-68 radiophone set for fighter planes. The TB-1 appeared after the VT-1 and -2 were in common use, so it is a mystery why "TB" was devised. In any case, there was no other "TB."

Another mystery: numerous later tubes never received VT-numbers, even though used widely in Army gear designed in the heyday of the VT-system. For example, the bread-and-butter 6A7 was used, with VT-76 and -77, in the BC-221A frequency meter, ca. 1936; but had no VT-number. Other examples: 1A3, 1L4, 3A4, 6C8, 6K5GT, 6N7GT, 6SH7, 7A8, 14AF7, 83V, and a large group of transmitting and WE types.

There is opportunity for confusion among VT-numbers issued by the Signal Corps, Royal Air Force, and British Post Office. These are unrelated! "Our" 807 was a VT-100 to the U. S. Army, a VT60 to the RAF, and a VT199 to the Post Office.

VT-1 [AKA WE 203A, q. v.].

VT-2 [AKA CW-931, WE 205B, q. v.]; "5-watt" xmtg. triode. Duovac also made some VT-2s, ca. 1932, in short ST bulbs. Unlike the WE product of 13 years earlier, they had fixed-voltage, not fixed-current, filaments.

VT-3 WE "P" rcvg. triode, WW I.

VT-4A [AKA WE 211A, q. v.].

VT-4B [AKA UV-203A, WE 211D, q. v.], 1922 vintage.

VT-4C [AKA CV620, 211 Spl, q. v.].

VT-5 [AKA WE 215A, q. v.]; rcvg. triode, VT-3 in "peanut" bulb.

VT-6 [AKA WE 212A, q. v.].

VT-7 [AKA WX-12, q. v.].

VT-8 [AKA UV-204, q. v.].

VT-10 Xmtg. triode, GE "P," WW I.

VT-11 Rcvg. triode, GE "G", WW I, GE version of VT-1, to interchange w/ it, fil. 4 V.

VT-12 Xmtg. triode, GE "T", 15 W, WW I, to replace VT-2.

VT-13 [AKA CG-890 (later version)]; ruggedized VT-11, prototype of UV-201.

VT-14 [AKA CG-1162]; xmtg. triode, GE "T," uprated VT-12, 1.74-A fil., prototype of UV-202.

VT-15 Experimental small rcvg. triode, post-WW I, low-current rcvg. type on "duckfoot" spade base.

VT-16 Xmtg. triode, WW I, GE "T" ruggedized.

VT-17 [AKA CV640, 860, q. v.].

VT-18 [AKA CG-1144]; xmtg. triode, GE "U", prototype of UV-203.

VT-19 [AKA CV641, 861, q. v.].

VT-20 Rcvg. triode, De F, WW I.

VT-21 Rcvg. triode, De F, WW I.

VT-22 [AKA CV2563, 204A, q. v.].

VT-23 Unknown - "obsol." early.

VT-24 [AKA CV2675, 864, q. v.].

VT-25 [AKA CV603, 10, q. v.].

VT-25A [AKA 2C25, 10Y, q. v.].

VT-26 [AKA 22, q. v.].

VT-27 [AKA CV604, 30, q. v.].

VT-28 [AKA CV936, 24 (A), q. v.].

VT-29 [AKA CV944, 27, q. v.].

VT-30 [AKA CV750, 01A, q. v.].

VT-31 [AKA CV931, 31, q. v.].

VT-32 Moorhead Labs WW I export item to Britain ("VT32"), on Br. 4-pin base. Not a Signal Corps procurement as such, but the number assignment was apparently protected.

VT-33 [AKA CV949, 33, q. v.].

VT-34 [AKA 207, q. v.].

VT-35 [AKA CV1752, 35/51, q. v.].

VT-36 [AKA CV1775, 36, q. v.].

VT-37 [AKA CV606, 37, q. v.].

VT-38 [AKA CV712, 38, q. v.].

VT-39 [AKA 869, q. v.].

VT-39A [AKA CV2723, 869A, q. v.].

VT-40 [AKA CV2501, 40, q. v.].

VT-41 [AKA CV2671, 851, q. v.].

VT-42 [AKA CV642, 872, q. v.].

U. S. ARMY
SIGNAL CORPS
Transmitting Vacuum Tube
VT - - 2

Radio Frequency Output 7.5 watts
Filament Potential - 7.0 volts
Filament Current - 1.35 amperes
D. C. Plate Potential - 350 volts
Maximum Safe Plate Dissipation - 15 watts
Amplification Factor - - - 8.0

Manufactured by
Duovac Radio Tube Corp.
BROOKLYN, N. Y.

Not all VT-2s were from WE.

VT-42A [AKA 872A (Spl fil.), q. v.].
VT-43 [AKA 845, q. v.].
VT-44 [AKA CV711, 32e, q. v.].
VT-45 [AKA CV596, 45, q. v.].
VT-46 [AKA CV32, 866, q. v.].
VT-46A [AKA 866A, q. v.].
VT-47 [AKA CV1772, 47, q. v.].
VT-48 [AKA CV608, 41, q. v.].
VT-49 [AKA CV1771, 39/44, q. v.].
VT-50 [AKA CV2533, 50, q. v.].
VT-51 [AKA 841, q. v.].
VT-52 [AKA 45 Spl, 2C45, q. v.].
VT-53 (Cancelled; was to be WL-436, repl. by VT-42A).
VT-54 [AKA CV1751, 34, q. v.].
VT-55 [AKA CV2676, 865, q. v.].
VT-56 [AKA CV611, 56, q. v.].
VT-57 [AKA CV612, 57, q. v.].
VT-58 [AKA CV613, 58, q. v.].
VT-59 [AKA CV2538, 59, q. v.].
VT-60 [AKA 850, q. v.].
VT-62 [AKA CV621, 801 (A), q. v.].
VT-63 [AKA CV2531, 46, q. v.].
VT-64 [AKA CV2657, 800, q. v.].
VT-65 [AKA CV582, 6C5, q. v.].
VT-65A [AKA CV581, 6C5G, q. v.].
VT-66 [AKA CV1186, 6F6, q. v.].
VT-66A [AKA CV1911, 6F6G, q. v.].
VT-67 ["30 Spl," AKA HY-24, q. v.].
VT-68 [AKA CV1891, 6B7, q. v.].
VT-69 [AKA CV1900, 6D6, q. v.].
VT-70 [AKA CV1915, 6F7, q. v.].
VT-72 [AKA 842, q. v.].
VT-73 [AKA CV639, 843, q. v.].
VT-74 [AKA CV1864, 5Z4, q. v.].
VT-75 [AKA CV614, 75, q. v.].
VT-76 [AKA CV615, 76, q. v.].
VT-77 [AKA CV616, 77, q. v.].
VT-78 [AKA CV2544, 78, q. v.].
VT-80 [AKA CV617, 80, q. v.].
VT-83 [AKA CV1773, 83, q. v.].
VT-84 [AKA CV619, 84/6Z4, q. v.].
VT-86 [AKA CV1942, 6K7, q. v.].
VT-86A [AKA CV1941, 6K7G, q. v.].
VT-86B [AKA CV1943, 6K7GT, q. v.].
VT-87 [AKA CV1951, 6L7, q. v.].
VT-87A [AKA CV1951, 6L7G, q. v.].
VT-88 [AKA CV1963, 6R7, q. v.].
VT-88A [AKA CV1962, 6R7G, q. v.].
VT-88B [AKA CV1964, 6R7GT, q. v.].
VT-89 [AKA CV833, 89, q. v.].
VT-90 [AKA CV1301, 6H6, q. v.].
VT-90A [AKA CV1931, 6H6GT, q. v.].
VT-91 [AKA CV1074, 6J7, q. v.].
VT-91A [AKA CV1937, 6J7GT, q. v.].
VT-92 [AKA CV588, 6Q7, q. v.].
VT-92A [AKA CV589, 6Q7G, q. v.].
VT-93 [AKA CV1894, 6B8, q. v.].
VT-93A [AKA CV1893, 6B8G, q. v.].
VT-94 [AKA CV1067, 6J5, q. v.].
VT-94A [AKA CV1932, 6J5G, q. v.].
VT-94B [AKA 6J5 Selected, q. v.].
VT-94C [AKA 6J5G Selected, q. v.].
VT-94D [AKA CV1934, 6J5GT, q. v.].
VT-95 [AKA CV1831, 2A3, q. v.].
VT-96 [AKA CV1957, 6N7, q. v.].
VT-96B [AKA 6N7 Selected, q. v.].
VT-97 [AKA CV1849, 5W4, q. v.].
VT-98 [AKA CV504, 6U5/6G5, q. v.].
VT-99 [AKA CV1917, 6F8G, q. v.].
VT-100 [AKA CV124, 807, q. v.].
VT-100A [AKA 807 modified].
VT-101 [AKA CV637. 837, q. v.].
VT-102 (Cancelled no.)
VT-103 [AKA CV1990, 6SQ7, q. v.].
VT-104 [AKA CV546, 12SQ7, q. v.].
VT-105 [AKA CV1969, 6SC7, q. v.].
VT-106 [AKA CV623, 803, q. v.].
VT-107 [AKA CV510, 6V6, q. v.].
VT-107A [AKA CV511, 6V6GT, q. v.].
VT-107B [AKA CV509, 6V6G, q. v.].
VT-108 [AKA 450TH, q. v.].
VT-109 [AKA CV1798, 2051, q. v.].
VT-111 [AKA CV836, 5BP4, q. v.].
VT-112 [AKA CV660, 6AC7, q. v.].
VT-114 [AKA 5T4, q. v.].
VT-115 [AKA CV208, 6L6, q. v.].
VT-115A [AKA CV1947, 6L6G, q. v.].
VT-116 [AKA CV70, 6SJ7, q. v.].
VT-116A [AKA CV590, 6SJ7GT, q. v.].
VT-116B [AKA CV866, 6SJ7Y, q. v.].
VT-117 [AKA CV1981, 6SK7, q. v.].
VT-117A [AKA CV1982, 6SK7GT, q. v.].
VT-118 [AKA 832, q. v.].
VT-119 [AKA CV248, 2X2, q. v.].
VT-120 [AKA CV1095, 954, q. v.].
VT-121 [AKA CV1059, 955, q. v.].
VT-122 [AKA WL-530, q. v.].
VT-123 [AKA (RCA) {A-5586}, q. v.]. Also see VT-128.
VT-124 [AKA CV756, 1A5GT, q. v.].
VT-125 [AKA CV1805, 1C5GT, q. v.].
VT-126 [AKA CV279, 6X5, q. v.].
VT-126A [AKA CV572, 6X5G, q. v.].
VT-126B [AKA CV574, 6X5GT, q. v.].
VT-127 [AKA 100TS, q. v.].
VT-127A [AKA 100TS, q. v., WL-534, 3-100D2]; VHF xmtg. triode; 100TL in tubular all-glass bulb w/ two plate pins out sides at top; two grid pins out sides at center; and two fil. fins out bottom; pulse-rated, at higher freqs. than 100TL; w/ bigger fil. (*QST*, Nov. 1946). Variants: Eimac 53A, 227A, 327A, and 327B.
VT-128 ["A5588A," AKA CV2715, 1630, q. v.].
VT-129 [AKA 304TL, q. v.].
VT-130 [AKA 250TL, q. v.].
VT-131 [AKA CV543, 12SK7, q. v.].
VT-132 [AKA CV24, 12K8Y, q. v.].
VT-133 [AKA CV700, 12SR7, q. v.].
VT-134 [AKA CV525, 12A6, q. v.].
VT-135 [AKA CV535, 12J5GT, q. v.].
VT-135A [AKA CV534, 12J5, q. v.].
VT-136 [AKA CV659, 1625, q. v.].
VT-137 [AKA CV1755, 1626, q. v.].
VT-138 [AKA CV1756, 1629, q. v.].
VT-139 [AKA CV216, 0D3, q. v.].
VT-140 [AKA 1628, q. v.] (repl. by VT-228).
VT-141 [AKA WL-531, q. v.].
VT-142 [AKA Bell Labs 39DY1, q. v.]. Experimental 200-MHz radar pulse osc.; never produced.
VT-143 [AKA CV625, 805, q. v.].
VT-144 [AKA CV26, 813, q. v.].
VT-145 [AKA CV275, 5Z3, q. v.].
VT-146 [AKA CV37, 1N5GT, q. v.].
VT-147 [AKA CV2, 1A7GT, q. v.].
VT-148 [AKA CV1811, 1D8GT, q. v.].

VT-149 [AKA 3A8GT, q. v.].
VT-150 [AKA CV1966, 6SA7, q. v.].
VT-150A [AKA CV1967, 6SA7GT, q. v.].
VT-151 [AKA CV16, 6A8G, q. v.;].
VT-151B [AKA CV580, 6A8GT, q. v.].
VT-152 [AKA CV207, 6K6GT, q. v.].
VT-152A [AKA CV1938, 6K6G, q. v.].
VT-153 [AKA CV128, 12C8Y, q. v.].
VT-154 [AKA CV629, 814, q. v.].
VT-155 to 157 "Signal Corps Labs Special" (Zahl/VT-155 variants?); no known use or CV no.; never produced?
VT-158 ["Special"]; Zahl packaged-in-glass 600-MHz 250-kW oscillator used in AN/TPS-3 radar {WX3092} (EI, ML, HK). Also see 1636.
VT-159, -160 "Signal Corps Labs Special." See VT-155.
VT-161 [AKA CV537, 12SA7, q. v.].
VT-162 [AKA CV697, 12SJ7, q. v.].
VT-163 [AKA CV1896, 6C8G, q. v.].
VT-164 [AKA CV723, 1619, q. v.].
VT-165 [AKA 1624, q. v.].
VT-166 [AKA (WE) 371A, q. v.].
VT-167 [AKA CV18, 6K8, q. v.].
VT-167A [AKA CV1944, 6K8G, q. v.].
VT-168A [AKA CV205, 6Y6G, q. v.].
VT-169 [AKA CV128, 12C8, q. v.] (a "VT-153A"!)
VT-170 [AKA CV45, 1E5GP, q. v.].
VT-171 [AKA CV782, 1R5, q. v.].
VT-171A "Loctal version of 1R5" - apparently abandoned; no CV no. or known use. Probably was 1LC6, which was on 1944 JAN "preferred" list.
VT-172 [AKA CV124, 1S5, q. v.].
VT-173 [AKA CV35, 1T4, q. v.].
VT-174 [AKA CV183, 3S4, q. v.].
VT-175 [AKA CV655, 1613, q. v.].
VT-176 [AKA CV66, 6AB7, q. v.].
VT-177 [AKA 1LH4, q. v.].
VT-178 [AKA CV3, 1LC6, q. v.].
VT-179 [AKA CV781, 1LN5, q. v.].
VT-180 [AKA CV187, 3LF4, q. v.]; VT-no. cancelled despite the CV-no.
VT-181 [AKA CV252, 7Z4, q. v.].
VT-182 [AKA CV228, 3B7, q. v.].
VT-183 [AKA CV91, 1R4, q. v.].
VT-184 [AKA 0B3, q. v.].
VT-185 [AKA CV187, 3D6, q. v.].
VT-186 "Special tube"; no CV-no. or known use.
VT-187 [AKA 575A, q. v.].
VT-188 [AKA CV891, 7E6, q. v.].
VT-189 [AKA CV893, 7F7, q. v.].
VT-190 [AKA CV75, 7H7, q. v.].
VT-191 [AKA CV683, (WE) 316A, q. v.].
VT-192 [AKA CV1770, 7A4, q. v.].
VT-193 [AKA CV76, 7C7, q. v.].
VT-194 [AKA CV22, 7J7, q. v.].
VT-195 [AKA CV2874, (CK) 1005, q. v.].
VT-196 [AKA CV279, 6W5G, q. v.].
VT-197A [AKA CV1268, 5Y3GT, q. v.].
VT-198A [AKA CV1926, 6G6G, q. v.].
VT-199 [AKA CV70, 6SS7, q. v.].
VT-200 [AKA CV686, 0C3, q. v.].
VT-201 [AKA CV220, 25L6, q. v.].
VT-201C [AKA CV551, 25L6GT, q. v.].
VT-202 [AKA CV664, 9002, q. v.].
VT-203 [AKA CV665, 9003, q. v.].
VT-204 [AKA CV745, 3C24, q. v.].
VT-205 [AKA CV109, 6ST7, q. v.].
VT-206A [AKA CV729, 5V4G, q. v.].
VT-207 [AKA CV529, 12AH7GT, q. v.].
VT-208 [AKA CV22, 7B8, q. v.].
VT-208A [AKA 7B8LM, q. v.]; no known use; never made in quantity.
VT-209 [AKA CV694, 12SG7, q. v.].
VT-210 [AKA CV181, 1S4, q. v.].
VT-211 [AKA CV1978, 6SG7, q. v.].
VT-212 [AKA CV650, 958, q. v.].
VT-213A [AKA CV151, 6L5G, q. v.].
VT-214 [AKA CV916, 12H6, q. v.].
VT-215 [AKA CV239, 6E5, q. v.].
VT-216 [AKA CV724, 816, q. v.].
VT-217 [AKA CV628, 811, q. v.].
VT-218 [AKA CV2552, 100TH, q. v.].
VT-219 [AKA 8007, q. v.]; "8007" and VT-no. cancelled; also see 807.
VT-220 [AKA CV2589, 250TH, q. v.].
VT-221 [AKA CV185, 3Q5GT, q. v.].
VT-222 [AKA CV647, 884, q. v.].
VT-223 [AKA CV104, 1H5GT, q. v.].
VT-224 [AKA CV18, 2C34, q. v.].
VT-225 [AKA CV2612, (WE) 307A, q. v.].
VT-226 [AKA CV817, 3EP1, q. v.].
VT-227 [AKA 7184, q. v.]; no known use.
VT-228 [AKA CV662, 8012, q. v.].
VT-229 [AKA 6SL7GT, q. v.].
VT-230 [AKA (WE) 350A, q. v.].
VT-231 [AKA 6SN7GT, q. v.].
VT-232 [AKA E1148, q. v.].
VT-233 [AKA 6SR7, q. v.].
VT-234 [AKA HY-114B, q. v.].
VT-235 [AKA HY-615, q. v.].
VT-236 [AKA CV636, 836, q. v.].
VT-237 [AKA 957, q. v.].
VT-238 [AKA 956, q. v.].
VT-239 [AKA 1LE3 SP, q. v.].
VT-240 ["710A," AKA 8011, q. v.].
VT-241 [AKA CV890, 7E5, q. v.].
VT-243 [AKA CV2705, 7C4, q. v.].
VT-244 [AKA CV276, 5U4G, q. v.].
VT-245 [AKA 2050, q. v.
VT-246 [AKA CV2692, 918, q. v.].
VT-247 [AKA CV207, 6AG7, q. v.].
VT-248 [AKA 3CP1, q. v.].
VT-249 [AKA (CK) 1006, q. v.].
VT-250 [AKA EF50, q. v.].
VT-251 ["441," AKA 2J32, q. v.].
VT-252 [AKA 923, q. v.].
VT-254 [AKA 304TH, q. v.].
VT-255 [AKA 8021, q. v.].
VT-256 ["GL-486," AKA 3J26, q. v.].
VT-257 ["K-7," AKA 2J30, q. v.].
VT-259 [AKA CV632, 829, q. v.].
VT-260 [AKA 0A3, q. v.].
VT-264 [AKA 3Q4, q. v.].
VT-266 [AKA 1616, q. v.].
VT-267 ["WL-578," AKA 8020, q. v.].
VT-268 [AKA 12SC7, q. v.].
VT-269 [AKA (WE) 717A, q. v.].
VT-277 ["417," AKA WL-417A, q. v.]; not *WE* 417A.
VT-279 ["1278-GY2," AKA D-161-831, q. v.]. No known use.
VT-280 ["C7063," AKA 1P24, q. v.; VT-no. cancelled; no known use.
VT-281 [AKA HY-145ZT/HY-115, q. v.]; cancelled; no known use.
VT-282 [AKA ZG489, q. v.]; no known use.
VT-283 [AKA {QF-206}, 2E27, q. v.]; cancelled.
VT-284 [AKA 2B24, q. v.]; VT-no. cancelled.
VT-285 [AKA {QF-200C}, 2C27, q. v.]; cancelled; no known use.
VT-286 [AKA 832A, q. v.].
VT-287 [AKA CV2663, 815, q. v.].
VT-288 [AKA CV922, 12SH7GT, q. v.].
VT-289 [AKA 12SL7GT, q. v.].

2. NAVY TYPE NUMBERS

Before the institution of the Joint Army-Navy numbering plan in 1942, the U. S. Navy designated tubes three ways: by four-digit numbers that it assigned and prefixed with a contractor code (e. g., "CG" for General Electric, "CRC" for Radio Corporation of America); by Bureau of Steam Engineering (SE-) numbers; or by a supply code in the 38xxx area, also prefixed with a contractor code. All three coding systems appear in the documentation on early Navy equipment. The following provides a conversion to industry numbering.

SE-1444 [AKA CW-933, CG-890]; "Navy no. prior to 1932."
SE-2566A [AKA 10Y]; Navy spec. RE 13A 312.
SE-2935 81
SE-3069 866
SE-3070 872
SE-3071 869, B
SE-3072 857B
SE-3119 860; Navy spec. RE 13A 353; IPA and PA in XK xmtr; two as IPA in XE xmtr., 1928.
SE-3124A 861; Navy spec. RE 13A 358; final in XE xmtr., 1928.
SE-3222 22 [AKA 38022]; Navy spec. RE 13A 381.
SE-3294 865; Navy spec. RE 13A 406.
SE-3373 UY-227; Navy spec. RE 13A 425.
SE-3382 112A; Navy spec. RE 13A 407.
SE-3572 866A
SE-3675 862A; Navy spec. RE 13A 410.
SE-3717 858; Navy spec. RE 13A 414.
SE-3720 40; Navy spec. RE 13A 424.
SE-3820, A 38309, A
SE-3841 80
SE-3864A 864; Navy spec. RE 13A 411.
SE-3871 50; Navy spec. RE 13A 428.
SE-3872 30
SE-3873 32; Navy spec. RE 13A 432.
SE-3918 874
SE-3921 45; Navy spec. RE 13A 430.
SE-3924 24A; Navy spec. RE 13A 427.
SE-3945 845; Navy spec. RE 13A 434.
SE-3982 868
SE-4102 871
SE-4171 71A
SE-4261 FG-17
SE-4374 Sel. UX-201A ("for Marshall rcvrs.") [AKA 38001]; Navy spec. RE 13A 377.
SE-4378 38304
SE-4465 35
SE-4610 36
SE-4611 37
SE-4612 38
CW-186 WE 201A, q. v.
CG-886 Receiving triode (WW I).
CG-890 [AKA SE-1444, q. v.]; GE "G," (early version); Navy version of VT-11.
CG-916 GE "P" (UV-204 prototype), WW I, ca. 250 W max. out.
CW-931 (WE) 205A, q. v. [AKA VT-2]; superseded by 205F.
CW-933 (WE) 203A, q. v. [AKA SE-1444, VT-1].
CG-1144 GE "U" (see VT-18).
CG-1144A UV-203 [AKA 38111].
CG-1162 [AKA VT-14]; xmtg triode, WW I.
CW-1344 (WE) 215A; Navy spec. RE 13A 219. Three in RG rcvr., 1925.
CG-1353 UV-208
CG-1787 Triode, WW I; "obsol. no.; nearest equiv. SE-4374."
CW-1818 (WE) 211A, q. v. [AKA VT-4A].
CW-1818A (WE) 211D; Navy spec. RE 13A 298; xtal osc. in XE xmtr, 1928.
CW-1819 WE 212A
CW-1819B (WE) 212D; Navy spec. RE 13A 131.
CG-1860, A UV-204, A; Navy spec. RE 13A 291.
CG-1861 UV-218
CW-1887 (WE) 220B; Navy spec. RE 13A 220.
CG-1971 UV-207; Navy spec. RE 13A 309.
CG-1984 UV-211
CG-2010 UX-210
CW-2110 38111A
CG-2172 851; Navy spec. RE 13A 310.
CW-2354 WE 225A
CG-2933 UV-206 [AKA 38106]; Navy spec. RE 13A 303.
CG-2935 UX-213B
CG-2935A 81
CG-2936, A UV-217, C
CG-2937 UV-219
CG-2938 UV-214
CG-2967 See 219.
38001 01A
38012 12A/112A [AKA SE-3382].
38015 215A
38019 19
38022 22 [AKA SE-3222].
38024 24A [AKA SE-3924].
38027 27 [AKA SE-3373].
38030 30 [AKA SE-3872].
38031 31
38032 32 [AKA SE-3873].
38032A 1B4P
38033 33
38034 34
38035 35 [AKA SE-4465].
38036 36 [AKA SE-4610].
38037 37 [AKA SE-4611].
38038 38 [AKA SE-4612].
38039 39
38040 40 [AKA SE-3720].
38041 41
38042 42
38045 45 [AKA SE-3921].
38047 47
38050 50 [AKA SE-3871].
38053 - 59 53 - 59
38064 864 [AKA SE-3864A].
38071 71A [AKA SE-4171].
38075 - 79 75 - 79
38085 - 89 85 - 89
38101, A 801, A
38102 12A/112A
38103 203A
38104 204A [AKA CG-1860A].
38106 UV-206 [AKA CG-2933].
38107 UV-207 [AKA CG-1971].
38110 10Y [AKA SE-2566A].
38111 [AKA CG-1984]; xmtg. triode, repl. by 38111A.
38111A, WA Xmtg. triode, low-capacitance 211 for crystal oscillator use; "use 211 for other applications."

38112	(WE) 212E [AKA CW-1819B].
38114	UV-214 [AKA CG-2938].
38116	4B26
38117	UV-217C [AKA CG-2936A].
38118	UV-218 [AKA CG-1861].
38119	UV-219 [AKA CG-2937].
38120	(WE) 220B [AKA CW-1887].
38138	838
38142	2C45
38143	843
38145	845 [AKA SE-3945].
38146	846
38149	849
38150	850
38151	851 [AKA CG-2172].
38152	852
38157	857 [AKA SE-3072].
38157A	857A
38157B	857B
38158	858 [AKA CG-3717].
38160	860 [AKA SE-3119]. Osc. in TAJ xmtr., 1932.
38161	861 [AKA SE-3124A]. Driver and final in TAJ xmtr., 1932.
38162	862 [AKA SE-3675].
38165	865 [AKA SE-3294].
38166	866 [AKA SE-3069].
38166A	866A
38169	869A [AKA SE-3071].
38169A	869B
38170	870
38171	871 [AKA SE-4102].
38172	872 [AKA SE-3070].
38172A	872A
38180	80
38181	81 [AKA CG-2935A].
38182	82
38183	83
38184	84/6Z4
38192	892
38193	893
38205	0C3
38211	211
38213	2A3
38215	2A5
38217	FG-17
38220	120
38222	3B23 (RK-22)
38227	2B7
38233	2C21 (RK-33)
38236	1C6
38250	0D3
38255	25Z5
38266A	RK-21
38267	1616
38268	868 [AKA SE-3982].
38274	874 [AKA SE-3918].
38275	0A3
38276	876
38278	"Obsol. no." (1278?)
38282	(WE) 282B
38286	886
38290	0B3
38296	896
38304	[AKA SE-4378]; "obsol. no."
38309, A	[AKA SE-3820, A]; "obsol. no."
38401	3B21
38402	4B22
38403	4B23
38404	4B24
38405	3B22
38412	312A
38565J	6J5
38566H	6H6
38567K	6K7
38567R	6R7
38568K	6K8
38593	5Z3
38616	6A6
38617	6A7
38627	6B7
38636	6C6
38646	6D6
38660	1951
38665E	6E5
38667	6F7
38674	5B24; "obsol. no.; use 4B25."
38674A	4B25 [AKA EL6CF].
38717E	1E7G
38765J	6J5G
38766Y	6Y6G
38768F	6F8G
38803	803
38807	807
38808	808
38814	814
38833	833
38837	837
38842	842
38853	1853
38884	884
38897	4B27
38954	954
38955	955
38956	956
38958	958
38959	959

3. REL TYPES

Research Enterprises Ltd. (REL) was a manufacturing arm of the National Research Council of Canada, located in Toronto (Leaside). It was established at the start of WW II to make radar and related equipment. For the Canadian forces it produced radars coded CDX, GL IIIC, RAW, RWG, RWM, and RXF/268, and the RCA-designed AYF radar altimeter. For the U. S. Army it made radars coded SCR-520, SCR-521, SCR-588, SCR-602A (AN/TPS-3), and SCR-636. It produced versions of some of these for export to Britain, other Commonwealth countries, and the USSR. The radar sets, and the most critical tubes in them, began as emergency adaptations of British types, but soon led to Canadian designs.

There was a line of radar tubes coded "REL." They were made by Canadian and U. S. firms, plus some CRTs from REL itself. Some were experimental, never appearing in actual equipment, but others turn up in today's collector market, in cartons explicitly carrying REL numbers. Some REL numbers may have been temporary, pending assignment of industry or military numbers. It has been possible to find information on a few of these, mainly as used in equipment supplied to the U. S. Here is what is currently known about these historic devices, many of which were "leading-edge technology" in their day.

The following prefixes appear in the equivalents: CV - Common Valve (British forces); VT - Valve, Transmitting (RAF); NT - Naval Transmitting (Admiralty);1B24, 2B27, and 4C27 - RMA-registered types; ZP - General Electric developmental.

There must have been numerous other "REL" tubes. For example, the SCR-636 radar used three RAF VU111 rectifiers (made by Canadian GE) and an RAF VR78 diode, along with REL 7s, an REL 21, and an REL 38. Surely the VU111 and VR78 had REL numbers.

REL1 [AKA VT90 (Br.), 8011, q. v.]. (NE).

REL3B [AKA "310A, 8310A"]; this is doubtless *not* the WE 310A.

REL3BX [AKA 3BHP]; cavity magnetron.

REL3C [AKA UK3108, 315B, E1198, CV38.]; early cavity magnetron.

REL3D [AKA UK310P, 315D, E1189, W2510, NT98]; quasi-experimental 3-GHz magnetron giving 12 kW out, brought to U. S. as {E1189} by the 1940 British scientific mission.

REL3DX [AKA 3DHP, E1325, CV56]; magnetron dev. from E1189, giving 80-100 kW out at 3 GHz; the first "strapped" type in British production.

REL5 [AKA (Br.) 10E/224, VT98]; copper-glass UHF xmtg triode dev. from RAF VT58, used in Chain Home Low air- defense radar and MB2 mobile set. Copper anode, side connection to grid; T-T fil. 12.6 V @ 58 A. Two tubes gave 30 kW peak at 600 MHz max. Major production in 1941. Two were used in xmtr of SCR-588 CHL/Ground Controlled Intercept set.

REL7 [AKA 4C27, q. v.].

REL8A-D [AKA RT66]; "rcvg UHF osc. (Can.)"; klystron?e

REL10 [AKA CW10, q. v.].

REL16 [AKA 579, 878, q. v.].

REL19 [AKA (GL, ZP) 436, q. v.].

REL20 "Watson thyratron." No further information.

REL21 [AKA VI507, 10E/467]; spark-gap (pulser?) tube; tungsten electrodes in argon-mercury atmosphere, like U. S. 1B42. Glass body is 1-7/8" long and 3/4" dia., w/ electrodes out ends and sealing tip on the side. British used it in CHL radar and Admiralty Type 291 air-surface warning set. REL applied it in BC-1188A unit of SCR-636A. Made by Rogers Radio Tubes Ltd.

REL22, Spl "Obsol. Can. no."

REL23 "Canadian number, Sutton." Early 3-GHz reflex klystron (NR89) from late 1940, named for Br. developer. Used as local osc. in Admiralty Type 271 radar and Br. Army GL3 gun-laying radar - and possibly in REL's GL IIIC.

REL25 [AKA ZP477, 12DP8, q. v.].

REL29 "Soft Sutton (8B)." TR cell used in Br. AI and H2S (bombing) radars; used glass structure and resonator of NR89 Sutton klystron; three variants for 2800, 2970, and 3300 MHz, w/ a "keep-alive" electrode - the first microwave TR switch. TR tubes were gas-filled, therefore "soft." This tube is *suspected* of being the same as the WE 709A, which was a TR switch comprising a WE 711A or 712A klystron cavity with gas added.

REL30 [AKA TR3SW, 1B24, q. v.]; "obsol. Can. no."

REL31 "F. C. M. sim. to 723A/B (Can.)." A puzzler: "F. C. M." presumably means "field-crossed magnetron," but 723A/B was a popular *klystron*.

REL32 [AKA BM313]; "F. C. M. magnetron, obsol. (Can.)"; see REL31.

REL33 [AKA 12HP3, q. v.] (REL-made).

REL33A [AKA 12HP1, q. v.] (REL-made).

REL34 [AKA 446, 2C40, q. v.].

REL35 [AKA ZP447, 2B27, q. v.].

REL36 Triode, 955 w/ three grid pins for G-G use, in BC-1082 rcvr. of SCR-602A radar.

REL37 [AKA 12HP7, q. v.] (REL-made).

REL38 [AKA 1923]; Br./Can."trigatron" cold-cathode spark-gap radar pulser, dev. in 1942. Not a *vacuum* tube: used argon under pressure in a 6-inch glass envelope, w/ a trigger electrode to fire it. Gave 170 kW in Modulator 64 in Br. Mk VIII AI and H2S radars.

REL38A [AKA 1923F, CV85]; trigatron.

REL38B [AKA 1924, CV100]; trigatron.

REL38C [AKA 2027, CV125]; trigatron.

REL38D "Special-based 2027 spark-gap mod. (Can.)."

REL39, 41 Can. 6AC7 selected for microphonics.

REL40 "CRT (Canadian)."

REL42 [AKA E1368, CV90, q. v.].

REL43 [AKA 9MP9, q. v.].

REL44 [AKA KRN2A, CV129]; klystron.

REL45 [AKA 12FP7, q. v.].

REL46 [AKA S25A, CV16]; disc-seal G-G triode.

REL47 [AKA E1273, CV58, q. v.].

REL48 [AKA CV82, q. v.]; disc-seal rcvg. triode.

REL49 [AKA CV88]; disc-seal rcvg. triode.

REL50 [AKA CV95]; RF measuring lamp..

REL51 RF measuring lamp (Can.). Also see PM3.

REL52 [AKA "878A, 8013A," q. v.]. Cannot have been *both* an 878A and an 8013A; may have been a third type similar to both.

REL53 [AKA NRC001M]; "xmtg UHF osc. (Can.)" (klystron?); "NRC": National Research Council?

REL54 [AKA NRC 002M] (klystron?).

REL55 [AKA CV108]; 9-GHz magnetron, 30 kW out.

REL56 [AKA 1365, CV114]; 9-GHz TR cell.

REL57 [AKA E1415, CV117]; ATR tube .

REL64 From inspecting a sample, made by Canadian GE: a disc-seal TR cell sim. to WE 721A but slightly larger (3" long, 1-1¼" dia. less sealing tip or cap connector, w/ two 1-7/8" connecting discs; has cap at end for keep-alive electrode.

DE FOREST

The De Forest Radio Company in the early '30s offered a variety of tubes: general-trade receiving, transmitting, and movie-sound. Some of these were kept on in RCA's line after purchase of the De Forest company in 1933. The numbering of these tubes generally relates to "industry" numbers by addition or subtraction, but there are numerous exceptions. De Forest did a modest amount of innovation, as in UX-199s with oxide-coated filaments said to avoid the failings of the tungsten version. The 575 rectifier seems to have originated with De Forest - an industry "standard" which considerably outlived the firm.

401A [AKA 01A, q. v.]; rcvg. triode.

402 "Theatre Audion" triode, interchanges w/ WE 102 series. As of 4-1-34, RCA - as successor to De Forest - had a 402-D {A225} under development.

405, D "Theatre Audion" power triode, interchanges w/ WE 205 series. As of 4-1-34, RCA offered a 405-D {A227}, "not listed for distributor sales."

410, 412A [AKA 10, 112A, q. v.]; power triodes.

420 [AKA UX-120, q. v.]; rcvg. triode; later, UX-120 w/ oxide-coated fil.

422, A [AKA 22, q. v.]; rcvg. tetrode. "A" is 22 w/ oxide-coated fil.

424, A [AKA 24, A, q. v.]; rcvg. tetrode.

426, 427 [AKA 26, 27, q. v.]; rcvg. triodes.

430 [AKA 30, q. v.]; rcvg. triode, tubular 199-style bulb.

431 [AKA 31, q. v.]; power triode, tubular 199-style bulb.

432 to **440** [AKA 32 to 40, q. v.]; rcvg. types.

445, 450 [AKA 45, 47, q. v.]; power triodes.

447 [AKA 47, q. v.]; power tetrode.

451 [AKA 35/51, q. v.]; rcvg. tetrode.

464A "Theatre Audion" triode, replacing WE 239A and 264A. Claimed to be quiet, for movie sound-head preamps. As of 4-1-34, RCA offered a 464-A {A231}, "not listed for distributor sales."

471B [AKA 71A, q. v.]; power triode.

480 to **483** [AKA 80 to 83, q. v.]; rectifiers.

499 [AKA UX-199, q. v.]; rcvg. triode; later, UX-199 w/ oxide-coated fil.

503A [AKA 203A, q. v.]; xmtg. triode.

504, A [AKA 204, A, q. v.]; xmtg. triode.

507 [AKA 207, q. v.]; xmtg. triode. Not strongly recommended for audio. Derated above 1.5 MHz.

510 [AKA 210, q. v.]; power triode.

511 [AKA 211, q. v.]; xmtg. triode.

511E "Theatre Audion" power triode, equiv. of WE 211E.

519W "Theatre Audion" power triode, replacing WE 219D, as in 10-A amplifier system.

520B, M Xmtg. RF triode, grid-out-side, fil. 22 V @ 34 A, μ 16, 5 kW, water-cooled via ext. cooler, "approximately interchangeable" w/ WE 228A (DeF, GE). "M" is modulator version w/ μ 10.

521 Xmtg. triode, two-kW version of DeF 520B, same format and fil., μ 25.

530 [AKA 830, q. v.]; xmtg. triode.

542A "Theatre Audion" power triode; T-T fil.; interchanges w/ 242A in "all" WE amp. systems. As of 4-1-34, RCA offered a 542-A, "not listed for distributor sales."

545 [AKA 845, q. v.]; xmtg. triode. "Primarily an audio tube."

548 [AKA 848, q. v.]; xmtg. triode for modulators, 10 kW (not derated to 6 kW as w/ 848); also see 563. Not strongly recommended for RF use.

549 [AKA 849, q. v.]; xmtg. triode.

550 [AKA 250, q. v.]; power triode.

551 [AKA 851, q. v.]; xmtg. triode; 600 W - not rated same as "industry" 851 at 750 W.

552 [AKA 852, q. v.]; xmtg. triode.

557 [AKA 857, q. v.]; mercury rectifier; seems to equal 857A; fil. 5 V @ 42 A (> 857B), 20 kV PIV, 20 A peak.

558 Mercury rectifier. See 558 in general list.

560 to **569** [AKA 860 to 869, q. v.]; xmtg. types.

571 Xmtg. triode, 750 W out; ("a 561 without the screen, or a big 552"); fil. 11 V @ 10 A, μ 16, 500 W, μ 16. Possibly same as RCA 831.

572 [AKA 872, q. v.]; mercury rectifier.

575, A Xmtg. mercury rectifier, intermediate between 872 and 869; spherical-tubular bulb, fil. 5 V @ 12.5 A, 12.5 kV PIV, 2.5 A peak curr. "A" has stepped-tubular bulb w/ cap, super-jumbo (A4-29) base; ratings are: fil. cut to 10 A, PIV uprated to 15 kV, 1.5 A out. 575A has same ratings as 673, but differs in base.

601 "Vis-Ion" neon lamp for De Forest disc-TV kit, tubular bulb on UX base, takes 100 mA, 1-½" square anode.

602 Phototube for De Forest disc-TV kit, spherical bulb, potassium-hydride cathode, 3.5-7.5 μA/-lumen.

603A Cesium phototube, round bulb, base w/ leads and phone tips, built like W. E. 3A photocell and replacing it; claimed to be quieter. Said to have good frequency response, implying a vacuum type. As of 4-1-34, RCA offered a 603A photocell, "not listed for distributor sales" {C714}.

604A De Forest crater lamp.

668, -S [AKA 868, q. v.]; cesium photocell, rated at only 35-77 μA/lumen, for use in RCA Photophone eqpt. "-S" is variant for use in Pacent sound eqpt; replaces potassium-hydride cells.

EIMAC

Eitel-McCullough Inc. began making tubes in 1934, starting with a line of transmitting triodes for the amateur and medium-power commercial market. These figured heavily in the SCR-299/399 mobile radio sets of WW II and in several important VHF radars. The company rescued its sales from the postwar glut of surplus tubes by introducing an innovative line of tetrodes in 1946. Besides developing a wide line of tube types extending to the megawatt range, over the years it took over product lines from Heintz and Kaufman, Lewis and Kaufman, Penta Laboratories, and eventually Machlett. Acquired by Varian Associates in 1965, the company has recently become a subsidiary of Communications and Power Industries.

1K015CA Reflex klystron, tunes 5.4-6.0 GHz, heater 6.3 V @ 800 mA, 350 V, 25-100 mW out, coaxial output.

1K015CG Reflex klystron, 1K015CA w/ WG output.

1K015XA Reflex klystron, 1K015CA for 8.7-9.6 GHz.

1K015XG Reflex klystron, 1K015XA w/ WG output.

2-01C UHF planar test-probe diode, heater 5.0 V @ 0.31-0.39 A, 1 kV PIV, 1 mA, resonates at 2 GHz. In HP 410B/C VTVMs. Also see EA53.

2-15A [AKA 15R, 4B30, q. v.]; triode.

2-25A [AKA 25R, q. v.]; rectifier.

2-50A [AKA 35R, q. v.]; rectifier.

2-66 Rectifier, "medical-surgical."

2-100A [AKA 8020, q. v.]; rectifier.

2-150A [AKA 152-R, q. v.]; rectifier.

2-150D [AKA 152-RA, q. v.]; rectifier.

2-240A Vacuum rectifier, fil. 7.5 V @ 12.0 A, 40 kV PIV, 500 mA {X461}.

2-250A [AKA 250-R, q. v.]; rectifier.

2CL40A Vacuum rectifier, fil. 6.0 V @ 2.1 A, 8 kV PIV (16 kV liquid-cooled), 120 mA.

2X1000A Vacuum rectifier, fil. 26.5 V @ 2.1 A, 35 kV PIV, 1.25 A, forced-air cooled.

2X3000F Vacuum rectifier, external anode, flying leads for fil., fil. 7.5 V @ 51 A, 25 kV PIV, 3.0 A.

2-450A Vacuum rectifier, fil. 7.5 V @ 26.5 A, 30 kV PIV, 1.0 A.

2-2000A Vacuum rectifier, T-T fil. 10.0 V @ 25.0 A, 75 kV PIV, 750 mA; probably a gridless 2000T.

3-15A3 [AKA 4C30, q. v.]; triode.

3-25A3 [AKA 25T, q. v.]; triode.

3-25D3 [AKA 3C24, q. v.]; triode.

3-50A4 [AKA 35T, q. v.]; triode.

3-50D4 [AKA 35TG, q. v.]; triode.

3-50G2 [AKA 834, UH50, q. v.]; triode.

3-50J4 [AKA 53A, q. v.]; triode.

3-50M2 [AKA UH51, q. v.]; triode.

3-75A2 [AKA 75TL, q. v.]; triode.

3-75A3 [AKA 75TH, q. v.]; triode.

3-100A2 [AKA 100TL, q. v.]; triode. In Hafstrom Technical Products LH-2000 linear.

3-100A4 [AKA 100TH, q. v.]; triode.

3-100D2 [AKA VT-127A, q. v.]; triode.

3-100E3 [AKA 227A, q. v.]; triode.

3-100G3 [AKA 327A, q. v.]; triode.

3-100H3 [AKA 327B, q. v.]; triode.

3-150A2 [AKA 152TL, q. v.]; triode.

3-150A3 [AKA 152TH, q. v.]; triode.

3-200A3 [AKA 592, q. v.]; triode.

3-250A2 [AKA 250TL, q. v.]; triode. Improved over 250T via non-emitting grid, lower-temperature plate, and new fil. structure.

3-250A4 [AKA 250TH, q. v.]; triode. Also see 3-250A2.

3-300A2 [AKA 304TL, q. v.]; triode; improved post-war version.

3-300A3 [AKA 304TH, q. v.]; triode.

3-300G4 [AKA 527, q. v.]; triode.

3-400Z [AKA 8163, q. v.]; triode.

3-450A2 [AKA 450TL, q. v.]; triode.

3-450A4 [AKA 450TH, q. v.]; triode.

3-500Z VHF zero-bias triode, all-glass w/ cap, robust 8163/3-400Z; fil. 5.0 V @ 14.5 A, 500 W, μ 160, for grounded-grid and/or linear use. One in Heathkit SB-1000 linear amp. Two each in other linears: Heathkit SB-220, Swan Mk II, Swan Mk 6B 6-m., Raytrack Horizon VI 6-m., Raytrack DX-2000-L, and Drake L-4B.

3-750A2 [AKA 750TL, q. v.]; triode.

3-1000A4 [AKA 1000T, q. v.]; triode.

3-1000H RF-heating triode, variant of 8164 w/ μ 25.

3-1000Z [AKA 8164, q. v.]; triode.

3-1500A3 [AKA 1500T, q. v.]; triode.

3-2000A3 [AKA 2000T, q. v.]; triode.

3CPN10A5 [AKA 7815, q. v.]; triode.

3CPX100A5 [AKA 7815R, 8745, q. v.]; triode.

3CPX800A7 Cermet focused triode, elevenar base, heater 13.5 V @ 1.5 A, μ 200, 800 W, 500 MHz.

3CPX1500A7 Cermet focused triode, spl. 7-pin base, heater 5.5 V @ 11.2 A, μ 200, 1500 W, for pulse modulators or pulse regulators.

3CPX3500U7 Cermet focused triode, coaxial base, heater 15.0 V @ 13.0 A, μ 200, 3.5 kW, 500 MHz.

3CPX5000A7 Cermet focused triode, mounted via grid flange, heater 15.0 V @ 14.0 A, μ 200, 5 kW, 110 MHz.

3CV30,000A1 Triode, vapor-cooled variant of 8158, uprated to 30 kW, same use.

3CV30,000A3 Triode, vapor-cooled variant of 3CX-

15,000A3, uprated to 30 kW, same use.

3CV30,000H3 Triode, vapor-cooled variant of 3CX-15,000H3, uprated to 30 kW and 100 MHz, same use. Two as Class B modulators in Gates VP50 50-kW AM xmtr.

3CW5000A3 [AKA 8242, q. v.]; triode.

3CW5000F1 [AKA 8241, q. v.]; triode.

3CW5000H3 Triode, water-cooled variant of 3CX2500H3, uprated to 5 kW.

3CW20,000A1 Triode, water-cooled variant of 8158, uprated to 20 kW, for audio amps and shaker-table drivers.

3CW20,000A3 Triode, water-cooled variant of 8159, uprated to 20 kW, same uses.

3CW20,000A7 Triode, water-cooled variant of 8160, uprated to 20 kW, same uses.

3CW20,000H3 Triode, water-cooled variant of 3CX-10,000H3, uprated to 20 kW, same use.

3CW25,000A3 RF-heating/xmtg. triode, water-cooled variant of 3CX15,000A3, uprated to 25 kW.

3CW30,000H3 Triode, water-cooled variant of 3CX-15,000H3, uprated to 30 kW, same use.

3CW40,000H3 Triode, water-cooled variant of 3CX-20,000A3, uprated to 40 kW, derated tp 90 MHz, for RF heating.

3CX100A5 [AKA 7289, q. v.]; triode.

3CX100F5 [AKA 8250, q. v.]; triode.

3CX400U7 [AKA 8961, q. v.]; cermet beam triode.

3CX600U7 Cermet focused triode, coaxial base, heater 6.0 V @ 5.4 A, μ 200, 600 W, 1 GHz.

3CX800A7 Cermet focused triode, large elevenar 11-pin base, heater 13.5 V @ 1.5 A, μ 200, 800 W, 350 MHz.

3CX800U7 Cermet focused triode, coaxial-based variant of 3CX800A7, raised to 1 GHz.

3CX1000A3 Cermet RF-heating triode, bipin base, T-T fil., 7.5 V @ 31 A, μ 24, 1 kW, 100 MHz.

3CX1000A7 [AKA 8283, q. v.]; xmtg. triode.

3CX1200A7 Cermet high-μ triode, T-T fil. 7.5 V @ 21.0 A, μ 200, 1.2 kW, 110 MHz, for linears.

3CX1200D7 Cermet triode, T-T fil. 6.3 V @ 25.0 A, μ 200, 1.2 kW, 110 MHz, for linears {YU-121}. Radiator different from 3CX1200A7.

3CX1200X7 Cermet triode, 3CX1200D w/ flanged base {YU-181}.

3CX1500A7 [AKA 8877, q. v.]; triode.

3CX1500C7 [AKA 8938, q. v.]; triode.

3CX1500U7 [AKA 8962, q. v.]; triode.

3CX2500A3 [AKA 8161, q. v.]; triode.

3CX2500F3 [AKA 8251, q. v.]; triode.

3CX2500H3 RF-heating triode, variant of 8251 w/ flying leads for fil. and ring terminal for grid, 75 MHz (EI, SV).

3CX3000A1 [AKA 8238, q. v.]; triode.

3CX3000A7 Triode, VHF zero-bias variant of 8238, μ 200, 75 MHz, intended for grounded-grid and/or Class B linear use; also for pulse and voltage-regulator uses. One in Continental 315R-1 5-kW AM xmtr, with another as PDM switch modulator. One in McMartin BF-3.5K 3.5-kW FM xmtr. (EI, SV).

3CX3000F1 [AKA 8239, q. v.]; triode.

3CX3000F7 [AKA 8162, q. v.]; triode.

3CX3000H7 Triode, HF zero-bias variant of 8238, flying leads for fil., for general use incl. voltage regs., μ 160, uprated to 4 kW, 30 MHz (ca. 1976).

3CX4000F3 Cermet RF-heating triode, flying leads for fil. and grid, T-T fil., 7.5 V @ 51.5 A, μ 20, 4 kW, 110 MHz {YU-199}.

3CX5000A3 Cermet med.-μ VHF triode, fil. 7.5 V @ 78 A, μ 18, 5 kW, 90 MHz, special base (grid contact flange can be bolted to grid deck); for RF heating.

3CX5000A7 Cermet focused triode, mounted via grid flange, cathode 10.0 V @ 17.5 A, μ 5 kW, 110 MHz.

3CX5000U7 Cermet focused triode, 5 kW, 1 GHz.

3CX5000H3 Triode, 3CX5000A3 w/ flying leads for fil.; same use. Two in RCA BTA-5L 5-kW AM xmtr.

3CX10,000A1 [AKA 8158, q. v.]; triode.

3CX10,000A3 [AKA 8159, q. v.]; triode.

3CX10,000A7 [AKA 8160, q. v.]; zero-bias triode.

3CX10,000B7 Cermet focused triode, 10 kW, 110 MHz.

3CX10,000H3 RF-heating triode, variant of 8159 w/ flying leads for fil., derated to 10 kW, 90 MHz (EI, SV).

3CX10,000U7 Cermet focused triode, 10 kW, 250 MHz.

3CX15,000A3 Cermet medium-μ triode, fil. 6.3 V @ 152-168 A, 15 kW, grid 500 W, 100 MHz, forced-air cooled, coaxial base, for RF heating or RF amps.

3CX15,000A7 Triode (EI, SV).

3CX15,000B7 Xmtg. cermet focused triode, 15 kW, 110 MHz (EI, SV).

3CX15,000H3 RF-heating/xmtg. triode, variant of 3CX-15,000A3 w/ flying leads for fil. and slightly different radiator, 90 MHz. One as final, one as switch modulator, in Gates MW-10 10-kW pulse-duration-modulated AM xmtr. Two in RCA BTA-10L 10-kW AM xmtr. (EI, SV).

3CX20,000A3 Cermet RF heating/xmtg. triode, coaxial base, fil. 10 V @ 160 A, 20 kW, grid 750 W, 110 MHz, for uses incl. G-G FM amps.

3CX20,000A3 Triode, variant of 3CX20,000A3 w/ flying leads for fil., same ratings, for RF heating and plate-mod. AM xmtrs.

3K2500LX Amplifier klystron, 980-1200 MHz, 7 kV @ 600 mA, 2.5 kW, 1.3 kW CW output, forced-air cooled.

3K2500SG Amplifier klystron, 1.7-2.4 GHz, 25 dB gain,

1.3 kW output, 7 kV @ 600 mA, 3 kW, forced-air cooled. In Leventhal Electric Co. 74T xmtr.

3KM2500LT Amplifier klystron, for 955-1270 MHz, three external cavities, modulating anode, 30 kW pulse, 35 dB gain, for air-navigation systems like VORTAC {X676}.

3K3000LQ Amplifier klystron, 610-985 MHz, 1.3 kW output, 10 kV @ 750 mA, 3 kW, air-cooled. In mid-'50s REL troposcatter system.

3KM3000LA Amplifier klystron, 375-540 MHz, 2.0 kW output, 9 kV @ 750 mA, 3 kW.

3K20,000LA-LK Amplifier klystron, for UHF TV, three external cavities, 20 dB gain, 13 kV @ 1.7 A, 20 kW, 5.5 kW peak output, water-cooled.

3K20,000LA For TV chans. 14-32.

3K20,000LF For TV chans. 33-55.

3K20,000LK For TV chans. 56-83.

3K50,000LA-LQ Amplifier klystron, for UHF TV and radar use, three external cavities, 30 dB gain, 19.5 kV @ 2.6 A, 50 kW, 12 kW peak output, water-cooled.

3K50,000LA For TV chans. 14-32.

3K50,000LF For TV chans. 33-55.

3K50,000LK For TV chans. 56-83.

3K50,000LQ For 750-1000 MHz.

3KM50,000PA Amplifier klystron, three cavities, 20 kV @ 2.5 A, 50 kW, 20 kW CW out, water-cooled.

3W5000A3 Triode, 3X2500A3 water-cooled via integral cooler, uprated to 5 kW.

3W5000F3 Triode, 3X2500F3 water-cooled via integral cooler, uprated to 5 kW.

3W10,000A3 VHF triode, water-cooled, integral cooler, coaxial construction, bombardment cathode w/ fil. 7.0 V @ 25 A and bombardment 1550 V @ 1.6 A, 10 kW, 220 MHz, promoted for TV xmtrs.

3X100A11 [AKA 2C39, q. v.]; triode.

3X150A3 [AKA 3C37, q. v.]; triode.

3X2500A, A1-A3 VHF triode, glass coaxial ring-seal construction, T-T fil. 7.5 V @ 51 A, μ 20, 2.5 kW, forced-air cooled. Can give 5 kW peak through TV Ch. 13. ("A3": EI, F). One in Gates BC-5P 5-kW AM xmtr, modulated by two. Two in BC-10P, HF-5, and HF-10 10-kW AM and SW xmtrs, modulated by two. Four in BC-20B and HF-20 20-kW AM and SW xmtrs, modulated by four. Six, in G-G, in Collins 205E (AN/FRT-22) 50-kW CW xmtr. Eight in parallel in Rixon Electronics 60-kW ionoscatter xmtr, 60 MHz.

3X2500F3 Triode, variant of 3X2500A series; same electrical ratings. incl. fil. (EI, F).

3X3000A1 Triode, variant of 3X3000F1, same electrical ratings incl. fil.

3X3000F1 HF triode, glass external-anode construction, forced-air cooled, fil. 7.5 V @ 51 A, μ 5, 3 kW, 30 MHz, flying leads for fil. and grid, for general use incl. linears.

3X5000A VHF triode; forced-air-cooled.

3X12,500A3 VHF triode, forced-air-cooled, four units sim. to (identical to?) 3X2500A3 mounted in parallel in large retaining rings, fil. 7.5 V @ 192 A, 12.5 kW, μ 20, 110 MHz, promoted for 50-kW FM broadcast xmtrs.

3X20,000A3 Triode, forced-air cooled, six triode units sim. to (identical to?) 3X2500A3 in parallel, fil. 7.5 V @ 288 A, 20 kW, μ 20, g_m 120 mS (ca. 1948).

4-65, A ["A" is AKA 8165, q. v.]; tetrode.

4-125A [AKA 4D21, q. v.]; tetrode.

4-250A [AKA X-117, 5D22, q. v.]; tetrode.

4-400A [AKA 8438, q. v.]; tetrode.

4-400C [AKA 6775]; tetrode, update of 4-400A.

4-750A Tetrode, discontinued (1947) in favor of 8166/4-1000.

4-1000A [AKA 8166, q. v.]; tetrode.

4CM300,000G [AKA 9000, q. v.]; tetrode.

4CN15A Tetrode, variant of 8167, less the anode radiator, 15 W, for pulse use.

4CPW1000KA [AKA 9009, q. v.]; tetrode.

4CPW10,000R [AKA 9016, q. v.]; tetrode.

4CPW300,000A [AKA 9008, q. v.]; tetrode.

4CPX250K [AKA 8590, q. v.]; tetrode.

4CPX800A Tetrode (EI, SV).

4CS250H, A Tetrode, conduction-cooled variant of 8321 (uses cooling contact block on anode), derated to 250 W.

4CV1500B Cermet external-anode vapor-cooled HF tetrode, heater 6.0 V @ 11.0 A, 1.5 kW, 30 MHz, breech-lock base, for AF and RF linear amps.

4CV8000A Tetrode, vapor-cooled version of 8169, uprated to 8 kW, fil. 10.0 V @ 43.5-48.5 A.

4CV20,000A tetrode, vapor-cooled version of 8170, uprated to 20 kW.

4CV35,000A Cermet vapor-cooled VHF tetrode, fil. 6.3 V @ 152-168 A, 35 kW, 110 MHz, for general use.

4CV50,000E Cermet vapor-cooled tetrode, reentrant anode, fil. 12 V @ 220 A, 50 kW, for general use. Has metal-wick cooler drawing water from boiler assembly. One as final, one as switch modulator, in Gates SW-100 100-kW pulse-duration-modulated SW xmtr.

4CV75,000A Tetrode, scaled-down 8351, derated to 75 kW.

4CV100,000C [AKA 8351, q. v.]; tetrode.

4CV100,000E Cermet vapor-cooled tetrode, fil. 16 V @ 230 A, 100 kW, freq. not stated, grounded-grid, pulse, and regulator use.

4CV250,000A Cermet HF tetrode, fil. 12 V @ 660 A, 250

kW, 30 MHz, vapor-cooled, for general use. Two in Continental 319D 250-kW AM xmtr.

4CV250,000V Tetrode, 4CV250,000A supplied w/ ion vacuum pump.

4CW800B Cermet UHF tetrode, heater 6.0 v @ 4.7 A, 800W, 800 MHz, liquid-cooled via internal cooler, for distributed (linear) amps., 5-pin base, related to 8809.

4CW800F Tetrode, 4CW800B w/ heater 26.5 V @ 1.25 A.

4CW2000A [AKA 8244, q. v.]; tetrode.

4CW10,000A [AKA 8661, q. v.] tetrode.

4CW20,000A [AKA 8173, q. v.]; tetrode.

4CW25,000A Tetrode, water-cooled version of 8281, uprated to 25 kW; promoted for regulator and pulse-modulator use. Two as cathode-follower modulators in Continental 320F 500-kW AM xmtr; three in 323C 1-MW and D323C 2-MW, same use. One also as driver in D323C.

4CW50,000E Cermet VHF tetrode, reentrant anode, fil. 12 V @ 220 A, 50 kW, 110 MHz, water-cooled via integral cooler, for general use.

4CW100,000D Cermet water-cooled HF tetrode, fil. 10.0 V @ 310A, 100 kW, 30 MHz, for general use.

4CW100,000E Cermet tetrode, fil. 16 V @ 230A, 100 kW, for pulse use or distributed amps., cooled by air and water via integral cooler.

4CW250,000A Cermet VHF tetrode, water-cooled, fil. 12.0 V @ 640 A, 250 kW, 50 MHz.

4CW250,000V Cermet tetrode, 4CW250,000A, w/ integral ion vacuum gauge.

4CX125C Cermet tetrode, variant of 8167 derated to 125 W, w/ horizontally-finned radiator for transverse air cooling or liquid immersion.

4CX125F Cermet tetrode, variant of 8167, heater 26.5 V @ 590-700 mA.

4CX150 Tetrode, cermet variant of 7034/4X150.

4CX150G Cermet UHF beam tetrode, heater 2.5 V @ 6.3 A, 150 W, forced-air cooled.

4CX250B [AKA 7203, q. v.]; tetrode.

4CX250BC, BM [AKA 8957, q. v.]; tetrode.

4CX250F [AKA 7204, q. v.]; tetrode.

4CX250FG [AKA 8621, q. v.]; tetrode.

4CX250K [AKA 8245, q. v.]; tetrode.

4CX250M [AKA 8246, q. v.]; tetrode.

4CX250R [AKA 7580W, q. v.]; tetrode.

4CX300A [AKA 8167, q. v.]; tetrode.

4CX300Y [AKA 8561, q. v.]; tetrode.

4CX350A [AKA 8321, q. v.]; tetrode.

4CX350AC tetrode (EI, SV).

4CX350F [AKA 8322, q. v.]; tetrode.

4CX350FJ [AKA 8904, q. v.]; tetrode.

4CX400A Tetrode (EI, SV).

4CX600B Cermet UHF beam tetrode, 5-pin special base, conductive-cooled, heater 6.0 V @ 4.3 A, 600 W, 800 MHz, forced-air cooled, for wideband (distributed) amps. and linears. In pulse use, gives 20 kW out @ 430 MHz.

4CX600F Tetrode, 4CX600B w/ heater 26.5 V @ 930 mA, same uses.

4CX600J [AKA 8809]; tetrode.

4CX600JA [AKA 8921, q. v.]; tetrode.

4CX800A Tetrode (EI, SV).

4CX1000A [AKA 8168, q. v.]; tetrode.

4CX1000K [AKA 8352, q. v.]; tetrode.

4CX1500A Cermet VHF tetrode, breech-block base, fil. 5.0 V @ 38-43 A, 1.5 kW, 110 MHz, forced-air cooled, rugged, for general use including linears. Two in Continental 416D 10-kW SW xmtr, 1966 (EI, SV).

4CX1500B [AKA 8660, q. v.]; tetrode.

4CX1600B, U Tetrode (EI, SV).

4CX3000A [AKA 8169, q. v.]; tetrode.

4CX3500A Tetrode (EI, SV).

4CX5000A [AKA 8160, q. v.]; tetrode.

4CX5000J [AKA 8909, q. v.]; tetrode.

4CX5000R [AKA 8170W, q. v.]; tetrode.

4CX10,000D [AKA 8171, q. v.]; tetrode.

4CX12,000A [AKA 8989, q. v.]; tetrode.

4CX15,000A [AKA 8281, q. v.]; tetrode.

4CX15,000J [AKA 8910, q. v.]; tetrode.

4CX20,000A [AKA 8990, q. v.]; tetrode.

4CX20,000D [AKA 9015, q. v.]; tetrode.

4CX35,000A Tetrode. In Continental 318C 150-kW AM xmtr.

4CX35,000C [AKA 8349, q. v.]; tetrode.

4KP40,000SQ Amplifier klystron, internal cavities, 2485-2865 MHz, WG, 10 mW pulse output.

4K50,000LQ Amplifier klystron, four external cavities, for 755-985 MHz, 20 kV @ 2.5 A, 50 kW, 6 kW CW output, water-cooled. For troposcatter xmtrs.; in REL mid-'50s tropo. system.

4KM50,000LA Amplifier klystron, four cavities, for 400-610 MHz, 10 kW out, for UHF TV (EEV, EI).

4KM50,000SG Amplifier klystron, "LA" for 1.7-2.4 MHz. In AM-1671/FRC-56 troposcatter set.

4KM70,000LA Amplifier klystron, 400-610 MHz, 25 kW CW output, 50 dB gain, BeO windows, for troposcatter and UHF TV.

4KM70,000LK Amplifier klystron, "LA" for 610-985 MHz.

4PR60A Pulse tetrode; see 8252 for ratings.

4PR60B [AKA 8252, q. v.]; pulse tetrode.

4PR60C [AKA 8252W, q. v.]; pulse tetrode.

4PR65A [AKA 8187, q. v.]; pulse tetrode.

4PR125A [AKA 8247, q. v.]; pulse tetrode.

4PR250C [AKA 8248, q. v.]; pulse tetrode.

4PR400A [AKA 8189, q. v.]; pulse tetrode.

4PR1000A [AKA 8189, q. v.]; pulse tetrode.

4PR1000B [AKA 8189W, q. v.]; pulse tetrode.

4W300B [AKA 8249, q. v.]; tetrode.

4W1250A VHF tetrode, water-cooled, glass body, integral cooler, fil. 5.0 V @ 13.5 A, 1250 W, for

TV and other VHF uses.

4W20,000A, F VHF beam tetrode, coaxial construction, water-cooled via integral cooler, bombardment cathode w/ fil. 10 V @ 30 A and bombardment at 1650 V @ 1.8 A, 20 kW, 220 MHz, promoted for TV xmtrs; gives 25 kW peak through Ch. 13. "F" has radial finger stock surrounding plate, screen, and grid rings.

4X100A Tetrode, ext. anode w/ radial fins, forced-air cooled, loktal-style base, fil. 6.0 V @ 2.8 A, 100 W, μ 4.5 (EI, ca. 1948).

4X150A [AKA 7034, q. v.]; tetrode.

4X150D [AKA 7035, q. v.]; tetrode.

4X150G [AKA 8172, q. v.]; tetrode.

4X150R [AKA 8296, q. v.]; tetrode.

4X150S [AKA 8297, q. v.]; tetrode.

4X250B UHF tetrode, ceramic w/ glass base, 9-pin spl base, fil. 6.0 V @ 2.6 A, 250 W, 400 MHz, forced-air-cooled, brazed external anode; interchanges w/ 7203 (AX, EI); interchanges physically w/ 7034.

4X250F Tetrode, airborne version of 4X250B, fil. 26.5 V @ 560 mA; interchanges w/ 7204 (AX, EI).

4X300A Tetrode. Two in Cosmophone 1000 xcvr.

4X500A [AKA 6F59R]; VHF beam tetrode, glass body (EI, PL, RCA) or ceramic (AX), fil. 5.0 V @ 13.5 A, 500 W, 120 MHz (Class C) or 220 MHz (Class B), forced-air-cooled, external-anode, promoted for FM and TV use (AX, EI, NEC, RCA). In REL 715 150-MHz police base xmtr.

4X500F Tetrode, variant of 4X500A, slightly larger, minor changes in some parameters (same electrical ratings incl. fil.).

5-75A [AKA 4E27A, q. v.]; tetrode.

5-500A Radial-beam power pentode, fil. 10.0 V @ 10.2 A, 500 W.

5CX1500A Cermet VHF power pentode, 3-½" dia. by 4-7/8" high, breech-block base, sim. physically to 4CX1000A, fil. 5.0 V @ 43 A, 1.5 kW, 110 MHz, forced-air cooled, for AF-RF linears. One each in following xmtrs: Gates FM-2H3 2-kW and FM-2.5K 2.5-kW FM, Continental 814R-1 and -2 2.5- and 1.25-kW FM, and Wilkinson FM2500E 2.5-kW (EI, SV).

5CX1500B Pentode (EI, SV).

5CX3000A [AKA 8966, q. v.]; power pentode.

5K210,000LQ Amplifier klystron, 755-985 MHz, six internal and external cavities, 75 kW out.

6K50,000LQ Amplifier klystron, six cavities, 720-890 MHz, 16 kV @ 1.62 A, 30 dB gain, 10 kW CW output, water-cooled. In Federal FTL-12-1 tropo system for AT&T, Florida-to-Cuba, 1957.

15E [AKA 4C30, q. v.]; triode.

15G Diode gas rectifier.

15R [AKA 4B30, q. v.]; rectifier.

25R [AKA 2-25A]; vacuum rectifier, T-T fil. 6.3 V @ 3.0 A, 25 kV PIV, 50 mA.

25T [AKA 3-25A3]; triode, fil. 6.3 V @ 3.0 A, μ 25, 25 W.

Twin 30 Duo-triode, UX base, top pins for grids and plates, quick-heating fil. 6 V @ 4 A, μ 32, 30 W total.

35R [AKA 2-50A]; vacuum rectifier, T-T fil. 5.0 V @ 4.0 A, 30 kV PIV, 75 mA.

35T [AKA 3-50A4]; triode, UX ceramic base w/ plate pin, fil. 5 V @ 4 A, μ 39, 35 W, 100 MHz (EI, GE, PE) (7-26-43). Plate cherry red @ 35 W. In AN/ART-5.

35TG [AKA 3-50D4]; triode, 35T w/ grid out side; nearest equiv. is 4C25 (7-26-43). In AM-18/APT.

UH35 Transmitting triode.

50T Triode, superseded by 100TL.

UH50 [AKA 3-50G2, 834, q. v]; VHF triode.

UH51 [AKA 3-50M2]; VHF triode, UX base w/ grid and plate caps, fil. 5.0 V @ 6.5 A, μ 10.6, 50 W.

53A [AKA 3-50D4]; VHF triode, VT-127 format w/ smaller electrodes.

53AX [AKA 5C23, q. v.]; triode.

53B Xmtg. "special-purpose."

75T Triode, renumbered as 75TH.

75TH [AKA 3-75A3]; triode, UX base w/ cap, fil. 5 V @ 6.25 A, μ 20, 75 W (7-26-43).

75TL [AKA 3-75A2]; triode, 75T w/ μ 12 (7-26-43).

100-R [AKA 8020, q. v.]; rectifier.

100TH [AKA 3-100A4, VT-218]; triode, UX base w/ cap, grid-out-side, fil. 5 V @ 6.5 A, μ 30, 100 W (sim. to 4C22, HF-100, RK-38, CV2552 and HK-254) (EE, EI, GE, GN, ICE) (7-26-43). In BC-610.

100TL [AKA RK-36, 3-100A2]; triode, 100TH w/ μ of 12; supersedes 50T (7-26-43). Promoted for RF heating.

100TS [AKA WL-534, VT-127, q. v.]; triode. In BC-407A and BC-621.

127A [AKA VT-127A, q. v.]; VHF triode.

150T VHF triode, fil. 5.0 V @ 10 A, μ 13, 150 W, 75 MHz. Eimac's first tube, 1934; superseded by 250TL.

152H, L Triode, renumbered as 152TH, TL.

152-R, RA [AKA 2-150D]; vacuum rectifier, fil. 5.0 V @ 13.0 A, 30 kV PIV, 150 mA; appears to be gridless 152TL, in tubular bulb.

152TH [AKA 152H, 3-150A3]; triode (twin paralleled 75THs), fils. 5 V @ 6.2 A each, μ 20, 150 W (7-26-43).

152TL [AKA HK-252L, 3-150A2]; triode, variant of 152TH w/ μ reduced to 12 (7-26-43).

227A [AKA 5C27, q. v.]; VHF triode. Four in 200-

MHz ring osc. in SC-1.

250-R [AKA 2-250A]; rectifier, fil. 5.0 V @ 10.5 A, 60 kV PIV, 250 mA; appears to be 250TH less grid (EI, PL, W); in SCR-268C.

250TH [AKA 3-250A4, VT-220, RK-63, HK-454H, CV2589]; triode, fil. 5 V @ 10.5 A, μ 32, 250 W (EI, ICE) (7-26-43). In BC-610.

250TL [AKA HK-454L, 3-250A3, VT-130]; triode, 250TH w/ μ 13, supersedes 150T, promoted for RF heating; sim. to uprated 806 (7-26-43). In BC-436.

253 Vacuum rectifier, fil. 5.0 V @ 10 A, 15 kV PIV, 350 mA.

254W Triode, Eimac version of HK-254; said to be common in base-station xmtrs. (1969).

300T Triode, fil. 7.5 V @ 12 A, μ 16, 300 W.; replaceable w/ 450TL (and probably abandoned in favor of it).

304TH [AKA 3-300A3, VT-254, 304H, WL-535, CV2611]; RF-heating/xmtg. triode, four sets of 75TH elements in parallel, all-glass tubular construction plus ceramic base ring, top cap, grid-out-side, fils. 5 V @ 12.5 A each, 300 W, μ 20, 40 MHz (EI, HK, RCA) (7-26-43) In T-8/CPN-3, BC-912, and SG.

304TL [AKA 3-300A2, VT-129, (HK) 304L, WL-525]; radar-pulser triode, fils. 5 V @ 12.5 A each, 300 W, μ of 12 (EI, HK) (7-26-43). In AN/ART-3, BC-409, et al.

327A [AKA 5C37]; UHF triode. Four in 175-225 MHz ring osc. of SC-2.

327B [AKA 3-100H3]; triode, 327A w/o plate fins.

450T Triode; see 450TH, 450TL.

450TH [AKA WL-450, VT-108, 3-450A4, 854H]; triode, fil. 7.5 V @ 12 A, μ 38, 450 W, 40 MHz, radiation-cooled; FCC-listed for broadcast xmtrs. (EI, F, W) (7-26-43). In BC-402B.

450TL [AKA 3-450A2, 300T, HK-854L]; triode, 450TH w/ μ of 18 (7-26-43).

500T Triode, fil. 7.5 V @ 21 A, 500 W; interchanges w/ 750TL.

527, A [AKA 3-300G4]; VHF triode, long tubular bulb, four plate lugs out sides, one large plate lug out side, dual-sleeve fil. bosses at bottom. Four in AN/TPS-18, giving 1 MW out at 210-225 MHz.

750TH Triode, high-μ variant of 750TL; μ 30.

750TL [AKA HK-1054L, 3-750A2]; triode, fil. 7.5 V @ 21 A, μ 15, 750 W, 100 MHz; interchanges w/ 500T; FCC-listed for broadcast use (7-26-43). Two in Collins 231D-11 (Navy TDH).

1000T [AKA 3-1000A4]; low-VHF triode, fil. 7.5 V @ 17 A, μ 35, 1000 W, 50 MHz, forced-air cooled (EI, GE, W) (7-20-43).

1000UHF VHF triode, fil. 7.5 V @ 16 A, 1000 W, forced-air cooled, 150 MHz. Dropped before 1945; replaceable by 1000T.

1500T [AKA 3-1500A3]; triode, fil. 7.5 V @ 24 A, 1500 W, μ 24, forced-air cooled. Two in Zenith W9XZR 1-kW prewar FM xmtr. (7-26-43).

2000T [AKA 3-2000A3]; triode, fil. 10 V @ 25 A, 200 W, μ 23, forced-air cooled (7-26-43).

X578 Tetrodes, airborne, heaters 26.5 V, for missile telemetry and airborne comm. (1960).

X578G Variant of 4CX300A, 1.65" long, 2.5" high, breech-block base, 300 W, 500 MHz, forced-air cooled.

X578H Variant of 4CX125C, 1.25" long, 2.5" high, breech-block base, w/ ringed anode for transverse air cooling.

X578J Variant of 4CN15A, 0.9" long, 2.5" high, breech-block base, plain unfinned anode, 15 W w/o cooling).

X2159 Water-cooled tetrode, 2 MW (originally 1.25 MW). Two each in Continental 323C 1-MW and D323C 2-MW AM xmtrs., 1973.

X2170 Water-cooled tetrode, 650 kW. Two in Continental 320F 500-kW AM xmtr.

X2099 Cermet tetrode, quasi-octal base, for linear amps., push-to-talk use w/ 250-msec. warmup time, fil. 2.5 V @ 10 A.

Y-503 Planar triode, 7855 w/ threaded stud on anode for mounting on cooler.

Y-676A [AKA 9013, q. v.]; tetrode.

YC-130 [AKA 9019, q. v.]; tetrode.

An old Eimac standby: the 100TH. *Photo: Bob Dennison, W2HBE.*

GENERAL ELECTRIC

General Electric made tubes beginning in WW I. Most of the company's production of entertainment and transmitting tubes in the '20s was numbered and sold through RCA; however, GE also sold tubes directly to the industrial market.

GE applied the following terms to its products as late as the '50s: a *kenotron* was a vacuum rectifier, a *phanotron* was a gas or mercury-vapor rectifier, and a *pliotron* was a gridded vacuum tube. *Thyratron* (a gridded gas or mercury-vapor tube) was used throughout the industry, as was *ignitron* (a mercury-arc pool-cathode device with an igniter electrode).

B-27 [AKA 876, q. v.]; ballast.

B-41 [AKA 886, q. v.]; ballast.

FA-6 [AKA 5627, q. v.]; vacuum switch.

FA-13 Resistance vacuum gauge, two-lead glass bulb with sensing tube, ca. 7 ohms resistance, covers 0-600 microns pressure; uses FA-14 compensator tube.

FA-14 Compensator tube for FA-13, UX-199 format, same filament as FA-13 but sealed-off in vacuum, for use in bridge ckt.

FA-15 [AKA 5626, q. v.]; vacuum switch.

FB-49 [AKA UX-897, q. v.]; ballast.

FB-50 [AKA UV-896, q. v.]; ballast.

FG-15 [AKA UV-857, q. v.]; mercury rectifier.

FG-19, A [AKA 872, A, q. v.]; mercury rectifier.

FG-20 Mercury full-wave rectifier.

FG-26 Mercury rectifier, 1 kV PIV.

FG-27A Gas-mercury triode thyratron for relay control, 1929, "S" bulb w/ cap, UX base, fil. 5.0 V @ 4.5 A w/ CT, 1 kV PIV, 2.5 A (GE, HY).

FG-28 Mercury rectifier, 5-1/16" dia. by 14-5/8" high, heater 5.0 V @ 17.5 A, 3.5 kV PIV, 12.5 A.

FG-29 Thyratron rectifier, heater 5.0 V @ 17.5 A, 3.5 kV PIV, 12.5 A avg., superseded by GL-44.

FG-32 [AKA (GL) 5558, q. v.]; mercury rectifier.

FG-33 Thyratron.

FG-37 Thyratron, 115-V heater (equals 91?).

FG-41 [AKA 5830, q. v.]; mercury triode thyratron.

FG-42 Gas rectifier, 10-¼" dia. by 25-½" long, heater 5.0 V @ 65 A, 16 kV PIV, 75 A.

FG-43 Thyratron.

FG-44 Gas rectifier, 10¼" dia. by 33½" long, heater 5.0 V @ 80 A, 15 kV PIV, 75 A.

FG-45 Triode thyratron, 10¼" dia. by 34-1/8" long, heater 5.0 V @ 80 A, 15 kV PIV, 75 A.

FG-52 Mercury rectifier, ca. 1932.

FG-53 Triode thyratron, 8-3/4" dia. by 25-½" long, heater 5.0 V @ 65 A, 1.5 kV PIV, 100 A.

FG-57 [AKA WL-631, 5559, q. v.]; thyratron.

FG-64 Mercury rectifier.

FG-65 Triode thyratron rectifier, fil. 2.5 V @ 2 A, 1 kV PIV, 100 mA.

FG-67 [AKA 5728, q. v.]; thyratron.

FG-81A Triode gas thyratron, ST bulb w/ cap, UX base, fil. 2.5 V @ 5.0 A, 500 V PIV (originally 180 V), 500 mA (GE, W).

FG-95 [AKA 5560, q. v.]; thyratron.

FG-97 Mercury tetrode thyratron, special conical bulb w/ cap, UX base, grid-out-side, fil. 2.5 V @ 5.0 A, 1 kV PIV, 500 mA (GE, HY).

FG-98A Mercury tetrode thyratron, ST bulb w/ cap, UX base, grid-out-side, fil. as for FG-97, derated to 500 V PIV.

FG-104 [AKA 5561, q. v.]; thyratron.

FG-105 [AKA WT-210-0074, WT-210-0079]; mercury tetrode thyratron, tapered or tubular bulb w/ cap, special 4-pin base, grid-out-side, heater 5.0 V @ 10 A, 2.5 kV PIV, 6.4 A (GE, HY, RCA, W) (5-10-46).

FG-118 Triode thyratron, grid-controlled rectifier, 5-1/16" dia. by 17-5/8" long, heater 5.0 V @ 20 A, 10 kV PIV, 12.5 A.

FG-119A Full-wave gas rectifier, 5" dia. by 19" long, heater 5.0 V @ 20 a, 1 kV PIV, 25 A.

FG-139 Ignitron, glass-bulb type, 1934; GE's first commercial ignitron; replaced by steel-body types; 4-¼" dia. by 15-¼" long, 900 V PIV, 15 A.

FG-154 [AKA 1254]; tetrode gas thyratron, special 3" dia. conical bulb w/ cap, UX base, grid-out-side, fil. 5.0 V @ 7.0 A, 500 V PIV, 2.5 A.

FG-157 Thyratron, cold-cathode, 2-¼" dia. by 5-5/8" long, 10 mA.

FG-166 [AKA 2252, WL-786]; mercury rectifier, steel case w/ mounting bracket and flying leads out both ends, fil. 2.5 V @ 100 A, 1.5 kV PIV, 20 A, similar physically to 870A.

FG-172 [AKA 1672, WT-210-0038, WT-210-0078]; mercury tetrode thyratron for continuous or welder use, steel case 2-¼" dia. by 8-3/8" long w/ mounting brackets and flying leads out both ends; fil. 5.0 V @ 10 A, 2 kV PIV, 6.4 A (GE, HY, RCA, W) (ca. 1936). Also see GL-414.

FG-178, A [AKA 0178]; gas triode thyratron, T10 tubular bulb w/ cap, UX base, heater 2.5 V @ 2.25 A, 500 V PIV, 125 mA.

FG-179 Ignitron, glass bulb 10-¼" dia. by 24" high, 1 kV PIV, 25 A (1934). GE's largest glass ignitron; replaced by steel-body types; nearest equivs. FG-235 and FG-271. "A" is derated to 900 V but raised to 35 A.

FG-190 [AKA 1290]; full-wave gas (not mercury) rectifier, steel case w/ mounting bracket, flying leads out bottom, grid-out-side, fil. 2.5 V @ 12 A, 175 V PIV, 1.25 A; seems to be a thyratron version

of FG-280.
FG-194 Ignitron, 25 A, 800 V PIV.
FG-235A [AKA 5552, q. v.]; ignitron.
FG-238, B [AKA ZG-238]; ignitron, glass body, cooling coil inside w/ tubing brought out base, flying lead for anode, lug for igniter, 1.5 kV PIV, 250 A, 1936. "B" [AKA 5555, q. v.] was successor.
FG-253 Ignitron, superseded by GL-415.
FG-258A [AKA 5553, q. v.]; ignitron.
FG-259B [AKA 5554, WL-679, q. v.]; ignitron.
FG-259R Ignitron.
FG-271 [AKA 5551, q. v.]; ignitron.
FG-280 Mercury rectifier, steel case w/ flying leads at both ends, heater 5.0 V @ 10.0 A, 2 kV PIV, 6.4 A. Also see FG-172.
FG-402 Ignitron, superseded by GL-415.
FH-11 Split-anode magnetron, tubular bulb, large 4-pin brass base, long dual plate pins, fil. 5.0 V @ 5.0 A, 1.5 kV, 50 mA, external magnet w/ field of 700 G; oscillates to 400 MHz; gives 25 W out @ 150 MHz or 35 W @ 60 MHz (*Proc. IRE*, Nov. 1931).
FJ-31 "Selenium tube" (photocell).
FJ-76 Phototube.
FJ-405 Vacuum phototube, tubular bulb, UX base, S-6 response, lab meas. device, 12 μA/lumen.
FM-500 [AKA UY-224, q. v.]; rcvg. tetrode.
FP-1 [AKA UX-852, q. v.]; xmtg. triode.
FP-2 "Pliotron, 500 W, short wave"; see FP-1.
FP-3 Pliotron.
FP-52 Vacuum rectifier, 2-5/16" dia. by 7-7/8" long, fil. 10.0 V @ 3.25 A, 3.5 kV PIV, 200 mA.
FP-53 CRT.
FP-54 [AKA 5740, q. v.]; electrometer tetrode.
FP-59 [AKA 30, q. v.]; triode.
FP-60 [AKA 31, q. v.]; power triode.
FP-61 [AKA 32, q. v.]; tetrode.
FP-62 Triode ionization gauge, flying leads at bottom, glass sampling tube at top, fil. 4.5 V @ 1.48 A, sensitivity 40 μA per micron pressure.
FP-78 Vacuum rectifier.
FP-84 Vacuum rectifier, 4-1/8" dia. by 14-7/8" long, fil. 10.0 V @ 14.5 A, 125 kV PIV, 100 mA peak.
FP-85, A [AKA 5741, q. v.]; vacuum rectifier.
FP-92 Vacuum rectifier, 8-1/8" dia. by 25-1/8", fil. 10.0 V @ 14.5 A, 140 kV PIV, 300 mA peak.
FP-126A Barkhausen UHF oscillator tube, ca. 1933.
FP-140 Vacuum rectifier, 2-5/16" dia. by 8-½" long, fil. 10.0 V @ 3.25 A, 7.5 kV, 600 mA peak.
FP-141 Vacuum rectifier, 4-1/16" dia. by 14-3/8" long, fil. 10.0 V @ 14.7 A, 11 kV, 750 mA peak.
FP-163 "Timing tube (inactive)."
FP-164 Rectifier filament timing control.
FP-197 RF-heating HF triode, replaceable by 211.
FP-198 RF-heating UHF triode osc.
FP-252, A [AKA NU200]; RF-heating triode, replaceable by HF-200, FV20.
FP-265 RF-heating HF triode, super-jumbo 4-pin base w/ cap, fil. 10 V @ 5.2 A, μ 75, 160 W, 15 MHz, suited to raw-AC plate feed.
FP-270 Xmtg. triode.
FP-285 RF-heating HF triode, jumbo 4-pin base, fil. 10 V @ 3.25 A, μ 12, 100 W, 20 MHz; replaceable by 311CT, 211C.
FP-400 Diode rectifier, "to be used in the study of electrical discharges in high vacuum"; ST-14 bulb, UX base, pure tungsten fil., 4.0 V @ 2.25 A, max. plate voltage 125 @ 25 mA, apparently for classroom-demo and teaching-lab use.
FQ-5 Tetrode, water-cooled.
FR-101 Pliotron, for vacuum-tube voltmeters.
FR-300 [AKA UX-864, q. v.]; audio triode.
FR-301 [AKA UX-245, q. v.]; power triode.
FR-503 [AKA UY-227, q. v.]; rcvg. triode.
FT-1-B [AKA UV-862, q. v.]; xmtg. triode.
FT-6 [AKA UV-858, q. v.]; xmtg. triode.
GL-34 (Long) [AKA 24154, q. v.]; vacuum switch.
GL-34 (Short) [AKA 24163, q. v.]; vacuum switch.
GL-34A [AKA 1S21, q. v.]; vacuum switch.
GL-75/30 [AKA 0A3, q. v.]; voltage regulator.
GL-90/30 [AKA 0B3, q. v.]; voltage regulator.
GL-105/30 [AKA 0C3, q. v.]; voltage regulator.
GL-146 Xmtg. triode, for zero-bias Class B amps., industrial 4-pin base w/ cap, T-T fil. 10 V @ 3.25 A, μ 75, 125 W, 15 MHz. Also see GL-152.
GL-150/30 [AKA 0D3, q. v.]; voltage regulator.
GL-152 [AKA P0016B]; xmtg. triode, GL-146 for Class B and C amps., μ reduced to 25.
GL-159 Xmtg. triode, large special base w/ cap, for Class B or C uses, promoted for police base stations, T-T fil. 10 V @ 9.6 A, μ 20, 250 W, 15 MHz, convection cooled. Also see GL-169.
GL-169 Xmtg. triode, GL-159 w/ μ of 85, same uses.
GL-299 Thyratron.
GL-411 Vacuum rectifier, stepped tubular bulb; smaller, lower-rated version of KC-4; fil. 10 V @ 14.5 A, 100 kV PIV, 300 mA peak out.
GL-414 [AKA WL-414, q. v.]; thyratron.
GL-415 [AKA 5550, q. v.]; ignitron; also see 1R21.
GL-418 Triode vacuum control tube.
GL-421 Xmtg. pentode control tube.
GL-427 Ignitron, glass bulb, small classroom-demo and lab type, 5 A, 1940; superseded by 5779.
GL-429 Thyratron.
GL-434, A [AKA ZP434]; xmtg. UHF triode. In BC-677.
GL-436 [AKA ZP436, REL19]; xmtg. UHF triode.
GL-441 [AKA CE61A/B, 59DV2]; vacuum phototube, 868 format, S-4 response, 45 μA/lumen, for measurement and relay uses.
GL-442 Thyratron.
GL-446, A, B [AKA REL34, ZP620]; "lighthouse" UHF planar rcvg-xmtg. triode, first "soldered can" cathode

{ZP446, A2179A}, 2-41. Superseded by "A," "B," then 2C40. Major improvement in radars: NF @ 1 GHz is 10 dB, vs. 23 dB for WE 368A. In BC-1057A and rcvr preamps of ASB-7, FC-FD, SC-2, and SCR-602T8 (AN/TPS-3). "A" in R-36/TPS-2 et al. "B" has 1656 as variant.

GL-451 [AKA ZP451, 8020, q. v.]. Four as bridge rectifier in xmtr of SC-1.

GL-452 Xmtg. triode superseded by 892.

GL-455 [AKA ZP455, 9LP7, q. v.]; CRT.

GL-464, A [AKA ZP464]; expanded, power version of 446; superseded by "A," then by 2C43, q. v. Also see GL-570. In BC-981A.

GL-471 [AKA ZG471, 1B21, A, q. v.]; TR tube. "Obsolete, no replacement needed - break glass and leave air gap." In AN/TPS-3 and SC-1.

GL-475 [AKA ZP475, 5FP7, q. v.]; CRT.

GL-481 [AKA ZA481]; diode gas switching tube.

GL-483 Special 851, dropped in favor of 851.

GL-484 [AKA ZP484, 3J24, q. v.]; magnetron.

GL-485 [AKA ZP485, 3J25, q. v.]; magnetron.

GL-486 [AKA ZP486, 3J26, q. v.]; magnetron.

GL-488 [AKA ZP488, 12GP7, q. v.]; CRT. In BC-986A.

GL-500 Phototube.

GL-502, A [AKA ZP502, WT-210-0137]; thyratron, metal 2050 (GE, RCA, W). Original shell was short and thick (6L6-dia.); later standard metal style.

GL-505 [AKA 3J35, q. v.]; magnetron.

GL-506 [AKA 6228, q. v.]; ignitron.

GL-507 [AKA 5564, q. v.]; ignitron.

GL-509 [AKA 889, q. v.]; xmtg. triode.

GL-509R Xmtg. triode superseded by 889R, q. v.

GL-510 Xmtg. mercury rect., supersed. by 857B, q. v.

GL-512, A [AKA 575A, q. v.]; mercury rectifier.

GL-515 Thyratron modulator, superseded by 4C35.

GL-516 [AKA ZP516, ZJ516, 1P24, q. v.]; phototube.

GL-522 [AKA ZP522, 3C22, q. v.]; xmtg. triode. In T-22/APT-5.

GL-529 [AKA ZG529, H350, 4C35, q. v.]; hydrogen thyratron.

GL-531 [AKA ZG531]; diode gap; in SCR-615.

GL-541 [AKA ZG541]; TR tube; in SCR-615.

GL-546 Min. gas tetrode thyratron, heater 6.3 V @ 150 mA, 500 V PIV, 20 mA. Also see 5696.

GL-553 [AKA 1P23, q. v.].

GL-559 [AKA 2B22, q. v.]; lighthouse planar diode (GE, RCA). Two in AN-91A, probably as TR switch.

GL-562 [AKA F-660]; vacuum rectifier; water-cooled.

GL-564 [AKA 1P24, q. v.]; phototube.

GL-570 [AKA 2C44, q. v.]; lighthouse triode, amp. version of 464 {ZP570}.

GL-572 [AKA ZP572, 2C39, q. v.].

GL-577 [AKA ZP577]; rcvg. multi-element UHF osc.

GL-579 [AKA ZP579, 19F4, 5J29, q. v.]; magnetron.

GL-584 [AKA ZP584, 5J31, q. v.]; magnetron.

GL-590 [AKA ZP590, 5J30, q. v.]; magnetron.

GL-594 [AKA ZP594]; magnetron.

GL-595 [AKA ZP595]; magnetron.

GL-597 [AKA ZP597]; magnetron.

GL-598 [AKA ZP598]; magnetron.

GL-599 [AKA ZP599]; magnetron.

GL-605 [AKA ML-101, q. v.]; xmtg. rectifier.

GL-615 [AKA ZP615]; magnetron.

GL-1367 Thyratron rectifier.

KC-1 Vacuum rectifier, 100 kV PIV.

KC-3 Vacuum rectifier, 150 kV PIV.

KC-4 [AKA 5625, q. v.]; HV rectifier.

KM-1 Vacuum rectifier, 30 kV PIV.

KP-1 Vacuum rectifier, 100 kV PIV in oil.

KP-2 Vacuum HV rectifier, 75 kV PIV.

KR-1 Vacuum rectifier, 125 kV PIV (not to be confused with Ken-Rad rcvg. KR1!).

PJ-1 Gas phototube.

PJ-2 [AKA RJ574; initially numbered TC-3]; "train control amp." high-μ voltage-amp. triode, UV base, fil. 4.5 V @ 1.1A, μ 30.

PJ-3 Mercury-vapor rectifier.

PJ-4 [AKA, initially, as TC-4]; "train control" power triode, UV base, fil. 4.5 V @ 1.1 A, μ of 8.5.

PJ-5 Vacuum phototube.

PJ-6 Vacuum rectifier.

PJ-7 [AKA 5742, q. v.]; triode.

PJ-8 [AKA 5556, q. v.]; power triode.

PJ-9 [AKA 842, q. v.]; power triode.

PJ-10 Triode, "special PT-201-A," μ of 8.

PJ-11 Triode vacuum control tube, "special PT-240."

PJ-12 Triode, "special PR-11-A," q. v.

PJ-13 [AKA UX-210, q. v.]; power triode.

PJ-14, B Vacuum phototube, 1.08 μA/lumen, 2-½" dia.

PJ-15 [AKA UX-867, q. v.]; phototube.

PJ-16, 18 Vacuum phototubes.

PJ-17, 19 Argon phototubes.

PJ-20 Phanotron, "AC gas-discharge regulator."

PJ-21 [AKA 5743, q. v.]; triode.

PJ-22 [AKA CE-1V, WL-734]; phototube.

PJ-23 [AKA 868, q. v.]; phototube.

PJ-24 Mercury rectifier, 2.5 V @ 12.5 A, 1.5 A peak, 20 kV PIV.

PJ-25 Gas lithium-cathode phototube, quartz bulb (UV-sensitive?).

PJ-26, A [AKA UV-869, A, q. v.]; mercury rectifier.

PJ-27 [AKA 865, q. v.]; tetrode.

PJ-28, A [AKA 866, A, q. v.]; rectifier.

PJ-29 Gas phototube, quartz bulb (UV-sensitive?).

PJ-30 Vacuum phototube, quartz window (UV-sensitive?).

PR-1 [AKA UV-201, q. v.]; triode.

PR-1-A [AKA UV-201A, q. v.]; triode.

PR-2A, B [AKA UV-202, A, q. v.]; xmtg. triode.

PR-3-A, B [AKA UV-203, A, q. v.]; xmtg. triode.

PR-4-A, B [AKA UV-204, A, q. v.]; xmtg. triode.

PR-10-A [AKA UX-210, q. v.]; power triode.

PR-11-A [AKA UV-211, q. v.]; xmtg. triode.
PR-51-A [AKA UV-851, q. v.]; xmtg. triode.
PR-845 [AKA 845, q. v.]; xmtg. triode.
PR-861 [AKA 861, q. v.]; xmtg. triode.
PT-210 [AKA 10, q. v.]; power triode.
PT-841 [AKA 841; q. v.]; xmtg. triode.
PT-842 [AKA 842; q. v.]; xmtg. triode.
PT-860 [AKA 860; q. v.]; xmtg. triode.
ZG530 [AKA 1B30, q. v.]; gas switching tube.
ZG547 [AKA H45, 3C45, q. v.].
ZG569 [AKA GK50, 5C22, q. v.]; hydrogen thyratron.
ZG621 [AKA 1B52, q. v.]; ATR tube.
ZG625 [AKA 1B55, q. v.]; TR tube.
ZG626 [AKA 1B35, q. v.].
ZG627 [AKA 1B36, q. v.]; ATR tube.
ZG629 [AKA 1B37, q. v.]; ATR tube.
ZG645 [AKA 1B38, q. v.]; pre-TR tube.
ZP2 Pliotron.
ZP6B [AKA 2C43, q. v.]; xmtg. triode.
ZP94 VHF triode, dev., water-cooled, ca. 1933.
ZP423 Xmtg. UHF triode, bombardment cathode; predecessor of ZP430 and then 8012.
ZP430 Xmtg. triode, fil.-type disc-seal (not dev. fully).
ZP432 Xmtg. UHF osc.
ZP433 Xmtg. UHF triode; large (4" dia.) disc-seal prototype giving 300 W out at 440 MHz.
ZP435 Xmtg. UHF triode.
ZP447 [AKA 2B27, q. v.]; lighthouse diode. Two in AN-137-A (as TR switches?).
ZP448 Lighthouse tetrode, ca. 3-41; no known commercial version, dev. from ZP464.
ZP449 [AKA 3C29]; xmtg. UHF lighthouse power triode, superseded by 3C22.
ZP453 [AKA 5BP4, q. v.].
ZP468 [AKA 7BP7, q. v.]; CRT.
ZP469 Xmtg. UHF triode.
ZP470 [AKA (WE) 316A, q. v.]; xmtg. triode.
ZP477 [AKA 12DP8, q. v.]; CRT.
ZP480 [AKA 5BP1, q. v.]; CRT.
ZP517 [AKA 5CP1, q. v.]; CRT.
ZP518 [AKA 5CP4, q. v.]; CRT.
ZP526 [AKA 3DP1, q. v.]; CRT.
ZP527 [AKA 6C21, q. v.]; xmtg. triode.
ZP540 [AKA 3FP7, q. v.]; CRT.
ZP548 [AKA 12FP7, q. v.]; CRT.
ZP549 [AKA 9GP7, q. v.]; CRT.

A whole generation of radio fans "got KFI" on GE-made UV-200 detectors.
Photo: Bob Dennison, W2HBE

ZP556 [AKA 4AP10, q. v.]; CRT.
ZP600 Xmtg. UHF osc.
ZP602 [AKA 2C41, q. v.]; lighthouse triode.
ZP612 [AKA 6J21, q. v.]; CW magnetron.
ZP614 Xmtg. UHF osc.
ZP616 Magnetron.
ZP620 [AKA 2C40, q. v.]; lighthouse triode.
ZP646 [AKA 5J32, q. v.]; magnetron.
ZP650 [AKA 2B23, q. v.]; diode.
ZP655 Xmtg. triode, 889 w/ disc electrodes; led to 7C21 and 9C24.
ZP666 Magnetron, 5J30 capacitance-loaded for lower freq. operation; obsol.; eqpt. mod. to use 5J32.
ZP676 [AKA 5J33, q. v.]; magnetron.
ZP679 Water-cooled triode.
ZP682 [AKA 2C49, q. v.]; lighthouse triode.
ZP693 Submin. type.

RAYTHEON

Raytheon produced specialized tubes in several areas: cold-cathode types, subminiatures, microwave types, and (mainly in the '30s) a line of transmitting tubes aimed primarily at the amateur market. This section covers those Raytheon devices that were numbered as "RK," "CK," and "QK" types. These are, of course, in addition to those with RMA/EIA registered numbering, including ordinary types whose numbers were prefixed with, say, "RK" for brand identification.

RK-CODED TYPES

RK-10 Power triode, 10 w/ Isolantite base, upped to 15 W.

RK-11 [AKA TZ40]; xmtg. triode, fil. 6 V @ 3 A, 25 W. Also see RK-12. Replaceable by 1623.

RK-12 [AKA HY30Z]; xmtg. triode, 809 w/ μ raised to 100. Also see RK-11.

RK-15 Rcvg. power triode, 4-pin base, grid out top, 10 W, equiv. to triode-conn. 46.

RK-16 Rcvg. power triode, 5-pin base, equiv. to triode-connected 59.

RK-17 Xmtg. pentode, "S" bulb w/ cap, 2A5 on 5-pin base w/ heater raised to 2 A.

RK-18 Xmtg. triode, stepped-tubular bulb on UX base w/ cap, fil. 7.5 V @ 3 A, μ 18, 35 W.

RK-19 Xmtg. full-wave vacuum rectifier, special base, heater 7.5 V @ 2.5 A, 3.5 kV, 200 mA.

RK-20, A Xmtg. pentode, stepped-tubular bulb w/ shrouded cap on Insolatite UX base (1935). Sim. to 804, q. v., but lower-rated. Per RA, "the original RF pentode for amateur service." Variants: RK-46, RK-47. "A" [AKA CK70] has fil. 3.25 A, 40 W.

RK-21 [AKA 38266A]; vac. rectifier sim. to 836, UX base w/ cap, heater 2.5 V @ 4 A, 3.5 kV, 200 mA.

RK-22 [AKA 38222]; full-wave rectifier, sim. to 3B23 w/ fil. current raised to 8 A, 3.5 kV PIV, 200 mA.

RK-23 Xmtg. pentode, w/ cap, 7-pin Isolantite base; third grid brought out separately for suppressor modulation, heater 2.5 V @ 2.0 A, 10 W. Variants: RK-25, RK-45.

RK-24 [AKA HY-24, q. v.]; triode.

RK-25, A, B Xmtg. pentode, w/ cap, 7-pin base, RK-23 w/ heater 6.3 V @ 900 mA; third grid brought out separately for suppressor modulation, replaceable by 802. "B" has phenolic base.

RK-28, A Xmtg. pentode, sim. to 803, q. v., except fil. current raised to 8.0 A and 100 W max. dissipation instead of 125.

RK-30 [AKA 800, q. v.]; xmtg. triode.

RK-31 Xmtg. triode, for class B modulators, UX base w/ cap, fil. 7.5 V @ 3 A, μ 170, 40 W.

RK-32 [AKA 834, q. v.]; xmtg. VHF triode.

RK-33 [AKA 2C21, q. v.]; duotriode.

RK-34 [AKA 2C34, q. v.]; duotriode; originally on Isolantite base. In BC-647A and CPR-60AAK.

RK-35 Xmtg. triode, pear-shaped bulb on UX base w/ grid-out-side and plate cap, Ta plate, fil. 7.5 V @ 3.25 A, μ 9, 35 W, 30 MHz. Variant: RK-37.

RK-36 Xmtg. triode, Ta plate, UX base w/ cap, grid-out-side, fil. 5 V @ 8 A, μ 14, 100 W; 100TL ratings w/ slightly greater μ and fil. current. Variant: RK-38.

RK-37 Xmtg. triode, RK-35 with μ raised to 30.

RK-38 Xmtg. triode, variant of RK-36, apparently vari-μ for modulators.

RK-39 Xmtg. pentode, non-beam version of 807, 25 W. Variant: RK-41.

RK-41 Xmtg. pentode, RK-39 w/ heater 2.5 V @ 2.4 A.

RK-42 Rcvg. triode, fil. 1.5 V @ 60 mA; 30 char. otherwise - RA claimed this and RK-43 were the first 1.5-V low-drain tubes.

RK-43 Rcvg. duo-triode, fil. 1.5 V @ 120 mA; for Class B use; μ higher than for RK-42.

RK-44 [AKA 837, q. v.]; xmtg. pentode.

RK-45 Xmtg. pentode, RK-23 w/ heater 12.6 V @ 450 mA.

RK-46 Xmtg. pentode, RK-20 w/ heavier bulb, fil. 12.6 V @ 2.5 A.

RK-47 Xmtg. beam tetrode, beam version of RK-20A. Also see 814.

RK-48, A [AKA HK-257B, 4E27, q. v.]; xmtg. beam tetrode, fil. 10 V @ 5 A, 100 W (vs. 125 for 4E27).

RK-49 [AKA T-21, q. v.]; beam tetrode, aligned-grid version.

RK-51 [AKA NU51A]; xmtg. triode, UX base w/ cap, fil. 7.5 V @ 3.75 A, μ 20, 60 W; variant: RK-52.

RK-52 [AKA NU51Z]; xmtg. triode, RK-51 w/ μ raised to 150, for zero-bias use.

RK-56 Xmtg. beam tetrode, 5-pin base w/ cap, heater 6.3 V @ 0.6 A, 8 W. Also see HY60.

RK-57 [AKA 805, q. v.]; xmtg. triode.

RK-58 [AKA 838, q. v.]; xmtg. triode.

RK-59 Xmtg. duo-triode, ceramic UX base, quick-heating fil., 6 V @ 1 A, μ 25, 15 W total.

RK-60 [AKA (GL) 1641]; full-wave vacuum rectifier, UX base w/ twin caps; fil. 5 V @ 3 A, 2120 V PIV, 250 mA. In Collins 32RA.

RK-61 Submin. gas triode, 0.55" dia., fil. 1.4 V @ 50 mA {QF-6}; see RK-62 for use.

RK-62 Rcvg. gas triode, prewar tube for superregen. receivers for remote control of models, fil. 1.4 V @ 50 mA, 45 V B+.

RK-63, A Xmtg. triode, fil. 5.0 V @ 14 A, μ 37 - 250TH parameters but rated at only 200 W. "A" has fil. 6 V @ 14 A.

RK-64 Xmtg. pentode, 5-pin base w/ cap, heater 6.3 V @ 500 mA, 6 W.

RK-65 [AKA 5D23, q. v.]; xmtg. tetrode.

RK-66 [AKA (HY, NU) 69]; beam tetrode, quick-heating

variant of 807, fil. 6.3 V @ 1.5 A.

RK-72, R [AKA RKR72, RX232, CE-220]; xmtg. vac. rectifier for radars; T-T fil. 2.5 V @ 3.0 A, 22 kV PIV, 20 mA (CE, HY, RA), replaceable by 3B24 (3-17-43). In T-2/APS-15.

RK-73, A, R [AKA RKR73]; HV rectifier (clipper diode), w/ cap, fil. 2.5 V @ 4.2 A, 13 kV PIV, 30 mA (3-17-43). In T-11/APN-2. "A" is cathode variant.

RK-75 [AKA (WE) 307A, q. v.]; xmtg. pentode.

RK-100 Xmtg. triode *mercury-vapor* amp., heater 6.3 V @ 900 mA, μ 50, cath. curr. 250 mA, 10 W; originally intended as output stage in AC-DC receivers, to develop substantial plate current and output power w/ limited supply voltage. Could put out RF power of 12 watts w/ only 110-V B+.

CK-CODED SUBMINIATURES

CK501, X AX, LX Hearing-aid pentode, ca. 3-40, tipped tubular bulb 1-5/8" long, 5-pin "peanut" base and metal shell, fil. 1.25 V @ 30 mA, g_m 0.33 mS, 30 V B+, zero bias. "X" has tinned leads, no shell.

CK502, A, X, AX, AY Hearing-aid power pentode, CK501 format; same ratings as CK501 but g_m 0.4 mS. "X" has tinned leads. "AX" has 6 mW out, is sim. to 2E35. "AY" is [AKA 2E36, q. v.].

CK503, AX, X [AKA 5854, q. v.]; hearing-aid power pentode. "X" has tinned leads.

CK504, AX Hearing-aid pentode, CK501 format, same ratings as CK501 but g_m 0.35 mS.

CK505, AX, X Hearing-aid pentode, CK501 format, fil. 0.625 V @ 30 mA, 30 V B+, zero bias, intro. 8-40. "X" has tinned leads.

CK506AX Hearing-aid power pentode, fil. 1.25 V @ 50 mA, 25 mW out.

CK507AX Hearing-aid pentode.

CK509AX Hearing-aid triode.

CK510AX Hearing-aid space-charge duo-tetrode, fil. 0.625 V @ 50 mA. One space-charge grid w/ two control grids and plates; gain of 250 for both sections.

CK511X Submin. AF "reentrant pentode," acts as two triode amps. in tandem, gives gain of 500 w/ 45 V B+. μ is 17 for both sections. *Electronics*, 6-46.

CK512AX Submin. low-microphonics pentode, fil. 0.625 V @ 20 mA, gain 39 X.

CK515BX Submin. rcvg. type.

CK518AX Hearing-aid power pentode, spray-shielded, fil. 1.25 V @ 30 mA, 10 mW out.

CK520AX Submin. power pentode.

CK521CX Submin. power pentode.

CK522AX [AKA 6088, q. v.]; hearing-aid power pentode.

CK523AX Hearing-aid power pentode, fil. 1.25 V @ 30 mA, 2.2 mW out.

CK524AX Hearing-aid power pentode, CK523AX char.

CK525AX Hearing-aid power pentode, fil. 1.25 V @ 20 mA, 2.2 mW out.

CK526AX Hearing-aid power pentode, fil. 1.25 V @ 20 mA, 4 mW out.

CK527AX Hearing-aid power pentode, fil. 1.25 V @ 15 mA, 0.8 mW out.

CK528AX Submin. hearing-aid pentode, spray-shielded, fil. 1.25 V @ 20 mA, 1.2 mW out.

CK529AX Hearing-aid power pentode, spray-shielded, fil. 1.25 V @ 20 mA, 1.6 mW out.

CK531DX Hearing-aid power pentode, fil. 1.25 V @ 20 mA, 1.6 mW out.

CK532DX Hearing-aid power pentode, fil. 1.25 V @ 15 mA, 1.8 mW out.

CK533AX Hearing-aid power pentode, CK532DX char.

CK534AX Submin. pentode, fil. 0.625 V @ 15 mA.

CK535AX Hearing-aid power pentode, CK531DX char.

CK536 Hearing-aid power pentode, CK532DX char.

CK537 Hearing-aid power pentode, spray-shielded, fil. 1.25 V @ 20 mA, 4 mW out.

CK538DX Hearing-aid power pentode, fil. 0.625 V @ 15 mA, 4 mW out.

CK539DX Hearing-aid power pentode, fil. 1.25 V @ 20 mA, 2.2 mW out.

CK541DX Hearing-aid power pentode, fil. 1.25 V @ 15 mA, 1.4 mW out.

CK542DX Hearing-aid power pentode, fil. 1.25 V @ 15 mA, 4 mW out.

CK546 [AKA DF651]; submin. voltage-amp. pentode, fil. 0.625 V @ 10 mA, g_m 0.068 mS.

CK549 [AKA DL651]; submin. power pentode, fil. 1.25 V @ 10 mA, g_m 0.45 mS

CK551AXA Submin. diode-pentode, fil. 1.25 V @ 30 mA, g_m 0.24 mS.

CK553AXA Submin. RF pentode, fil. 1.25 V @ 50 mA, g_m 0.55 mS.

CK556AX [AKA 5676, q. v.]; VHF triode.

CK568AX [AKA 5677, q. v.]; VHF triode.

CK569AX [AKA 5678, q. v.]; RF pentode.

CK570AX [AKA 5697, q. v.]; submin. electrometer tube.

CK571AX [AKA 5886]; submin. electrometer pentode, fil. fil. 1.25 V @ 10 mA, w/ grid lead segregated.

CK573AX [AKA 6029]; submin. triode, VHF osc., fil. fil. 1.25 V @ 200 mA, g_m 2.0 mS.

CK574AX Submin, RF pentode, spray-shielded, fil. 0.625 V @ 20 mA, g_m 0.16 mS.

CK600AX Submin. triode UHF osc.

CK602 Submin. UHF pentode, double-ended, heater 6.3 V @ 200 mA, spray-shielded T3 bulb; like CK604; has integral screen bypass cap.

CK603 Submin. double-ended UHF med.-μ triode, heater 6.3 V @ 200 mA.

CK604A Submin. RF pentode, spray-shielded T4-½ bulb, heater 6.3 V @ 200 mA, g_m 5 mS; effectively a submin. 6AK5.

CK605CX [AKA 6540, q. v.]; submin. pentode.

CK606BX [AKA 5704, q. v.]; submin. diode.

CK608CX [AKA 5703, q. v.]; submin. triode.

CK611 [AKA 6N4, q. v.]; min. triode.

CK612 [AKA 6AN5, q. v.]; min. pentode.
CK619, CX [AKA 5744, q. v.]; submin. triode.
CK626CX T4 submin. rectifier, heater 6.3 V @ 300 mA, 850 V PIV, 45 mA.
CK628 [AKA 6247, q. v.]; submin. triode.
CK1037 [AKA 6437, q. v.]; submin. voltage regulator.
CK1038 Submin. voltage regulator, 900 V @ 5-15 μA.
CK1039 [AKA 6438, q. v.]; submin. voltage regulator.
CK1045 [AKA 6542, q. v.]; submin. voltage regulator.

OTHER CK-CODED ITEMS

CK70 [AKA RK-20A, q. v.]; xmtg. pentode.
CK108 Rcvg. low-noise pentode; replaceable by 77.
CK113 Rcvg. rectifier and pentode.
CK1003 [AKA 0Z4A, q. v.]; gas rectifier.
CK1004 Rcvg. full-wave gas rectifier.
CK1005 [AKA QMG159, VT-195, CV2874]; rcvg. octal full-wave gas rectifier, fil. 6 V @ 100 mA, 450 PIV, 70 mA (11-2-42).
CK1006 [AKA VT-249]; rcvg. full-wave gas rectifier, ST14 bulb on 4-pin base, graphite anodes, fil. 2.5 V @ 2.0 A, 1600 V PIV, 200 mA (1-20-43).
CK1007 Rcvg. octal full-wave gas rectifier, fil. 1 V @ 1.2 A, 980 V PIV, 110 mA (5-15-45).
CK1008 Rcvg. full-wave gas rectifier.
CK1009 [AKA BA]; rcvg. full-wave gas rectifier.
CK1010 Rcvg. full-wave gas rectifier.
CK1011A [AKA 1B48, q. v.]; HV rectifier.
CK1012 Rcvg. full-wave gas rectifier, ST14 bulb on 4-pin base, 1.2 kV PIV, 300 mA, w/ aux. filament (1.75 V @ 2 A) to help starting. Can be used directly heated or ionically heated.
CK1013 [AKA 5517, q. v.]; rectifier.
CK1017 [AKA 5962, q. v.]; voltage regulator.
CK1018-21 Geiger-Muller counter tubes.
CK1022 Min. corona volt. reg. w/ cap, 1 kV @ 5-55 μA.
CK1023 Geiger-Muller counter tube.
CK1024 Full-wave cold-cath. octal rect., 1 kV PIV, 160 mA.
CK1026 Geiger-Muller counter tube, min. dual-ended, Auqadag coat on outside, 800-950 V.
CK1027 HV rectifier sim. to CK1013.
CK1028 Rectifier, heater 6.3 V @ 550 mA.
CK1029 Geiger-Muller counter tube.
CK1042 [AKA 6659]; cold-cath. rect., 2.8 kV PIV, 8 mA.
CK1050 [AKA 7979, q. v.]; submin. indicator thyratron. Also see 7323.
CK1089 Tetrode cold-cathode visual-glow thyratron, in fluorescent-starter bulb, flying leads, two starter electrodes, 225 V, 15 mA. Intro. 4-45.
CK1090 Visual-glow thyratron, CK1089 w/ one starter electrode, sim. to WE 395A.
CK1091 [AKA 2B25, q. v.]; rectifier.
CK1800 [AKA 2C50, q. v.]; duotriode.

QK-CODED TYPES

QK26 [AKA 707B, q. v.]; klystron.
QK28 [AKA 2K28, q. v.]; klystron.
QK30 Pirani vcuum gauge.
QK33 [AKA 2K33, q. v.]; klystron.
QK36 [AKA 3B26, q. v.]; rectifier.
QK47 [AKA 2J55, q. v.]; magnetron.
QK50 [AKA 5C22, q. v.]; hydrogen thyratron.
QK51 Magnetron.
QK59-64 [AKA 4J60-4J65, q. v.]; magnetrons.
QK70-73 CW magnetron series, supersede QK59-62, q. v.
QK74A Magnetron, superseded by QK155.
QK75 Magnetron.
QK77 CW magnetron.
QK78 Magnetron, extension of QK69-73 series.
QK85 Klystron, 2K28 w/ integral cavity.
QK95 Rectifier, improved 3B26.
QK98 [AKA 4B31, q. v.]; rectifier.
QK111 Reflex klystron, replaceable by QK202.
QK143 Klystron.
QK151, 155 Magnetron.
QK159 Reflex klystron, 2950-3275 MHz, IC, heater 6.3 V @ 650 mA, 300 V, 150 mW output.
QK172, 181 Magnetron.
QK174A CW magnetron, tunes 1990-2110 MHz, 70 W.
QK201-202 Klystrons.
QK205 [AKA 5721, q. v.]; reflex klystron.
QK206 Magnetron.
QK217A CW RF-heating magnetron, fixed at 2450 ± 25 MHz, 1.5 kW, water-cooled.
QK221 Magnetron.
QK228 Magnetron; superseded by QK-254.
QK235, 241 Magnetron.
QK243 [AKA 6037, q. v.]; klystron.
QK246 Klystron.
QK247 Magnetron.
QK253 [AKA 6695, q. v.]; magnetron.
QK254 Magnetron.
QK269 Klystron.
QK288 Magnetron.
QK299 [AKA 6229, q. v.]; magnetron.
QK312 CW RF-heating magnetron, fixed at 2450 ± 25 MHz, 1.5 kW.
QK329 Beam-deflection tube, square-law multiplier, tubular domed bulb on diheptal base.
QK338A Magnetron, 5 MW, 2.7-2.9 GHz. In AN/FPS-6.
QK349 [AKA 6249, q. v.]; magnetron.
QK351 [AKA 6115, q. v.]; klystron.
QK358 Magnetron, tunes 1280-1350 MHz, 1 MW pulse output. In AN/FPS-8.
QK366 [AKA 6967, q. v.]; magnetron.
QK447 Magnetron, 5.5-5.8 GHz, 250 kW pulse output. In AN/MPQ-10C and AN/MPQ-12C.
QK496 CW magnetron, tunes 2350-3600 MHz, 150 W output. In AN/ALT-83.
QK685 [AKA 7571, q. v.]; storage tube.
QK797 [AKA 7521, q. v.]; magnetron.
QK915A, B, C Magnetrons; "A" is AKA 4J59, q. v.]; "B" is [AKA 4J58, q. v.] "C" is [AKA 4J57, q. v.].

RCA

RCA introduced special-purpose tubes under numbers in the 1600s in the early '30s. Many of these were low-runner variants of regular types, or specially selected production. The 1600 number series was used for another ten years, and other makers like GE offered the more popular types. A few even received RMA registration. Types 1637-1639 are of interest because they represent U. S. production of exclusively British types during WW II.

Early television-related tubes were numbered in the 1800s, beginning in the mid-'30s.

RCA coded tubes in the low 2000s, here and there, from the '50s through the early '70s. Its 4000-series devices were a major product line, developed and made at the plant at Lancaster, Pennsylvania: photosensitive devices, image converters, specialized CRTs, and transmitting tubes. These appeared between roughly 1955 and 1975.

1600-SERIES

1601 Power triode in T18 bulb, 4-pin jumbo base, fil. 7.5 V {A242} - probably based on 211A.

1602 [Ex - 10 Spl.]; power triode, low-micro. 10, "especially selected for NBC," uprated to 15 W (GE, RCA).

1603 [AKA CV652]; low-hum, non-microphonic 6C6; braced structure (GE, RCA). "Not listed for distributor sales" as of 1937. Widely used in broadcast audio gear; three in W. S. No. 33 [Aust.]).

1604 Rcvg. pentode, selected low-noise 78, used in RCA Victor PA-110-A amp. "Not listed for distributor sales" as of 1937.

1605 Rcvg. triode, ST9 bulb, spl. small 5-pin base, fil. 1.25 V, used in "noise-measurement" equipment, otherwise unknown {A208}. Also see 1606. 1605-1606 were "not listed for distributor sales" as of 1937.

1606 RF pentode, 1605 format/use, fil. 1.25 V {A471}.

1607 Power triode in ST16 bulb, ceramic base, fil. 7.5 V. 2A3 construction with sprayed micas and special stem leads {A248B}. Also see 1608.

1608 Xmtg. triode, 801 with μ raised to 20, quick-heating fil. 2.5 V @ 2.5 A, and ceramic (later Micanol) base (GE, RCA) {A248A}. Also see 1607.

1609 Non-microphonic audio pentode, ST12 bulb, 5-pin base, fil. 1.1 V @ 0.25 A {A4030}; like a pentode version of 864 (GE, RCA). "Not listed for distributor sales" as of 1937.

1610 Power tetrode; selected, xmtr-rated 47 for xtal oscs., 6 W (GE, RCA) {A4053}.

1611 [AKA CV653]; power pentode, 6F6 selected for cutoff {A4016}. Used (only?) in RCA Radiomarine AR-8600 auto-alarm rcvr.

1612 [AKA CV654]; pentagrid mixer, selected non-microphonic 6L7 (GE, RCA, W) {A5500}.

1613 [AKA VT-175, CV655]; power pentode, selected xmtr-rated 6F6, 7 W, 45 MHz (GE, RCA) {A4016}.

1614 Beam power tube, selected, xmtr-rated 6L6, 21 W, 80 MHz (GE, RCA) {A494}.

1615 "Crystal-oscillator pentode," ST14 bulb, med. 6-pin base, heater 6.3 V {A453}.

1616 [AKA 38267, VT-266, CV2679]; quick-heating vacuum rectifier, T16 bulb w/ cap, UX base, fil. 2.5 V @ 5.0 A, 130 mA, 6 kV PIV (GE, HY, RCA) (9-24-43). Appears to be a derated, rebulbed 836.

1617 Rcvg. pentode "similar to" 6AC7; not directly interchangeable.

1619 [AKA VT-164, CV723]; beam power tube, quick-heating 2.5-V fil.-type 6L6, slightly derated to 15 W, 45 MHz, introduced in 1938 but used mainly in BC-604/BC-684 FM tank xmtrs (9-24-43) (GE, RCA).

1620 [AKA CV657]; rcvg. pentode, selected non-microphonic 6J7 (metal 1603); common in broadcast eqpt. (GE, RCA, W).

1621 Power pentode, 6F6 sel. for long life (GE, RCA, W).

1622 [AKA CV658]; beam power tube, 6L6 selected for long life in theater sound systems, et al.; derated to 15 W (GE, RCA, W).

1623 Xmtg. triode, 809 w/ μ reduced to 20, ceramic or Micanol base, promoted for use in oscillators (GE, NU, RCA).

1624 [AKA VT-165]; beam power tube, quick-heating 2.5-V fil.-type 807, 25 W, 60 MHz (GE, RCA) {A4154D} (9-24-43). Used in same eqpt. as 1619; also as osc. and mod. in RCA ET-8043 xmtr.

1625 [AKA VT-136, CV659]; variant of 807 on 7-pin base w/ separate suppressor lead, heater 12.6 V @ 450 mA (GE, HY, NU, RA, RCA, TS) (9-24-43). Common in WW II xmtrs.: SCR-274, AN/ARC-2, AN/ARC-5, TCS, etc. As made, had beam-forming plates tied to cathode. However, amateur lore had it that non-RCA 1625s had a separate leadout wire for the beam plates, and could used better in linears by opening the base and separating that lead.

1626 [AKA VT-137, CV1755]; xmtg. triode for stable VFOs; has Micanol base and internal bracing, heater 12.6 V @ 250 mA, μ 5, 5 W (GE, HY, NU, RCA, TS) (9-24-43). For oscillators; "not particularly suited to other applications." Variant: SD838.

1627 Xmtg. triode, 810 w/ fil. 5 V @ 9 A (GE).

1628 [AKA VT-140]; VHF power triode, 8012 format, fil. 3.5 V @ 3.25 A, μ 23, 40 W. Variants: 8012, 8025.

1629 [AKA VT-138, CV1756]; electron-ray tuning ind., octal version of 6E5 w/ 12.6-V heater, intro. 1-41 (GE, RCA) (9-24-43).

1630 [AKA VT-128, 12H, CV2715]; orbital-beam hexode VHF amp. in "big acorn" pkg.; electron-multiplier preamp in early SCR-270 radar. {A-5588A} (2-18-43).

1631 Beam power tube, highly uniform 12-V version of 6L6, derated to 17.5 W (GE, RCA).

1632 Beam power tube, highly uniform metal variant of 12L6GT (GE, RCA).

1633 Rcvg. duotriode, 6SN7GT w/ heater 25 V @ 150 mA, sections matched (GE, RCA). In BC-1199A.

1634 Rcvg. duotriode, 12SC7 w/ sections matched (GE, RCA).

1635 Rcvg. duotriode, low-distortion, low-heater-current, Class B high-μ type, heater 6.3 V @ 600 mA, μ 65, 3.3 W (RCA, W) (9-24-43).

1636 Beam-deflection heptode balanced mixer for 600 MHz. {A5594A} (11-3-44). In AN/TPS-3 radar; see VT-158 for xmtg. tube in same set.

1637 [AKA VT52]; power pentode, Br. CV1052 (MU EL32) {A4273}.

1638 [AKA VR54]; full-wave rectifier, Br. CV1054 (MU EB34) {R6139}.

1639 [AKA VR55]; duodiode-triode, Br. CV1055 (MU EBC33) {A4270}.

1640 [AKA 6405, q. v.]; phototube.

1641 [AKA RK-60]; full-wave xmtg. rectifier, ST bulb, UX base base w/ two caps, fil. 5 V @ 3 A, 2.1 kV PIV, 250 mA (GE). In BC-1239A.

1642 [AKA 2C21, q. v.]; duotriode.

1643 Power pentode, otherwise unknown.

1644 Dual power pentode, 12L8GT w/ sections matched.

1645 Vac. phototube w/ capacitance-balancing electrode. Also see 5652.

1649 Rcvg. pentode, selected non-microphonic 6AC7W, used in ATJ and BC-1211 airborne TV cameras.

1650 Acorn triode, low-leakage 955.

1651, A Vacuum xmtg. rect., UV-204A format. "Inactive" as of 4-1-34. "Purchased outside but merchandized by RCA Mfg. Co." as of 1-20-38.

1652 [AKA 520B, 820B, F3562A]; water-cooled xmtg. triode, UV-207 format, derated to 5 kW, fil. 14.5 V. Active item as of 3-25-37. "Purchased outside but merchandized by RCA Mfg. Co." as of 1-20-38.

1653 "Obsolete."

1654 Min. low-drain HV rect. w/ plate pin at top, fil. 1.4 V @ 50 mA, 7 kV PIV, 1 mA (HY, RCA, TS) {R6193} (9-28-45). Powered 1P25 image tube in WW II Snooperscope and Sniperscope.

1655 [AKA CV2716]; duotriode, 6SC7 w/ sections matched.

1656 Lighthouse triode, selected 446B.

1657 Thyratron, 2050 w/ Pins 1 & 4 connected.

1658 Low-leakage 2-volt fil. triode {A205D}.

1659 [AKA Boonton 101A, 105A]; rcvg. duodiode-triode, selected 2A6 for use in Q-meters {A4006B}.

1660 Vacuum diode rectifier.

1661 [AKA C6A, q. v.]; triode thyratron.

1662 Rcvg. pentode, selected 3A4.

1663 "Small, special purpose."

1664 [AKA 12C8Y, VT-153]; rcvg. duodiode-pentode, 12C8 w/ Micanol base.

1665 Thyratron, special 2050.

1666 Min. diode-pentode, selected 1S5.

1672 [AKA FG-172, q. v.]; tetrode thyratron.

1674 [AKA 38674, 5B24]; full-wave mercury rectifier, replaceable by 4B25.

1675 Water-cooled triode, fil. 22 V @ 33.5 A, μ 25, 5 kW {A243}. Informal RCA report: "1675 is 520B construction with 521 grid in a 507 anode." Probably inherited by RCA on purchase of De Forewt.

1800-SERIES

1800 CRT, 9", P3 phosphor, electrostatic focus and 45° magnetic deflection, 20-3/8" long, 6-pin base w/ anode cap, heater 2.5 V @ 2.1 A, 7 kV max.

1801 CRT, 5", P3 phosphor, electrostatic focus and 39° magnetic deflection, 15-1/8" long, 5-pin base w/ anode cap, heater 2.5 V @ 2.1 A, 3 kV max.

1802P1 [AKA 2525A, 55BP1, q. v.]. In main indicator of FC-FD.

1802P3 [AKA 5BP3, q. v.]; CRT.

1802P4 [AKA 2525B5, 5BP4, q. v.]; CRT; (NU, RCA).

1802P7 [AKA 5BP7, q. v.]; CRT.

1803 [AKA 12AP4, q. v].

1803P4 [AKA 1803, 12AP4]; CRT (NU, RCA).

1804P4 [AKA 9AP4, q. v.]; CRT (NU, RCA).

1805P1 [AKA 5AP1, q. v.]; CRT.

1805P4 [AKA 5AP4, q. v.]; CRT.

1806P1 [AKA 3EP1, q. v.]; CRT.

1806P4 [AKA 3EP4, q. v.]; CRT.

1808P1 [AKA 3CP1, q. v.]; CRT.

1808P4 [AKA 3CP4, q. v.]; CRT.

1808P7 [AKA 3CP7, q. v.]; CRT.

1809P1 [AKA 9JP1, q. v.]; CRT.

1810P1 [AKA Br. VCR139A]; CRT (electrostatic focus and deflection) {C7428}.

1811P1 [AKA 7CP1, q. v.]; CRT.

1812P5 [AKA 5FP5, q. v.]; CRT.

1812P7 [AKA GL-475, 5FP7, q. v.]; CRT.

1813P1 [AKA 7BP1, q. v.]; CRT.

1813P7 [AKA 7BP7, q. v.]; CRT.

1814P1 [AKA 2AP1, q. v.]; CRT.

1814P5 [AKA 2AP5, q. v.]; CRT.

1840 Orthicon, for cameras and film chains, 4-1/16" dia. body stepped to 1-7/16" dia. by 18-1/8" long, octal base, four caps on faceplate, heater 6.3 V @ 600 mA (GE, RCA). 1947 price: $1050.

1846 Iconoscope {C781}.

1847 Iconoscope, 2" dia.; promoted ca. 6-40 to amateur market, electrostatic-deflection, suitable for 120-line use, heater 6.3 V @ 600 mA, 600 V, in format of 2" CRT, octal base w/ signal plate connector at front (GE, RCA) {C778A}.

1848 Iconoscope for portable cameras, octal base w/ two caps, heater 6.3 V @ 600 mA, 1.2 kV, superseded by 1846 (GE, RCA) {C787A}.

1849 Iconoscope, for film-chain cameras, classic 6-½" barrel-bulb format w/ angled gun, 6-pin base, four caps on bulb, heater 6.3 V @ 600 mA, 1 kV, superseded by 1850A {C785} (GE, RCA).

1850, A Iconoscope, 1849 for live-view cameras (GE, RCA) {C791} (2-15-51).

1854 [AKA 6474, q. v.]; image orthicon.

1855 [AKA 6896, q. v.]; "graphechon" image converter.

1898 Monoscope, 3", for test pattern (GE).

1898 Monoscope, girl's-head test pattern, 3" electrostatic-deflection, CRT format, 10-7/8" long w/ cap in center of faceplate, 7-pin base, heater 2.5 V @ 2.1 A, 1.3 kV max.

1899 Monoscope, Indian-head test pattern or custom test pattern (e. g., station callsign) per customer order, 5" magnetic-deflection, CRT format, 12-11/16" long w/ caps in center of faceplate and on bell, 6-pin base, heater 2.5 V @ 2.1 A, 1.7 kV; repl. by 2F21.

2000-SERIES

2000 [AKA 4B26, q. v.]; rectifier, replaceable by CE-226 (HY, RCA).

2005 Digital readout.

2013 [AKA 1216, q. v.]; duotriode.

2014 Video pentode, premium 6CL6.

2020 Photomultiplier, for scintillation counters, 10-stage, S-11 response, head-on bulb 2" dia. by 4-7/8" high, diheptal base (ca. 10-63).

2022 Vacuum phototube, S-1 response, 20 μA/lumen.

2040 [AKA 1C21, q. v.]; cold-cathode triode.

2048A Vidicon, shock- and vibration-resistant, 7697 format shortened to 5.1", heater 6.3 V @ 95 mA (ca. 1-62).

2052 Tetrode thyratron rectifier.

2053 Display-storage tube, 5", ruggedized, guns for writing and viewing, bulb stepped from 5.6" dia. to 2.4" dia, 13.6" long, diheptal base w/ 3 caps, electrostatic deflection.

2054 Xmtg. cermet double-ended triode, coaxial electrodes, fil. 3.6-4.5 V @ 6800-7200 A, 300 kW, 450 MHz, gives 5 mW pulse, water-cooled via integral cooler; for radars and particle accelerators {A2346} (ca. 6-66). Also see 7835.

2055 Photomultiplier, 931A format.

2060-65 Photomultiplier, variants of other types w/ removable diheptal base on the flying leads and revised performance limits. Base was to be removed after testing.

2060 Variant of 6199 (ca. 10-66).

2061 Variant of 6342A (ca. 6-66).

2062 Variant of 6655A (ca. 10-66).

2063 Variant of 8053 (ca. 6-66).

2064B Variant of 8054 (ca. 6-66).

2065 Variant of 8055 (ca. 6-66).

2067 Photomultiplier, 10-stage, ruggedized, S-11 response, domed head-on T12 bulb, 12 flying leads on removable duodecal base, 1.56" dia. by 2.8" high.

2076 [AKA 5R4GYB, q. v.]; full-wave rectifier.

2081 Min. triode-pentode, computer-rated 6AW8A.

2082 Min. 9-pin duotriode, computer-rated 12AY7.

2300 Image converter.

2342A, B Phototube, ultraviolet (modified 935).

2346 Xmtg. pulse cermet beam triode, water-cooled via integral cooler {A2346E} (1959). In RCA G-G long-pulse 400-MHz xmtr., in AN/FPS-50 BMEWS radar.

2399 Power duotriode, med.-μ pass tube for voltage regulators.

4000-SERIES

4007 TWA, 2.0-4.0 GHz, IM, CO, 3 kV, 80 mW output, 23 dB gain.

4009 TWA, 2.0-4.0 GHz, IM, CO, 6.3 V @ 1.8 A, 700 V, 25 mW output, 35 dB gain. Usable in tandem w/ 4010.

4010 TWA, 2.0-4.0 GHz, IM, CO, 6.3 V @ 1.3 A, 1.2 kV, 1 W output, 35 db gain, 30 dB NF. Usable as power amp. following 4009.

4011A Magnetron, tunes 8.8-9.6 GHz, heater 13.8 V @ 3.2 A, 23 kV @ 28 A, 215 kW pulse output (RCA, W). In guidance radar for Bomarc missile.

4015 TWA, 8.0-12.0 GHz, IM, CO, 3 kV, 1 W output, 33 db gain.

4016 TWA, 7.0-11.0 GHz, IM, 3 kV, 30 dB gain, 1 W out.

4017 TWA, 2.0-4.0 GHz, IM, CO, 6.3 V @ 500 mA, 1 kV, 28 dB gain, 17 dB NF, 10 mW output.

4019, /V2 TWA, 1.0-2.0 GHz, 6.3 V @ 600 mA, 1 kV, 10 mW output, 28 db gain, 17 dB NF (16 dB for "V2").

4020 TWA, 4.0-8.0 GHz, IM, CO, 6.3 V @ 1.8 A, 1.1 kV, 28 dB gain, 18 dB NF, 10 mW output.

4021 TWA, 1.0-2.0 GHz, IM, CO, 6.3 V @ 1.8 A, 1.2 kV, 1 W output, 32 dB gain.

4028A Xmtg. cermet pencil high-μ triode, for plate-pulsed G-G use to 4 GHz, heater 6.3 V @ 300 mA, μ 70, 1 kW typical pulse out; also usable as CW amp. w/ output of 6 W or so (9-65).

4032 [AKA 7977, q. v.]; digital readout.

4037, A Xmtg. glass pencil high-μ triode on octal base, to replace 2C40; heater 6.3 V @ 140 mA, μ 56, 6.25 W, 1.7 GHz, built as regular pencil triode w/ metal-shell octal base at cathode end (3-62). "A" has lower μ of 30, intended to replace 2C40A. Construction is close to traditional lighthouse style (7-65).

4041 TWA, 8-12 GHz, integral magnet, heater 6.3 V @ 700 mA, collector voltage 3 kV, 1-10 mW out, TNC input connector, waveguide out (12-68).

4042 Xmtg. glass pencil triode, pulsed osc. giving 425 W pulse at 1.7 GHz.

4043 Xmtg. glass pencil triode, amp., 5 W out @ 500 MHz.

4053 TWA, 1-2 GHz, integral magnet, heater 6.3 V @ 1.75 A, collector voltage 3 kV, 5 W in, 13 W out, forced-air cooled, "N" connectors in and out (5-65).

4054 TWA, 1.7-2.7 GHz, integral magnet, heater 6.3 V @ 1.75 A, collector voltage 3 kV, 20 mW in, 20 W out, N connectors in and out (12-68).

4055 Xmtg. triode, cermet pencil type, heater 6.3 V @ 295 mA, μ 70, 10 W, 4 GHz, giving 1 kW pulse out at 1 GHz (9-67).

4058 Xmtg. glass pencil triode, heater 6.0 V @ 300 mA, μ 40, 8 W, gives 800 W pulse out at 3.3 GHz, construction sim. to 5675 (2-66).

4060 Xmtg. pencil triode packaged as 1090-MHz osc. for IFF transponders, for use w/ 4061 as amp., heaters 6.3 V @ 660 mA total, 18 W total, 1100 V, 500 W pulse out, package includes tuning screws and coax output fitting (9-67).

4061 Packaged amp.; see 4060.

4062A Xmtg. cermet pencil triode, heater 6.3 V @ 295 mA, μ 100, 10 W, 4 GHz, 600 W pulse out at 1090 MHz (7-67).

4063 Packaged pencil-triode pulsed osc. for 1090-MHz airborne transponders, 1967. Also see 4064.

4064 Packaged pencil-triode G-G pulsed amp. for use in tandem w/ 4063, 500 W output, contains one tube.

4065 Xmtg. triode, cermet pencil type, giving 1.3 kW pulse @ 3 GHz.

4068 Xmtg. pencil triode packaged as 1090-MHz osc., heater 6.3 V @ 330 mA, 500 W pulse out, package has tuning screw and coax output (2-70).

4070 Xmtg. pencil triode packaged as 1090-MHz osc. for IFF transponders, slight revision of 4060-4061 (8-70).

4071 Packaged pencil-triode amp.; see 4070.

4072 Xmtg. pencil-triode packaged as 1090-MHz osc. for IFF transponders, electrically close to 4070, different case design (8-70).

4401, V1, V3 Image orthicon, 5655 format, S-10 response, for low-light color use and X-ray linkage systems, can replace 5820, 6474, and 7513 (ca. 10-60).

4402 Cadmium-sulfide photocell.

4403 Cadmium-sulfide photocell.

4404 Cadmium-sulfide photocell.

4407 CRT, for flying-spot scanners, 5", P37 phosphor, magnetic focus and 53° deflection, 11-¼" long, 5 kV.

4409 Phototube, S-4 response, 929 format.

4412 Display-storage tube, 10", electrostatic deflection, 80-sec. persistence, 9 kV, military.

4413 Cadmium-sulfide photocell.

4415 Image orthicon, 5655 format, for color use in studios lighted at B&W levels; matched set consists of 4415s for R and G, and one 4416 for B. Repl. by 8674.

4416 Image orthicon; see 4415.

4420 Germanium P-N photojunction sim. to 7467, S-14 response, 0.35" dia. by 1.1" long.

4423 Cadmium-sulfide photocell.

4424 Cadmium-sulfide photocell.

4425 Cadmium-sulfide photocell.

4438 Photomultiplier, for measurement of nuclear radiation, etc., 10-stage, head-on T12 bulb 1.5" dia. by 3.2" high, 12 flying leads, S-11 response (ca. 8-63).

4439 Photomultiplier, 4438 w/ removable duodecal base for preuse testing (ca. 5-70).

4440 Photomultiplier, S-11 response, 10-stage, head-on bulb 1-½" dia. by 3.5 " high, shortened duodecal base (ca. 8-63).

4441, A Photomultiplier, 4440 w/ flying leads (ca. 8-63). "A" has specs on risetime and dark current.

4445 Vidicon, for single-tube color cameras, 7697 format, heater 6.3 V @ 600 mA (ca. 6-72). Integral dichroic filter stripes optically encode color information.

4448 Cadmium-sulfide photocell.

4449A Image converter, S-11 response, for use as nanosecond high-speed light shutter in fast photography, body 4" dia. by 9.9" long (ca. 6-64).

4453 Cadmium-sulfide photocell.

4454 Display-storage tube, 5", 34° magnetic deflection, 20-sec. persistence, 8.5 kV. Also see 4457.

4456 Photomultiplier, 5" head-on, S-20 response, 7.7" long.

4457 Display-storage tube, erasable variant of 4454.

4459 Photomultiplier, 12-stage, S-20 response, for ruby-laser detector systems, photometry, etc.; head-on bulb 2" dia. by 5.5" long, bidecal base (ca. 6-64).

4460 Photomultiplier, 10-stage, S-11 response, for nuclear instrumentation, head-on bulb 3/4" dia. by 2.9" long, flying leads (ca. 6-64).

4461 Photomultiplier, ruggedized 4440 w/ base flange and flying leads (ca. 5-65).

4463 Photomultiplier, 10-stage, S-20 response, head-on bulb 2" dia. by 4.9" high, diheptal base (ca. 6-64).

4464 Photomultiplier, 4463 w/ bulb flared to give 3" dia. faceplate, 6.3" long (ca. 8-64).

4465 Photomultiplier, 4463 w/ bulb flared to give 5-¼" dia. faceplate, 7.7" long (ca. 8-64).

4470 Image-intensifier orthicon, for low-light TV cameras, S-20 response, 5" dia. body stepped to 2" dia. by 22" long, diheptal base and flying leads (ca. 9-67). "First (RCA) low-light-level tube."

4471 Photomultiplier, 931A w/ added red response, 5% min. red-to-white ratio (ca. 5-65).

4472 Photomultiplier, 931A w/ added red response, 7% min. red-to-white ratio (ca. 5-65).

4473 Photomultiplier, 1P21 w/ added red response, 7% min. red-to-white ratio (ca. 5-65).

4478 Vidicon, 7735B for noncritical uses, lower resolution (600 lines at center).

4486 Projection kinescope, 7NP4 w/ fitted HV anode cable in place of cap (ca. 12-66). Also see 4583-4585.

4488 Vidicon, 7735B for noncritical uses, lower resolution (625 lines at center).

4490 CRT, 7", electrostatic focusing and defl., accelerator type, P31 phosphor, 14.6" long, diheptal base w/ anode cap, heater 6.3 V @ 600 mA, 8 kV max on accelerator, 4 kV on first anode (ca. 10-66).

4491 CRT, 8" variant of 4490, 16-½' long (ca. 10-66).

4492 Image orthicon, 4-½", for luminance channel of 4-tube color camera, 4-½" dia. bulb stepped to 3-1/8" by 19-3/8" long, diheptal base (ca. 5-70).

4493 Vidicon, 4-½", sulfide target, for red chroma channels of 3-tube color cameras for live-pickup use; also for CCTV (ca. 1-68). Also see 4494 and 4495.

4494 Vidicon, green-sensitive counterpart of 4493.

4495 Vidicon, blue-sensitive counterpart of 4493.

4499 CRT, electrostatic focusing and defl., P1 phosphor, 12" long, duodecal base, heater 6.3 V @ 600 mA, 2.8 kV max. (ca. 12-65).

4500 Vidicon, 7697 format, for televising scenes having limited motion (ca. 12-66).

4501, /V3, /V4 Photomultiplier, 12-stage head-on, 2" dia. by 5.8" long, 21-pin base, for liquid scintillation counters (ca. 11-73).

4503A Vidicon, sulfide-target, 7697 format shortened to 5.1", vibration-resistant, for CCTV (ca. 12-68).

4506 CRT, magnetic focusing and 42° defl., 6", P1 aluminized phosphor, 22-1/8" long, duodecal base w/ anode cap, heater 6.3 V @ 600 mA, 25 kV max. (ca. 7-71). for computer-output microfilm uses.

4507 Photomultiplier, 12-stage, head-on body 2.1" dia. by 5.7" high, 21-pin base (ca. 10-71).

4510 CRT, 5", P7 phosphor, accelerator type, electrostatic focus and defl., P7 phosphor, 12" long, special duodecal base w/ anode cap, heater 6.3 V @ 600 mA, 6 kV max. on accelerator, 3 kV max. on first anode (ca. 10-66).

4513 Image orthicon, repl. by 8673.

4514 Vidicon, 1" sulfide-target, ruggedized, for CCTV.

4516 Photomultiplier, 10-stage, bialkali response, 3/4" dia. by 3.9" high, flying leads (ca. 9-67).

4517 Photomultiplier, 4516 expanded to 1.56" dia. by 3.9" high, duodecal base (ca. 9-67).

4518 Photomultiplier, 4516 expanded to 2" dia. by 4.9" high, diheptal base (ca. 5-68).

4520 Camera tube {C23193}.

4521 Photomultiplier, 10-stage, bialkali response, head-on bulb w/ aluminum oxide window, 3" dia. necked down to 2", 4.9" high, 13 flying leads to removable diheptal base (ca. 2-70).

4522 Photomultiplier, 14-stage, bialkali response, head-on bulb 5" dia. stepped to 3" dia. by 12" long, 21-pin base, for scintillation counters, 3-ns risetime (ca. 6-68).

4523, /V1 Photomultiplier, 10-stage, bialkali response, for scintillation counters, head-on body 2" dia., 4.9" high, diheptal base (ca. 2-67).

4524, /V4, /V5 Photomultiplier, 4523 w/ body expanding to 3" faceplate, 5-3/8" long (ca. 2-67).

4525, /V1 Photomultiplier, 4523 w/ body expanding to 5" faceplate, 6.8" long (ca. 2-67).

4526, /V1 Photomultiplier, 1-½" "dormer-window" bulb 3" long.

4531 CRT, 5", for airborne weather-radar display, electrostatic focus, magnetic deflection, P7 aluminized long-persistence phosphor, 9-13/16" long, neck offset from bulb, trapezoid faceplate for cockpit installation, 9-pin base w/ anode cap, heater 6.3 V @ *300* mA, 12 kV max. (ca. 11-72).

4532, A, B Vidicon, silicon-target, high sensitivity, wide spectral response, based on 8507A; for CCTV; low-power heater, 6.3 V @ 100 mA; replaces 7735 or 8507 w/ minor camera changes (ca. 6-73). Developmental variants of 4532 are C23143, C23161, C23162, C23174, and C23195. "A" and "B" have progressively tighter blemish limits.

4536 Image orthicon, 4-½", 4492 format, for luminance channel TK42 and TK43 color cameras; replaces 4492 family (ca. 12-68).

4542 Vidicon, 7697 format, for slow-scan and signal-storage uses (ca. 12-68).

4543 Vidicon, 1" sulfide target, for color cameras and film chains; also for CCTV {C23087}.

4547 Display-storage tube, 5", P20 aluminized phosphor, TV-capable, electrostatic defl., guns for writing and viewing, 11" long, 7-pin min. and 8-pin "neoditetrar" bases plus flying leads, heaters 6.3 V @ 600 mA each, 9 kV max. (ca. 2-71).

4549 Image intensifier, 8606 w/ automatic brightness control, integral power converter and voltage multiplier, S-20 response, cylindrical package 3.7" dia. by 12" long, supply voltage 7.0 @ 20-45 mA (ca. 2-72).

4550 Image-intensifier, 18-mm fiber-optic type, 3 stages, automatic brightness control, 2" dia. by 5.9" long.

4551 Image-intensifier, variant of 4550.

4552 Photomultiplier, bialkali cathode, peak response at 400 nm, 10-stage, Compactron-style side-on bulb 1.2" dia. by 2.6" high, 12-pin duodecar base (ca. 7-71).

4555 Photomultiplier, 4552, no detected difference (ca. 10-71).

4557 CRT, for TV monitors, P4 phosphor, 12" rectangular, 16-3/4" long, electrostatic focus and magnetic 70° defl.

4560 Monoscope, for alphanumeric character generation in computer displays, w/ pattern custom-made from buyer-supplied stencil pattern, electrostatic focus and deflection, 2" dia. by 11-½" long, diheptal base w/ anode caps, heater 6.3 V @ 600 mA, 2.5 kV max. (ca. 5-69).

4565-68 Color projection CRTs, 6", 50° deflection, 16" long, 50 kV max.

4565	Blue.	4567	Red.
4566	Green.	4568	White.

4569 Vidicon, 8541A w/ fiber-optic faceplate for coupling to image intensifiers or fiber-optic cables.

4572 CRT, 5", P-1 phosphor, 15.1" long, 2 kV.

4583 Projection kinescope, color variant of 4486 (or, equivalently, of 7NP4); blue unit of matched set w/ 4584 and 4586 (ca. 11-72).

4584 Projection kinescope, green counterpart of 4583 (ca. 11-72).

4585 Projection kinescope, red counterpart of 4583 (ca. 11-72).

4588 Vidicon, 7697 format, for low light and limited motion {C23042}.

4589 Vidicon, 8507A w/ fiber-optic faceplate (ca. 8-70) {C23112B}.

4591 "Vistacon" vidicon, 1", PbO target, 4593 format, for color CCTV. See 4592 for variants.

4592 "Vistacon" vidicon, 1", PbO target, 4593 format, variants /R, /G, /B, /L, for 3- or 4-tube cameras.

4593/R "Vistacon" vidicon, 1", PbO target, 1.2" dia. by 7-½" long, 7-pin base, for red channel of TK44 and TK45 color cameras, heater 6.3 V @ 300 mA {C23160}.

4594/G, /L, /R "Vistacon" vidicons, variants of 4593; "/G" for green channel of TK44 and TK45; full series for use in other cameras {C23253}.

4598 Graphechon scan-conversion storage tube, small and ruggedized, 3.7" dia. by 15" long, integral yoke, 6-pin bases for write and read guns plus flying leads (ca. 2-71).

4600A Xmtg. cermet beam tetrode, for voltage-reg. use, heater 5.5 V @ 17.3 A, 1750 W, forced-air cooled (ca. 9-62).

4604 Beam power tube, quick-heating variant of 6146A, fil. 6.3 VCT @ 650 mA (ca. 1-65).

4616 Xmtg. cermet beam tetrode, fil. 0.95 V @ 495 A, μ 7, 30 kW, 600 MHz, gives 275 kW long-pulse or 2 MW short-pulse output; water-cooled via integral cooler (ca. 6-66).

4617 Xmtg. cermet beam triode, fil. 1.5 V @ 1800 A, 150 kW, 450 MHz, gives 8 MW pulse output at 425 MHz; water-cooled via integral cooler (ca. 6-66).

4618 Xmtg. cermet beam tetrode, appears to be RF variant of 4600A, heater 5.5 V @ 17.3 A, 1500 kW, 1215 MHz;, forced-air cooled (ca. 9-67).

4621 Xmtg. cermet beam tetrode, for pulsed RF amps., heater 6.3 V @ 3.2 A, 125 W, 1215 MHz, forced-air cooled (ca. 9-68).

4622 Xmtg. cermet beam tetrode, conduction-cooled variant of 7649, typical power output 4.5 kW pulse at 1215 MHz (ca. 6-64).

4624 Xmtg. cermet beam tetrode, for distributed amps. up to 500 MHz or UHF TV xmtrs, heater 6.3 V @ 3.5 A, 400 W, forced-air cooled (ca. 4-65). Has internal 600-pF screen bypass cap. Variant: 4636.

4626 Xmtg. cermet beam tetrode, for pulse modulators, heater 5.5 V @ 17.3 A, 1.5 kW, 17 kV, 30 A, conduction-cooled (ca. 9-68). Variant: 4638.

4628 Xmtg. cermet beam tetrode, for SSB linears and general use, also for UHF TV translators using Y1246 cavity, fil. 4.5-4.75 V @ 125 A, 10 kW, forced-air cooled (ca. 12-65).

4629 Xmtg. cermet type, fil. 5.5 V @ 17.3 A, forced-air cooled.

4630 Xmtg. cermet beam tetrode, for pulse modulators, heater 22 V @ 12.6 A, 7.5 kW, 22 kV, 40 A, conduction-cooled (ca. 9-68).

4631 Xmtg. cermet beam tetrode, for RF linears, heater 26.5 V @ 540 mA, 150 W, 1215 MHz, forced-air cooled (ca. 9-68).

4632 Xmtg. cermet beam tetrode, for general RF use including UHF TV, heater 5.5 V @ 31 A, 5 kW, 900 MHz, forced-air cooled (ca. 7-67).

4633 Xmtg. cermet beam tetrode, for linear amps, conduction-cooled by large cylindrical BeO insulator leading into rectangular finned radiator. Gives 240 W PEP. Has grid ring and pin base for misc. leads.

4634 Xmtg. cermet beam tetrode, for pulse modulators, promoted as having nanosecond switching time, heater 6.0 V @ 7.3 A, 600 W, switches 7 kV @ 8 A peak, forced-air cooled (ca. 9-68).

4635 Xmtg. cermet beam tetrode, for general RF use, heater 22 V @ 12.6 A, 10 kW, 500 MHz, forced-air cooled (ca. 9-68).

4636 Xmtg. cermet beam tetrode, variant of 4624, heater 6.3 V @ 3.5 A, 1 kW, 900 MHz, liquid-cooled via integral cooler (ca. 9-68). Typical coolant is Coolanol-35, silicate-ester-based liquid, at -50 to 135° C.

4637 Xmtg. cermet beam tetrode, for general UHF use, heater 6.3 V @ 3.2 A, 300 W, 1215 MHz, conduction-cooled (ca. 9-68).

4638 Xmtg. cermet beam tetrode, variant of 4626, uprated to 20 kV (ca. 9-68).

4647 Xmtg. cermet UHF tetrode, fil. 3.7 V @ 840 A, 20 kW, 1 GHz, 250 kW CW or 500 kW pulse output, liquid-cooled via integral cooler (ca. 10-71).

4648 Xmtg. cermet UHF tetrode, uprated 4647, fil. 3.7 V @ 1600 A, dissipation and power outputs doubled (ca. 10-71).

4651 Xmtg. cermet UHF beam tetrode for linear amps., for VHF TV use in Y1170 cavity; heater 5.5 V @ 17.3 A, 1.5 kW, 400 MHz, forced-air cooled (ca. 2-69). Possibly developed from 4626. Also see 4661 and 8792.

4652 Beam tetrode, variant of 8042 w/ lower cost and slightly longer heating time (1.25 sec.) (ca. 2-70).

4658 Multicavity amplifier klystron, 4.4-5.0 GHz, heater 6.5 V @ 7.6 A, beam voltage 8.5 kV, 1.3 kW CW out, forced-air cooled (ca. 4-71). Variants: 4659, 4660, 8811.

4659 Multicavity amplifier klystron, appears to be a pulsed variant of 4658, 5 kW pulse output (ca. 2-71)

4660 Multicavity amplifier klystron, appears to be 4659 with cooling thermostat added (ca. 2-71).

4661 Xmtg. cermet UHF beam tetrode, heater 5.5 V @ 17.3 A, 1.5 kW, 400 MHz, forced-air cooled (ca. 2-69). Appears to be a variant of 4551.

4662 Xmtg. cermet UHF tetrode, general RF use, heater 13.5 V @ 1.3 A, 400 W, 500 MHz, forced-air cooled (ca. 8-70).

4675 Xmtg. cermet beam tetrode, for pulse modulators, 1.5 kW, 20 kV, 30 A, conduction-cooled (ca. 9-68).

4681 Xmtg. cermet UHF beam tetrode, heater 9.0 V @ 140 A, 15 kW, 400 MHz, forced-air cooled (ca. 2-75).

4682 Xmtg. cermet UHF beam tetrode, FM and TV use, heater 5.5 V @ 112 A, 8 kW, 400 MHz, forced-air cooled (ca. 2-75).

4802 Photomultiplier, for scintillation counters, 12-stage, S-11 response, 3/4" dia. by 3.5" long, flying leads on removable bidecal base {C31005B} (ca. 6-72).

4804, A, B Vidicon, 16-mm silicon-intensifier target, low-light, fiber-optic faceplates, mod. S-20 response; body 1.5" dia. by 7.5" long, ditetrar 8-pin base; repl. C23165D or E intensifier vidicons w/ minor changes (ca. 6-72). Variants are 4804/PRE ("premium"), 4804/PRO (tight blemish limits), 4804/SUR (surveillance), and 4804/SCI (loosened limits at lower price). "A" and "B" are controlled to progressively fewer blemishes. "P2" and "A/P2" are potted; "/P5" subvariant is potted w/o voltage divider for focus electrode.

4807, V1, A, A/V1 35/35 mm Image Isocon, for low-light use, S-20 response, fiber-optic faceplate, body 3" dia stepped to 2-¼" dia. by 16" long, flying leads, ruggedized, military version w/ 1:1 aspect ratio (ca. 11-72). "V1" has 20-pin base in place of flying leads. "A" is TV version w/ 4:3 aspect ratio {C21095C}.

4809, /B Vidicon, 7697 format, sulfide target, for color film chains and CCTV, "dark" heater, 6.3 V @ 100 mA {C23158}. "/B" is tested for sensitivity for use as blue channel.

4810 [AKA 8480, q. v.]; vidicon.

4811 [AKA 8134, q. v.]; vidicon.

4816 "Vistacon" vidicon, PbO target, 1", for CCTV and X-ray linkage systems {C23157A}.

4817 "Vistacon" vidicon, PbO target, 1", for CCTV and X-ray linkage systems.

4818 [AKA 1P28B]; phototube, 1P28 w/ bialkali cathode, peak response @ 400 nm {C7045G}.

4820 Projection kinescope, B & W, greatly modernized 5TP4, P45 aluminized phosphor, 5" dia. by 12" long, duodecal base w/ flying anode lead, heater 6.3 V @ 600 mA, 42 kV max., air-cooled above 9 W anode power (ca. 11-72).

4825, A Camera tube {C23250, A}.

4826 Vidicon, 25-mm silicon-intensifier target, low-light, fiber-optic faceplate. Also see 4804.

4827 35/35 mm Image Isocon, low-light ("moonlight"), clear-glass faceplate, resolution > 1000 lines, 1:1 aspect ratio.

4827A Image Isocon, 4827 w/ 4:3 aspect ratio {C21127}.

4828, A 35/35 mm Image Isocon, low-light ("deep twilight"), clear-glass faceplate, resolution > 1000 lines, 1:1 aspect ratio. "A" has 4:3 aspect ratio {C21093A}.

4831 Photomultiplier, 1.56" dia. by 3.9" high {C7151U}.

4832 Photomultiplier, 9-stage, 931A format, GaAs photocathode, response near-flat from UV to near-IR {C31025C} (ca. 3-72).

4833, A Vidicons, 2/3" size, silicon-target, CCTV, 0.77" dia by 4.3" long, 7-pin base, heater 6.3 V @ 100 mA {C23228A}. "A" is limited to fewer blemishes.

4836 Photomultiplier, extended red response for detection of light from He-Ne and ruby lasers, 10-stage, head-on 3/4" dia. by 3.8" long, flying leads on detachable 12-pin base {C70042R} (ca. 3-73).

4837 Photomultiplier, 9-stage side-on, 931A format, for spectrometer use {C7045J} (ca. 3-73}.

4846 Vidicon, 1" sulfide-target, for CCTV {C23265}.

4848 Vidicon, 2/3" sulfide-target, only 4.2" long, for CCTV.

4849, A Low-light camera tube, comprises 8857/V1 image intensifier tube fiber-optically coupled to 4804 silicon intensifier-target camera tube, bulb 2-¼" dia stepped to 1.2" by 9.5", S-20 response, ditetrar 8-pin base and flying leads. "A" is controlled to fewer blemishes.

4855 Photomultiplier, for low-energy spectrometer use, 10-stage, head-on, 1-½" dia. by 4.4" long, 12-pin base {C31061}.

4861-63 Projection color kinescopes, 5", 12-¼" long, 40 kV.
4861 Blue {C24001}.
4862 Green {C24286}.
4863 Red {C24288}.

4869 CRT, 4870 use and characteristics, drum-type front body instead of flared {C24290} (ca. 9-74).

4870 CRT for photorecording, 5", magnetic focus and deflection, narrow line width, flat faceplate, G533 purple-blue phosphor, 15" long, for computer-output microfilm systems, heater 6.3 V @ 450 mA, anode voltage 20 kV, duodecal base and flying lead for anode {C24317} (ca. 9-74).

4873 CRT, for flying-spot scanners, 5", G591 phosphor, electrostatic focus and 50° magnetic defl., 11.2" long, 25 kV {C24322}.

4874 CRT, 4873 w/ flying lead for anode {C24322A}.

SYLVANIA

Most tubes bearing 1200-series numbers were special-purpose types from Sylvania. Some of these (e. g., 1232/7G7) were apparently interim self-assigned codes, used until the RMA could issue a regular identification - analogous to Sylvania's use of interim letter codes like XXD or XXL. Other numbers designated low-volume specials, regular types selected for special characteristics, variants on regular designs, etc. Because of Sylvania's promotion of the loktal format, most of these types are loktals. National Union seems to have been the only significant second-source for these types. The numbers were assigned semi-randomly; while the 1200-series items dated from 1939 to the early '50s, it is difficult to determine vintage from a particular code.

Many of Sylvania's developmental types of the '40s were publicized before regular registration and/or production. A list of these is included for reference.

1201, A — [AKA 7E5, q. v.]; rcvg. triode (NU, SY) (3-26-42).

1203A — [AKA 7C4, q. v.]; rcvg. triode (NU, SY) (3-26-43).

1204 — [AKA 7AB7, q. v.]; rcvg. pentode (NU, SY).

1205 — Rcvg loktal VHF sharp-cutoff pentode, heater 6.3 V @ 300 mA, g_m 3.1 mS, usable to 200 MHz; dual cathode pins {SD673D}.

1206 — [AKA 7G8, q. v.]; twin sharp-cutoff tetrode {SD706} (NU, SY).

1207 — Rcvg loktal UHF duo-triode (common plates and cathodes), for converter use to 600 MHz; heater 6.3 V @ 300 mA, μ 5.5, g_m 3 mS {SD698C}.

1216 — Min. long-life computer duotriode, heater 6.3 V @ 300 mA, μ 27 <7BF>.

1217 — Min. long-life computer heptode, heater 6.3 V @ 300 mA, sim. to 6BE6 but characterized for lower plate voltage <7CH>.

1218A — [AKA 8334, q. v.]; min. triode, variant of 6AN4 <7DK>.

1219 — Min. 9-pin duotriode, variant of 5670.

1220 — Min. pentode, variant of 5654 w/ ferrite isolator to prevent parasitic oscillation [!].

1221 — Rcvg. pentode, low-microphonics variant of 6C6, ST12 bulb <6F> (NU, SY). Also see 1603.

1222, A — Beam power tube, premium 6L6GA w/ pinout scrambled (NU, SY). "A" is uprated to 21 W.

1223 — Rcvg. pentode, ST12-bulbed octal variant of 1221 <7R> (NU, SY).

1225 — Loktal receiving type, otherwise unknown.

1227 — Electrometer tube.

1229 — [AKA 32E, 932]; rcvg. tetrode, electrometer version of 32 in ST12 bulb w/ low grid current <4K>. Possibly same as Beckman Instruments 931 (SY-made), which has Aquadag external coating and clear insulating lacquer at top of ST12 bulb.

1230 — Rcvg. triode, electrometer version of 30 w/ low grid current <4D>.

1231 — [AKA 7V7, q. v.]; rcvg. pentode <8V>. (Not to be confused w/ HY1231Z.)

1232 — [AKA CV894, 7G7, q. v.]; rcvg. pentode (1-30-40). Widely used in 1940-model Philco home rcvrs.

1236 — Loktal bright-emitter temperature-limited diode; fil. 1.9 V @ 450 mA, 330 V, 0.63 mA. Voltage-sensing tube in '50s Superior Electric electronic (magnetic-amplifier) AC voltage regulators.

1237 — Argon full-wave rectifier, fil. 2.5 V, 3 A. In American Flyer No. 14 and 16 "electronic rectiformer" for toy trains, 1947-50. Promoted as late as 1955.

1238 — Dual beam tetrode, special 28D7, uprated to 3.3 W <8BS>.

1245 — T3 hearing-aid pentode, fil. 0.7 V @ 30 mA, 1945.

1246 — T3 hearing-aid power pentode, fil. 1.25 V @ 30 mA, 1945.

1247 — T3 subminiature diode w/ plate lead, ca. 1948, fil. 0.7 V @ 65 mA, 300 V in, 400 μA. Used in AC test probe of SY Polymeter VTVM; kept on into the '60s.

1252 — Min. triode-pentode, premium 6U8A/6AX8.

1254 — [AKA 5559, q. v.]; tetrode thyratron (non-SY type).

1257 — [AKA FG-57, q. v.]; triode thyratron. Also [AKA VC-1257, q. v.]; hydrogen thyratron (non-SY types).

1265 — Voltage regulator, 0B3 w/ starting voltage raised from 125 to 135 <4AJ>.

1266 — Voltage regulator, 0B3 formulated to regulate at 70 V; same pinout <4AJ>, less internal jumper.

1267 — [AKA Z300T]; cold-cathode triode, GT version of 0A4G <4V>.

1273 — Pentode, non-microphonic 7C7 <8V>.

1274 — Full-wave rectifier, variant of 6X5GT w/ heater cut to 500 mA <6S>.

1275 — Full-wave rectifier, variant of 5Z3 w/ fil. current cut to 1.75 A <4C>.

1276 — Power triode, ST16-bulb version of WE 205D; has 205D's short-pin base and 4.5-V fil., but draws only 1.14 A. Presumably intended for replacement use in WE PA and theater amps., broadcast eqpt., aircraft radios, etc. <4D>. Rated at 16.5 W.

1278 — [AKA 38278]; otherwise unknown.

1280 — Rcvg. pentode, non-microphonic 14C7 <8V>.

1282 — [AKA 7W7, q. v.]; rcvg. pentode <8BJ>.

1284 — Loktal VHF receiving pentode, heater 12.6 V @ 150 mA, g_m 2 mS <8V> (NU, SY).

1285 — Beam power tube, similar to 25L6G.

1288 — Rcvg. duotriode, special 1291/3B7.

1289 Loktal VHF receiving triode, otherwise unknown.

1290 [AKA FG-190, q. v.]; full-wave gas rectifier.

1291 [AKA 3B7, q. v.]; loktal duotriode (2-1-43) (NU, SY) <7BE>.

1292 Loktal twin triode, superseded by 3B7/1291 <7BE>.

1293 Loktal VHF rcvg. triode, fil. 1.4 V @ 110 mA, μ 14; sim. to ½ 3B7 <4AA>.

1294 [AKA 1R4, q. v.]; rcvg. diode (2-1-43) (NU, SY).

1295 [AKA 5560, q. v.]; mercury tetrode thyratron.

1296 Loktal VHF power tube, otherwise unknown.

1297 [AKA 4B27, q. v.]; full-wave gas rectifier.

1298 Rcvg. loktal, otherwise unknown.

1299 [AKA 3D6, q. v.]; beam power tube (2-1-43) (NU, SY).

DEVELOPMENTAL TYPES

SA584 See 1563 under Western Electric.

SC650 Submin. diode, T4 bulb, heater 6.3 V @ 150 mA.

SC651 [AKA 7AB7, q. v.]; rcvg. pentode.

SC673D [AKA 1205, q. v.]); rcvg. pentode.

SD698C [AKA 1207, q. v.]; rcvg. duotriode.

SD705 Med.-μ power duotriode, heater 12.6 V @ 300 mA, μ 12, 7.5 W.

SD706 [AKA 7G8, q. v.]; duotetrode.

SD712 Loktal UHF rcvg triode, fil. 1.5 V @ 135 mA, dual pinouts for grid and plate.

SD719 [AKA 1AB5, q. v.]; rcvg. pentode.

SD742 Rcvg. pentode, 1LN5 w/ double screen ("obsol.").

SD759A Loktal accelerometer, covers 1 to 100 G.

SN770 Submin. ("obsol.").

SA780 [AKA 2C28, q. v.]; submin triode.

SA781A [AKA 2E29, q. v.]; submin pentode.

SA782B [AKA 2D29, q. v.]; submin. tetrodethyratron.

SD828 Submin. pentodes.

SD828A [AKA 5638]; heater 6.3 V @ 150 mA.

SD828E [AKA 5634, q. v.].

SD828F [AKA 1W5, q. v.];.

SD829 [AKA 1B43, q. v.]; gas switching tube.

SD834 [AKA 6K4, q. v.]; submin. triode.

SD835 Reflex klystron.

SB836 Diode gas switching tube.

SD836 Submin. pentode.

SD838 Power triode, loktal variant of 1626, heater 25.0 V @ 130 mA.

SB846 "Rocket" SHF triodes.

SB846 [AKA 2C36, q. v.]; .

SB846A [AKA 2C37, q. v.].

SB846B "Probeless 2C36 for Navy application."

SB846D [AKA 5764, q. v.].

SB846H [AKA 5767, q. v.].

SD847A [AKA 1B40, q. v.]; TR tube.

SD849B [AKA 3J22, q. v.]; magnetron ("Army").

SR886 Rectifier, loktal variant of 117Z6GT.

SD917 T3 submin. high-μ triode, heater 6.3 V @ 150 mA, μ 70, 1.1 W, for use to 1.2 GHz.

SD917A [AKA 5637, q. v.]; submin. triode.

SN944 [AKA 5633, q. v.]; submin. pentode.

SN946 [AKA 5647, q. v.]; submin. diode.

SN947, C [AKA 5640, q. v.]; submin. beam power tube.

SN948 [AKA 5644, q. v.]; Submin. voltage reg.

SN949, B, C [AKA 5643, q. v.]; submin. thyratron.

SN953 [AKA SN974]; submin. pentode; see 5639.

SN953A, D Submin. pentode; see 5639.

SN954 [AKA 5641, q. v.]; submin. rectifier.

SN955B [AKA 5635, q. v.]; submin. duotriode.

SN956, A, B [AKA 5642, q. v.]; submin. rectifier.

SN957, A [AKA 5645, q. v.]; submin. triode.

SN966 Submin. low-microphonic pentode.

SC968 Low-μ triode.

SB970 Twin pentode RF osc.

SA971 Low drain secondary-emission pentode.

SN972 [AKA SN944]; submin. pentode, long-life 5633.

SN973 Submin. pentode, long-life 5634.

SN973B [AKA 5840, q. v.]; submin. pentode.

SN974 Submin. pentode, long-life 5639.

SN975 Submin. duodiode-triode.

SN976 Submin. pentode, long-life 5640.

SN977 Submin. rectifier, long-life 5641.

SN978 Submin. diode, long-life 5647.

SN979 Submin. triode, long-life 6K4.

SN980 Submin. triode, long-life 5637.

SN981 Submin. voltage regulator, long-life 5644.

SN982 Submin. thyratron, long-life 5643.

SN983 Submin. RF pentode.

SN984 Submin. mixer.

SN985 Submin. RF pentode.

SN986 Submin. mixer sim. to 6AS6.

SA987 Low drain secondary-emission pentode.

SA988 Submin. low-μ triode.

SA989 Submin. duodiode.

SA990 Submin. remote cut-off pentode.

SA991 Submin. AF power amp.

SA992 Submin. high-μ triode.

SA993 Submin. low-μ triode.

SA994 Submin. duodiode.

SA995 Submin. remote cut-off pentode.

SA996 Submin. sharp cut-off pentode.

SD997 High-μ triode.

SD998 High-μ triode w/ feedback probe.

SN1006 [AKA 5646, q. v.]; triode (version of SD917A).

SN1007A [AKA 5636, q. v.]; submin. pentode.

SN1016A T3 submin. sharp cut-off pentode, heater 6.3 V @ 150 mA.

SN1039A Submin. sharp-cutoff pentode.

SD1066A [AKA 1D3, q. v.]; submin. triode.

SN1096A Submin. RF pentode.

WESTERN ELECTRIC

The Western Electric Company made tubes over a span of 75 years, initially for telephone repeaters, then for military communication in WW I, then for broadcasting, theater sound, and airborne/land-mobile communications. WE's design bureau, Bell Telephone Laboratories, developed tubes for production by Northern Electric and overseas licensees like STC and NEC. During WW II Bell Labs figured in the development of radar tubes - first the "700" series, then types numbered per the RMA registration system. Although tending mainly to Bell System needs, WE later made specialized magnetrons, etc., for missile systems like the Nike series. Renamed AT&T Technologies in 1984, the company kept making a few types (mainly the 300B audio triode) until 1989, and reportedly resumed production of 300Bs for the audio market in 1995.

WE was perhaps the world's biggest specialty tube house: its line included, at one time or another, "everything" except ignitrons and entertainment types. Although WE's main production volume was for telephone use, the company had a disproportionate effect on the rest of the industry. One or another "WE" special-purpose tube were made or distributed at one time by all of the major producers, not to mention the foreign makers. This included Amperex, Bomac Laboratories, Chatham Electronics, Continental Electric Co., De Forest, Du Mont, Electronic Enterprises, Federal/IT&T, General Electric, General Electronics, Heintz & Kaufman, Hytron, Jenkins & Adair, Ken-Rad, Lewis & Kaufman, Litton, National Electronics, National Union, Raytheon, RCA, Sylvania, Taylor, Tung-Sol, United Electronics, and Westinghouse.

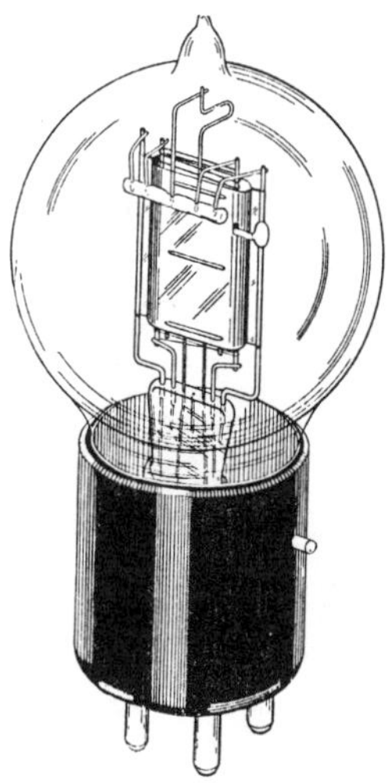

WE offered a line of CRTs in the '30s, including a highly sophisticated three-beam type. Despite full-page ads promoting this product line, it never "took off" - Western wasn't in the oscilloscope or TV business - and WE CRTs are rare today.

WE's receiving tubes reflect a stand-alone approach: with no need to interchange with other makers' products, early indirectly heated tubes had odd heater voltages of 2.0 or 10.0 V. Types with 6.3-volt heaters were needed mainly in connection with later audio equipment. Many of these tubes have unique pinouts wholly different from the RMA "standard" diagrams.

Tubes numbered by WE with its three-digit codes are listed here. "Industry" (RMA/EIA-registered) tubes originated by BTL and/or WECo include the 5531, 5541, 5649, 5650, 5681, 5780, 5795, 6891, 7208, 7400, 7535, 7589, 7950, 7976, 8021, 8079, and 8123.

Radar/military types prominent in WW II were 1B22, 1B23, 1B29, 1B31, 1B42, 1B62, 2A21, 2C45, 2J48-51, 2J53, 2J55, 2J56, 2K22, 2K23-26, 2K28-29, 2K40, 2K45, 2K48-50, 2K54-56, 3B24, 3J21, 4J21-30, 4J42, 4J45-47, 4J50-52, 4J78, 5D21, 5J23-24, 6AJ5, 6AK5, 6AR6, 6AS6, VT-25A, and VT-52.

The 700-series, besides types like 719A and 723A/B that were common long after the war, served as prototypes for radar tubes into the '60s. They were critical material in their time: data sheets were classified CONFIDENTIAL for the 700A-D, 701A, 702A, 706-714-718 series, 707B, 708A, 710A, and 719A. The 707A was rated SECRET.

Several WE transmitting types of the '30s were shared with the rest of the industry. These were 260A (≈ 860), 295A (≈ 203A), 304A (≈ 834), 318A (≈ 837), 319A (≈ 872A), 322A (≈ 803), and 331A (≈ 805). It is not clear what firm actually designed them.

In 1948, WE abandoned production of large transmitting tubes, which were then in substantial demand because of a large installed base of WE broadcast and communications transmitters. The product line was transferred to Machlett. The affected types included the 212E, 220C, 220CA, 222A, 228A, 232B, 233A, 236A, 240B, 241B, 251A, 279A, 298A, 298B, 315A, 321A, 340A, 341AA, 342A, 343A, 343AA, 357B, and 379A. Types in advanced development, like the 5531 and 5541, were part of the transfer.

Western Electric repeater tubes had a reputation for unusually long life, meaning 100,000+ hours in some cases. Not to take credit away, but repeater service was about the gentlest possible use for tubes: heater voltages were regulated, equipment ran continuously, tubes were safe from shock and vibration, and the repeaters did not exceed the tubes' ratings. Still, WE's general line of tubes made generous use of life-extending features like gold-plated grids.

Unlike practice in the rest of the industry, letters appended to a WE number do not necessarily mean compatibility with an earlier lettered type. The 274A and B, for example, are the same electrically but have different bases; likewise the 350A and B. The 254A and B are in the same physical format but quite different electrically.

WE was unusual in one respect: it was straightforward about the radioactive content of its cold-cathode tubes, including giving directions for safe disposal. The other tube makers were silent on the subject.

In the ratings below: most early "repeater bulbs" and transmitting tubes operated on a constant-current basis (e.g., 1.0 A) rather than a fixed voltage. "Quasi-UV base" means a short-pin four-pin base with precious-metal contact tips. The "cold-cathode" types listed here, specifically the ones having starter electrodes, were officially considered usable as rectifiers, relay tubes, or voltage regulators.

Developmental types not made in quantity are not generally listed; they were BTL Model Shop, not WE, types. However, three developmental types not reported before are included at the end.

Besides its usual three-digit numbering, WE assigned "D-specification" numbers to special tubes: preliminary-production types, modified or selected versions, low-usage types, etc. A list of these is included for reference.

101-TYPE Triodes for telephone repeaters, carrier telephone systems up to ca. 40 kHz, etc.

101A In S20 tennis-ball bulb, cast-brass base, fil. 4-5 V @ 1.0 A, μ 5-6.5 {M} <4D>.

101B Longer-life variant of 101A, S20 tennis-ball bulb, fil. 5.0 V @ 1.3 A, μ 5 {L}.

101D, DW [AKA 3A/141A, 4101D]; "tennis-ball" or ST14 bulb, quasi-UV base, replacing D-86326, fil. reduced to 1.0 A @ 4.5 V, μ 6.2, g_m 0.9-1.3 mS; repl. 101A, B (NEC, STC, WE). "DW" was for commercial sale. Variants: D-86326, D-92437.

101F [AKA 4101F]; "half-current" version in S17 tennis-ball bulb (later ST14), quasi-UV base, fil. reduced to 500 mA @ 4.15 V, μ 6.5, 2 W, controlled as to microphonics (NEC, STC, WE) <4D>. ST-bulbed version could replace 101J. Variants: D-91143, 440A.

101FA ST14-bulbed 101F w/ μ raised to 9, for slightly higher gain in voice-frequency repeaters.

101G [AKA 4101G, 4019A, 4022A, TB-600A] (NEC, STC, WE).

101H Electrically same as 101D, "for foreign use." May carry different patent marks than domestic tubes.

101J 101F in S14 bulb with increased insulation between elements.

101L Low-current 101F, ST14 bulb, quasi-UV base, fil. reduced further to 250 mA @ 4.15 V, μ 6.5, g_m 1.1-1.2 mS, 2 W.

101M Quick-heating 101L; fil. 3.75 V @ 150 mA.

102-TYPE Triodes for telephone repeaters, etc.

102A S20 tennis-ball bulb, fil. 4.0 V @ 1.3 A, μ 30 {V} <4D>.

102D, DW [AKA 3A/142A]; S17 tennis-ball bulb (later ST-14), quasi-UV base, fil. 2.1 V @ 1 A, μ 30, g_m 0.5-0.56 mS (NEC, STC, WE) <4D>. "DW" was for commercial sale.

102E "Quiet 102D"; superseded by 102G. Variant: D-85792.

102F [AKA 4102F]; redesigned, half-current 102D, ST-14 bulb, fil. reduced to 500 mA @ 2.1 V, controlled for microphonics (NEC, WE). Variants: D-90278, D-90279, D-91144, 441A.

102G [AKA 4102G, 4020A, TB-601A]; 102D w/ reduced sputter noise, less fluctuation noise, and better insulation between elements; S17 tennis-ball bulb; replaced by 102D (NEC, STC, WE).

102H Export 102G analogous to 101H.

102L Quarter-current 102D, fil. reduced further to 250 mA @ 2.1 V, μ 30, g_m 0.65-0.8 mS, 500 mW.

103A Triode, S19 tennis-ball bulb, fil. 10.0 V @ 1.3 A, μ 1.5 {H}. Actual production unknown.

104-TYPE Power triodes for repeaters and general use.

104A S20 tennis-ball bulb, fil. 5.5-7.0 V @ 1.3 A, μ 2.2-2.8 {O} <4D>.

104D, DW [AKA 3A/144A, 4104D]; ST14 bulb w/ quasi-UV base, reduced noise, improved fil. 4.5 V @ 1.0 A, μ 2.5, g_m 0.9-1.4 mS; replaces 104A and D-86327 (NEC, STC, WE). "DW" is for commercial sale.

104F [AKA 4104F, 4021A, 4022A, TB-602A, TB-603A] (NEC, STC, WE); improved version.

104G [AKA 4104B, 3A/144]; variant of 104D.

104H D-86327 for export use, analogous to 101H.

105A, AW Power triode, repl. by 205F.

112A, AW Xmtg. triode, repl. by 212E.

113A, AW Xmtg. triode, repl. by 242C.

115AW Rcvg. triode, repl. by 215A.

117AW Rectifier.

201A, B [AKA CW-186]; S19 tennis-ball bulb on 3-pin Navy base, fil. 2.6-3 V @ 1.3 A; μ 37-44; replaced by 102D {D}. "B" is on 4-pin base. Unrelated to UX-201A.

202A Early version of "E" (205-type) w/ larger (G18) bulb and 4-pin nub-style base, fil. 8.0 V @ 1.3 A, μ 6 {K}.

203A, B, D [AKA CW-933, VT-1]; rcvg. triode, WW I; T10 tubular bulb, fil. 2.0-2.5 V @ 1.1 A, μ 5-6.5 {J}. "B" has same electrical specs as "A." "D" is a general-purpose triode, fil. 2.3-3.0 V @ 1.0 A.

204A, B Xmtg. triode, fil. 8 V @ 3 A, μ 1.5 {S}. "B" is variant {W}. Both were used in 1915 Arlington xmtr.

205A, B Xmtg. triode, "5-watt," oxide fil. 1.35 A @ 6.5-7.5 V, μ 7 {E} <4D>. "B" [AKA VT-2, CW-931] has S19 tennis-ball bulb, UV-style base w/ pin rotated 45°, WW I production, same electricals as 205A; see VT-2 re 1932 production by Duovac. Variant: D-79509. "D" [AKA 3B/151A, 4205D, 1276, q. v.] has S17 tennis-ball bulb; fil. 1.6 A @ 4.5 V, μ 7, g_m 1.7-2.5 mS, 14 W, 15 MHz (AX, STC, WE). Replaced by 205F. Variants: D-97822, D-156734; also 1276 and graphite-anode "D" from SY. Osc. and final amp. in 11A aircraft xmtr. "E" is "D" treated for reduced sputter noise. "F" and "FA" are "D"s in ST14 bulb. Variant: 442A.

206A Rcvg. triode, same bulb and electricals as 203A.

207A Xmtg. triode, S19 tennis-ball bulb, same electricals as 205A. Not to be confused with "industry" 207!

208A, C Triode, 101D for commercial sale; replaced by 205F. "C" was replaced by 101D.

209A Triode, 102D for commercial sale, repl. by 102D.

210A Triode, 104D for commercial sale, repl. by 104D.

211A, D, E [AKA VT-4, CW-1818]; xmtg. triode, fil. 3.4 A @ 9-10 V, μ 12 {G}. "A," "B," and "C" are graded by fil. current. Derivative: 214A. "D" [AKA CW-1818A] has constant-voltage fil., 10 V @ 3.25 A, 120 W (AX, WE); also see 242C. Variants: D-80050, D-86737. "E" is "D" w/ parasitic-suppressor chokes built into grid and plate leads (JA, WE). AX sold a T-T, graphite-plate version as of 1936.

212A, D, E [AKA VT-6, CW-1819]; power triode, general-purpose incl. Class B broadcast linears, T28 bulb, special 4-pin base with pins in "cross" configuration; fil. 13 V @ 6.265 A, μ 16, 250 W {I}. "A" and "D" [AKA CW-1819B] are superseded by "E." Derivative: 219A. Variants: 241A, D-88057, D-88526. "E" [AKA 38112, (F, UN) 312E, 2RA3, NL615, V.1505] has T-T fil., 14 V @ 6 A, g_m 8.5 mS, 7.5 MHz (in 1934; uprated to 275 W, but derated to 1.5 MHz in 1936). Each tube stamped "1" to "4" to indicate static plate current for purposes of matching (AX, F, ML, RE, WE). Variants: 241B, 308B, q. v.

214A, D, E Vacuum rectifier, gridless 211A w/ close-spaced plate. "D" is superseded by 214E. "E" is improved, rated fil. 10 V @ 3.25 A, 750 V in, 150 mA; officially replaceable by diode-connected 242C.

215A [AKA 38015, VT-5, CW-1344, R215A]; rcvg. AF-RF "peanut" triode, repackaged VT-3; T5 bulb w/ special nub-contact base, silver contacts, fil. 1.0 V @ 250 mA, μ 5.7, g_m 0.35-0.42 mS, controlled for microphonics <4D>. Variants: D-80039, D-98677. Prototype for Br. Wecovalve, on four-pin flared base.

216A Power triode, wide-tolerance 101D w/ VT-1-style elements, fil. 5-6 V @ 1.0 A, μ 6, 100-150 V. Variant: 217A.

217A Vacuum rectifier, gridless 216A w/ close-spaced plate, 150 V in, 25 mA.

219A, D Rectifier, gridless 212A w/ close-spaced plate. "D" has improved fil., 14 V @ 6.0 A, 1500 V in, 250 mA.

220A, B, C Xmtg. triode, pure-tung. fil. 21.5 V @ 41 A, water-cooled via external cooler. "B" [AKA F320B, 188, 38120, CW-1887]; is rated at μ 40, 10 kW, 3 MHz; is quite close to UV-207. Replaced by 343A. (F, WE). "C" [AKA F-320C] is "B" w/ more rugged fil. structure, hard glass, short-pin fil. connectors in place of clasp-type, new internal structure, and short grid-out-side lead rather than vertical glass stalk; uprated from 12 kV to 15 kV and from 3 to 4 MHz (AX, F, ML, WE). Also see 342A (ML). "CA" is radiator version of "C," derated to 10 kW (ML, WE). AX offered such a variant as of 1938.

221D Equivalent of UX-201A for production by NE.

222A, B [AKA F322A]; vacuum rectifier, gridless 220B, PIV 25 kV, 5 A (F, ML, WE). "B" was nominally replaced by "A," which outlasted it in production by 11 years.

223A Tel. repeater triode, fil. 4-5 V @ 1.55 A, μ 2.5; intermediate in power between 104D and 205D.

224A, B, C CRTs, 5", 5-pin base, gas-focused (argon fill), UX base, platinum-ribbon fil., 2.0 V @ 1.0 A, 300 V, electrostatic or magnetic defl. Variant: D-94341.

225A [AKA CW-2354]; triode.

227A (Number used by Eimac: 227A VHF xmtg. triode, q. v.).

228A [AKA F328A]; xmtg. triode, pure-tung. fil. 21.5 V @ 41 A, μ 16, 5 kW, 3 MHz, water-cooled via external cooler (ML, WE). AX offered a radiator version (228R) as of 1938.

229D Rcvg. triode, fil. 3 V @ 60 A; production unknown.

230D Rcvg. triode, 231D w/ UV-199 style base wired like 101-type. Variant: D-94230.

231D Rcvg. triode, T9 bulb w/ UX-199 style base, fil. 3.1 V @ 60 mA, μ 8.5, g_m 0.38-0.56 mS, controlled for microphonics <4D>. In other words: a UX-199 w/ μ raised 25%. Variants: D-87722, D-94230.

232A, B, BA [AKA F332A]; xmtg. triode, pure-tung. fil. 20 V @ 61 A, 20 kW, water-cooled via external cooler. In WOR 50-kW xmtr., 1935, w/ 266A rects. Replaced by 343A. "B" [AKA F332C] has grid-outside bulb and pin-style connectors, μ 40, uprated to 25 kW, 3 MHz (ML, WE). Variants: 340A, 342A. "BA" is radiator version of "B." "C" has more rugged fil. structure than predecessor. AX also offered a "radiator" version as of 1938.

233A, B Xmtg. vacuum rectifier, gridless 232A, fil. reduced to 21.5 V @ 41 A (possibly by use of thoriated tungsten), 20 kV applied, 2 A (ML, WE). "B" has enlarged glass section, PIV 50 kV, 5 A.

234A Xmtg. vacuum rectifier, T-T fil. 11 V @ 3.85 A, 2 kV applied, 1 A; xmtrs. converted to 249B instead.

235D Rcvg. triode, equivalent of UX-201A; origin of NE DX-235.

236A Xmtg. triode, pure-tung. fil. 21.5 V @ 41 A, μ 40, 20 kW, 3 MHz, water-cooled via external cooler (ML, WE).

237A [AKA F337A]; xmtg. rectifier, gridless 232A, 50 kV PIV, 3 A, water-cooled via external cooler (F, WE).

238A Phototube, long tubular (10") bulb; gas fill; alkali (cesium?) cathode; replaceable by 1A Photocell. (Most WE phototubes were officially designated "Photocells" rather than "Vacuum Tubes."

239A Rcvg. triode used in condenser-mic. preamps, etc., quiet 215A, small 199-style 4-pin base, fil. changed to 1.05-1.25 V @ 270 mA <4D>.

240A, B Xmtg. triode, mid-anode, double-ended, pure-tung. fil. 21.5 V @ 41 A, μ 40, 10 kW, 20 MHz, water-cooled via integral cooler, repl. by 240B. Variant: D-85716. "B" is uprated from 10 kV to 12 kV (ML, WE).

241A, B Xmtg. triode, 212D w/ cap, on special three-pin base, for HF use. "B" is 212E w/ same cap and base, uprated to 7.5 MHz (ML, WE).

242A, B, C [AKA 3B/850A]; WE version of 211, T-T fil., derated to 85 W (originally 50 W) and 6 MHz (AX, GE, STC, WE) <4E>. Variant: 284D. "B" is carbon-plate version, uprated to full 211 rating of 100 W (AX, GE, WE) "C" [AKA C342C, 4242C];

is redesigned "A" w/ metal plate, same ratings as 211 (AX, GE, STC, NEC, WE).

243A Xmtg. triode, double-ended, fil. 10-11 V @ 41 A, μ 40, 2 kW, water-cooled. Variant: D-85715.

244A Rcvg. triode, S14 bulb on UY base w/ locating pin; heater 2.0 V @ 1.6 A, μ 10, g_m 0.5-1.2 mS, controlled for microphonics and sputter noise; suited to audio preamps <5A>. In other words: a re-voltaged, controlled-quality UY-227. Variants: 247A, 285A.

245A Rcvg. tetrode, ST14 bulb w/ cap, UY base w/ locating pin, heater 2.0 V @ 1.6 A, g_m 0.75 mS, controlled for microphonics <5E>. In other words: a re-voltaged, controlled-quality UY-224. Variants: D-90319, 246A, 259A.

246A Rcvg. tetrode, filament variant of 245A, fil. 3.3 V @ 100 mA, g_m 0.39-0.51 mS <4K>. Repl. by "outside manufactured" KS-4750.

247A Rcvg. triode, higher-μ variant of 244A on UY base w/o pin, μ 15, g_m 0.52-1.2 mS <5A>. Later production had ST14 bulb.

248A [ex-D-86737]; 211D w/ reduced grid spacing, presumably for higher g_m. In low-level stages of 5C broadcast transmitter.

249-TYPE Mercury rectifiers, 2.5-V oxide filament.

249A Fil. 7.0 A, 6.3 kV PIV, 1.1 A peak.

249B [AKA (T) 249B]; T21 bulb w/ small ceramic-ringed cap, UX base, fil. 7.5 A, 7.5 kV PIV (per AX and WE; 10 kV per T), 640 mA. Variant: 258B.

249C "B" w/ large (0.552" dia.) ceramic-ringed cap (CE, HY, WE). HY uprated to 8.5 kV; WE did not. In BC-315 and BC-460A.

251A [AKA 4251A, TC-517-A]; xmtg. triode, all-glass double-ended stepped T48 bulb 22" long overall, grid-out-side, bipin ceramic base, T-T fil. 10 V @ 16 A, μ 10.5, 1 kW, 30 MHz, radiation-cooled (ML, NEC, STC, WE). In BC-315 and WE 9-type police-airline xmtr. Variant: 278A.

252A Power triode, S21 bulb on UX base, fil. 5.0 V @ 2.0 A, μ 5.2, g_m 2.2-3.5 mS, 38 W <4D>. Modulator in 10A airport xmtr. and 11A aircraft xmtr. Variant: 275A.

253A [AKA 38171]; mercury rectifier, S17 bulb w/ cap on bipin base; oxide fil. 2.5 V @ 3.0 A, 3.5 kV PIV, 250 mA out. Variant: 280A.

254A, B Xmtg. tetrode, S19 bulb w/ cap, UX base, T-T fil. 5 V @ 3.25 A, μ 80, 20 W, 10 MHz. Final in 10A airport xmtr. Variant: 268A. "B" is considerably redesigned: T-T fil. 7.5 V @ 3.25 A, μ 100, 25 W, 15 MHz. Not to be confused w/ HK-254A or HK-254B.

255A, B Mercury rectifier, T41 spherical-tubular bulb w/ ceramic-ringed cap; no base but strap assembly holding fil. pins; graphite anode; oxide fil. 5.0 V @ 21 A, 20 kV PIV, 2 A out. "B" has redesigned internal structure, fil. reduced to 19 A (AX, WE).

256A [ex-D-90279]; argon triode thyratron, grid-controlled rectifier; ST14 bulb on 5-pin base w/ locating key, heater-type 102F with cathode tied to one heater pin; heater 2.3 V @ 1.7 A, 325 V PIV, 53 mA <5A>.

257A Rcvg. triode, 231D w/ grid cap, also controlled for microphonics.

258B Mercury rectifier, 249B on bipin base (AX, CE, HY, T, WE).

259A, B [ex-D-92321]; rcvg. tetrode, 245A w/ higher g_m of 1.3-1.5 mS, also controlled for microphonics. Later production in ST14 bulb, still w/ locating pin <5E>. Variant: 283A. "B" is lower-noise variant for audio preamps.

260A [AKA 860, q. v.]; WE production of RCA UV-860 (NEC, WE).

261A [AKA (UE) 361A]; xmtg. triode sim. to 276A, 242C, etc.; T-T fil., 10 V @ 3.25 A, μ 12, 100 W <4E>. (AX rated at 120 W) (AX, UE, WE).

262A, B Rcvg. triode, low-noise audio preamp, S14 bulb w/ cap, UX base w/ silvered pins, heater 10 V @ 320 mA, μ 15, g_m 0.6-1.2 mS <4BY>. Superseded by 262B. "B" is lower-noise variant w/ extra heater-cathode insulation, internal hum shield, etc. Later production in ST12 bulb on 4-pin base. Variant: 347A.

263A, B Full-wave low-voltage argon rectifier, double-ended S26 spherical-tubular bulb 10-3/4" long, bipin bases at both ends, fil. 2.5 V @ 15 A, 100 V PIV, 3.2 A. "B" is uprated to 5 A.

264A, B, C Rcvg. triode, T9 bulb, small 4-pin base, fil. 1.5 V @ 300 mA, μ 7.2, g_m 0.56-0.64 mS, controlled for noise <4D>. Variant: D-94044. "B" has silver-plated pins (AX, WE). "C" has slightly lower input capacitance.

265A Xmtg. triode, double-ended, 45" long, fil. 22 V @ 180 A, μ 32, 100 kW, wate-cooled, ca. 1934.

266A, B, C Mercury rectifier (F, GE, WE). "B" has T56 spherical-tubular bulb w/ ceramic-ringed cap, 7-1/8" max. dia. by 19-½" long (857B size), right-angle fil. pins out bottom, graphite anode, fil. 5 V @ 42 A (WE; GE rates at 30 A), 5 A, 22 kV PIV (GE). (AX, F, GE, WE). (857B is rated at 10 A; 266B at 10 A in quadrature operation, but there seems to be some specmanship here.) "C" has 8-½" 857-style flying leads for filament.

267B [AKA F267B]; mercury rectifier, 319A on bipin base (AX, F, WE).

268A Xmtg. *triode*, screenless variant of 254A, μ 5, uprated to 25 W and 30 MHz <3Q>.

269A [ex-D-90278, q. v.]; argon triode thyratron, medium-voltage grid-controlled rectifier (modified 102F), S17 tennis-ball bulb (later ST14) on 4-pin base, fil. 2.2 V @ 550 mA, 275 V, 14 mA <4D>.

270A Xmtg. triode, all-glass double-ended stepped-tubular bulb, T-T fil. 10 V @ 9.75 A, μ 16, 350 W, 7.5

MHz, radiation-cooled. In 100-W broadcast xmtrs. (AX, WE)

271A [AKA TB-627-A]; power triode, like scaled-down heater-type 252A, S19 bulb on 5-pin base, heater 5.0 V @ 2.0 A, μ 8.4, g_m 2.3-3.5 mS, 27 W (NEC, WE) <5A>.

272A Rcvg. triode, general-purpose, ST14 bulb, 5-pin base w/ locating pin, heater 10 V @ 320 mA, μ 6.0, g_m 0.76-0.95 mS, 2.0 W <5A>.

273A Rcvg. detector triode, fil. 2.0 V @ 1.6 A, in receiver developed for Army. Commercial production unknown.

274A, B [AKA 4274A]; full-wave rectifier, ST16 bulb on UX base, fil. 5.0 V @ 2 A, 1.5 kV PIV, 225 mA (5R4 fil., 5U4 output) (EE, STC, WE) <4C>. "B" [AKA 4274B] is variant on octal base <5T>. Six in FD radar.

275A Power triode, half-power variant of 252A, ST17 bulb on 4-pin base, fil. 5.0 V @ 1.2 A, μ 2.7, g_m 1.0-3.8 mS, 17 W (STC, WE) <4D>.

276A Xmtg. triode, another WE version of 211, same format and ratings. Differs from 242C in having less interelectrode capacitance; full rating to 30 MHz <4E>. Replaces 261A (GE, WE).

277A Thyratron, heater 5.0 V @ 2.8 A, 350 V PIV, 500 mA peak; replaceable by 287A <5A>.

278A Xmtg. tetrode, tetrode variant of 251A w/ three-pin base; same ratings.

279A [AKA 4279A, TC-522-A]; xmtg. triode, 251A scaled-up, fil. uprated to 21 A, dissipation uprated to 1.2 kW, derated to 20 MHz (HK, ML, NEC, STC, WE). "Largest radiation-cooled tube on the market" as of 1932. Two in 71A 1-kW broadcast final amp. Variant: 379A.

280A Mercury rectifier, 253A on UX base.

281A Power audio tetrode, aligned-grid type, S21 bulb on 5-pin base, fil. 5.0 V @ 1.6 A, g_m 1.5 mS, 11 W <5C>.

282A, B Xmtg. tetrode, S19 hard-glass bulb w/ off-center cap on UX base, T-T fil. 10 V @ 3.0 A, g_m 1.4 mS, 70 W, 30 MHz (T, WE). Three in 13A aircraft xmtr. "B" [AKA 38282] has Isolantite base and internal bracing; is in BC-315. "282" version is one of the few tetrodes Taylor ever made.

283A Rcvg. tetrode, variable-μ variant of 259A.

284A, B, D Audio power triode, variant of 242A, T-T fil. 10.0 V @ 3.25 A, g_m 2.5 mS, 50 W out from two tubes, superseded by "D" <4E>. "B" is apparently a version w/ plate cap <3N>. "D" is a low-μ variant of "A," μ 4.8, w/ lower interelectrode capacitances like those of 276A.

285A Rcvg. pentode dev. from 244A triode, same format and other ratings, g_m 0.86-1.02 mS <5F>.

286A Rcvg. pentode., variable-μ, S14 bulb w/ cap on 6-pin base, heater 2.0 V @ 1.6 A, g_m 1.2 mS. For 13A radio receiver, etc.

287A [AKA 1287, CE302]; mercury triode thyratron sim. to 5557, S16 (later ST16) bulb w/ ceramic-ringed cap on 5-pin base, oxide fil. 2.5 V @ 7.0 A, 2.5 kV PIV, 640 mA (CE, HY, WE). Replaces 277A. Variant: 323A.

288A Mercury tungar rectifier, same as 4B26 except for lower PIV rating of 300 V vs. 375; replaced by it.

289A [AKA 4B26, q. v.]; mercury rectifier.

290A Rcvg. variable-μ pentode, ST14 bulb w/ cap on 6-pin base, heater 10.0 V @ 320 mA <6F>.

291A Pentagrid conv., ST-12 bulb w/ cap on 7-pin base, heater 10.0 V @ 320 mA.

292A Duodiode-triode, ST14 bulb w/ cap on six-pin base, heater 10 v @ 320 mA, μ 13.5. Same pinout as 75 or 85 <6G>; μ intermediate between those of 75 and 85; higher heater voltage. Superseded by 352A. Variant: 303A.

293A Rcvg. power pentode, ST14 bulb on 6-pin base, heater 10 V @ 320 mA, g_m 1.14-1.34 mS, 5 W <6B>. Has format and pinout of 42 but lower ratings. Variant: 294A. Also see 336A.

294A Rcvg. power pentode, 293A w/ cap on 5-pin base <5F>. In C-40/APQ-10. Variant: D-96412.

295A Xmtg. triode, WE version of 203A derated to 6 MHz <4E>.

296A Xmtg. triode, T32 bulb, graphite anode, T-T fil. 11.0 V @ 6.0 A, ca. 250 W.

297A Argon triode thyratron, 264A format, fil. 1.75 V @ 350 mA, 250 V PIV, 10 mA <4D>.

298A, B Xmtg. triode, audio or RF use, mid-anode, double-ended, 53-3/4" high, used w/ corona-grading rings at grid end, pure-tung. fil. 27 W @ 225 A, μ 32, 100 kW, 4 MHz, water-cooled via integral cooler (ML, WE). Two in WE 50-kW broadcast xmtr. "B" is RF-only variant w/ μ raised to 57.5 (ML, WE).

299A Gas (mercury?) rectifier ("not active"), 870-sized.

300A, B [AKA 4300]; audio power triode, ST19 bulb on UX base w/ locating pin in normal position, fil. 5.0 V @ 1.2 A, μ 3.85, g_m 5.5 mS, 40 W <4D>. Promoted to the amateur-radio market in a special catalog booklet issued in 1936. Replaced by 300B (STC, WE). "B" [AKA 4300A] is "A" w/ locating pin rotated 43° (yes, 43°!) toward Pin 3 (STC, WE).

301A Full-wave mercury rectifier, 83-oid in bigger (ST19) bulb, uprated on PIV (1.8 kV vs. 1.55) and peak current (1000 mA vs. 675 mA) <4C>.

302A CRT, early (1934).

303A Duodiode-triode, 292A w/ heater 2.0 V @ 1.6 A.

304A, B [AKA 834, q. v.]; graphite anode <2D>. Two in Navy experimental radar, 200 MHz, 1936. "B" [AKA (STC) 4304B] has metal anode, claimed to give greater output at VHF (STC, WE).

305A Xmtg. tetrode, hard-glass format like 304A, w/ screen, plate, and fil. CT pins out top of bulb, S19 bulb on UX base, T-T fil. 10 V @ 3.1 A, μ 56, 60 W, 50 MHz. In 16A VHF base-station xmtr.

306A Xmtg. pentode, ST16 bulb w/ cap on five-pin base w/ locating pin, oxide-coated fil. 2.75 V @ 2.0 A, g_m 3.7-4.1 mS, 15 W, 50 MHz (NU, WE). Variant: 307A.

307A [AKA RK-75, VT-225, CV2612]; xmtg. pentode, 306A w/ quick-heating fil. 5.5 V @ 1 A, 15 W, third grid brought out for use in suppressor modulation, derated to 40 MHz (KR, NU, RA, SY, WE). In 19A aircraft xmtr., 1935, and in BC-315.

308A, B Xmtg. triode, low-μ audio variant of 212E, μ 8, derated to 2250 V and 250 W. "B" is further derated to 1.5 kV.

309A Rcvg. pentode, variable-μ, ST12 bulb w/ cap on 5-pin base, pins silver-plated; heater 10 V @ 320 mA, g_m 1.1 mS, controlled for noise <~5E>. Like a re-voltaged 39/44; same pinout.

310A, B [AKA 5A/150A, 4310A]; rcvg. pentode, ST12 bulb w/ cap on 6-pin base, silvered pins, gold-plated grid, heater 10 V @ 315 mA, g_m 1.8 mS (RA, NEC, STC, WE) <6F>. Variants: 328A, 348A. "B" is lower-hum version w/ heater-cathode voltage derated from 150 to 30 V.

311A, B [AKA 4311A]; rcvg. pentode, ST12 bulb w/ cap on 5-pin base, heater 10 V @ 640 mA, g_m 2.9 mS, 9 W (NEC, STC, WE) <~5E>. Variant: 329A. Differences in, and production of, "B" are unknown.

312A [AKA 38412]; xmtg. pentode, T18 bulb w/ cap tapered heavily to fit 6-pin base, T-T fil. 10 VCT @ 2.8 A, g_m 3.8 mS, 50 W, 20 MHz, third grid wired out separately for suppressor modulation.

313-TYPE Cold-cathode neon triode, T9 bulb on 4-pin base. Later production has a coat of deep-red light-shield lacquer. RaBr activator added in 1952; changed to Kr-85 in 1960.

313A Superseded by 313C.

313AA Variant superseded by 313CA.

313B Variant superseded by 313AA.

313C [AKA CV75, KB-758-B, 4313C]; rated at 185 V, 35 mA. Variants: 333A, 346 family, 353A, 430 family (NEC, WE).

313CA 313C uprated to 200 V, derated to 25 mA.

313CB 313C derated to 170 V and 25 mA.

313CC 313C derated to 150 V and 25 mA.

313CD 313CA w/ viewing window in lacquer coat.

314A Full-wave medium-voltage mercury rectifier, ST19 bulb, fil. 5.0 V @ 5.0 A, 300 V PIV, 2.5 A peak (ca. 800 mA avg.) <4C>.

315A [AKA F315A]; mercury rectifier, 321A on bipin base (AX, F, ML, WE).

316A [AKA CV683, VT-191, (STC) 4316A]; all-glass "doorknob" xmtg. UHF triode, pins out bottom in "T" configuration, T-T fil. 2.0 V @ 3.65 A, μ 6.5, 30 W, 500 MHz; oscillates to 600 MHz (GE, STC, TS, WE) {ZP470}. Intro. 1936. In BC-645 and T-13/APN-5. Local osc. in FC-FD radars.

318A Xmtg. pentode equiv. to 837A; interchanges w/ it.

319A [AKA 4H72]; mercury rectifier, 872A format, fil. 5 V @ 6.75 A, 7.5 kV PIV, 1 A. Effectively same as 872A, but with different internal structure (F, NEC, WE) <4AT>. Variant: 267B.

320A Xmtg. triode, mid-anode/dual-ended format, 94" high by 12" dia. overall, pure-tung. triple fil. 35 V @ 435 A, μ 30, 150 kW, 2 MHz (250 kW), water-cooled via integral cooler. In XER border-blaster xmtr. ca. 1939.

321A Mercury rectifier, sim. to 575A or 673, spherical-tubular bulb 3-7/8" dia. by 9-23/32" high w/ plate pin out nippled top, on jumbo 4-pin base; fil. 5.0 V @ 10 A, 12.5 kV PIV, 1 A <4AT>. Ratings are somewhat below 575A (AX, ML, WE). Variant: 315A.

322A [AKA 803, q. v.] <5J>. Variant: 332A.

323A, B Triode thyratron, 287A with argon-mercury fill for increased temperature stability, PIV derated to 500 V (CE, TS, UE, WE). "B" is uprated to 1250 V PIV, 1.25 A (per HY; WE rates at 1.5 A) (CE, HY, NL, TS, UE, WE).

324A [AKA D-96413]; rectifier, tall T9 bulb, fil. 5 V @ 3 A, 12 kV PIV, for oscilloscope beam supplies.

325A, B, C CRTs, 4", electrostatic focusing and def., accelerator type, conical bulb 4-5/8" tapering to 1-7-16" by 16-½" long, five-pin regular and five-pin ring bases, heater 5 V @ 550 mA, 5 kV on accelerator, 1.5 kV on first anode. "A" has P1 phosphor; "B" has long-persistence screen like P7; "C" has blue actinic phosphor sim. to P5 or P11.

326A, B, C CRTs, 7" versions of 325-type; bulbs 7-1/8" dia. tapering to 1-7/16" by 22" long. Other details, phosphors, and ratings same. Some later 326As (for Navy radar indicators) made by DuM.

327A [AKA GE12X825, WL-966626]; mercury Tungar rectifier, single-ended; 4B35 uprated to 275 V PIV from 200. Replaceable by GE 12X825. (Not to be confused w/ the EI 327B xmtg. triode.)

328A, B Pentode, 310A w/ heater 7.5 V @ 425 mA.

329A, B [AKA 4329A]; pentode, 311A or B w/ heater 7.5 V @ 850 mA (STC, WE).

330A, B, C CRTs, 7" three-gun, bulb 7-1/8" tapering to 3-5/8" dia. and stepped to 2-17/32" dia. by 23-1/16" long, octal base and 13-pin ring base, heater 5 V @ 1.65 A. Other details, phosphors, and ratings same as for 325-type.

331A [AKA 805, q. v.]; vari-μ triode for Class B modulators; typical value for μ is 40 <3N>.

332A Xmtg. pentode, variant of 322A not intended for suppressor modulation: third grid connected to center of filament.

333A Cold-cathode neon triode, 313C w/ molded mounting bracket in place of base, covered w/ corrugated paper protective sleeve, and w/ 11-7/32" flying leads; rated at 150 V and 35 mA. Later production has deep-red lacquer coat. Used as ringer isolator in telephones. Variant: 372A.

334A Mercury thyratron for 24-V/48-V regulated rectifiers, replaceable by 354A w/ circuit change.

335A Argon-mercury thyratron, variant of 334A; replaceable by 355A w/ circuit change.

336A Rcvg. power pentode, ST12 bulb on six-pin base, gold-plated grid; heater 10 V @ 640 mA, g_m 4.1-4.4 mS, 12 W <6B>: A revoltaged, higher-g_m 42. Variant: 349A.

337A Rcvg. variable-μ pentode, ST12 bulb w/ cap on six-pin base, gold-plated grid, heater 10 V @ 320 mA, g_m 1.7 mS <6F>; effectively a revoltaged 78.

338A Argon triode thyratron, T12 bulb on five-pin base, heater 10 V @ 500 mA, 325 V PIV, 100 mA <5A>. For grid-controlled rectifiers and control circuits; a re-voltaged 885.

339A Xmtg. pentode, for Class B linears, S19 bulb w/ cap, 5-pin metal-ceramic base, oxide fil. 5.0 V @ 1.2 A, g_m 4.8, 45 W <5J>. Same pinout as 2E22 but 50% higher dissipation.

340A Xmtg. triode, variant of 232B, T-T fil. raised from 60 to 73 A, lower capacitances, uprated from 3 to 10 MHz (ML, WE).

341A, AA Xmtg. triode, 232B re-filamented to 40 A instead of 60 (by use of thoriated tungsten?). "AA" is for Class B audio, pure-tung. fil. 21.5 V @ 57.5 A, μ 9, 5 kW, radiator type, roughly sim. to 891R (ML, WE).

342A Xmtg. triode, variant of 232B, fil. raised from 60 to 67 A, uprated to 4 MHz (appears to be a lower-freq. version of 340A w/ lower grid dissipation) (F, ML, WE).

343A, AA Xmtg. triode, replacing 220B and 220C, grid-outside, fil. upsized to 21.5 V @ 57.5 A, μ 40, 5 kW, 4 MHz, water-cooled by external cooler (F, ML, WE). "AA" [AKA F343R] is radiator version, w/ corona grading ring, derated to 5 kW.

345A [AKA 4345A]; rcvg. full-wave rectifier, ST12 bulb on 5-pin base, heater 6.3 V @ 1 A, 1250 V PIV, 100 mA <5D>: an uprated 84/6Z4 (STC, WE). Variant: 351A.

346A, B, C Cold-cathode neon triode relay, 313C w/ molded screw-down mounting bracket in place of base, carrying soldering studs; includes optional 120 K resistor in series with starter electrode; 225 V, 30 mA; superseded by "B." "B" has different mounting base carrying soldering lugs, starter resistor reduced to 110 K. "C" has shorter mounting base.

347A Rcvg. triode, 262B w/ heater 6.3 V @ 500 mA, ST12 bulb, and octal base <5M>.

348A Rcvg. pentode, 310B w/ heater 6.3 V @ 500 mA <7R>; sim. to 6J7G.

349A Rcvg. pentode, 336A w/ octal base, heater 6.3 V @ 1.0 A <7S>: a small-bulbed, higher-heater-current 6F6G.

350A, B [AKA 5B/250A, RK-49, T21, VT-230, CV2629]; aligned-grid beam power tube, ST16 bulb w/ cap, on 5-pin base, gold-plated grid, heater 6.3 V @ 1.6 A, g_m 7.1-8.5 mS, 30 W (NEC, STC, WE) <5AW>. Comparable to slightly scaled-up 807. In BC-748A. "B" has octal base <7AC> and is derated to 27 W; comparable to scaled-up 6L6G. In O-1/ASQ-1.

351A [AKA 4351A]; 345A on octal base (STC, WE) <6S>. WE's 6W5G.

352A Duodiode-triode, 292A in ST12 bulb on octal base w/ silver-plated pins <6G>.

353A, B Cold-cathode neon triode relay, 346C rated at 150 V and 35 mA. "B" is uprated to 165-225 V.

354A Mercury triode thyratron, for grid-controlled rectifiers like 48-V battery chargers, S25 bulb w/ cap on 4-pin industrial base, fil. 2.5 V @ 16 A, 1.5 kV PIV, 4 A. Replaces 334A w/ circuit change. Variant: 355A.

355A Thyratron, 354A w/ argon-mercury fill for increased temperature tolerance, PIV derated to 350 V; same use.

356A, B [AKA 5T33]; xmtg. high-μ triode, all-glass T18 domed bulb w/ cap and ceramic base ring holding four pins; T-T fil. 5.0 VCT @ 5.0 A, μ 50, 50 W, 100 MHz, replaceable by 356B (NEC, WE). Variant: 378A/705A. "B" [Ex-D-159781]; is has redesigned grid structure and Zr-coated anode; is uprated to 60 W.

357A, B VHF high-μ xmtg. triode, all-glass T41 domed hard-glass bulb w/ cap, dish base w/ three large pins (sim. to 701A); T-T fil. 10 V @ 10 A, μ 30, 350 W, 100 MHz, to be forced-air cooled above 40 MHz (ML, WE). "B" has Zr-coated anode; is uprated to 400 W. Two "B"s as drivers in Continental Electronics 105B 1-MW AM xmtr. (see 5682).

358A Neon indicator, visual telephone ringer, T4 bulb on molded mounting base, fires at 65 V; carries 18 mA.

359A Neon-argon cold-cathode triode for relay or indicator use, T4 bulb 1-3/4" long, flying leads, 165 V, 40 mA. Variant: 395A.

360A Hearing-aid pentode. Variant: D-164696.

361A Hearing-aid voltage-amp. pentode, fil. 1.4 V @ 20 mA, g_m 250 mS (not to be confused w/ UE 361A).

362A Hearing-aid power pentode, fil. 1.4 V @ 50 mA, g_m 0.6 mS, 9 mW out.

363A VHF aligned-grid xmtg. pentode, pentode version of 357A, same format and fil. but w/ six pins; 350 W, 85 MHz, forced-air cooled.

364A VHF high-μ xmtg. triode, T20 domed bulb w/ dual heavy plate caps, all-glass construction w/ ceramic ring holding five large wide-spaced pins, T-T fil. 5.0 VCT @ 5.0 A, μ 50, 50 W, 125 MHz, forced-air cooled.

365A Power pentode, single-ended octal base, heater 6.3 V @ 650 mA. Model shop only. Variant: D-159511.

366A Rcvg. pentode, T12 bulb on octal base, heater 6.3 V @ 650 mA, gm 10.3 mS. Model shop only. Variant: D-159511.

367A Beam power tube, 350A adapted to video frequencies by use of special pinout w/ wide grid-plate spacing, in odd spherical-tubular ST16 bulb, derated to 25 W.

368A, AS [AKA (STC) 3B/250A]; xmtg. SHF triode, 2" dia. ridged doorknob all-glass hard-glass construction, tungsten plate and grid pins at top, fil. pins at bottom, graphite plate, T-T fil. 1.2 V @ 4.0-4.5 A, μ 8-10, 20 W, will oscillate to 1750 MHz (STC, WE). In T-52/APQ-1. Variant: 388A. "AS" [AKA 703A, D-159764] is single-ended; will oscillate to 1400 MHz. In AN/APT-2.

369A [AKA 5H69A]; xmtg. triode, 25 kW, water-cooled, for AM xmtrs. (NEC, WE).

370A Xmtg. triode, fil. 10 V @ 32 A, forced-air cooled, for 3-kW xmtrs. in prewar (42-50 MHz) FM band.

371A, B [AKA VT-166]; xmtg. vacuum rectifier, 319A format but in nippled bulb, fil. 5.0 V @ 10.3 A, 25 kV, 250 mA. In RA-38. "B" [AKA 221B, 2K71] is ruggedized (EE, GN, NEC, NU, UE, WE). In BC-1037A et al.

372A Cold-cathode neon triode relay, 333A w/ leads shortened from 11-7/32" to 5".

373A Rcvg. pentode, T11 bulb on octal base w/ shortened pin for screen grid, button stem, fil. 2.0 V @ 250 mA, g_m 1.3 mS, 2 W. Short pin is to reduce clicks from in-service changing of tubes paralleled w/ others, i. e., the screen is the first element disconnected. Variant: 374A.

374A Rcvg. power pentode, 373A w/ fil. 3.0 V @ 530 mA, uprated to 4.8 W and 3.3 mS.

375A Rcvg. beam power tube, T11 bulb 4-5/16" high on octal base, heater 20 V @ 320 mA, g_m 4.8-6.2 mS, 6 W, optimized for 48-volt plate supply <7AC>.

376A, B, C Cold-cathode triode, T9 bulb on octal base, w/ RaBr activator to aid firing. "B" [AKA 5589, q. v.] is rated at 275 V, uprated to 20 mA, fires at 80 V, has deep-red lacquer coating. "C" differs in use of Kr-85 in place of RaBr; fires at 75 V.

378A [Ex-D-159780; AKA 705A, q. v.]; vacuum HV rectifier (manufactured as 705A).

379A Xmtg. triode, 279A w/ grid-out-side (ML, WE).

380A [Ex-D-157653]; rcvg. VHF diode, mushroom bulb w/ flying leads and sealing tip at bottom, heater 6.3 V @ 150 mA, 500 V PIV, 5 mA.

381A [Ex-D-159508]; rcvg. diode, 380A on octal base.

382A [Ex-D-157591]; rcvg. triode, small unbased mushroom style {177YQ}. In RA-1095 preamp for 640AA condenser mic. Variant: 383A.

383A [Ex-D-159509]; rcvg. triode, 382A on octal base.

384A [Ex-D-157654]; rcvg. VHF pentode, mushroom bulb w/ flying leads at top and bottom, heater 6.3 V @ 150 mA, g_m 2.6 mS {HQ201}. Variants: 385A, 386A.

385A Rcvg. VHF pentode, 384A w/ grid cap on octal base {HQ201}.

386A [Ex-D-158796]; rcvg. VHF pentode, 384A w/ higher g_m of 4 mS {HQ201}, forebear of 717A and 6AK5. Variant: 387A.

387A [AKA M16-T-63870, D-159512] {HQ201}; rcvg. VHF pentode, 386A on octal base w/ cap.

388A [Ex-D-156548]; xmtg. UHF triode; 368A scaled-up to 2-9/16" dia., fil. 1.5 V @ 9.0-9.25 A, μ 6, 50 W, oscillates to 900 MHz. In T-9/APQ-2. Variant: D-159769.

389A, AA Xmtg. triode intended for prewar FM band. "AA" is radiator variant; has pure-tung. fil. 11.0 V @ 150 A, μ 22, 7.5 kW, 50 MHz (42-50 MHz).

393A [AKA 4G93]; argon-mercury triode thyratron for regulated rectifiers, ST16 bulb w/ ceramic-ringed cap, octal base, fil. 2.5 V @ 6.25 A, 1250 V PIV, 1.5 A out, sim. to 3C23 (CE, GE, HY, NEC, NL, TS, UE, WE) <5AV>. (Some 393As distributed by WE under its name came from by GE, and are marked accordingly.) Variant: 7410.

394A Argon-mercury triode thyratron, "half-current" 393A: same format, fil. 2.5 V @ 3.25 A, 1250 V PIV, 640 mA (CH, GE, KL, WE).

395A [AKA 727A]; neon triode indicator/thyratron, variant of 359A, flat-ended T4 bulb 1-3/4" long, 140 V, 13 mA (TS, WE). Variant: D-168535.

396A [AKA 2C51, q. v.] {1491} <8CJ>.

397A [AKA 2K56, q. v.]; reflex klystron, 723A/B format, original local osc. in TE microwave rcvr.; tuning range of 397A, w/ osc. 65 MHz above received signal, restricted rcvr. to 3890-4170 MHz. Replacement by 438A gave full coverage of the 3700-4200 MHz band. Variant: D-178461.

398A [AKA 5603, q. v.] <6BD>.

399A, B Min. fil.-type pentode, for M1 carrier telephone system, sim. to 1T4 exc. for fil. voltage. "B" is heater version, heater 1.15 V @ 650 mA, g_m 1.1 mS <6AR> plus cathode on pins 1 and 5.

400A Min. fil.-type pentagrid converter, for M1 carrier telephone system, sim. to 1R5 exc. for fil. voltage. "B" is heater version, heater 1.15 V @ 650 mA, same use as "A"; <7AT> plus cathode on Pins 1 and 5.

401A [AKA 5590, q. v.]; min. rcvg. pentode. Same original use as 399A.

402A Xmtg. amp. klystron, glass disc-seal, small three-pin base, dual external cavities and magnet, finned radiator at collector end, forced-air cooled, heater 6.3 V @ 500 mA, 500 mW out at 4 GHz, for 1948 TDX trial; later used in TD-2 test sig. gen.

403A, B [AKA 6AK5, q. v.]; min. rcvg. pentode {1381HQ} <7BD>. "B" [AKA 5591, q. v.]; has heater reduced to 150 mA.

404A [AKA 5847, q. v.]; min. pentode (LME, WE) <9X>.

405A Min. cold-cathode triode, repl. by, 426A.

406A Reflex klystron sim. to 2K54, original transmitter tube in TE portable microwave system; gave 200 mW output @ 3.7-4.2 GHz. Also see 431A.

407A, B Min. duotriode, 396A w/ heater for 20 V @ 100 mA or 40 V @ 50 mA (RCA, SY, TS, WE). In N1 carrier repeater, where heater and plate current were supplied on a series-string basis via cable pairs over a 3-5 mile distance. Variant: 7370. "B" is uprated to 1.6 W, 130 V from heater to cathode, and g_m 6.0 mS.

408A [AKA 6028, q. v.]; (NEC, RCA, SY, WE) <7BD>.

409A [AKA 6R-R8, 6AS6, q. v.]; min. pentode (NEC, WE) <7EN>.

410A Mercury triode thyratron, steel case 2-½" dia. by 10-3/4" long w/ mounting bracket, anode lug at top, other lugs at bottom w/ finned radiator, fil. 2.5 V @ 18 A, 1.5 kV PIV, 8 A.

411A Cold-cathode relay tube, replaced by 425A.

412A [AKA 6754, q. v.] <9ET>.

413A, B Cold-cathode symmetric neon-argon diode, T9 bulb on molded screw-down mounting w/ soldering lugs, fires at 180 V, 70 mA. "B" has Kr-85 instead of RaBr to aid firing.

414A Min. pentode sim. to 6AJ5.

415A Min. pentode, 409A w/ heater reduced to 150 mA.

416A-D Xmtg. glass SHF planar triode, double-ended gold-plated body 1.07" dia. by 1.88" high, glass seals, anode pin at top, threaded grid ring at upper center, four pins at other; heater 6.3 V @ 1.18 A, μ 200, g_m 50 mS (4 GHz), forced-air cooled {1553}. Used in three-stage transmitting amp. in TD-2 and TD-3D microwave systems, later replaceable by 660-type GaAs microwave ICs. "B" [AKA 6280, q. v.] is uprated to 7.5 W, gives 500 mW output at 4 GHz. "C" is uprated to μ 250, g_m 65 mS, and 15 W; gives 2 W output at 4 GHz. "D" is selected "C," later w/ BeO seals, for 5 W output at 4 GHz.

417A [AKA 6R-H2, 5842]; 9-pin min. grounded-grid triode, heater 6.3 V @ 300 mA, μ 43, g_m 24 mS, 4.5 W {1547} <9V>. Introduced as 70-MHz IF preamp in receiver of TD-2 microwave system (NEC, LME, SY, WE).

418A [AKA 6B-P16]; power tetrode, short Compactron-style T9 bulb on 9-pin base w/ large (0.050" dia.) pins and sealing tip out base, heater 6.3 V @ 600 mA, g_m 26.5 mS, 8.5 W (NEC, WE) {1546}.

419A Klystron, improved 2K45 for operation w/ reflecgor supply of only 250 V {1522}. "419B" [AKA 2K39, q. v.] seems to be non-WE tube.

420A [AKA 5755, q. v.]; min. duotriode {1554}.

421A Power duotriode, pass tube in voltage regulators, ST16 bulb on octal base, heater 6.3 V @ 2.4 A, μ 6.0, g_m 18-20 mS, 26 W total <8BD>. 6AS7G w/ slightly higher μ.

422A Full-wave vacuum rectifier, ST16 bulb on octal base, heavy plate structure, fil. 5.0 V @ 3.0 A, 1800 V PIV, 400 mA <5L, but w/ cathode on pin 2, not 8>.

423A, B, C [AKA 6140, q. v.]; voltage reference <9BY>. "C" has Kr-85 activator, possibly in place of RaBr. Variant: 453A.

424A Pentode, 409A w/ redesigned grid. Production unknown.

425A Cold-cathode tetrode, min. 9-pin bulb on molded screw-down mounting bracket w/ flying leads; deep-red lacquer coating; has starter electrode with its own cathode, so that starting circuit is normally isolated from the main circuit. Includes 18 K resistor in starter cathode and 82 K resistor in starter cathode. Rated 180 V, 5 mA. Ringer isolator in 500-type telephones, replacing 333A. Variants: 426A, 451A.

426A Cold-cathode relay, triode variant of 425A, same use.

427A [AKA 6141, q. v.]; min. voltage regulator w/ starter. Variant: 446A.

428A Reflex klystron, 723A/B format and pinout, tunes 6.4-7.2 GHz, heater 6.3 V @ 460 mA, 330 V. An oddity: officially intended as xmtr. osc. and local osc. in "the 7 KMC microwave television relay." 7 GHz is not a common-carrier (Telco) band, suggesting intention perhaps to sell to broadcasters.

429A Rcvg. beam power tube, taller 418A format, heater 20 V @ 140 mA, g_m 6.6 mS, 12 W. Used in pairs to key 60-mA telegraph local loops until replaced by plug-in solid-state switch.

430A, B, C Cold-cathode relay, 313C w/ faster firing time via addition of RaBr; fires in 70 μsec, vs. 6 msec; 185V, 35 mA. "B" has activator changed from RaBr to Kr-85.

431A Metal reflex klystron, not in 723A/B format, IC, CO, heater 6.3 V @ 440 mA, 475 V, replacing 406A in TE microwave transmitter, giving 300 mW <1: shell and resonator; 2 & 7, heater; 4, coax output line; 6: reflector; 8: cathode>.

432A, B Min. 9-pin voltage reference, w/ starting electrode, 100 V @ 4-8 mA. <9BY>. "B" is high-stability version, claimed to be free from parasitic oscillations.

435A [AKA 6R-R21]; min. 9-pin rcvg. tetrode, gold-plated frame grid, heater 6.3 V @ 300 mA, g_m 15.5 mS (NEC, WE).

436A Rcvg. tetrode, 418A format, gold-plated frame grid, heater 6.3 V @ 450 mA, 30 mS, 4.5 W. Also see 7150.

437A Rcvg. triode, 418A format and heater, μ 41, g_m 46 mS, 7 W.

438A Reflex klystron, 723A/B format and pinout. See 397A for use.

439A [AKA 6167, q. v.]; counter.

440A Triode, 101F in ST14 bulb on 4-pin push base.

441A Triode, 102F on 4-pin push base.

442A Power triode, 205F on 4-pin push base, uprated to 14 W.

443A [AKA 6388, q. v.], cold-cathode triode.

444A TWT, external mounting and magnet, glass tube w/ flying leads, heater 6.3 V @ 920 mA, anode voltage 3.5 kV, 80 W, 33 dB gain, 8 W out @ 6 GHz. Replaceable by VA VTJ2686B2. Variant: 463A.

445A [AKA VRX7204B]; reflex klystron, 10.7-11.7 GHz, octal base, 723A/B pinout, waveguide output, cooling fins, heater 6.3 V @ 880 mA, resonator 600 V, gives 400 mW output; convection-cooled. In TJ microwave system as xmtg. osc. and local osc.

446A Min. voltage reg., 108-V variant of 427A <9BZ>.

447A Min. 9-pin cold-cathode neon voltage reference and indicator, 82 V, 8 mA.

448A, B [AKA 6B-R22]; rcvg. tetrode, 436A format and heater, 31.5 mS, 6 W (NEC, WE). "B" has inside of bulb coated w/ Aquadag to suppress fluorescence. Variant: 454A.

449A Reflex klystron, 5925-6425 MHz, four pins in same format as on 416A, WG, heater 6.3 V @ 880 mA, resonator 525 V, 150 mW out. Variants: 450A, 459A-459B. "B" replaces "A," has tuner assembly added to waveguide. Variant: 450A.

450A Reflex klystron, version of 449A for 6090-6175 MHz. Variant: 459A.

451A Cold-cathode tetrode, 425A in 9-pin min. form; lacquered deep red.

452A Reflex klystron, version of 2K25 for 3.9-4.46 GHz, gives 70 mW output.

453A Voltage reference, 83-V variant of 423A <7BY>.

454A Rcvg. tetrode, 448A w/ plate pin shortened for use in a "special test fixture" for TH microwave system.

455A-F Rcvg. pentodes, T9 bulb w/ ceramic base ring and flying leads, heater 7.5 V @ 305 mA, μ 5.9 mS, B+ 45 V. Tubes were graded in manufacture into the six codes, for use in undersea repeaters of the SD submarine-cable system. Originally coded F-51887 thorough F-51883.

456A, B Gas protector diodes for cable repeaters (see 455A); short T9 bulb and ceramic base w/ flying leads. Fires at 160 V; carries 450 mA. "A" is for undersea use; "B" is for shore application. Variants: 458A, B.

457A [AKA VRX7204A]; reflex klystron, 10.7-11.7 GHz, octal base, WG, heater 6.3 V @ 880 mA, resonator 600 V, 10 mW out. Xmtg. klystron and local osc. in TL microwave system.

458A, B Gas protector diodes, 456A and B w/ firing voltage lowered to 50-95 V; same uses.

459A, B, D Reflex klystron, 450A adapted for vapor-phase cooling, 100 mw out. "B" gives 80 mW for local-osc. use. Both replaceable by VA VRJ7203A. "D" is replaceable by VA VRX7203A.

460A Cermet gas protector diode, 0.39" dia. by 5/8" long, fires at 200 V min., takes 2000-A surges {2185}.

461A TWT, packaged type w/ integral magnet, heater 7.5 V, helix voltage 2900, gives 5 W output at 4 GHz. Replaceable by 660E-660F GaAs microwave integrated circuits.

462A TWT, packaged type w/ integral magnet, heater 7.0 V, helix voltage 2300, 4 W out at 6 GHz. Replaceable by Calif. Microwave CX64/TM-1 solid-state kit.

463A TWT, 444A w/ collector voltage 1.2 kV, 90 W, 1 W output at 6 GHz {4095}. Replaceable by VA VTJ2686B3.

464A TWT, packaged type w/ integral magnet, heater 7.5 V, helix voltage 3.6 kV, gives 12 W output at 6 GHz {4095}.

466A Vidicon, for PICTUREPHONE sets, non-burn-in silicon-target, packaged with yoke and IR filter, rectangular package 2.1" by 2.2" by 8.1", flying leads going to ribbon connector, aspect ratio 1.1:1, heater 8.0 V, supply voltage 500 V.

467A Cermet three-element gas protector, 0.39" dia. by ½" long, fires at 200 V min., takes 250-A surges.

468A Camera tube, for PICTUREPHONE sets, 1", Si target, packaged w/ focus magnet and deflection ckts., fil. 8 V @ 105 mA, 500 V.

469A Gas diode protector tube for SG submarine cable system, T5½ bulb in plastic mount, 75 V, 3 mA.

471A Surge protector.

472A Surge protector sim. to 467A.

473A TWT, highly linear type for AR-6A SSB microwave system, 5925-6425 MHz. Replaceable by KS-22469.

700A-D, T Family of fixed-frequency external-magnet coaxial-output cavity magnetrons, all later replaceable by the tunable 4J42. Heater 13.5 V @ 1.85-2.0 A, 12 kV, 10 A, 40 kW pulse out, forced-air cooled. Four frequency codes: "A," 715 MHz; "B," 705 MHz; "C," 695 MHz; "D," 685 MHz, each ±5 MHz. In, and probably unique to, FA-FB-FC-FC radars. "T" is informal designation for tunable version, AKA 4J42, q. v.

701A [AKA D-160810]; xmtg. tetrode radar modulator, format like 357A but reduced to T33, four base pins, interior contains four sets of 350A electrodes w/ common Mo anode, gold-plated grid and screen, fil. 8.0 V @ 7.5 A, 12.5 kV, 100 W, 10 A pulse; in FC-FD and BC-717A.

702A, B TR tube, disc-seal cell for external cavity, hydrogen - water vapor fill, body 1-3/16" dia. by 2" long; gaps break down at 500 V. In BD-106A and FC-FD. "B" is improved version.

703A [AKA 368AS, D-159764]; xmtg. UHF triode, 368A with cutoff limits (TS, WE). In BC-716A. Receiver mixer in FD radar; in T-26/APT-2 and AN/APQ-2.

704A [Ex-D-159765]; submin. UHF diode mixer, T2 body 1-1/16" long w/ heater leads; plate-out-side; heater 4.5 V @ 500 mA, 1.5 kV PIV, 10 mA {225TQ}. In BD-106A and wavemeter unit of FC-FD.

705A, WA [AKA 1K75, (WE) 378A, D-159780]; HV rectifier, gridless 356A w/ plate pin instead of cap; fil. 5.0 V @ 5.0 A, 30 kV PIV, 100 mA. Charging diode and/or HV rectifier in FC-FD, BC-740A, MD-2/TPS-1, MD-36/MPG-1, SG, SJ, O-18/CPN-2A, AN/CPN-6, et al. Variant: 8021, q. v. (NEC, RA, TS, WE). "WA" (UE) has graphite anode, plate cap on glass stalk.

706, T1 Family of fixed-frequency cavity magnetrons w/ external magnets and coaxial output; heater 6.9 V @ 1.5 A, 22 kV, 22 A, forced-air cooled. Original codes were "A"-"C," 25 kW pulse output, 3019-3100 MHz; later versions were "AY"-"GY," 200 kW at 2914-3100 MHz [AKA WL-443 family]. In xmtr. of SG; "FY" and "GY" in SJ. Also see 714AY and 718 family. "T1" is informal designation for a tunable version [AKA 2J61].

707A, B Ext.-cavity disc-seal reflex klystron, body 1-¼" dia. stepped to 13/16" dia. by 5" long, octal base, unusual 3-sided glass press for leadout wires, cap for reflector; 2.5-3.75 GHz (or wider range, depending on clamp-on cavity). Heater 6.3 V @ 650 mA, resonator voltage 300 V. First widely used U. S. klystron; superseded by 707B. In BC-740A, SE, SG, SH, SJ, Mk. VIII, et al. AKA "McNally" tube. Input power is limited by heating of grids, which normally run bright yellow. "B" [AKA QK26] gives 80 mW out, usually replaceable by 2K28 (RA, HK, SY, WE). Differences from 707A minor; mainly 5 mW more output (WE). In RF-7/MPN-1, AN/APN-7, and SO. Variant: 2K28.

708A Rcvg. UHF G-G triode mixer; flanged steel body 1-1/32" dia. by 23/64" high, w/ grid connected to it, three coaxial feed-throughs for fil. and plate leads out bottom; prominent seal-off tube out top; fil. 1.7 V @ 1.85 A, μ 10, 15 W, dev. ca. 1940 {1272Y}. In initial 1-3 GHz radars: BC-740A, early SE, SG, early SH, early SJ, early Mk. VIII. Variants: D-160127, D-160206. AKA "Shepherd-Pierce" converter.

709A Three-electrode TR tube. In BC-740A and Navy radars.

710A [Ex-D-160810; AKA VT90 (Br.), 8011, q. v.]; xmtg. triode. In BC-747A.

711A Osc. (Br.). Suspected of being a derivative of the Br. NR89 3-GHz klystron, incidental source of the 709A TR tube. Cancelled, probably before production.

712A "Superseded by 707A." Cancelled, probably before production.

713A Rcvg. pentode, 717A on bakelite base (TS, WE) <8BK>. In BC-716A, SE, SH, SJ, and Mk. VIII.

714A, AY Cavity magnetron, fixed at 3300 ± 20 MHz, "A" gave 20 kW pulse out; "AY" gave 125 kW (WE, W). See 706 family for details. In BC-1050B.

715-TYPE Tetrode radar modulators, for airborne use, 356A format slightly expanded. Has quadruple cathode, single gold-plated grid and screen, single Mo-Zr plate.

715A Original version, heater 27.0 V @ 2.15 A, 14 kV, 10 A peak, 60 W, superseded by "B" and "C." In T-2/APS-15 and MD-5/APS-3.

715B Improved version, has bigger heat radiating fins on grid rods, fil. changed to 26.5 V @ 2.1 A, uprated to 15 kV and 15 A; superseded by "C." In T-25/TPS-2 and AN/APS-3.

715C [AKA P552/1E]; improved "B," all-glass construction, T-20 bulb w/ cap, ceramic wafer "jumboid" base; sim. to 5D21 (RA, RCA, STC, WE). May not have been made in quantity until postwar.

716A Thyratron, cancelled, probably before production. Superseded by 722A.

717A [AKA VT-269, CV3594]; rcvg. RF pentode (713A w/ bantal micanol base), heater 6.3 V @ 175 mA, g_m 4 mS (TS, WE) <8BK>. In BC-695, SCR-717, and SJ. Derives from 384A; led to 6AK5.

718AY-EY Family of fixed-frequency cavity magnetrons, long-life versions of 706 family, q. v. "AY" - "EY" covered 2720-2890 MHz, giving 136 kW pulse output. In BC-947A.

719A Radar HV rectifier, 356A format slightly expanded; fil. 7.0 V @ 7.0 A, 25 kV, 500 mA. In BC-1039A and AN/MPG-1. Postwar version (LK) had Zr-coated plate and no internal insulators.

720AY-EY Family of fixed-frequency cavity magnetrons, external magnet, 53.5-ohm coaxial output, heater 13.9 V, 27 kV, 70 A, 1 MW pulse output, forced-air cooled (WE) (ca. 7-42). Original codes were "A" - "C"; later versions "AY" -"EY." "AY," "BY," and "CY" are AKA 4J45, 4J46, and 4J47 respectively. "CY" was used in SCR-615B.

721A, B TR-ATR cell, disc-seal bulb 1-1/8" dia. by 3" long, w/ cap for keep-alive electrode, fixed-tuned, for use w/ external cavity at 2.7-3.3 GHz, superseded by 721B (BL, WE). In RF-7/MPN-1, SCR-545, and SL. "B" is improved version.

722A [AKA 287A, CE302]; triode thyratron sim. to 287A; w/ cap; fil. 3.6 V; modulator (w/ 813) in SJ.

723A, A/B Reflex klystron, tunable 8702-9548 MHz. Has

metal body; is tuned by slightly flexing internal cavity. Original design was for 9375 MHz only, then expanded to full band. Has coaxial output via small tubing line extending out octal base. Heater 6.3 V @ 440 mA, 25 mW out (AX, WE) {1439XQ} <1: shell & resonator; 2 & 7, heater; 4, coax output line; 8: cathode>. Replaced by 723A/B. In RT-5/APS-4 et al. "A/" is improved, 330 V, 33 mW out (AX, KR, MU, NEC, RA, TF, VO, WE). Sim. to REL31. Repl. by 2K25 late in WW II. Variants: 2K25, D-169618. In AN/-MPG-1, AN/APS-15, SCR-717, etc.

724A, B TR-ATR tube, disc-seal, fixed-tuned, tubular body 0.37" dia. by 2" long, w/ cap at one end for keep-alive electrode, for use w/ external cavity at 8.6-9.7 GHz; replaced by 724B; {1331} (WE). In RT-10/APS-3, AN/APS-4, and AN/MPG-1. "B" has temperature compensation; in AN/APS-1.

725A Magnetron, fixed at 9375 ± 30 MHz, EM, output waveguide parallel to axis of cathode, heater 6.3 V @ 1 A, 16 kV @ 16 A, 60 kW pulse output, interchangeable w/ 2J21 and replaceable by 2J53 {1330M} (AX, KK, MU, NE, PH, RA, TSU, VO, RA, WE). In AN/APS-3 (interchangeably w/ 2J21A), RT-5/APS-4, AN/APS-15, and AN/SPN-18. Also see 730A.

726-TYPE Reflex klystrons, 723A/B format and pinout, IC, CO, heater 6.3 V @ 440 mA, 390 V, 180 mW out.

726A Tunes 3.2-3.4 GHz (NEC, RA, SFR, THF, WE). In BC-1901A.

726B Tunes 2.9-3.2 GHz, 150 mW output (KK, NEC, RA, THF, WE). In SJ.

726C Tunes 2.7-3.0 GHz (KK, NEC, RA, THF, TSU, WE).

727A Cold-cathode triode sim. to 359A. Also see 313CA.

728AY-JY, T Family of fixed-frequency cavity magnetrons, external magnet, coaxial-output, giving 275 kw at 900-970 MHz. "T" is informal designation for tunable version [AKA 4J51, q. v.] {1443M}.

729A [AKA 1B23, q. v.]; TR tube.

730A Magnetron, 725A w/ waveguide outlet rotated 90° {1384M} (KK, NEC, WE). In BC-1276A and SCR-717.

731A [AKA 6AK5, q. v.]; pentode {1381HQ}. No known manufacture under this number.

732A [AKA 3B24, q. v.]; HV rectifier {429T, 1299X2}. No known manufacture under this number.

DEVELOPMENTAL TYPES

240H Doorknob duopentode, 368A format, common screen and suppressor, fil.-type, 10 W out @ 150 MHz (*Electronics*, 12-36 p. 60)

1563 Five-element cold-cathode device, short *loktal* 7F8-like ruggedized; has what appears to be a pair of starter electrodes, a regular cathode, and two anodes. Also carries number SA584, which, along with the loktal bulb, suggests a collaboration w/ SY. Carries usual lettering WESTERN ELECTRIC / MADE IN U. S. A. / BELL TEL. LABS / NOT STANDARD / FOR EXPERIMENTAL USE ONLY.

39DY1 [AKA VT-142]; xmtg. triode 200-MHz radar pulse osc.; lab only; never made in quantity but VT-no. was not cancelled.

PHOTOCELLS

WE catalogued its phototubes (other than 238A) as "Photocells" rather than "Vacuum Tubes."

1A Phototube, short bulb w/ long insulating stems for leads.

3A [AKA 1P31, G10, CE3A/B, CE3D, 79A4]; gas type; also see 1P31. Cesium cathode, central anode, S17 spherical bulb w/ flying leads, sensitivity 100 μA/lumen.

5A Vacuum type, central anode, S11 spherical bulb, cesium cathode, 4-pin base, sensitivity 40 μA/-lumen.

6A [AKA CE-26]; gas type, central anode, S17 spherical bulb, 4-pin base, cesium cathode, sensitivity 100 μA/lumen (HY, WE).

7A Gas typesim. to 3A, spherical bulb, central anode, S17 spherical bulb, flying leads, sensitivity 150 μA/lumen.

8A [AKA CE-1]; gas type, S17 spherical bulb, central anode, cesium cathode, 4-pin base, sensitivity 100 μA/lumen (HY, WE).

9A Gas type, partitioned cathode, two anodes, S17 spherical bulb, 4-pin base, sensitivity 150 μA/-lumen.

D-SPECIFICATION TUBES

D-79509 Spherical ionization manometer derived from rebased 205B, 8" glass tube added as measurement inlet.

D-79510 Spherical ionization manometer (soft glass); wire leads out bottom and inlet tubing out top; total length 7-¼".

D-79511 Ionization manometer, hard-glass variant of D-75509.

D-79512 Ionization manometer (D-79510 w/ Nonex hard glass), pear-shaped.

D-80039 [AKA VT-5, 215A, q. v.]; rcvg. triode. Bell Labs listing ES-452954 is in error on this item, reversing 211D and 215A/D-80039.

D-80050 [AKA VT-4B, 211D, q. v.]; xmtg. triode. Bell Labs listing ES-452954 is in error on this item, reversing 211D and 215A/D-80039.

D-85715 Xmtg. triode, 243A w/ different lead arrangement and lead glass instead of Pyrex.

D-85716 Xmtg. triode, 240A w/ changes as for D-85715.

D-85792 Rcvg. triode, high-insulation 102E, superseded by 102G.

D-86326 Rcvg. triode, high-insulation 101D, disc. after improved production of 101D. In line amps of C2 carrier telephone system.

D-86327 Rcvg. triode, high-insulation 104D, disc. after improved production of 104D. Used optionally in 22A2 telephone repeater.

D-86737 Xmtg. triode, 211D w/ closer grid spacing, recoded as 248A, q. v.

D-87491 Diode voltage regulator, unbased, tungsten fil. 7.75 V @ 1.65 A.

D-87722 Rcvg. triode, unbased 231D, used in 1A Pilot Channel ('20s tel. transmission-level regulator).

D-88057 Xmtg. triode, 212D w/ half as many grid turns.

D-88526 Xmtg. triode, 212D w/ fewer (23/39) grid turns.

D-90278 Thyratron, recoded as 269A, q. v.

D-90279 Thyratron, recoded 256A, q. v.

D-90319 Rcvg. tetrode, 245A (q. v.) shortened to fit 8A aircraft rcvr.

D-91143 Rcvg. triode, 101F w/ UX base and "S" or "ST" (not tennis-ball) bulb, repl. by 440A.

D-91144 Rcvg. triode, 102F w/ UX base, repl. by 441A.

D-92321 Rcvg. tetrode, recoded as 259A.

D-92437 Rcvg. triode; "quiet" version of D-86326, which itself derives from 101D.

D-94044 Rcvg. triode, selected 264A.

D-94230 Selected 231D with UX-sized bayonet-style base bearing 215A-style studs, ca. 1925-28.

D-94341 Selected 224A CRT.

D-95521 Rcvg. triode, for program amps.

D-95954 "AEO-lite" ("alkaline earth oxide") [photocell] used by Movietone News.

D-96412 Pentode, replaced by 294A.

D-96413 HV rectifier sim. to 324A.

D-96415 Space-charge electrometer tube, "S" bulb w/ grid cap on 1-½" glass stalk for added insulation; roughly sim. to GE FP-54. Replaced by 5674.

D-96475 Space-charge electrometer tetrode, grid cap on 1-½" glass stalk for insulation, fil. 0.98 V @ 270 mA.

D-96820 [AKA 38111]; low-capacitance 50-W power triode, sim. to 211D and 242B.

D-96987 Xmtg. triode; see 845.

D-97004 Stroboscope flash tube.

D-97822 Selected 205D.

D-98677 215A as made for Navy as CW-38015.

D-99022 Cold-cathode tube.

D-152455 TR switch, 9 GHz, gold-plated cast body to clamp between waveguide sections.

D-155023 Unbased phototube, 1" dia. and 1-½" high, w/ 360° annular field of view. May have been used in BTL photoelectric VT fuze for shells and rockets (*Bell Labs Record*, Feb. 1947).

D-156547 Rcvg. diode, repl. by D-157653/380A.

D-156548 Doorknob tube used in prewar 1A radio altimeter, recoded 388A, q. v.

D-156734 205D w/ silver-plated studs, repl. by 442A.

D-157591 Rcvg. triode {HF175}, recoded 382A, q. v.

D-157653 Rcvg. diode recoded as 380A, q. v.

D-157654 Rcvg. pentode, recoded as 384A, q. v.

D-158643 Rcvg. pentode, repl. by D-158643/384A.

D-158796 Rcvg. pentode, recoded as 386A, q. v.].

D-159076 Photomultiplier, 6-stage.

D-159508 Rcvg. diode, recoded as 381A, q. v.

D-159509 Rcvg. triode, recoded as 383A, q. v.].

D-159511 Pentode, 366A (q. v.) w/ grid cap.

D-159512 [AKA 387A, q. v.]; xmtg. tetrode.

D-159764 [AKA 368AS, 703A, q. v.]; xmtg. triode.

D-159765 Submin. diode, recoded 704A, q. v.

D-159769 Xmtg. triode, selected 388A, q. v.

D-159778A Thyratron.

D-159780 HV rectifier, recoded 378A, then 705A, q. v.].

D-159781 Xmtg. triode, recoded 356B, q. v.

D-160052 Br. E1189 magnetron, as copied for developmental purposes; 30 or so made.

D-160127 Rcvg. triode sim. to 708A (experimental).

D-160206 Rcvg. triode, recoded 708A, q. v.

D-160573 Early flanged triode sim. to 708A {1301M}.

D-160810 Xmtg. triode, recoded 710A, q. v.].

D-161145 "Xmtg. UHF osc." {1325M} (also D-161458).

D-161714 Thyratron.

D-161831 [AKA 1278-GY2, GY2, VT-279]; thyratron, characteristics and production unknown.

D-162836 Si diode, superseded by 1N21.

D-163876 Thyratron {1374GY}.

D-164694 Small klystron, EC, EM, flying leads, superseded by 2K40, q. v. {1280CT}.

D-164696 Hearing-aid pentode, ruggedized 360A.

D-164713 "Xmtg. UHF triode osc."

D-164714 [AKA 7C22, q. v.; duotriode packaged osc., made in small numbers.

D-166713 External-anode air-cooled VHF duotriode capable of 600-1000 kW pulses at 200 MHz, not made in quantity; loosely related to 7C22. Two-ended construction w/ anode radiator at center.

D-167158 "UHF osc." {1289CT}.

D-168535 Cold-cathode triode, ruggedized 359A w/ main-gap voltage 150.

D-169618 Klystron, modified 723A/B.

D-170135 Radar gap, recoded 1B42, q. v. {1438GT}.

D-170573 Klystron, sim. to 5649, q. v.].

D-170924 Xmtg. triode, 343A w/ spiral fil. for longer life.

D-171089 Duotriode VHF osc. sim. to 7C22; shorter in height.

D-171225 Emergency duplicate of captured Siemens & Halske spray-shielded repeater tube: 1000 copies made in three weeks in 1944 to restore telephone repeaters in France.

D-171561 Ge diode, high PIV.

D-171612 Crystal diode for AN/APS-3A, DC restorer, high PIV.

D-172733 Tunable magnetron like QK185.

D-176627 Klystron, registered as 5650, q. v.

D-178461 Selected 397A/2K56 for operation at 250 V; repl. by 419A.

WESTINGHOUSE

Like General Electric, Westinghouse numbered and sold its mass-market tubes in the '20s through RCA, but served the industrial market separately.

Numbers assigned by W to special-purpose types follow a simple numerical sequence, with various prefixes (mainly WL; also KU, RO, etc.). Developmental devices are prefixed "WX."

GS-10 Grid-glow thyratron, relay or grid-controlled rectifier, S17 bulb, 500 mA, 1930.

GS-12 Grid-glow thyratron, relay or grid-controlled rectifier, S16 bulb, 100 mA, 1930.

GS-14 Grid-glow thyratron, relay or grid-controlled rectifier, spherical-tubular bulb ca. 3" dia., 6 A, 1930.

SR-17 Vacuum phototube.

GS-18 Grid-glow relay thyratron, S14 bulb, 20 mA, 1930.

WC-21 Vacuum rectifier, fil. 10 V @ 16.75 A, 15 kV PIV, 220 mA.

SR-50 Cesium vacuum phototube, tubular bulb w/ side-on cathode, industrial 4-pin base, S-1 response, 15 μA/lumen.

SR-51 Phototube, SR-50 w/ violet response.

SR-53 Phototube, SR-50 w/ sensitivity raised to 25 μA/lumen.

SR-58 Vacuum phototube, T17 stepped tubular bulb, 4-pin industrial base, large cathode (1-5/8" x 3-½"), ultra-violet response, 15 μA/lumen.

SK-60 Gas phototube, SR-50 format, S-1 response, 60 μA/lumen.

SK-61 Phototube, SK-60 w/ violet response, 60 μA/-lumen.

SK-62 Phototube.

SK-63 Phototube, SK-60 w/ sensitivity raised to 125 μA/lumen.

SK-68 Phototube, gas version of SR-58, S-1 response, 60 μ/lumen.

WL-108 Reflex klystron, 12.4-18 GHz, heater 6.3 V @ 1.2 A, 625 V, 70 mW output.

WL-112 Power triode, prototype of UX-112, q. v.

WL-195 RF-heating HF triode, 852 format, fil. 10 V @ 3.25 A, 125 W, giving 155 W @ 50 MHz (1-10-47); also see WL-196.

WL-196 RF-heating HF triode, WL-195 w/ higher μ of 35 (1-10-47).

AW-200 Xmtg. triode, double-ended, external anode w/ tubular glass extensions at fil. and grid ends, fil. 14.5 V @ 153 A, μ 30, ca. 15 kW, water-cooled, giving 5 kW output @ 50 MHz, promoted as power source for cyclotrons.

AW-220 Xmtg. triode, 4" dia. by 63" long, double-ended, fil. 30 V @ 320 A, 150 kW, water-cooled anode and grid, fluted anode, ca. 1929. Six in experimental 300-kW power amp. at KDKA, 1932, modulated by six more.

WL-386 HV rectifier ("medical and surgical").

WL-402 "Small special-purpose."

WL-404 Xmtg. and RF-heating HF triode.

WL-405, 6 Xmtg. triode.

WL-407 Xmtg. tetrode.

WL-409-12 Xmtg. triodes.

WL-413, 15 Xmtg. tetrode.

WL-414 [AKA GL-414]; mercury tetrode thyratron for continuous or welder use, steel case w/ mounting bracket and flying leads out both ends, grid-out-side; fil. 5.0 V @ 19 A, 3 kV PIV (2 per W), 5 A {WX3187} (GE, W). Also see FG-172.

WL-417, A, B [AKA VT-277]; klystron, 2650-3330 MHz, IC, octal base, input-output via threaded coax fittings; cap for repeller; mechanically tuned by clamp-on assembly. Input 75 W max. w/ forced-air cooling; output 250 mW (SP, W). In BC-1086 and R-6/CPN-1. "B" is low-grid-emission variant; see 2K41.

WL-418 Xmtg. triode.

WL-419 Xmtg. triode, superseded by 891.

WL-420 Xmtg. triode UHF osc.

WL-421 [AKA 2K21]; klystron.

WL-422 [AKA 2J21, q. v.].

WL-424 Xmtg. rectifier, superseded by 869B.

WL-427 Vacuum rectifier, bulb 6-1/8" dia. by 24-3/8" high, fil. 10 V @ 50 A, 150 kV PIV, 120 mA, superseded by WL-612. Also see WL-444 and WL-612.

WL-432 Xmtg. vacuum rectifier.

WL-433 Xmtg. diode vacuum rectifier.

WL-435 [AKA 40, q. v.]; rcvg. triode.

WL-436 Xmtg. diode gas rectifier (obsol.); was to be SigC VT-53.

WL-440 Triode amp.

WL-441 [AKA 2J32, q. v.]; magnetron.

WL-442 [AKA 2J21, q. v.]; magnetron.

WL-443 [AKA 706AY, q. v.]; magnetron.

WL-444 Vacuum rectifier, bulb 6-1/8" dia. by 24-3/8" high, fil. 10 V @ 50 A, 110 kV PIV, 240 mA, superseded by WL-612. Also see WL-427.

WL-450 [AKA 450TH, q. v.]; xmtg. triode.

WL-455 RF-heating triode, 110-W tungsten fil., for VHF diathermy machines w/ output controlled only by fil. curr.; derived from WL-680; grid and plate caps out opposite sides. Gave 120 W @ 50 MHz.

WL-456 Vacuum rectifier, bulb 5-1/8" dia. by 18-5/8" high, fil. 11 V (earlier 10 V) @ 20 A, 140 kV (earlier 150 kV) PIV, 60 mA.

WL-460 RF-heating HF triode, tubular bulb w/ cap, grid-out-side, jumbo 4-pin base; fil. 10 V @ 4.1 A, μ 18, 150 W; gave 250 W out @ 50 MHz; rated at higher freq. and voltage than WL-468; sim. to FP-252A or HV18 (1936). Also see HF-200 and WL-464.

WL-463 RF-heating HF triode, fil. 10 V @ 5 A, μ 22, 200 W,sim. to HF-300.

WL-464 RF-heating HF triode, WL-460 w/ higher μ of 65.

WL-468 RF-heating HF triode, fil. 10 V @ 4.1 A, 150 W; also see WL-460.

WL-469 [AKA FP-285]; RF-heating HF triode, 211 and WL-471 ratings.

WL-471 [AKA 8003, q. v.]; RF-heating triode.

WL-472 RF-heating triode.

WL-473 [AKA 5736, q, v.]; RF-heating/xmtg. triode.

WL-477R Xmtg. VHF tetrode, fil. 5.0 V @ 23 A, 700 W, 120 MHz, forced-air-cooled, promoted for FM xmtrs.

WL-478R Xmtg. VHF tetrode, fil. 5.0 V @ 70 A, 3 kW output, 120 MHz, forced-air-cooled, promoted for FM xmtrs.

WL-479R Xmtg. VHF tetrode, fil. 5.0 V @ 206 A, 6 kW , 120 MHz, forced-air-cooled, promoted for FM xmtrs.

WL-481, B Vacuum rectifier, fil. 2.5 V @ 5 A, 25 kV PIV, 30 mA. Comparable to 3B24 in construction and ratings.

WL-488 Triode vacuum control tube.

RH-507 Electrometer triode, ST bulb, 4-pin industrial base, fil. 2 V @ 60 mA, μ 0.8.

DRJ521 Rcvg. interstage amp.

RJ-524 Low-microphonic triode, fil. 1.1 V @ 250 mA, μ 7.8 (864 ratings). Variant: RJ-544.

WL-525 [AKA 304TL, q. v.]; xmtg. triode.

RJ-526 Triode "interstage amp.," fil. 1.25 V @ 920 mA, μ 8.2.

DRJ528 Low-filament-current amp.

WL-530, A [AKA GL-530, VT-122]; xmtg. radar power triode, cylindrical bulb, grid-out-side, T-T fil., water-cooled (10-2-47). In BC-400B of SCR-270.

WL-531 [AKA ML-531, VT-141]; xmtg. vacuum rectifier, fil. 11.5 V @ 20 A, 50 kV PIV, 290 mA; external anode w/ cooling fins cast into anode; flying fil. leads out bottom; in RA-39 of SCR-270 (1-27-43).

WL-532, A [AKA 1B32, q. v.]; TR tube. In BC-961 and T-25/TPS-2.

WL-534, A [AKA VT-127A, q. v.]; xmtg. triode.

WL-535 [AKA 304TH, q. v.]; xmtg. triode.

WL-536 [AKA VT31 (Br.), CV1031]; xmtg. tetrode.

WL-537 [AKA VT114 (Br.)]; xmtg. HF tetrode.

WL-538 [AKA VT90 (Br.), 8011, q. v.]; xmtg. triode.

WL-539, A ["A" is AKA 1B39, q. v.]; gas switching tube.

WL-533 UHF triode, sim. to Br. VT98.

RJ-544 Low-microphonic triode, fil. 1.1 V @ 250 mA, μ 35 (RF-524 w/ higher μ).

RJ-547 [AKA PJ-2]; non-microphonic triode.

RJ-550 Triode, "low-grid-current" photocell amp., etc., smaller/cheaper RJ-553, heater 2.5 V @ 920 mA, μ 8.5. Also see RJ-553, -557, -572.

RJ-551 Triode, low-μ "current amplifier," fil. 7.5 V @ 1.25 A, μ 8. Also see RJ-563.

RJ-553 Triode, "low-grid-current" photocell amp., etc., heater 6.0 V @ 1.05 A, μ 3.8.

DRJ555 Osc. and current amp.

RJ-557 Triode control tube, "low-grid-current," heater 2.5 V @ 920 mA, μ 10.7.

DRJ559 Low-grid-current amp.

DRJ562 Current amp.

RJ-563 Triode (low-μ "current amplifier"), fil. 2.5 V @ 3.0 A, μ 4. Also see RJ-551.

RJ-571 [AKA WL-571]; space-charge tetrode, ST12 bulb, heater 2.5 V @ 1.0 A, g_m 1.2 mS.

RJ-572 "Low-grid-current" amp.

WL-577 Voltage regulator.

WL-578 [AKA 8020, q. v.]; rectifier.

WL-579, B Vacuum rectifier, T16 bulb w/ anode pin, super-jumbo 4-pin base, fil. 2.5 V @ 6 A, 20 kV PIV, 25 mA (originally 2.5 V @ 7 A, 15 kV PIV, 3 mA) (9-5-46) (RCA, W).

DRO580 Rcvg. vacuum rectifier.

RO-581 Vacuum rectifier, bulb 2-3/16" dia. by 6-1/8" high, fil. 5.0 V @ 2.2 A, 6.5 kV PIV, 3 mA.

RO-583 Vacuum rectifier, bulb 2-9/16" dia. by 7-½" high, fil. 5.0 V @ 4.5 A, 25 kV PIV, 8 mA.

RO-585 Vacuum rectifier, bulb 1-9/16" dia. by 4-3/8" high, fil. 4.5 V @ 1.1 A, 1.5 kV PIV, 3 mA (originally 5.0 V @ 1.1 A, 2 mA).

RO-586 Rcvg. full-wave rectifier, bulb 1-9/16" dia. by 4-3/4" high, fil. 2.5 V @ 1.5 A, 1.2 kV PIV, 50 mA.

RO-587 Rcvg. full-wave rectifier, bulb 2-3/16" dia. by 6-¼" high, fil. 2.5 V @ 5 A, 1.5 kV PIV, 290 mA.

DRO588 Xmtg. vacuum rectifier.

KI-605 Gas rectifier, 1-7/16" dia. by 5" high, fil. 2.5 V @ 2.0 A, 5 kV PIV, 200 mA.

WL-608 Vacuum rectifier, bulb 6-1/8" dia. by 24-3/8" high, fil. 10 V @ 10 A, 60 kV PIV, 60 mA.

KU-610 [AKA WT-210-0120]; triode gas thyratron, fil. 2.5 V @ 6.5 A, 500 V PIV, 100 mA. In RM-7A.

WL-612 Vacuum rectifier, fil. 10 V @ 50 A, 150 kV PIV, 240 mA (WL-427 w/ current doubled).

WL-613 Vacuum rectifier, fil. 11 V (originally 10 V) @ 10 A, 140 kV PIV, 60 mA.

WL-614 Vacuum rectifier.

WL-615 Vacuum control tube.

WL-616 [AKA 5625, q. v.].

KU-618 Tetrode grid-glow thyratron, 800 V, 15 mA, 2-3/16" dia. by 5-3/4" high.

WL-619 Xmtg. air-cooled type {WX3260}.

KI-620 Gas rectifier, 3-¼" dia. by 9" high, fil. 5.0 V @

11.5 A, 5 kV PIV, 4 A.

KU-621 Thyratron.

KU-623H Mercury triode thyratron, 869 w/ grid added, fil. 5.0 V @ 20 A, 10 kV PIV, 2.5 A (1937).

KI-624 Gas rectifier, 7-1/8" dia. by 19-3/4" high, fil. 5.0 V @ 40 A, 20 kV PIV, 25 A.

WL-624 Tetrode mercury thyratron, stepped-tubular bulb w/ cap, fil. 5.0 V @ 10 A, 2.5 kV PIV, 6.4 A.

KI-625 Gas rectifier, 5-1/8" dia. by 14-3/8" high, fil. 5.0 V @ 20 A, 20 kV PIV, 10 A.

KI-626 Mercury rectifier, S-19 bulb w/ cap, industrial 4-pin base, fil. 2.5 V @ 6.0 A, 5 kV PIV, 300 mA.

KU-627 [AKA CE309]; mercury triode thyratron, S19 or (later) ST19 bulb w/ cap, 4-pin industrial base, fil. 2.5 V @ 6 A, 2.5 kV PIV, 640 mA (HY, RCA, W), sim. to CE309 (9-5-46). Intro. pre-1936.

KU-628 Mercury triode thyratron, S25 bulb w/ cap, industrial 4-pin base, fil. 5 V @ 12.3 A, 2.5 kV PIV, 2 A. Introduced pre-1936.

WL-629 [AKA M16-T-66290]; triode thyratron sim. to 885, ST12 bulb, 5-pin base, fil. 2.5 V @ 40 mA, 350 V PIV, 40 mA (RCA, W) (9-5-46).

WL-630, A [AKA 730]; thyratron, replaceable by 2050.

WL-631 [AKA 5559, q. v.]; thyratron.

WL-632, A, B ["A" is AKA 5560, FG-95, q. v.]; mercury tetrode thyratron, T18 stepped bulb w/ cap, grid-out-side via cap on base ring, UX base, heater 5 V @ 5 A, 1.5 kV PIV, 2.5 A (RCA, W). "B" is AKA WT-210-0057, WT-210-0116.

KU-633 Mercury triode thyratron, ST13 bulb w/ cap, industrial 4-pin base, fil. 2.5 V @ 2.0 A, 1 kV PIV, 125 mA. Introduced pre-1936.

KU-634 Mercury triode thyratron, S28 globe bulb w/ cap, industrial 4-pin base, fil. 5 V @ 11.5 A, 7.5 kV PIV, 25 A. Introduced pre-1937.

KU-635 Triode grid-controlled rectifier, bulb 3-¼" by 9" high, fil. 5.0 V @ 14 A, 750 V PIV, 1.0 A.

KU-636 Triode thyratron, fil. 2.5 V @ 7 A, 350 V PIV, 100 mA.

KU-637 Ignitron, glass bulb, 5-1/8" dia. by 11-5/8" long, plug-in, 750 V PIV, 20 A, forced-air cooled (1937).

KU-639 Ignitron, glass bulb, 10-1/8" dia. by 20-5/8" long, w/ terminals, 750 V PIV, 50 A, forced-air cooled (1937).

KX-642 [AKA WL-642]; 3-electrode voltage arrester (protector tube), sphrical bulb, 4-pin base; fires at 300-500 V.

KU-651 Ignitron, metal body, 4-¼" dia. by 15" high, w/ terminals, 360 V PIV, 85 A, water-cooled (1937).

WL-651/656 [AKA 5552, q. v.]; ignitron.

WL-652/657 [AKA 5551, q. v.]; ignitron.

WL-653/658 [AKA 5553, q. v.]; ignitron.

WL-653B [AKA 5555, q. v.]; ignitron.

WL-654/659 Ignitron, 4" dia. by 10-½" high, 900 V PIV, 12.5 A (10-20-43).

WL-655/658 [AKA 5553, q. v.]; ignitron.

WL-656 [AKA 5552, q. v.]; ignitron.

WL-657 [AKA 5551, q. v.]; ignitron.

WL-658 [AKA 5553, q. v.]; ignitron.

WL-659 WL-654/659, FG-259B]; ignitron.

WL-660 Vacuum rectifier, bulb 6-1/8" dia. by 33-5/8" high, fil. 10 V @ 10 A, 230 kV PIV, 30 mA.

KI-662 Diode gas rectifier, bulb 3-3/4" dia. by 10" high, fil. 5.0 V @ 6 A, 10 kV PIV, 2.5 A.

KI-663 Full-wave gas rectifier.

KI-664 Full-wave mercury rectifier, fil. 2.5 V @ 14 A, 750 V PIV, 2.5 A.

KI-666 Full-wave mercury rectifier, bulb 2-1/8" dia. by 7" high, fil. 2.5 V @ 6.4 A, 750 V PIV, 1 A.

WL-669 Full-wave mercury rectifier, fil. 2.5 V @ 12 A, 1 kV PIV, 2 A.

WL-670, A Full-wave mercury rectifier, fil. 2.5 V @ 17 A (originally 24 A), 725 V PIV, 6.4 A {WX-3072}. "A" is AKA 4B25, q. v.].

KU-671 Ignitron, glass bulb, 10-1/8" dia. by 23-½" long, w/ terminals, 750 V PIV, 64 A, water-cooled (1937).

WL-674 [AKA 831]; xmtg. triode.

WL-675 Xmtg. triode, carries 203A ratings, difference from 203A unknown if any.

KU-676 [AKA WL-676, WT-210-0058]; mercury triode grid-glow thyratron, for welding controllers, ST30 bulb w/ cap, 4-pin industrial base, fil. 5 V @ 9.5 A, 2.5 kV PIV, 6.4 A {WX3141} (RCA, W) (9-5-46). Introduced pre-1936.

KU-677 [AKA WT-210-0059]; mercury thyratron, KU-676 uprated to 10 kV but derated to 4 A (RCA, W) (9-5-46).

WL-677 Thyratron?; "ST" bulb w/ cap.

WL-678 [AKA WT-210-0134]; mercury triode thyratron, for grid-controlled rectifiers, T20 stepped bulb w/ ceramic-ringed cap, 4-pin (special?) base, fil. 5 V @ 7.5 A, 15 kV PIV, 1.6 A (RCA, W) (9-5-46).

WL-679 [AKA 5554, q. v.]; ignitron.

WL-680 Xmtg. triode, 203A w/ lower μ of 12, superseded by 835. Also see WL-455.

WL-681/686 [AKA 5550, q. v.]; ignitron.

WL-682 Ignitron {WX3194}.

WL-688 Rectifier ignitron, 12-¼" dia. by 41" high, 2.1 kV PIV, 300 A, water-cooled.

WL-706 Voltage regulator.

WL-710Y, Z Ballast tube, med. screw base, 20-28 V @ 250 mA.

WL-711, Y Ballast tube.

WL-712X, Y Ballast tube.

WL-715 Neon glow lamp.

WL-728 Phototube.

WL-734 [AKA PJ-22, CE-1V/VC, CE-1VE, PG-94, V9]; vacuum phototube, 868 format, cesium cathode, S-1 response, 20 μA/lumen (GE, W).

WL-735 [AKA 868, q. v.]; gas phototube, cesium cathode, 868 format, visible wavelengths, 60 μA/lumen.

WL-737 [AKA CE-2]; gas phototube, 868 format, S-1 response, 60 μA/lumen.

WL-738 [AKA CE-3E, CE-1V]; vacuum phototube.

WL-739 [AKA CE5BB]; gas phototube.

WL-741 [AKA 71A, CE23D, 923, q. v.]; phototube.

WL-759 Cold-cathode relay, S-14 bulb w/ black light-tight coating, 4-pin base, trigger electrode on cap, 250-325 V, 4 mA, 1-3/16" dia. by 4" long (1936).

WL-762 Pirani pressure gauge, hard glass, 8" high, platinum fil. 0-300 mA.

WL-767 Vacuum phototube, UV-sensitive (2000-3200 Å), titanium cathode.

WL-769 Vacuum control tube.

WL-770 Vacuum phototube, cesium-magnesium cathode, tipped spherical bulb 3-¼" dia. by 6-½" high, UX base, cap for cathode, for visible wavelengths, 0.75 μA/lumen.

WL-773 Vacuum phototube, UV-sensitive (2000-3700 Å), thorium cathode.

WL-775 Vacuum phototube, UV-sensitive (2000-3000 Å), tantalum cathode.

WL-781 "Lenard-ray" electron-beam generator, built like X-ray tube, fil. 10 V @ 4 A, 250 kV, 200 W, ca. 1936.

WL-786 [AKA 2252, FG-166, q. v.]; mercury rectifier, fil. 2.5 V @ 100 A, 1.5 V PIV, 30 A.

WL-787 Classroom demo triode showing grid action on 1-½" by 3" P1 phosphor-coated plate (changing voltage varies width of glowing stripes between grid wires, and introducing a magnet bends the stripes), 2-5/8" dia. by 10" long, industrial 4-pin base, multiple fils., 6 V @ 1.6 A total, μ 2, 400 V. (W also built a huge - 27" - version for display at the New York World's Fair; see *Communications*, 5-39.)

WL-788 Ballast, med. screw base, 9-18 V @ 250 mA.

WL-789 Vacuum phototube, UV-sensitive (< 2000 Å), platinum cathode, 500 V.

WL-790 Thyratron.

WL-859483 [AKA 4B36, q. v.]; rectifier.

WX3072 Full-wave gas rectifier.

WX3074 [AKA 7C24, q. v.]; triode.

WX3089 Xmtg HF triode.

WX3090 Xmtg HF triode.

WX3092 [AKA VT-158, q. v.]; Zahl osc.

WX3102 Xmtg HF triode.

WX3116 Klystron multiplier.

WX3121 Klystron.

WX3141 Thyratron rectifier (superseded by KU-676).

WX3152 UHF osc.

WX3153 UHF osc.

WX3154 TR tube.

WX3158 [AKA 3J30, q. v.]; magnetron.

WX3160 Ignitron.

WX3161 UHF load lamp.

WX3164 UHF load lamp.

WX3165 UHF load lamp.

WX3168 UHF load lamp.

WX3169 [AKA C1A, q. v.].

WX3170 [AKA C6A, Navy Type C, q. v.].

WX3172 [AKA WL-672].

WX3178 UHF load lamp.

WX3194 [AKA WL-682, q. v.].

WX3199 [AKA 1B24, q. v.]; TR tube.

WX3210 [AKA 1B26, q. v.]; TR tube.

WX3225 [AKA WL-473, q. v.].

WX3230 [AKA WL-678, q. v.].

WX3239 TR tube.

WX3240B [AKA 1B45, q. v.]; TR tube.

WX3240D [AKA 1B41, q. v.]; TR tube.

WX3240G [AKA 1B49, q. v.]; TR tube.

WX3243 Klystron.

WX3243A [AKA 2K51, q. v.]; klystron.

WX3243B [AKA 2K52, q. v.]; klystron.

WX3243C [AKA 2K53, q. v.]; klystron.

WX3260 [AKA WL-619, q. v.]; xmtg. type.

WX3506 In large rectangular flat glass pkg., two flying leads out one side, five out opposite side.

WX3608 "Matchbox" pentode sim. to 6AG5, 2 leads out one side, 4 leads out opp. side, to be soldered onto PC board, heater 6.3 V @ 300 mA.

WX3683 "Matchbox" pentode sim. to 6BN4; two leads out each side; heater 6.3 V @ 200 mA. Likely source of 7430.

WX4073 "Matchbox" type in rectangular flat glass pkg., two flying leads out each side.

WX5669 Rotary spark gap (radar pulser?), not a tube.

TUBE USER'S GUIDE

Electronics restorer-collectors and audio enthusiasts are rather intimately involved with tubes. The following ideas may prove helpful in preserving and enjoying classic equipment.

1. THINK TWICE BEFORE REPLACING

It's surprising how rarely trouble in a piece of old gear derives from a bad tube. Even though tube makers in the old days promoted routine replacements of tubes, and even though radio repairmen got a major part of their profits from sales of tubes, the tubes in an old set are quite likely to be usable. (The main exceptions are audio power tubes, which have often at the ends of their lives, and 35Z5s, which commonly have the heater section between Pins 2 and 3 burned out.) Corroded-open inductors, leaky or dried-out capacitors, partly open resistors, and similar troubles are far more likely to be present - and it's a rare set that has only one trouble. Before replacing a tube, consider the following factors.

Loose Base. A tube may be loose from its base. This has no effect on performance, but there is some risk of the bulb twisting around enough for the wires to short-circuit in the base. Recementing (see below) is all that's needed.

Rattles. Bits of loose cement in the base have no effect on the operation of the tube. Likewise with glass particles inside the tube: as long as the vacuum is OK, bits of glass from the "press" are harmless. Loose metal parts like getters, are more serious: there is a risk of their falling into other elements and causing a short circuit.

Occasionally a large filament-type tube like a 5U4 or 2A3 has rectangular white flakes of filament coating loose in the bulb. This is suspicious, and calls for a test of emission.

Blue Glow. Tubes filled with gas (0C3, 0Z4G, 2A4G, etc.) or mercury vapor (82, 83, etc.) glow in vivid colors in normal use; lack of glow indicates failure. Some healthy power tubes (6CA7/EL34, etc.) show a small area of blue inside the glass at the top, the result of stray electrons striking the glass. However, blue glow *inside the elements* of a vacuum-type tube indicates a gassy tube which had better be replaced.

One Side Bad. Duodiode-triodes (75, 6Q7, 12AT6, etc.) often show weak emission on one or both diodes. This diode is only a detector; it's not worked hard, and most radios have the two diodes wired in parallel anyway. Here is a fine case not to replace unless a new tube produces improved volume. The same goes for pentagrid converters (6A7, 6A8, 12BE6, etc.), in which the oscillator section commonly tests weak. If the oscillator works on all bands, at both ends of the dial, it's OK. (For a more refined test, reduce the line voltage 10% with a Variac.) Full-wave rectifiers often show imbalance between the sides, in a radio which performs satisfactorily.

2. TESTERS AND TESTING

People naturally like to delegate hard "go - no go" decisions to a machine, but a tube tester is simply a guide to the usefulness of a tube. In earlier times it it was a handy sales tool, a scientific-appearing device helping convince the customer that a new tube was required. The "emission" type of tester checks for gross defects like failed emission - the usual problem with old-time tubes - and for tubesfor interelectrode leakage. The "transconductance" type of tester is more refined, and usually tests for gas as well. Regardless of type, a tester may "fail" a usable tube or pass a defective one. Any tester in use today is at least 20 years old, and its components have aged appreciably - "5%" composition resistors, for example, have almost universally aged upward, out of tolerance. The tester may not really simulate reality - a TV sweep tube may test "OK" for emission, yet fail miserably in a linear amplifier that works the tube harder than the tester. Because individual tester designs apply different test parameters, a given tube routinely yields different readings from one tester to another. The message: use one tester consistently, rely on it to catch the gross defects (leakage, gas, burnouts), but expect it to give no more than relative ratings of tube quality. *There is no other place in electronics where keep-or-replace decisions were commonly made, based on a drift-prone analog instrument lacking even a calibration control!*

Certain expensive 2.5-volt tubes (the 45 and 2A3) draw multiple amps of filament current from the tester. Before placing too much faith in the meter reading, it is wise to be sure the filament voltage is close to 2.5 volts under load - just put the probes of a digital meter on pins 1 and 4 of any of the unused sockets. The voltage often sags to 2.1 or so, guaranteeing a low reading. This is partly a fault, the result of using an inexpensive transformer. It is also partly a virtue, answering the need for current-limiting at times when the user sets up the tester to apply heater voltage to an internal jumper in the tube! Bottom-of-the-line testers may supply "2.5" volts, but don't even give setup data to test 2A3s - and test 5U4s at 4.6 volts.

The test instructions can be screwy, too. Example: on the Heathkit IT-17 and -21 testers (the best of the kit emission checkers), the test instructions for a 199 tube call for a filament voltage of 2.5 (vs. a proper 3.3). This lets only unusually good tubes read "OK." Turn the voltage up to 3.15, and the tube perks up considerably. The use of 2.5 V is probably a carryover from earlier models that lacked the 3.15-V setting.

3. FURTHER TUBE SCOOP

Multiple-Numbered Types. Far too many tubes carry multiple numbers, as the result of prototypes that were only triflingly different. Early examples were the 35/51, 39/44, and 47/PZ. Toward the end of the tube era this effect mushroomed. Thus, the owner of a 6BS8 automatically has a 6BQ7A and a 6BZ7. The best defense against this confusion is a tube-substitution book.

Improved Versions of Tubes. An "A" or "B" version of a tube is improved over the basic model, and will replace it (except for some Western Electric tubes which are incompatible from "A" to "B"). However, many "A" versions of the mid-'50s are improved only in that they warm up in a controlled length of time. This is important in series-string TV sets but meaningless in parallel-wired equipment. Some such types are:

6AM8A	6AU6A	6BK7B	6CL8A	6U8A
6AN8A	6AU8A	6BR8A	6CG8A	12AZ7A
6AQ5A	6AW8A	6SN7GTB	6CB6A	12BH7A
6AT8A	6BA8A	6S4A	6T8A	12BY7A

Improvements in other types may or may not matter, depending on the use: The 6AF4A is simply shortened physically; the 12AU7A is a low-microphonics type; the 12AX7A is a low-noise-low-hum variant.

Interchangeability Between Makers. In North America, the land of interchangeable parts, the tube industry was blessed with uniform standards. Makers produced tube parts *and complete tubes* for branding and sale by their competitors. An 80 made by RCA can be replaced confidently by one from Arcturus, GE, Sylvania, etc. This seems obvious, but the writer has seen an old-timer insisting that the only feasible replacement for his blackened National Union ripply-plate 80 was another NU.

Military Tubes. There are numerous ex-military tubes around. These lack prestige to some degree, having once been sold off as "war surplus" at low prices. However, they had to meet rigid requirements (at the time, the MIL-1-A series of specifications) and are a higher-quality product than ordinary civilian tubes. They are thus perfectly good for radio-electronic restoration, and carry some cachet in the tube-audio world. The military departments are still selling excess tubes in lots of thousands, so this source has not dried up.

Early Metal Tubes. Plain black metal tubes carrying steel-stamped brands and type numbers date from the first production in 1934-35. Looking like engraving, the stamping appears mainly on RCA-Cunningham and Raytheon tubes: the "original nine" (5W4, 6A8, 6C5, 6F5, 6F6, 6H6, 6J7, 6K7, 6L7) and the 6X5. Tubes in this style are "correct" replacements for 1935-season radios like the Atwater Kent 317 or GE A-81; by contrast, they are not really right in a later set.

Metal tubes with a protruding "getter bump" on the shell just above the header are early, dating pre-1936-38. Grid-cap types (6K7, etc.) having brown insulation under the grid cap are of pre-war vintage. Those with the type number steel-stamped into the bottom of the shell are from 1945 or later.

"MR" Tubes. Many old tubes are stamped "MR." These came from a 1943-45 program of manufacturing a few popular types for "maintenance and repair" of civilian radios, thus reducing a crippling shortage caused by WW II military needs.

"T9" Oldies. Interesting tubular replacement tubes appeared in the '50s: T9-bulbed versions of the 27, 37, 41, 80, 84, and 25Z5. They were a useful simplification to makers: put a 6K6GT bulb on a six-pin base, and a 41 results; put a 5Y3GT bulb on a four-pin base, and get an 80. It is easy to overlook these today because of their GT-sized cartons.

Substitutions. The beginning tube substituter can get good guidance from the wartime booklets produced by the tube manufacturers, or any edition of the Howard W. Sams substitution guide (still in print). There has been some wild substitution advice in the tube-audio world ("replace a 47 with an 814"), but the old-time sources are reliable. Speaking of substitutions: the little known industrial 5R4GY is a dandy plug-and-play replacement for the 5Y3G; it draws no more filament current, but has bigger plates and runs cooler.

Making Your Own. Many otherwise uninteresting tubes can be made into useful types by making an adapter. Equipment restorers are unnecessarily reluctant to make adapters, perhaps out of simple unfamiliarity, yet this was routine stuff in the tube shortage of WW II. For example, the 7591A audio tube is now scarce and pricey. The 7868 is common and the 6GM5 is more or less an orphan. Yet these are electrically the same - quite fit for use with adapters.

Changing or rewiring the base can work fine. As an example, consider the 6A4/LA: a 6.3-V filament-type five-pin power tetrode used in 1931 Motorola car radios. A lot of these have survived as new-old-stock, being useless anywhere else. The 6A4 becomes a good replacement for the 201A by changing from a five-pin to a four-pin base, triode-connecting the tube. At five volts, it draws the 201A's 250 mA and has ample emission. An Atwater Kent 35 plays nicely with six rebaseed 6A4s.

Rectifiers, which usually involve only four leads, are particularly easy to re-base. Some handy conversions among electrically identical (or better) types:

5R4GY to 80	5U4G to 5Z3	5Y3GT to 5Y4GT
5U4G to 5X4G	5V4G to 83-V	5Y3GT to 80

What You See Is (?) What You Get. In inspecting flea-market tubes, it is helpful to verify that the tube in the box matches the brand and type on the carton. However, RCA tubes are a special case: new-old-stock with the "new" (1969) logo often appears in cartons with the "old" logo. RCA must have had a large stock of obsolete boxes when the corporate symbol was redesigned. A variant tube should match the box - a "12AU7A" box should contain a 12AU7A - but the writer has seen a group of distributor-stock 6GT5s in 6GT5A cartons.

4. BUILDING A REPAIR STOCK

Restoring old equipment goes a lot easier with a good supply of "tubeware" on hand. The hints that follow will be helpful in assembling a stock for general repair use. The assumed starting point is, say, a bag of loose tubes fresh from a flea market or auction: begrimed and unknown as to condition. The rate of recovery is not bad, usually running 80-90%.

Clean-up. Inspection comes first, with filtering-out of the hopeless color-TV types. Then a wash under running water with dish detergent is effective on the grime, with a little steel-wool help on the power tubes where the dirt is baked on. Tubes should be generally be held upside down to keep water from running into the bases. Cardboard beer flats with paper towels on the bottom organize the finished product and aid drying. Tubes needing repairs get set aside for the Tube Hospital.

The exception to washing is miniatures and loctals, most of which are identified with water-soluble ink. Either the brand name disappears or the tube becomes wholly anonymous. Washing should be limited to the non-inked part of the tube. The ink even smudges when handled with moist hands.

Miniatures and compactrons need a check for bent pins. A pin straightener or fingernails are the preferred tools to correct any bending. The use of pliers is likely to crack the glass.

An important hint applies to big-pin (4-5-6-7-pin) tubes: if they've been stored outside a radio through humid summers, the heater pins may have corroded enough to give poor contact in the tester. This applies particularly to the 2.5-V types. These draw an amp or more, so a contact resistance of even 1/10 ohm will keep them from heating fully. Hand wire-brushing is insufficient. It usually is necessary to scrape the heater (big) pins vigorously with a single-edged razor blade. The same goes for loktals with their small pins: the heater pins (1 and 8) may look clean, yet be covered with an insulating film.

Testing. The game now is to speed up a basically slow task. The affordable luxury in this area is two testers of the same type. For multiple single-section tubes of the same type - 50L6s, say - the test goes left-right-left-right, warming up in one tester while reading on the other. For multiple-section tubes, each tester handles one-half of the tube.

Many testers require the user to cruise up and down the roll chart to get test instructions, and then to set up all pin connections before the heater voltage is applied. It is easy to make up a simplified card showing the setups for the most common audio and antique tubes. The setup sequence also can be changed so that the first thing set up is the heater; the tube is then warming up while the other settings are in process. This is a safe process because no known tester applies high voltage until the "test" button is pushed.

Regardless of the hardware, the idea is to group the octals together, the 7-pin miniatures, the loctals, and so on; then to bunch up tubes of the same or similar types. With the tester set up for a particular type, one can breeze through all units of that type. This idea leads immediately to grouping all types having the same pinouts, for example the 6SG7-6SH7-6SJ7-6SK7-6SS7 family or the 12AU7-12AV7-12AX7-12AY7-12AZ7 group. Here it is possible to go between types by changing only the potentiometer(s) on the tester, not the pin switches. Or one can group by heater voltage: test the 6BA6s, then do the 12BA6s by simply turning up the heater voltage. Wisdom requires testing the lowest-voltage types first, to avoid the bitterness of blowing a tube with excessive voltage. With practice, one can do a hundred mixed tubes an hour.

In the '50s, the counterman in a parts store who spent the day testing whole TV-sets-ful of tubes had a preheater: a chassis with multiple sockets wired to apply six volts to all heater pins. This shaved a few minutes off the time to check a load of tubes - not a bad idea for anyone expecting to test a lot of tubes.

Further Tests. Supplemental tests are sometimes necessary. 35Z5s and 35W4s may test "good" even when the pilot-lamp section of the heater is open. An extra check with an ohmmeter is needed on these. Big rectifiers like 5U4s that test OK may arc-over when used at full voltage; only set-testing finds these.

Mystery Tubes. Tubes of the two-digit variety may be numbered almost anywhere: on the crown, the side of the bulb, or the base. The numbers are often faint, especially on Sylvania and Philco tubes of the early '30s. Here the old repairman's trick of rubbing the tube on one's hair comes into play. The tube collects a light coat of oil, which one strokes with a finger into a microscopically thin layer. Holding the glass up to the light then reveals the number by contrast. The best test is to do this in bright sunlight, observing slowly and *patiently*.

With practice it becomes possible to guess the type pretty closely by inspection. (If it has a cylindrical structure with a zillion visible grids, a cap, and a seven-pin base, it's either a 2A7 or a 6A7. If it then tests OK as a 2A7, that's it.)

Storage. Boxing-up comes next. New white boxes are available from sources like Antique Electronic Supply. They sometimes appear in collector markets. Old boxes can be reused nicely by covering the type-number stamping with typist's correction tape of the 2/6" width, giving a fresh writing surface. (New-old-stock TV tubes often are auctioned off for much less than the value of their cartons.)

For storage, a cardboard box of the 11" x 17" size used for xerocopy paper can be cut down to tube-box height. It then takes 96 octals or 192 miniatures, with a convenient slip-on lid. With 3" x 10" spacers cut from old file folders to keep the rows straight, a tidy package for storage results.

To store tubes without individual boxes, cardboard beer flats can be cut down in height (to 1-¼" for T9s, etc.) and the tubes laid out. The flats will stack in xerocopy paper cartons. This is not as elegant as individual cartons but definitely beats jumbling the tubes in a box.

After sorting out the first few hundred items, tubes resolve themselves into three classes: the Prized Repair Supply, the Flea-Market Sale Stock, and the Embarrassing Glut. One quickly learns the low value of most TV-set "pulls." (The 6SN7 is an "audio" type today, but how many 17D4s can you <u>use</u> in a lifetime?) The flea-market stock can be sold off, if the tube boxes are labeled clearly and displayed in order. Miniatures have only minor value in the collector market, but have some draw with comm-gear restorers. Six-volt Compactrons have some (small) value for use in '60s amateur and audio gear.

5. TUBE HOSPITAL: REPAIR & REACTIVATION

In principle, tubes are not repairable, other than the rebuilding operations that are applied to big transmitting types. Still, there are numerous ways to salvage a troubled tube.

Loose Bases. To re-cement the tube, use clear nail polish - paint a ring around the base, let the polish soak into the old cement, recoat, and let dry overnight. Polystyrene "coil dope" works as well, and can be used to refill the nail-polish bottle. Solvents (acetone, etc.) are ineffective in softening the old base cement - the cement was baked hard in manufacture, and "nothing" dissolves it.

Broken Keys on Octals. These have no effect on performance but are unesthetic. Inspection will usually show where the locating rib on the key was located. If not, an ohmmeter check for the heater will reveal which pins are 2 & 7 or 7 & 8.

Key breakage is most common on big metal tubes (6L6, 5T4), whose weight stresses the key if the equipment is dropped. Fortunately, these are the easiest to repair, simply by replacing the base wafer. To do so, flex each pin carefully with pliers until the brass breaks, then unsolder the pin. Uncrimp the shell from the base wafer. Swap the wafer with a good one from a dud. With practice, this takes less than 15 minutes.

It is possible to cement new keys onto bases, but holding the key aligned until the cement (epoxy, etc.) cures is a problem. A wafer-style tube socket, pushed onto the pins upside down, makes a holding jig. Still, the key often comes out tilted.

Rebasing. Old tubes have sometimes been stored in wet locations where a pin has corroded to the point of breaking - the brass was cold-worked heavily in fabrication, and is sensitive to chemical contamination. High-value tubes like 201As or 45s are well worth rebasing. To get the old base off without worrying about the base cement, break the remaining pins and unsolder then. Cut a slit across the bottom of the base with a hacksaw. Insert a screwdriver and twist; this cracks the base off in two halves. Then solder on a new base and recement.

Missing Grid/Plate Caps. The artful restorer keeps a supply of caps from dud tubes. Cap replacement is straightforward if the leadout wire is intact and long enough - just scrape the wire with a razor blade before soldering. If the wire is a bit short, put the new cap on the bench right-side-up and place the tip of a Phillips screwdriver in its center. Tapping the screwdriver will dish-in the cap enough to meet the wire.

If the leadout wire is broken off at the glass, there is still hope. On tubes where the glass seal is long and tapered (e. g., RCA), it is feasible to crush *the very end* of the seal with pliers to expose a tiny amount of wire. On tubes with a "blob" or rounded seal (e. g. Arcturus), the trick is to use a vibrating engraver (Burgess Vibrograver, etc.) to chip away just a bit of glass, say, 0.020". Then take a few inches of hair-fine copper wire from a piece of zip cord. Loop the center of the wire around the exposed leadout - only a single turn is needed - and solder. This gives two pigtails that will reach up into the new cap.

"Open" Filaments. Oldies like 201As routinely have "open" filaments from corrosion inside the base pins. The best repair is to make a tight hairpin from a short piece of #20 tinned wire. This fills space inside the pin, and gives lots of surface for new solder to grip. Melt the old solder, insert the hairpin, and apply new solder liberally. This also applies to 2.5-volt tubes, which cannot take even small amounts of joint resistance. Resoldering was a standard trick of TV technicians in the days of soldered-base picture tubes.

Resoldering is a slow way to troubleshoot tubes. To determine quickly whether the "open" is real, connect a *current-limited* high-voltage transformer to the filament pins. If there is an arc visible inside the glass, the tube is hopeless. If not, the current has probably broken down the solder joint and is flowing harmlessly. Resoldering is then justified. A typical high-voltage transformer is the oil-burner type, putting out 10 kV but self-limiting to 23 mA. This current poses no danger to a 250-mA 201A. Neon-sign or bug-zapper transformers would probably work as well. Just use well insulated leads and a three-wire power cord, and stay clear of the high voltage.

Leaky Tubes. Old tubes often show interelectrode leakage. This is generally harmful to circuit operation, but a 201A with, say, a one-megohm leak from grid to filament will work fine in a '20s battery radio. Many such leaks (other than temporary ones from water getting in the base during washing) result from cathode material evaporating onto insulators inside the tube. These can often be cleared by a brief spritz from the oil-burner transformer - half a second will do. A leaky miniature tube can often be fixed with a vigorous rap on the bench. And there are special cases: a pair of EL34s with leakage from plate to heater (Pin 2 to 3) turned out to have arc tracks burnt into the base. (Someone must have turned the volume up with the speakers disconnected.) Scraping away the burnt material fixed them.

One nice "S"-bulbed UX-280 that had arcing between plate leads inside the base was saved by drilling a hole into the bottom of the base and pushing the wires apart. An old-time tipped BH cold-cathode rectifier that arced-over internally was cured by putting a pair of 1N4007 diodes in its base. This is one of the few cases where silicon diodes are unlikely to stress the rest of the circuit from overvoltage.

Gas. The word among old-time radio operators was that a gassy tube could be saved by applying normal voltages and heating the flashed getter deposit with a small flame. A couple of recent tries on gassy CX-245s failed to confirm this, but it may be worth trying.

Rejuvenation. Reactivating tubes was a well known practice in the '20s but not common later. There are three techniques involved, depending on the type of tube.

For *pure-tungsten* filaments (in the original 200, 201, and early transmitting types), the goal is simply to evaporate contaminants. Operate the tube at 110% of rated filament voltage for up to half an hour, then retest.

For *thoriated-tungsten* filaments (in the 10, 22, 71 - not 71A -, 120, 199, 200A, 201A, 216B, 240, and later transmitting types), the trace of thorium on the surface of the filament that provides most of the emission needs to be restored. The starter technique is to operate the filament at 135% of rating for 30 minutes. This makes fresh thorium diffuse to the surface. Test, and repeat once if necessary. If that doesn't succeed, take a deep breath and give the tube 350% of rated voltage for 30 seconds. (This usually wipes out whatever emission there was.) Then let the tube idle at normal voltage for a while - two hours, four hours, overnight - and the emission is usually normal.

For *oxide-coated* filaments or cathodes, it may be successful to run the tube for an hour before testing. Then go to 120% of rated voltage and monitor the emission for an hour or so. It should slowly increase, peak, and gradually decline. The trick is to go back to normal voltage just after the emission peaks, and let the tube run for at least four hours.

For *cathode-type* tubes, direct action in nothing-to-lose cases is to apply triple voltage for 30 seconds, let the tube cool, and retest. Tubes with 6.3-volt heaters rarely burn out at 19.6 volts. One or two cycles of this will improve - or impair - a lot of '30s and even "modern" types.

"Eye" Tubes. These are usually dim as-found; being nonessential to the operation of the radio, they were rarely replaced during normal servicing. Despite a couple of hopeful-sounding articles in the collector press, they rarely can be reactivated (this is the result of a test run on 25 "dims"). The villain is *not* low emission; it appears to be contamination of the willemite (green-glowing) target coating. The contamination is visible on the metal targets of really bad 6E5s and the like; it is easily visible on used 6AL7s, which use a coated mica target and readily show the blackening. Two partial fixes: (1) make sure the 1-meg resistor in the socket is good - an open resistor will dim the display; (2) rewire the plate lead directly to the rectifier output, thus giving it extra voltage. A small increase in voltage gives a disproportionate increase in brightness.

6. IDENTIFICATION AND DATING

MANUFACTURER CODES

Tubes are often marked with private brands. A set-maker (Du Mont, Emerson, Philco, Zenith, etc.) usually equipped its products with tubes made by the tube companies but marked with its own brand . . . then warned the owner to use only its brand for replacements. RMA-EIA codes on the the following list may be found on tubes in addition to the private brand. They show the real origin of the product, like "Emerson" tubes made by GE (Code 188), or "Philco" items from National Union (Code 247).

The codes do not distinguish between manufacturers and distributors; they appear on, say, imported tubes for which GE was the U. S. distribution agent.

This list can also be used on military codes. The letter codes in parentheses are military identifiers that often give the maker of an unidentified tube - for example, a JAN-CIM-2C39 would be an Eimac product. The services did not put up with third-party manufacture: the code on the tube represents the actual maker [1].

MANUFACTURER CODES

111:	Amperex (CEP)
117:	Arpin Mfg. Co. (CAPQ)
158:	Du Mont (CDU)
162:	Eimac (CIM)
170:	Electronic Tube Corp./ General Atronics (CVG)
177:	Farnsworth Tel. & Radio Corp. (CFN)
179:	Federal T. & R. Co. (CFT)
188:	GE (Ken-Rad) (CG, CKR)
189:	General Electronics (CDR)
210:	Hytron/CBS-Hytron (CHY)
212:	Ind'l & Comm'l Elect. (CIZ)
226:	Kuthe Labs (CADK)
231:	Machlett Labs (CAGD)
247:	National Union/- Lansdale Tube Co. (CNU)
260:	Philco
274:	RCA (CRC)
280:	Raytheon (CRP)
301:	Sperry Gyroscope Co. (CS)
308:	Stromberg-Carlson
312:	Sylvania (CHS)
322:	Tung-Sol (CTL)
323:	United Electronics (CUE)
336:	Western Electric (CW)
337:	Westinghouse (CWL)
354:	Lewis Electronics (CYN)
423:	North Am. Philips (CNY)
431:	Waterman Prod. Co. (CAGX)
557:	Electronic Products Co.
562:	Polarad Electronics
636:	Sheldon Electric Co.
653:	American Telev. Inc. (CAGE)
672:	Thomas Electronics (CBUP)
677:	Rogers Elect. Corp. (CQ)
713:	Taylor Tubes, Inc. (CTY)
738:	Lewis & Kaufman, Ltd.
744:	Hughes Aircraft Co.
749:	National Electronics (C)
771:	Penta Labs
781:	Vacuum Tube Products
787:	Sonotone Corp. (COZ)
803:	Microwave Associates
806:	Gordos Corp.
809:	Varian Associates
818:	Tel-O-Tube Corp.
879:	Litton Industries
884:	C. R. T. Electronics Corp.
886:	Calvideo Tube Corp.
935:	Electrons, Inc. (CEL)
940:	Bomac Labs. (CBNQ)
964:	Huggins Labs
1012:	Bendix Red Bank (CJEA)
1101:	Rauland (CIY, CBQZ)
1120:	General Electrodyn. Corp.
-	Arcturus (CAA)
-	Cable (Speed) (CRS)
-	Champion (CRS)
-	Daven (CDN)
-	Duovac (CBW)
-	Electronic Enterprises (CDZ)
-	Heintz & Kaufman (CKH)
-	Johnsonburg (CJR)
-	Majestic (CMW)
-	Northern Electric (CT)
-	Victoreen (CBBM)
-	Canadian Westinghouse (CK)
-	Chatham Electronics (CAHG)

DATE CODES

Most tubes of 'mid-'30s and later vintage carry date codes. The date was encoded in a scheme devised by the maker before about 1950. Afterward, it was usually in the form YY-WW, where YY is the last two digits of the year and WW is the week within the year (often rounded to 13, 26, 39, or 52 to indicate just the quarter).

The "date" is often ambiguous: it may represents the month of "manufacture" or the month that the maker's guarantee expired. (This is particularly messy with low-runner tubes, which might rest in warehouse stock for months or years before marking and shipment.) It is further confused by the use of different codes for original equipment manufacturer (OEM) vs. replacement sales, where the guarantee conditions were different. In any event, here is what is presently known in this area.

RCA. RCA tubes were coded by an elaborate and cryptic system of symbols, too complex to repeat, up into the mid-30s [2]. Afterward, things were simpler. The code [3] became simply a year letter:

U: 1937	R: 1939	Y: 1941	K: 1943	V: 1945
T: 1938	X: 1940	S: 1942	H: 1944	F: 1946

plus a month number:

1: Jan-Feb.	3: May-June	5: Sept.-Oct.
2: Mar.-Apr.	4: July-Aug.	6: Nov.-Dec.

plus the letter "E" for OEM tubes.

Transition to a semi-open code began in 1946: "7-26" indicates the 26th week of 1947. The dash applies to an OEM tube; no dash, a replacement. RCA went to the RMA-EIA scheme (YY-WW) in 1956. Somewhere in the process there were also two-letter date codes (AW, BF, etc.).

Sylvania. Sylvania tubes of the '30s can be dated as follows [4, 5]. The code consists of a letter and a digit, corresponding to the month of production and to the last digit of the year. From 1930 to 1935, the code was as follows:

C	January	D	April	A	July	B	October
H	February	J	May	D	August	F	November
M	March	N	June	K	September	L	December

This is coupled with the last digit of the year: "N5," stamped into the tube base at manufacture, indicated June, 1935.

In September 1935 Sylvania changed to a different set of letters, which lasted through 1938:

A	January	D	April	G	July	J	October
B	February	E	May	H	August	K	November
C	March	F	June	I	September	L	December

Up to mid-1938, the maker's logo and date code were hot-stamped into the base. After May or June, these were silk-screened in green ink. If another date code is etched or stamped on the glass, under the type number, it often means that another manufacturer made the tube.

The code changed in 1939 to letters Z-V-T-N representing quarters. An "R" following the code indicated "replacement," not OEM.

Code	Shipment Interval	Guarantee Ends
Z9	5-1-39 to 8-1-39	7-1-40
V9	8-1-39 to 11-1-39	10-1-40
T0	11-1-39 to 2-1-40	1-1-41
N0	2-1-40 to 5-1-40	4-1-41

From examining the dates above, the plan gave actual guarantee intervals ranging from 11 to 14 months.

. In 1947 Sylvania went to an three-digit open date code with only one digit for the year, the same as RCA's. The "week" digits were changed only every month, or even every quarter, as "13," "26," "39," and "52" are more common than others.

Some Zenith tubes of the late '30s and '40s were Sylvanias, with the same date codes.

Philco. Sylvania made most Philco tubes, and the codes were identical until 1935, so tubes in Philco radios can be dated the same way. The characters usually ran number-letter.

After 1935, Philco used only the four letters O-P-U-S, one per quarter of the year, chosen to be easy to remember and not overlapping the previous code. As with Sylvania tubes, in mid-1938 the logo and date code went from hot-stamping in the base to yellow silk-screening. This system was in force through 1949. In 1950 it changed to the EIA system.

Ken-Rad. This company had a prewar dating policy [6] that depended in part on type (battery vs. regular), but without explicitly stating how long the guarantee ran. The replacement scheme ran as follows. Defective tubes were to be replaced if returned before or during the month shown below:

Code	Month	Code	Month	Code	Month
H9	7-40	N0	1-41	V0	7-41
I9	8-40	P0	2-41	W0	8-41
J9	9-40	R0	3-41	X0	9-41
K9	10-40	S0	4-41	Y0	10-41
L9	11-40	T0	5-41	Z0	11-41
M0	12-40	U0	6-41	A1	12-41

For battery tubes (in the 1.5-volt and 2.0-volt series), the guarantee ended three months *sooner than* the month shown.

Ken-Rad tubes coded A7 to L7, M8 to Z8, and A9 to G9 were out-of-guarantee at the time, suggesting that the same code had been used for expiration dates back to Dec. 1937.

Raytheon. This maker's quarter-oriented guarantee was nominally a year, but actually ran up to 15 months. It expired one year after the listed shipped-before date. Its system was:

Code	Shipped Before	Code	Shipped Before	Code	Shipped Before
C9	3-31-39	L9	12-31-39	I0	9-30-40
F9	6-30-39	C0	3-31-40	L0	12-31-40
I9	9-30-39	F0	6-30-40	C1	3-31-41

Stewart-Warner. This receiver maker guaranteed tubes used in its sets (as did some other makers) for 12 months *after* the code date. It pointed out that the code date was set to allow for time in dealer stock: "Tubes sold to set manufacturers usually are coded ahead so that the warranty period actually begins several months after the date of manufacture, thus normally allowing sufficient time for the set to go from factory to distributor to dealer to customer. For example, a tube shipped to us in July might be coded August or September, and the warranty would then extend 12 months after the code marking."

Western Electric. WE used a two-letter date code on the bases of some of its repeater tubes beginning in mid-1936 [7]. The first letter was the year: "A," 1936; "B," 1937; "C," 1938. The second letter was the month:

A	Jan.	D	April	K	July	N	Oct.
B	Feb.	E	May	L	Aug.	P	Nov.
C	March	H	June	M	Sept.	S	Dec.

On ST-bulbed types, a date code was embedded into the circular WE marking at the top of the bulb. A mark above or below letters in the name WESTERN ELECTRIC indicated the year and month:

Letter	Year*	Month@	Letter	Year*	Month@
W	1938	Jan.	N	1944	July
E	1939	Feb.	E	1945	Aug.
S	1940	March	L	1946	Sept.
T	1941	April	E	1947	Oct.
E	1942	May	C	1948	Nov.
R	1943	June	T	1949	Dec.

* Index below letter.

@ Index above letter.

In the second quarter of 1947 the date code became the three-digit system used by RCA, less the dash and rounded to the quarter (13, 26, etc.). In the first quarter of 1956 WE went to the EIA system, again roundeed to quarter.

WE dome marking for Oct. 1943.

To sum up: one can date a tube, knowing the code, but one ought not try to be too precise about it.

REFERENCES

1. "Military Standard - Electron Tubes and Crystal Rectifiers," MIL-E-1C, 1955, p. 31.

2. Bro. Patrick Dowd, W2GK, "Dating the RCA (Cunningham) Composition-Base Radio Receiving Tubes - From Mid-1924 Thru 1941 (Start of WW II)," session handout, AWA Conference, Canandaigua, NY, Sept. 1978 (available from Antique Electronic Supply).

3. P. A. Keller, The Cathode-Ray Tube - Technology, History, and Applications (New York: Palisades Press, 1991), Appendix 6.

4. R. Ramirez, "Philco Tube Date Code," *Antique Radio Classified*, Vol. 10 No. 6 (June 1993), p. 7.

5. A. Douglas, "Dating Philco, Sylvania, and (Some) Zenith Tubes," Delaware Valley Historic Radio Club *Oscillator*, Vol. 4 No. 2 (Feb. 1996), p. 2.

6. L. A. Sibley, "More Date Codes for Tubes," *ibid.*, pp. 2-3.

7. A. R. Balaton, "Elements of Reference for Identifying and Dating Western Electric Electron Tubes," AWA Review, Vol. 5 (Holcomb, NY: Antique Wireless Assn., Inc., 1990), pp. 118-120.

TV SWEEP TUBES IN RF AND AUDIO USE

The octal beam-power tubes intended for horizontal-amplifier use in TV sets had only modest usage in other equipment in the '50s. However, the arrival of large-screen color television in the early '60s had helpful consequences: it caused the tube makers to develop new lines of sweep tubes, particularly in the all-glass novar and Compactron formats. These were essentially small transmitting tubes with continuous-duty dissipation ratings of 18 watts and up. (At 40 watts, the 6LF6 was 14% "tougher" than even the rugged 6550 audio tube.) Such powerful devices were needed to drive color picture tubes with their then-low light-generating efficiency.

The new generation of sweep tubes had strong appeal for use in single-sideband linear amplifiers, where they could be pushed to extraordinary power levels for short periods of time. Used in parallel, with forced-air cooling, they could displace much costlier transmitting tubes. The champion for tube use in this fashion was the Galaxy 2000 linear, which used ten sweep tubes in parallel. These tubes were most popular for new designs from roughly 1963 to 1973. Their level of intermodulation distortion was a bit high to meet the FCC limits for commercial SSB transmitters, but was acceptable under the amateur rules. Sweep tubes also found their way into *sub rosa* CB linears, into industrial applications like shaker-table drivers, and into output-transformerless audio amplifiers. It is ironic that tubes for sweep use, designed to put out slam-bang 16-kHz sawtooth signals with multi-kilovolt flyback pulses, adapted successfully to high-end audio use.

Today's equipment restorer may be interested in the higher-heater-voltage "cognate" versions of many of these tubes, which are in little demand and therefore relatively cheap. For example, the popular 6JB6 has related versions in the 12JB6 and 17JB6. Needing to re-tube a transceiver using several 6JB6s, one could save considerable money by using a less popular cognate version and adapting the heater circuit to higher voltage.

Many of these tubes went through "A", "B", etc. variants. The differences are usually trivial (sealing tip on the base, vs. on the top), but occasionally significant (increased dissipation rating). The differences are pointed out below.

It may also be useful to know what groups of tubes share a common pinout, for potential use in substituting types. A list is given later in this section. Even if a potential substitute has a different pinout, only one or two pins may have to be rewired.

By contrast, no non-TV use is known for the 6HS5, 6HV5, 6JH5, 6KV6, 6LJ6/6LH6A, or their higher-voltage cognates. These are special beam-voltage regulators that only *look* like sweep tubes.

In the data below, tubes are divided into octal, novar (9-pin), Compactron (duodecar 12-pin), and magnoval (large 9-pin). They are broken down further as single-ended or capped. In hope of making it easier to find substitutes for hard-to-locate types, cap sizes are given. Caps are listed as "min." (0.25" dia.) or "small" (0.36" dia.). Seated heights are cited; some compact equipment may not accept tubes taller than the originals. Basing diagrams are shown in brackets < >; these refer to the EIA-standard layouts found in most tube manuals and substitution books. Heater voltages are 6.3 unless noted. Transconductance figures are included for completeness, although the manufacturers considered numerous tubes to be equivalent despite nominal differences in g_m - an illustration of the fact that transconductance varies widely with test conditions.

Related non-TV power tubes include the 7867, 7868, 7984, 8149, 8156, 8233, 8417, 8505, 8608, and 8950. Some of these may be replaceable with sweep tubes - or vice versa.

In the collector market, old-time tube caddies are often offered for sale, filled with TV-type tubes - and occasionally a surprise in the form of one or more high-value sweep tubes.

SINGLE-ENDED OCTAL

6FW5 6L6GB format shortened by 3/8", heater 1.2 A, g_m 7.3 mS, 18 W <6CK>. Special pinout w/ plate pin isolated to stand 6.5-kV peaks.

CAPPED OCTALS

6AV5 GT, GA — Heater 6.3 V @ 1.2 A, 11 W (12-9-49) <6CK>. "GA" has T11 or T12 bulb.

6BG6G, GA — Variant of 6L6G w/ small cap, uprated to 20 W (RCA) (7-20-46) <5ET>. "GA" (4-2-55) is in T12 bulb, 4-¼" seated height.

6BQ6G, GA, GT, GTA, GTB — Heater 1.2 A, g_m 5.5 mS, 10 W (RCA) (5-23-49) <6AM>. "GTB" [AKA 6CU6] has button stem; is uprated to 11 W. One each in Bramco CD-10X and 2CD-20X xmtrs. Two in Gonset G-77 xmtr. Four in Sansui 500A stereo receiver.

6CB5, A — Early color-TV type, ST-16 or T14 bulb w/ small cap, seated height 4.6", heater 2.5 A, g_m 8.8 mS, 23 W (GE) (4-2-56) <8GD>.

6CD6G, GA — ST-16 bulb w/ small cap, seated height 5", heater 2.5 A, g_m 7.0 mS, 15 W (RCA) (12-5-49) <5BT>. "GA" is uprated to 20 W.

6CL5 T12 bulb w/ small cap, seated height 4.4", button stem, heat radiators on grid and screen, heater 2.5 A, g_m 6.5 mS, 25 W (GE, SY) (7-5-55) <8GD>.

6CN6 [AKA EL38]; heater 1.4 A, dist. by AX (rare) (3-27-53) <8EW>. Four each in Lakeshore P-400-GG and WRL LA-1 Globe Linear amps.

6CU6 [AKA 6BQ6GTB, q. v.].

6DN6 6CL5 format, heater 2.5 A, g_m 9.0 mS, 15 W (GE, SY) (7-18-55) <5BT>. SY version has grid and screen heat radiators; GE does not.

6DQ5 Early color-TV type, 6CL5 format, heater 2.5 A, g_m 10.5 ms, 24 W (RCA) (5-27-57) <8JC>. One each in Hallicrafters HT-40, Hammarlund HX-50. Two in Hallicrafters HT-44, Hammarlund HX-50A and Swan SW-240. Two (?) in Gonset GSB-100 linear.

6DQ6, A, B — T12 bulb w/ min. cap, seated height 3.5", heater 1.2 A, g_m 6.0 ms, 15 W (RCA) (5-2-55) <6AM>.

"A" [AKA 6GB6] is uprated to 6.5 mS and 18 W (3-5-56). "B" [AKA 6GW6] has button stem; is uprated to 7.3 mS (3-30-59). One each in Conar 400, Eico 723, Heathkit DX-20 and HX-11, International Crystal AOT-50, Knight-Kit T-60, and Lafayette HE-25. Two each in Eico 753 xcvr. and WRL Meteor and DSB-100 xmtrs. Three in Johnson Challenger.

6EX6 6CL5 format and construction, heater 2.25 A, g_m 7.7 mS, 22 W. Same base and outline as "audio" tube 7867, which is rated for 24, vs. 22, W (GE) 3-23-59 <5BT>. Variant: 21EX6.

6FH6 T12 bulb w/ small cap, seated height 3.7", heater 1.2 A, g_m 12 mS, 17 W (GE) (3-24-58) <6AM>.

6GB6 [AKA 6DQ6A, q. v.].

6GC6 6DQ6 format, button stem, heater 1.2 A, g_m 6.6 mS, 17.5 W (GE, SY) (11-23-59) <8JX>. GE and SY production is identical.

6GW6 [AKA 6DQ6B, q. v.] (9-10-60).

6LW6 T14 bulb, 4-3/4" seated height, small cap, button stem, thick support rods for grid and screen but heat radiator only on plate; heater 2.65 A (GE) <8NC>.

12DQ6, A, B [AKA 12GW6]; 12-V variant of 6DQ6 (RCA) (5-2-55). Two "B"s in Hallicrafters SR-150 xcvr.

12GW6 [AKA 12DQ6B, q. v.].

SINGLE-ENDED NOVARS

These had no known non-TV use but are included for completeness. All have T12 bulbs. "A" variants are tipless.

6GT5, A Repackaged 6DQ6B, seated height 3", uprated to 17.5 W and 7.1 mS (RCA) (9-25-61) <9NZ> (same pinout as 7868). Variants: 12GT5, 17GT5A. Also see 6GJ5A.

6JG6, A Repackaged 6JF6, seated height 3" (RCA) (3-9-64) <9QU>. Variants: 17JG6A, 22JG6A.

6JR6 Seated height 2.8", heater 1.6 A, 17 W; lower g_m than 6JG6A - 7 mS vs. 10 <9QU>. Variants: 17JR6, 22JR6, 33JR6.

6JT6, A Lower-peak-current variant of 6JG6A, seated height 2.8"; heater reduced to 1.2 A, g_m 7.1 mS, 17.5 W (RCA, SY) (4-20-64) <9QU>. Variants: 12JT6A (tipped), 17JT6A.

CAPPED NOVARS

All are in T12 bulbs.

6GJ5, A Repackaged 6DQ6B, w/ min. cap, seated height 3.04" (3.13" for "A"), uprated to 17.5 W (RCA) (11-27-61) <9QK>. Also see 6GT5A. Two in National NCX-3 xcvr.

6JB6, A Min. cap, seated height 2.9" (3.13" for "A"); heater 1.2 A, g_m 7.1 mS, 17.5 W (RCA) (12-3-62) <9QL>. Two each in National NCX-200 xcvr, Drake T-4XC xmtg. conv., and Drake T-4 series xmtrs. Three in Drake TC-6 6-m xmtg. conv. and TR-6 6-m xcvr. Four in Drake TR-4 xmtr. Variants: 12JB6A, 17JB6A. Note: unconfirmed scuttlebutt among Drake owners is that SY 6JB6s perform better than other brands; some others fail to operate at 28 MHz.

6JE6, A, B, C Small cap, seated height 4.22" (3.75" for "A," "B," and "C"); heater 2.5 A, g_m 10.5 mS, 24 W (SY) (12-31-62) <9QL>. "A" is uprated to 30 W (8-3-64). "B" is uprated slightly on plate current. "C" [AKA 6LQ6; also see 6MJ6] has controlled warmup, uprated slightly on plate current over "B"; heater current reduced to 2.3 A. Two in Swan 350C xcvr. and Knight-Kit T-175 linear; four in Amitron linear. Six in Sideband Engineers SB1-LA and SB-2LA linears. Eight in McIntosh Labs MC3500 audio amp.

6JF6 [AKA 6KM6, 6JU6, q. v.]; min. cap, seated height 3.17", heater 1.6 A, g_m 10 mS, 17 W (GE, RCA, SY) (3-22-65) <9QL>. Also see 6JG6A. Variants: 17JF6, 22JF6.

6JU6 [AKA 6JF6, 6KM6, q. v.]; g_m 7.0 mS, 17 W (RCA) (2-14-66). Variant: 22JU6.

6KM6 [AKA 6JF6, 6JU6, q. v.]; g_m 9.5 mS, 20 W (RCA) (6-21-65). Variant: 22KM6.

6LQ6 [AKA 6JE6C, q. v.]; g_m 7.5 mS, seated height 4". One each in Swan 260 and 270B xcvrs. Two in Kenwood (Trio) TS-511S xcvr., Kenwood (Henry Radio) TS-900, and Swan 500C xcvrs. and Knight-kit T-175 linear. Four in Swan Cygnet 1200 W linear. Also in National NCX-500 xcvr. Variant: 31LQ6.

6LZ6 Small cap, seated height 4", heater 2.3 A, g_m 11 mS, 30 W <9QL>. Variant: 6LZ6.

6ME6 6LZ6 format and ratings except for g_m of 9.6 mS (RCA) <9QL>.

6MJ6 [AKA 6JE6C (q. v.)]; has ruggedized cap and other detail improvements; has 200-W short-term rating (RCA) {A40806}. Introduced in 1975 - just before RCA gave up making receiving tubes - and rare today; GE made some.

12JB6, A Variant of 6JB6 w/ heater 12.6 V @ 600 mA (RCA) (12-3-62). Two in Lafayette HA-250 mobile linear; three in Drake TR-3 xcvr. Six in P & H LA-500M.

31LQ6 Variant of 6LQ6 w/ heater 31.0 V @ 450 mA (RCA). Six in Top Band Systems TBS-2000 linear.

SINGLE-ENDED COMPACTRONS

All but 6GE5 and 6HE5 have T12 bulbs; all use duodecar base.

6GE5 Seated height 2.5", heater 1.2 A, g_m 7.3 mS, 17.5 W (GE, RCA, SY) (10-9-61) <12BJ>. One in Heathkit HW-16 xmtr. Two each in Heathkit HW-12(A), -22(A), and -32(A) "single-bander" xcvrs. and HW-18 xmtr. Variants: 12GE5, 17GE5. GE and RCA versions differ internally.

6GF5 T9 bulb, seated height 2.5", heater 1.2 A, g_m 4.2 mS, 9.0 W (GE) <12BJ>.

6HB5, A Seated height 3", heater 1.5 A, g_m 9.1 mS, 18 W (GE, RCA, SY) <12BJ>. Eight in Futterman H-3 amp. Variant: 21HB5. GE and SY versions differ.

6HD5 Seated height 3.25", heater 2.25 A, g_m 10 mS, 24 W (GE) (11-19-62) <12ES>.

6HE5 [AKA 6JB5, 6JC5]; *vertical* sweep amp., T9 bulb, seated height 2.5", heater 800 mA, g_m 4.1 mS, 12 W

(GE) (10-15-62) <12EY>.

6HJ5 Seated height 3.25", heater 2.25 A, g_m 10 mS, 24 W (GE, SY) (5-27-63) <12FL>. GE and SY versions have different getter locations; otherwise same.

6JB5 [AKA 6HE5, q. v.].

6JC5 [AKA 6HE5, q. v.].

6JN6, A Single-ended variant of 6JM6, q. v.; seated height 2.5" (RCA, SY) (3-9-64) <12FK>. "A" is uprated on plate and screen current. Two in David Berning Co. EA-230 amp. Variants: 12JN6, 17JN6.

CAPPED COMPACTRONS

All are in T12 bulbs w/ duodecar base. All use small caps except 6GV5, 6GY5 and 6KN6.

6GV5 Min. cap, seated height 3.25", heater 1.2 A, g_m 7.3 mS, 17.5 W (GE, RCA, SY) (5-7-62) <12DR>. Variant: 17GV5.

6GY5 6GV5 format, heater 1.5 A, g_m 9.1 mS, 18 W (GE, RCA, SY) (12-31-62) <12DR>. Lower peak current than 6GV5 (550 mA, vs. 800). Variants: 16GY5, 21GY5.

6HF5 Seated height 3.75", heater 2.25 A, g_m 11.3 mS, 28 W (GE, RCA, W) (1-7-63) <12FB>. One in Drake 2-NT xcvr and Hallicrafters HT-46 xmtr. Two in Hallicrafters SR-400 xcvr.; Swan 350, 400 and 500 xcvrs.; and WRL Duo-Bander 84, Galaxy III, and Galaxy 300 xcvrs. Ten in WRL Galaxy 2000T linear. W version has heat radiators on grid and screen; RCA does not.

6JM6, A 6GV5 format, heater 1.2 A, g_m 7.3 mS, 17.5 W (3-9-64) <12FJ>; also see 6JN6. "A" (12-21-64) has "instantaneous plate knee" value of 55 V, vs. 60 for 6JM6. Variant: 17JM6A.

6JS6, A, B, C Seated height 3.75", heater 2.25 A, g_m 11.3 mS, 28 W (8-31-64) <12FY>. "A" has controlled warmup; same ratings (4-20-64). "B" has factory limit on emission from beam plates; same ratings. "C" is uprated to 30 W. Two in Henry Radio Tempo I and Yaesu FT-101B xcvrs. Variants: 23JS6A, 31JS6A.

6JZ6 6GV5 format, heater 1.5 A, g_m 9.0 mS, 18 W (SY) <12GD>.

6KD6 Seated height 4.25", heater 2.85 A, g_m 14 mS, 33 W (RCA, SY) {A40699A} (4-26-65) <12GW>. Some SY-made tubes use two parallel sections; are shortened by ½". One in Hallicrafters FPM-300 xcvr. Two each in Hallicrafters SR-400A, Yaesu FT-DX-400 and FT-DX-560; Swan 600-T; and Robyn 500 xcvrs. Four in Yaesu FL-2000 linear.

6KE6 6GV5 format, heater 1.5 A, g_m 9.1 mS, 18 W (GE) (2-22-65) <12GM>.

6KN6 Seated height 4", min. cap, heater 3.0 A, g_m 16 mS, 30 W (RCA, SY) (7-26-65) <12GU>. Some contain two units in parallel. Variant: 42KN6.

6LB6 Seated height 4", heater 2.25 A, g_m 16 mS, 30 W (RCA, SY) {A40700} <12JF>. Two in Galaxy V Mk. 3 and (WRL) Galaxy GT-550 xcvrs. Ten in WRL Galaxy 2000B linear.

6LF6 [AKA 6LX6]; seated height 4.6", heater 2.0 A, 40 W (RCA) <12GW>. Five in Top Band Systems TBL-2000 linear. Six in Futterman H-3aa OTL amp.; eight in Counterpoint SA4 amp. Variant: 20LF6.

6LG6 Seated height 3.75", heater 2.0 A, g_m 11.5 mS, 28 W (GE) (9-18-67) <12HL>.

6LR6 6LB6 format, heater 2.5 A, g_m 16 ms, 30 W (RCA) <12FY>. Variant: 35LR6.

6LV6 Heater 2.0 A; replaceable by 6LF6 <12GW>.

6LX6 [AKA 6LF6, q. v.].

6MH6 [AKA 6LF6, q. v.].

SWEEP TUBES WITH COMMON PINOUTS

(High-voltage variants not listed.)

CAPPED OCTAL

<5BT>:	6CD6, 6DN6, 6EX6, 7867
<5ET>:	6BG6
<6AM>:	6BQ6, 6BW5, 6CU6, 6DJ6, 6DQ6, 6FH6, 6GW6, 12DQ6
<6CK>:	6AV5
<8EW>:	6CN6
<8GD>:	6CB5, 6CL5, 6FN5
<8JC>:	6DQ5
<8JX>:	6GC6
<8NC>:	6LW6

SINGLE-ENDED NOVAL

<9NZ>:	6GT5, 7868
<9QU>:	6JG6, 6JR6, 6JT6

CAPPED NOVAL

<9QK>:	6GJ5
<9QL>:	6JB6, 6JE6, 6JF6, 6JU6, 6KM6, 6LQ6, 6LZ6, 6ME6, 6MJ6, 12JB6

SINGLE-ENDED COMPACTRON

<12BJ>:	6GE5, 6GF5, 6HB5
<12ES>:	6HD5, 6HJ5
<12EY>:	6HE5, 6JA5, 6JB5, 6JC5
<12FK>:	6JN6
<12FL>:	6HJ5

CAPPED COMPACTRON

<12DR>:	6GV5, 6GY5
<12FB>:	6HF5
<12FJ>:	6JM6
<12FY>:	6JS6, 6LR6
<12GM>:	6KE6
<12GU>:	6KN6
<12GW>:	6KD6, 6LF6, 6LX6, 6MH6
<12HL>:	6LG6
<12JA>:	6LX6
<12JF>:	6LB6

CAPPED MAGNOVAL

<9NH>:	6GB5
<9RJ>:	6KG6, 40KG6

CAPPED MAGNOVALS

For "neonoval" 9-pin socket.

6GB5 [AKA EL500]; T9 bulb w/ min. cap, seated height 3.75", heater 1.38 A, 17 W (RCA, SY) (4-17-61) <9NH>. Two in Sideband Engrs. SB-34 xcvr. Variants: 13GB5/XL500, 18GB5/LL500, 27GB5/PL500.

6KG6, A [AKA EL509]; T12 bulb w/ small cap, seated height 4.6", heater 2.0 A, g_m 9.5 ms, 34 W (RCA) (8-30-65) <9RJ>. "A" rated same. Variant: 40KG6A/PL509.

27GB5 [AKA PL500]; variant of 6GB5 w/ heater 27 V @ 300 mA. Two in Sideband Engineers SB-33 xcvr.

40KG6, A [AKA PL509]; variant of 6KG6 w/ heater 40 V @ 300 mA (8-30-65) <9RJ>. Two in Esoteric Audio Research EAR-509 mono amp.; four in EAR-549.

EQUIPMENT USING SWEEP TUBES

AUDIO

Counterpoint SA4 amp. (1983-88) .... 6LF6 (8).
David Berning Co. EA-230 amp. (1981-87) .... 6JN6 (2).
Esoteric Audio Research mono amplifiers
- EAR-509 (1981-86) .... 40KG6 (2).
- EAR-549 (1983-92) .... 40KG6 (4).

Futterman OTL amplifiers
- H-3 (1962-69) .... 6HB5 (8).
- H-3aa (1978-82) .... 6LF6 (8).

McIntosh Labs MC3500 amp. (1968-71) .... 6JE6 (8).
Sansui 500A stereo rcvr.: .... 6BQ6 (4).

RADIO

Amitron linear .... 6JE6 (4).
Bramco CD-10X and CD-20X xmtrs. (1952) .... 6BQ6 (1).
Conar 400 xmtr. (1965) .... 6DQ6 (1).
Drake
- 2-NT xmtr. .... 6HF5 (1).
- T-4 series xmtrs. (1965-80) .... 6JB6(A) (2).
- TC-6 6-m xmtg. conv. .... 6JB6 (3).
- TR-3 xcvr. .... 12JB6 (3).
- TR-4 xcvr. .... 6JB6 (4).
- TR-6 6-m xcvr. .... 6JB6 (3).

Eico 723 xmtg. conv. (1960-64) .... 6DQ6 (1).
Eico 753 xmtr. .... 6DQ6 (2).
Galaxy
- V Mk. 3 xcvr .... 6LB6 (2).
- 2000 linear (1965-68) 6HF5 (10)
- 2000B linear (1969) 6LB6 (10)

Gonset G-77 xmtr. (modulator) (1955-59) 6BQ6 (2)
Gonset GSB-100 linear (1958-62) 6DQ5s
Hallicrafters
- HT-40 xmtr. (1960-64) .... 6DQ5 (2).
- HT-44 xmtr. (1962) 6DQ5 (2)
- HT-46 xmtr. (1966-68) .... 6HF5 (1).
- FPM-300 xcvr. (1972-74) .... 6KD6 (1).
- SR-150 xcvr. (1961-63) .... 12DQ6 (2).
- SR-400 xcvr. (1965-69) 6HF5 (2)
- SR-400A xcvr. (1969) 6KD6 (2)
- HX-50 xmtr. .... 6DQ5 (2).

Hammarlund HX-50 xmtr. (1961-64) 6DQ5 (1)
Hammarlund HX-50A xmtr. (1965-68) 6DQ5 (2)
Heathkit
- DX-20 xmtr. (1957-60) .... 6DQ6A (1).
- HX-11 xmtr. (1961-63) .... 6DQ6A (1).
- HW-12, -18, -22A, -32A xcvrs. (variously, 1963-74) ... 6GE5 (2).
- HW-16 xcvr. (1967-76) .... 6GE5 (1).
- KL-1 linear .... 6DQ6 [screen clamp] (1).

Henry Radio (Kenwood) TS-900 xcvr. .... 6LQ6 (2).
Henry Radio Tempo I xcvr .... 6JS6 (2).
International Crystal AOT-50 xmtr. (1963) .... 6DQ6 (1).
Johnson Challenger xmtr. (1959-65) .... 6DQ6 (3).
Kenwood (Trio) TS-511S xcvr. .... 6JS6C (2).
Kenwood TS-900 xcvr. .... 6LQ6 (2).
Knight-Kit T-60 xmtr. (1961-62) .... 6DQ6 (1).
Knight-Kit T-175 linear .... 6JE6A (2).
Lafayette HA-250 mobile linear (1966-69) .... 12JB6 (2).
Lafayette HE-25 xmtr. (1960-61) .... 6DQ6 (1).
Lakeshore P-400-GG linear (1956-60) .... 6CN6 (4).
National
- NCX-3 xcvr. .... 6GJ5 (2).
- NCX-200 xcvr. 6JB6 (2)
- NCX-500 xcvr. .... 6LQ6s.

P & H LA-500M linear (1963-66) .... 12JB6 (6).
Robyn 500 xcvr .... 6KD6 (2).
Sideband Engineers
- SB-33 xcvr .... 27GB5 (2).
- SB-34 xcvr .... 6GB5 (2).
- SB-1LA linear (1964) .... 6JE6 (6).
- SB-2LA linear (1965-69) .... 6JE6 (6).

Swan
- 260, 270B xcvrs. .... 6LQ6 (1).
- 350, 400, 500 xcvrs .... 6HF5 (2).
- 350C xcvr. .... 6JE6 (2).
- 500C xcvr .... 6LQ6 (2).
- 600T xmtr. (1971-75) .... 6KD6 (2).
- Cygnet 1200-W linear (1969-72) .... 6LQ6 (4).
- SW-240 .... 6DQ5 (2).

Tempo I xcvr. .... 6JS6 (2).
Top Band Systems TBS-2000 linear (1971) .... 31LQ6 (6).
Top Band Systems TBL-2000 linear (1972-73) .... 6LF6 (5).
World Radio Labs.
- Duo-Bander 84 xcvr. .... 6HF5 (2).
- DSB-100 xmtr. (1958-59) .... 6DQ6 (2).
- Galaxy GT-550 xcvr .... 6LB6 (2).
- Galaxy 300, III xcvrs .... 6HF5 (2).
- Galaxy linears (see Galaxy).
- LA-1 Globe Linear .... 6CN6 (4).
- Meteor xmtr. (1962-64) .... 6DQ6 (2).

Yaesu
- FT-DX-400, 560 xcvrs. .... 6KD6 (2).
- FT-101B xcvr. .... 6JS6C (2).
- FL-2000 linear .... 6KD6 (4).

AUCTION PRICES FOR TUBES

The following are auction prices for tubes at the Rochester conference of the Antique Wireless Association (1993-95, plus selected 1991-92 items), the Dearborn (Ford Museum) sale of 1995, and the New Jersey Antique Radio Club events at Hightstown in 1994-95. These were large events with numerous collectors in attendance and unusual tubes on offer.

The "exponent" on each price (e. g., 55D95) gives the site ("R," "D," or "H") and year. Abbreviations used are: NIB, new in box; BBT, brass-based and tipped; CRT, cathode-ray tube; bklt, bakelite. Unless indicated otherwise, "good" refers only to filament/heater continuity. Some of these tubes had been checked for emission as well, as shown.

BY BRAND NAME

Airline 201As, gold-globe, NIB, pair ... 90R95
Amrad S ... 20R91
Arcturus blue
- 124 ... 11R93
- 127 (2 lots of 1) ... 12R93, 28D95
- 127s (4), and 201A ... 25R94
- 45 and 80, good ... 30R93
- Mixed (5) (2 lots) ... 105R93, 65D95
- Mixed (9, 8 of them good) ... 75R95
- PZ, NIB, pair ... 90R93
- PZ/47 S-bulb & 59 ST-bulb, good ... 45R95
- RA-1 Sonora, good ... 90D95
- RD-1 Sonora, NIB, nice ... 55R95
- Wunderlich Auto, good ... 105D95
- Wunderlich (1 reg. + 1 capped), good ... 100R95
- Wunderlich, 1 good, 1 open ... 80R94
- Wunderlich, good emission ... 22H95

AudioTron tubular
- 1 good fil. ... 43R94
- Cunningham (2 lots) ... 180R93, 95R95
- Open ... 55R95, 140D95

AudioTron-style tubular, both fils. open., no label; and unident. 6" spark-gap protector tube ... 16H95
Barber-Colman furnace-control tubes (3), NIB ... 10R94
Brightson blue, good (2) ... 125D95, 140D95
Cardon C-484 (4), 1 in box ... 11R91
CeCo AC Pentode ... 17R91
Cetron CE-201A (industrial), fil. good ... 15R93
Cooper Hewitt mercury rectifiers (4) 70R91, 90R91, 205R92, 100R93
Cooper Hewitt mercury rectifier, small dud ... 50D95
Crookes "railway" discharge tube on stand ... 450R94
Crookes CRT, repro, Maltese Cross target on stand ... 210R95
Crookes CRT, repro, vertical style, on stand ... 200R95
Cunningham
- C-11, WD-11, WX-12 ... 160R93
- C-12, BBT, good ... 100D95
- C-301, BBT, good ... 40D95
- C-301A, BB, in box ... 23R94
- C-301As, pair, BBT, good emission ... 40H95
- CX-200A, BBT, fil. good ... 29H95

Daven MU-20, good ... 35D95
De Forest
- "H," open, tip off ... 90R94
- 410s, pair, 1 good ... 25D95
- 426s, pair, NIB ... 95D95
- CF-185 ... fil. good, 250R92; rebased, open, 90R91
- DV2 or -5 & RCA UV-199 ... 14R93
- DV2, good ... 20R94
- DV2, good, DV5, open, both in cans ... 60R91
- DV3 ... 15R93
- DV3 (3), good ... 65D95
- DV3/3A, in can (3 lots) ... 175D95, 175D95, 250D95
- DV3s, open (3) ... 20R95
- DL4, DV5, DV7, DL14 (1 good); in cans ... 140R94
- DV5 ... w/ can, good, 85R91; new in can, 30R92
- Oscillion, 8" unbased, good ... 2100D95
- Singer rectifier, good ... 230D95
- Spherical Audions
  - Single-wing, both fils good, bulb repaired, vac. unk. ... 800D95
  - Single-wing, both fils good, short leads ... 1400D95
  - Single-wing, cond. unstated ... 900R94
  - Double-wing, early type, fils. good ... 850R91
  - One lead missing, open fil., broken tip ... 310R93
  - Open, 1 lead gone ... 210R91
- Tubular audion, good ... 900D95
- VT-20, good, and VT-21, open ... 75R91
- VT-20, open ... 60R93
- VT-21 (2 lots) ... good, 55H94; open, 55R92
- DeF-Marconi (Moorhead), open, tip off ... 10R94

Du Mont 9877EM electron multiplier ... 12R94
Eimac
- 75T, in box, pair ... 15R94
- 100TH (e) ... 21R92
- KY-21 (4) ... 25R94

Electrad Diode and Welsh V. T. 501, good ... 100R91
Electrad Diode, cond. unstated ... 60R91
Electrad Diode, open, w/box ... 70R95
Electrons Inc. 6CF, good ... 5R95
Farad early AC tubes (3), 2 w/ good fils ... 95R93
General Electric
- "U" tube, WW I ... 85R94
- 201A, BBT, rainbow colors, good ... 30R94
- CG-886, cond. unstated ... 240R93
- CG-886, good (2 lots of 1) ... 140R91, 110D95
- CG-890, good ... 120R91
- CG-1162, good (2 lots of 1) ... 25R94, 50D95
- CG-1787 (4) ... 40R94
- TB-1 (3) and UV-202, good ... 38R95
- TB-1 (4) w/ boxes ... 40R92
- TB-1 and 2 tipped light bulbs ... 35R95

H&K 24G, pair, NIB ... 5R93

- Kellogg
  - 401, good (2 lots) ... 20R93, 25D95
  - 401, short pins ... 20R93
  - 401s (4), good ... 25R94
  - 401s, pair ... 48R95
  - AC-401 and Crusader RX-401, good emission ... 20H95
- Machlett 861, good fil. ... 65R93
- Magnavox A gridless, good ... 70R94
- Magnavox A w/ wood carton, open ... 40R91
- Marathon AC-608A, pair, fils. good. ... 10H95
- Moorhead
  - A, good ... 90D95
  - A, VT, and ER (pair); 1 good ... 100R94
  - B, good ... 170R94
  - ER, NIB ... 90R91; open, 16R91
  - "Solenoid," fil. good ... 220R92
  - 2 (1 good) ... 75R94
- Myers RAC
  - Good, w/ box ... 30R91
  - Good (2 lots) ... 60D95; w/ socket, 60D95
  - Good, in UV adapter ... 70D95
  - Open ... 65D95; orig. box, 25R94
  - Open, pair, w/ sockets, ... 75H95
- National Union 01As (5), NIB ... 47R93
- Northern Electric 221 & DX-235, in crisp orig. boxes ... 23R95
- O & T Electric Co. large tubular, and UV-200; both BBT, good fils. ... 27H95
- O. T. xmtg. valve, BBT, good ... 165D95
- QRS Red Top ... 33R94; in box, 55R94
- QRS Red Top (5), 2 in orig. boxes, 3 loose ... 100R92
- QRS Red Top w/boxes (4), nice ... 200R95
- Raytheon
  - 4B31 w/socket ... 5R93
  - BH, NIB ... 11R94
  - RK-20A in box ... 6R94
  - RK-39, 830B ... 6R94
  - TV photocell ... 21R91
- RCA
  - "Original 9" metal octals on display stand 37R92; w/ copy of 1935 promo booklet, 55R95
  - RCA - De Forest 203A ... 10R95
  - RCA - De Forest 841, NIB ... 30R94
  - 10s (2), 1 50 in ST bulb, tested ... 60R93
  - 199s, good, pair ... 23R95
  - 201A, BBT, good ... 55D95
  - 203As, boxed, pair, UV-203A ... 32R93
  - 203As, pair new in sharp "world" boxes ... 80R95
  - 211, pair ... 20D95
  - 211, pair, alum. bases, good fils., sockets ... 33H94
  - 250, good emission ... 80D95
  - 811/811A/812s in boxes, pair (3) ... 20R95, 43R94, 35R94
  - 813 in box ... 12R94
  - 852, good ... 70D95
  - 866s (3) ... 10R94
  - 5820 image orthicon ... 50R93
  - 5826 image orthicon, pair ... 170R94
  - 8093A image orthicon ... 20R94
  - A-2402 beam-deflection tube ... 100R94
  - "Radiotron UX-225," as factory-engraved on base, S14 bulb, unique base w/ five pins in "W" formation, not in Tyne's Saga but fourth example seen, good ... 20H95
  - UV-196, cond. unk. ... 40D95
  - UV-199 (7 lots of 1 each) ... avg. 26 ea.D95
  - UV-200 & UV-201, w/ boxes, good ... 55R95
  - UV-202, BBT, open ... 7R94
  - UV-202, good (4 lots of 1 each) ... 20R94, 15D95, 85D95, 90D95
  - UV-202, NIB, crisp ... 25R95
  - UV-204, no base or cap, good ... 300D95
  - UV-204A, mounted w/socket ... 30R95
  - UV-204As, (3), good fils., 2 BB, 1 alum. base ... 16H95
  - UV-204As, four (3 good) ... 35D95
  - UV-207, "unknown" cond. ... 70R95
  - UV-207, pair ... 85D95
  - UV-216 and UV-201, open ... 13R93
  - UV-218, good ... 30D95
  - UV-849 ... 20R95
  - UV-851, good ... 80D95
  - UV-857 and GE FG-195 ... 115D95
  - UV-858 ... 25D95
  - UX-120 in box ... 15R94
  - UX-120, good (5 lots of 1 each) ... avg. 18D95
  - UX-120s, good, pair ... 25R95
  - UX-199, good, in plug-in shock-mount ... 15R95
  - UX-201A, BBT, 7 (4 good) ... 295D95
  - UX-201A, crisp box (new?) ... 20R94
  - UX-201As, BB (5) ... 85R93
  - UX-213, new with box ... 11R93
  - UX-852, good ... 5R95
  - WD-11, good (2) ... 100D95, 125D95
  - WX-12, good ... 15D95
- Rogers 30, pair, good ... 23R94
- Rogers 32 (3), good ... 380R94
- Schickerling S-4000, 2 in crisp boxes, 1 loose ... 85R94
- Schickerling S-700, pair, 1 good ... 75R94
- Sodion D-21, fil. open (2) ... 30R93, 80D95
- Speed 295, good ... 52R92
- Sylvania 6SQ7, odd round-top metal ... 5R95
- United Electronics UE-X80 radar clipper diode ... 23R94
- Vac-O-Bub detector, open ... 25R94
- Van Horne 3BP, good ... 15D95
- Volutron Type A, good ... 20R92
- Westinghouse
  - 857B, minor crack ... 29R92
  - 861, open fil. ... 44R93
  - Developmental diode, pre-WD-11 ... 33R94
  - UV-201, BBT ... good, 20R94; cond. unk., 160R94
  - WL-787 classroom demo triode ... 85R91
  - WR-21 Aeriotron, good ... 70R95
  - W-McCandless auto headlamp in box (2 lots of 1) ... 5R94, 5R94
- Western Electric
  - "P" (pre - VT-3) ... 160R94
  - 101B, cast base, good ... 170R91

101B, cast; 102A; arbors broken in both ... 20R91
101D & 101F (5 lots of 1 each) ... avg. 66 ea.D95
101D, 102F, 104D ... 85R93
101D ... w/ info sheet, NIB, 280R94; pair, 200R93
101F, fil. good ... 10R93
102A & 102E (3) ... avg. 47 ea.D95
102F, fil. good ... 52R93
102F, pair, 216A, VT-1, good ... 190R95
104D & 101F, good fils. ... 30H95
201A ... 110R94
202A, broken in tip area ... 200D95
205B, pair, good emission ... 125D95
205D ... cond. unk., 70R95; good, 75H94
205D, good emission (2) ... 65D95, 141H95
205Ds, cond. unstated (3) ... 160R93
205Ds, pair, VG emission ... 303H95
208A, good ... 115D95
211A, good ... 25D95
211D ... fil. good, 37R93; NIB, 380R94
211E ... good w/box, 22R93; good, 50D95
212A, lot of seven (!) ... 180D95
212D, fil. good, w/box ... 45R93
215A (2 lots of 1) ... 20D95, 30D95
215As, 2 good, 1 open ... 16R93
215A, NIB ... 35R93
215A, pairs ... w/ sockets, good, 125D95; open, 6R94
215A/VT-5s, 3 boxed, 1 loose ... 22R93
216A (2 lots of 3 each) ... 125R93, 210R95
216A, fil. good ... 38R93, 62R93, 110R94, 35R95
216A, pair, good emission (2) ... 78H95, 225D95
216A, pair, 1 w/ VG emission, 1 weak ... 90H95
216As, pair; 1 VT-1, good ... 110R95
219D ... 55R93
224A CRT ... 260R95
228A, cond. unstated ... 40D95
232A, cond. unstated, looks new ... 70D95
262A, 310A, 310B, 717A; pair of each ... 40R94
264C, good ... 10R95
270A, good ... 80D95
274A ... 5R93
300B (3), good ... 310R92
350Bs, pair ... 100R93
CW-105 ballast ... 38R95
D-86326, good fil. ... 40R93
Photocell 2A, NIB ... 280R94
Photocell 3A ... 13R93
Photocells 2A (4) & 9B (1) ... 38R95
VT-1, good (5 lots of 1) ... 55R93, 65R93, 43R94, 85D95, 125D95
VT-1, NIB, crisp (2 lots of 1) ... 75R95, 81R93
VT-1, open ... 32R93
VT-1s (3), 1 with paper patent-license sticker added. Good fils.; emission VG, fair, and poor. ... 48H95
VT-1s, (6 lots of 1), good emission ... avg. 100D95
VT-1s, pair ... 75R93
VT-2, good (6 lots of 1) ... 60R94, 80R94, 43R95, 35D95, 105D95
VT-2s, fil. good ... 56R93
VT-2s, pair ... 125R93
Collection of > 100 types ... 625R92
WW I era - CW-1344 in socket, VT-1, VT-2; good ... 150H94
Weagant Valve ... 220R94; open, 190R93, open, 150 R95
Welsh WT-501 ... cond. unk., 70R93; open, 50D95; good, 75R93
Welsh WT-501, box & info sheet only (3) ... 11R94, 10R94, 10R94

BOX LOTS

"AC" types in nice cartons, unusual brands incl. Rogers, Sparton (16) ... 80R95
Audio: 6550 (6), 5881, pair, KT88, pair, in boxes ... 60R94
Ballasts, tested/reboxed (40) ... 25R95
Big-pin types, *weak*, 12 incl. collectible cartons ... 30H94
Big-pin ST-bulb types - 01A, 10, 22, 24, 27, 36, 46, 47, 71A, 81s, 120, 485 - 21 total ... 180H94
Big-pin ST-bulb types (31) - 32, 33, 42s, 80, 83, 874, 2A5, *good emission*, plus 3 bright 1629s ... 60H94
Cartons, ca. 12 for display ... 13R94
Loktals, 100 ... 25R95
Loktals-acorns-Compactrons (79) ... 20R93
Min., 8 pounds ... 21R93
Min., ca. 2 gallons ... 3H95
Mixed incl. big-pin and eyes, 1+ cu. ft., fils. good ... 55R94
Mixed incl. big-pin, 1+ cu. ft., fils. good ... 43R94
Octals, "ST" bulb, 350 ... 45R94
Octals, large, 225 ... 43R94
Octals, 0.6 cu. ft ... 20R94
Octals-loktals (96), "radio" types, tested/boxed ... 33H95
Octals-loktals etc., plus 2 clean AX 6155Us with good fils. 16H95
Submin. (50+) ... 43R93
TV sweep types, 50 NIB ... 5R94
01As, good (45!), 45s (5), blue (5) ... 320R95
UV-200, WE 216A, 201 in boxes, 11 good ... 100R95
210s, 27s, 01As, 800, 3B24, 6L6 ... 25R95
01A BBT good, QRS dud, Myers dud ... 48R95
VT-2 good, 216As (2 good), De Forest DL-14, W KX-642; rest mostly duds (20) ... 250R95
WE 101D, 102F, & 104D; GE CG-1162; Ediswan Type C, open (rest good); many NIB (16 total) ... 1050R95
WE 335A, Amperex 575A, Federal F-128A ... 42R93
Xmtg, 3 Eimac 4-400As; 811A, 813, 814; fils. G; appear new 90H94
Xmtg., 866, UV-204A, good fils. ... 25R93
Xmtg: "50-watters": three GE FP-285s (1 bad); RCA UV-203/203A (four, 2 BB); 2 unident. BB; 1 unident. ... 25H95
Xmtg: "50-watters": WE 242A (BB), WE 211E (BBT), GE FP-285, unident. BBT, De Forest 511, RCA UV-203A (BBT), good fils. ... 58H95
WE 102F, Cunningham C-299, good emission; GR10 prototype mercury rect., good fil. ... 30H95
WE 212E, GE CG-886, 2 others ... 225D95

BULK LOTS

Unboxed, untested, "raw" condition, from multiple sales.

Big-pin types: ... 53 ¢ to $2.33 each, avg. $1.21.
Miniatures, TV types, NIB ... 3.6¢ to 60 ¢ each, avg. 21 ¢.
Miniatures ... 1.5 ¢ to 7.2 ¢ each, avg. 3.5 ¢.
Octals ... 3.1¢ to 80¢ each, avg. 29.3 ¢.

CRTs

2AP1 & 2-3BP1 .... 7H94
5BP1, pair, 5UP1, 5CP1, pair 20R94
Sylvania 8YP4 TV-test 2R93
Du Mont 54XH + 7 common 5" 38H94

DISPLAYABLE DUDS

Mostly BBT (7) - WD-11, UV-199, UX-199, blue 138, etc.. 42H94
Moorhead VT & De Forest 20 .... 65R95
WE, 13 tennis-ball, 3 VT-1 .... 300D95
BB (6) 45H95
"S"-bulb battery (20) (2 lots) 18H95, 20R95
Moorhead VT BBT, UV-201 BBT, UV-202 BBT, C-301A BBT, unk. spherical BBT, DeF D-01A (?), 2 UV-199s 30H95
199 (6) 10H95

FOREIGN TYPES

AEG K1 (WW I) repeater tube, good fil. 360R93
AEG K6 (WW I) w/sockets 150R94
ERECT rectifier valve 550R93
Fotos 1915, open 100R91
French xmtg, like SIF, good 400D95
French, like "R" valve (TM?), open 65D95
GEC KT88s in boxes (3) & 45 .... 350R95
GEC-Osram R5 w/box, good 180R94
Gecovalve, good 28D95
Gold Lion KT-66s (3), good 30R93
Hitachi 2M121A magnetron 4R92
Japanese UV-814, screw-on-top variant, open 65R95
Japanese xmtg. (3) 40R93
LE-244 (German), good 40R94
Loewe 2HF 240R91
Loewe 3NF, spray-shield flaking 110R94
Marconi "High Impedance Triode," DeF DV-2, 1 unk..... 160R94
Marconi-Osram
- LS2 in box, good 110R94
- NT54 silica .... bad seal/good fil., 210R92; 375R91; good, 240R92
- V24 .... good, 40R91; in display frame, 75R95

Mètal TM .... gassy, 80R91; open, 33R94
Mètal S-625, untested 75R91
Mullard 904V and U10 rectifier 13R93
Mullard, unident. '20s, big orig. box 18R95
Philips
- DV triode, good 130D95
- Miniwatt 19 425, good 8R92
- Miniwatt A21, good 9R92
- Miniwatt A409, good 7R92
- Miniwatt E415, good 8R92

"R" valve, open 125D95
Siemens GR xmtg 85D95
Star (Philips) UV-199, NIB 280R94
Telefunken
- Arcotron, spray-shield flaking 110R94
- EL-152 (8) 25R93
- EVE 173, fil. open 35H95
- Gleichrichter (rectifier), '20s 45R93
- RS-214 xmtg., on display base 100R95
- RV2D800, pair 14R92
- RV2P-style (5), good htrs 50R93
- RXV-275A, display stand 80R95

Telefunken & Valvo xmtg. 70R93
Valvo QB3/300 .... 7R93
WW II British types, NIB (18) 43R94
WW II radar tubes (British CV76, CV79, NT99; German LD15) 35R94

GENERIC OLDIES

KT-88, tested good 31R93
UV-199 (3); UV-120; good 35R94
UV-199, blue 40R94
UV-201, BBT, peach color, good 20R94
UV-201As, BBT, 4 good, 1 gassy 130R93
UV-203As, pair, good 30R94
UV-861, in display stand, gassy 22H95
UX-120s, pair, UX-199, pair, UV-199, good 480R94
UX-201As (9), tested 105R93
UX-201As, good (22) 180D95
UX-201C, Canadian (3), 201 (1) 50R93
UX-216B and UX-250, S-bulb, good emission 45H94
UX-222 (good emission) and GR10 rect. w/ good fil. 5H95
UX-852s: 6 good, incl. 1 DF, & 1 dud 143H95
VT-4C, new w/ socket .... 8 (!)R91
VT-11, BBT, good 50D95
VT-27 (civilian 30) (37), NIB 72R94
VT-158 Zahl tube 160R94
WD-11s, pair, good 60R94
WD-11s, pair, bklt/tipless 53R94, 60R94
WX-12, 3 BB, 1 bklt, good fils. 75R93
WX-12, BBT, orig. style, good emission 25H95
WX-12s (4), mixed mfgrs., good .... 20 (!)R94
2A3s (12), in boxes, good 130R95
2A3s (12, 4 of them in CeCo boxes), good 180R95
2A3s (5), good 43R94, 53R94
2A3s (6), tested (2 lots) 130R93, 150R93
45s (6 - 2 w/ "S" bulbs), tested 55R93
45s (6), ST bulb, tested 36R93
45s, Philco "S"-bulb, pair, good emission 20H94
45s, S-bulb, pair, 6B4G, 6L6s (4), good emission 35H94
4C29, pair untested 35R91
50, tests good (2) 95D95, 95D95
6550, pair, newish, good emission 40H95
6A3s (5), tested 60R93
810s (5, 1 open), RCA globe carton 30R94
Geissler tube, blue/yellow 90R94
Geissler-tube set, rotating, 3 tubes w/ induction coil....... 1550R94
Spectrum-glow tubes: O_2, He, Ne, etc. (5) & WE 216A. 100R95
Tuneon (blackened) 20R94
Tuning eyes (6), fils. good 30R93
X-ray, rotating-anode .... w/field coil, 10R93; CGR, 35R92

NEON TV LAMPS

Baird 110R94
De Forest 601 .... 180R91; arbor broken, 65R95
De Forest 601?, 4-pin, 7" high, tipped, works 115H95
Jenkins four-plate, not working 83H95
Raytheon Kino-Lamp (3 lots of 1) 150R91, 100R92, 130R95

INDEX

TUBE LORE
SUPPLEMENT 2
REPLACES SUPPLEMENT DATED 9-19-96

CORRECTIONS [1]

P. 3. Five more (final?) nuvistors are the RCA 2EG4, RCA 8059 (more info needed), Hitachi 2B-H5, Toshiba 3D-HH13, and Toshiba 6D-HH13.

P. 8, col. 2, 1st para. The "rocket" tube is 2C26.

P. 11, col. 2. Cunningham numbers parallel RCA's, "numbered 100 higher" (C-299 = UX-199; CX-301A = UX-201A; CX-220 = UX-120), with exceptions (C-112 = UX-112).

P. 12, col. 1. Ken-Rad Wunderlich tubes were KR20 and KR22.

P. 14, col. 1. The Magnavox Type A has two small rectangular grids, each surrounding its own filament strand, in a common plate.

P. 15, col. 2. For equivalents of "WLS" tubes, see p. 16.

P. 67, col. 1. The 6327 has a unique pinout, not covered by a standard EIA diagram.

P. 148. The following three types got lost in editing:

1851 Video pentode, top-cap variant of 6AC7 w/ 1/8" pin for grid cap; has same characteristics as 6AC7 except for capacitances, which are actually slightly higher.

1852 [AKA 6AC7, q. v.]; pentode.

1853 [AKA 6AB7, q. v.]; pentode.

P. 174, col. 1, third para. There are indeed a few tube testers that apply test voltages before the "test" button is pushed. An example is one of the Jackson emission types.

P. 175, col. 2, 2nd para. Revitalizing a BH rectifier involves putting a diode in series with each plate lead to guarantee rectifier action. The tube still conducts, thus giving its normal large voltage drop, which is ca. 100 volts.

P. 178. The 6AV5 lacks a top cap, and belongs under the heading "Single-Ended Octal."

P. 186. "HY"-prefixed tubes are listed on p. 123.

SUBMINIATURE TYPES (P. 2)

Numerous subminiature types carry dual numbering, depending on whether they have long leads for soldering-in or short leads for socket use. The following are some such duals. This may be handy in equipment restoration: clip the leads on a 1C8, and a 1E8 results.

Long	Short	Long	Short
1C8	1E8	2E31	2E32
1S6	1T6	2E35	2E36
1V5	1AC5	2E41	2E42
1V6	1V6/S	2G21	2G22
1W5	1AD5	6BF7	6BG7

As time went on, use of sockets for subminiature tubes became rare, and new types were introduced with only long leads - the military-grade types were considered reliable enough to be soldered in, and sockets were a reliability problem in their own right.

It is an indication of what a specialty item subminiature tubes were that about 200 types were registered, yet all are scarce today. There was negligible civilian use of these types: a few dictating machines, hearing aids, and a few portable radios.

DE FOREST TYPES (P. 11)

The De Forest 401-499 are "industry" types, that is, types known elsewhere in the tube industry as (UX)-201 through (2)80, etc. and (UX)-199.

Besides the other power types, there was a DFV464, an RF-heating triode equivalent to the W WL-464 [2].

ROGERS (P. 15)

New information on these Canadian types [3], plus some further research, yields the following.

EARLY TYPES

There were very early rectifiers, R100 and R200. These were basically gridless 201As.

"S" TYPES

Rogers' "S" spray-shielded types after 1934 were usually sprayed all over, including the bakelite base. Grounding was via contact springs that clasped the base.

Like Majestic, Rogers made the 56AS, 57AS, and 58AS, which are electrically similar to the 76, 77, and 78 respectively.

The 87S and 88S are pentodes like the 77 and 78 respectively but drawing 400 mA; same pinouts.

The 89RS is rectifier-duodiode and power-pentode, renumbered 6G7S in 1935; exact registration date not known. The diodes share an independent cathode which is internally wired to the screen of the pentode. It has seven pins plus a grid cap.

The 6H7S triode-pentode (there seems to have been no 6H7M) also has seven pins plus cap, roughly like the octal 6AD7. The 6G7S and 6H7S have odd pinouts with heater on Pins 3 and 4.

"M" TYPES

The octal "Metal Spray" types having grid caps had "small" caps such as are used on big-pin tubes like the 78.

The following types are similar to non-"M" standard versions: 1C7M, 1F7M, 1P5M, 75M/6B6M, 6F5M, 6F6M, 6J5M, 6J7M, 6J8M, 6K6M, 6K7M, 6R7M, 6T6M, 12A8M, 12B8M, 12K7M.

Other "M" types are related to more familiar types as follows:

34M = 1D5	80M/R80 = 80
41M = 6K6	85M = 6R7
75M = 6Q7	86M ~ 6C5, 6J5, or 6P5.
77M = 6J7	87M = 6J7

88M = 6K7 w/ 400-ma heater.

84M is a duodiode with the same pinout <6S> as the 6X5GT.

6A7M = 6A8

6B7M = 6B8 but w/ unique pinout.

6F7M = octal 6F7 w/ pinout differing from the 6P7 "octal 6F7."

6H7M ~ 6AD7 w/ unique pinout and a top cap.

12B6M = 12Q7

20J8M (see p. 33).

2X3 ≈ ½ 5Y3; fil. 2.5 V @ 2.0 A, 350 V in, 125 mA (11-4-37) <4X>. Has pinout of first section of 5Y3GT.

2Y3 - see 2X3.

6X6 - see p. 29. Same pinout <7AL> as 1629.

RECEIVING TYPES (P. 17 ff.)

It is appropriate to emphasize that this book does not attempt to list "all" registered receiving types, as that would bring in an ocean of uninspiring color-TV types. However, receiving types are being added as "interesting" ones appear - particularly obscure types not mentioned in the usual tube manuals. It has also been suggested that information on TV vertical-output triodes having potential audio applications would be helpful. (There was, for example, a *Radio News* article in the '50s on a small hi-fi amp using a 6BX7 as push-pull output stage.)

2A3 ("More Lore.") There are three 2A3s. The single-plate version, beloved of tube-audio enthusiasts, was introduced by RCA in early 1933. The next variant has a complex plate structure, in a ++ shape as viewed from above, with dual "M"-shaped filaments. ("Dual-plate," the usual name, is a misnomer - it's a single structure, and "dual-filament" would be more accurate.) A postwar version has two independent triodes in parallel. Of 18 2A3s inspected, three were single-plate, 13 were dual-plate, and two had the twin-independent construction. These bore an RCA-style etch and "A8M" date code on the bulb, but were base-stamped "CBS-Hytron" [date] "5243" and "Sylvania" [date] "326" respectively, i. e, 1952-53.

The single-plate 2A3 uses 21 tiny filament wires in parallel, similar to the multi-wire, low-voltage "harp cathode" introduced by Amperex for instant-heating mobile transmitting tubes 30 years later, in types like the 8042, 8300, or 8509. There's nothing new under the sun . . .

2EG4 Nuvistor high-μ triode, heater 1.7 V @ 600 mA, μ 68, g_m 12.5 mS, 1.5 W, for grounded-cathode neutralized RF amps in TV and FM receivers (RCA) <12AQ>.

6AF4, A Min. triode, dual pins for grid and plate, heater 6.3 V @ 225 mA, μ 16, g_m 7.5 mS, 2.25 W <7DK>. "A" has controlled warmup time and shorter bulb. Probably the first mass-market UHF tube - the local oscillator in UHF TV tuners, competing w/ SY 6T4.

6AW5 Full-wave rectifier, metal or spray-shielded, octal base, dual cathodes, heater 6.3 V @ 600 mA, 250 V in, 120 mA total, prewar Italian type (by FIVRE?) <7Q>.

6AY8G Duodiode-power-tetrode, octal w/ cap, heater 6.3 V @ 1.25 A, gm 9.5 mS, 4 W output, prewar Italian type (FIVRE). Similar to Philips EBL1. Detector and reflexed IF/audio stage in "Radio Roma" receiver.

6BL7GT, GTA Medium-μ octal duotriode, heater 6.3 V @ 1.5 A, μ 15, g_m 7.0 mS, 10 W per section, for vertical sweep (12 W total) (RCA, SY, TS) (6-27-50) <8BD>. "A" has longer-life first section; both sections have controlled plate current under zero bias.

6BN8 Duodiode-pentode, metal or spray-shielded octal w/ cap, heater 6.3 V @ 300 mA, gm 1.1 mS. Prewar Italian type; *not* the (SY) 9-pin 6BN8 duodiode-triode as registered 10-24-55.

6BX7GT Medium-μ octal duotriode, for vertical sweep, heater 6.3 V @ 1.5 A, μ 10, g_m 7.6 mS, 10 W per section (12 W total) (GE, RCA) (12-7-51) <8BD>.

6BY8 Duodiode-power-pentode, octal w/ cap, heater 6.3 V @ 1.25 A, gm 11 mS, 4.5 W output. Prewar Italian type; *not* the (SY) 9-pin 6BN8 duodiode-pentode as registered 5-21-56.

6CK4 Low-μ octal triode, for vertical sweep, heater 6.3 V @ 1.25 A, μ 6.6, g_m 5.5 mS, 12.0 W (GE, RCA, SY) (9-9-57) <8JB>. Promoted by SY for audio use; potentially a heater-type replacement for 6B4G.

6DY7 Dual beam power tube, T12 bulb on octal base, said to be a frame-grid type, heater 6.3 V @ 1.2 A, gm 6 mS, 15 W per section (GE, TS) (10-27-58) <8JP>. In Stromberg-Carlson ASP422 stereo amp.

6DZ7 Dual beam power tube, T12 bulb on octal base, heater 6.3 V @ 1.52 A, 400 V, 13.2 W per section (GE, RCA, SY, TS) (6-22-59) <8JP>. Apparently equals two 6BQ5s, uprated on plate voltage. In at least one Heathkit stereo amp.

6EA7 Dissimilar octal duotriode, for vertical sweep, heater 6.3 V @ 1.05 A; first section μ 66, g_m 2.2 mS, 1.0 W; second section μ 5.5, g_m 6.0 mS, 10 W (RCA, SY) (3-30-59) <8BD>. (SY 1968 manual lists incorrectly as min. 9-pin.)

6EM7 Dissimilar octal duotriode, for vertical sweep, heater 6.3 V @ 925 mA; first section μ 68, g_m 1.6 mS, 1.5 W; second section μ 5.4, g_m 7.2 mS, 10 W (RCA, SY) (9-14-59) <8BD>. Barely different from 6EA7; usually labeled "6EA7/6EM7."

6FY5 [AKA EC97]; min. frame-grid RF triode, heater 6.3 V @ 200 mA, μ 70, g_m 13 mS (MU) (5-9-60).

6S4, A Min. 9-pin med.-μ triode, for vertical sweep, heater 6.3 V @ 600 mA, μ 4.5, 8.5 W (GE, RCA, SY) (12-5-49) <9AC>. "A" has controlled warmup time.

6TH8 Triode-hexode, octal base, heater 6.3 V @ 700 mA, sim. to ECH3 (prewar European type).

12AC6 Min. remote-cutoff pentode, heater 12.6 V @ 150 mA, gm 0.73 mS, for 12.6-V B+ (TS) (3-26-56) <7BK>. In 1956 Motorola car radios w/ 12AD6, 12AE6, 12K5, and TS-176 transistor.

12AD6 Min. heptode, heater 12.6 V @ 150 mA, conversion gm 0.3 mS, for 12.6-V B+ (TS) (3-26-56) <7CH>. Also see 12AC6.

12AE6, A Min. duodiode-triode, heater 12.6 V @ 150 mA, μ 17, gm 1.3 mS, for 12.6-V B+ (TS) (3-5-56) <7BT>. Also see 12AC6.

12K5 Min. space-charge tetrode, heater 12.6 V @ 400 mA, for 12.6-V B+. Puts out 40 mW. Space-charge grid draws 75 mA @ 12.6 V (TS) (12-27-55) <7FD>. Also see 12AC6.

14A5 Loktal power pentode, like 12-V 7C5 but lower-powered; *not* a 12-V 7A5. Heater 12.6 V @ 150 mA, 7.5 W (GE) (10-24-40) <6AA>.

19A3 Rectifier, variant of 35W4 w/ nontapped heater, 19.0 V @ 150 mA, in Japanese-made AM-FM receivers ca. 1965; generally replaceable by 35W4.

21TH8 Triode-hexode, 6TH8 w/ heater 21.0 V @ 200 mA (prewar European).

117P7GT Diode rectifier - pentode, 117L7GT w/ simpler pinout (TS) (1-7-41) <8AV>.

800-SERIES TYPES (P. 35)

812-H Top-cap variant of the 812, with graphite anode, hard-glass bulb, and metal-ceramic base; rated at 85 W intermittent at 30 MHz and promoted for use in amateur transmitters (UE).

825 In addition to the two "825"s listed, there was a Sylvania version apparently having the same ratings as Taylor's except for a μ of 8.0 instead of 20.

853 Xmtg. triode, variant of UV-204A with pure-tungsten filament having current raised to 16.25 A [4].

854 Xmtg. triode, variant of UV-207 with pure-tungsten filament drawing 52 A at 14.5 V (instead of 22 V); also rated at 10 kW dissipation but giving lower output than the 207 because of lower filament emission [4].

855 Vacuum rectifier, gridless version of 854, rated at 50 kV PIV and 5 A peak [4].

874 (More info.) Originally filled with neon; later S-bulbed production and ST-bulbed had xenon or argon fill giving deep blue glow instead of orange.

EARLY REGISTERED SPECIALS (P. 41)

1B102 "Thyrode" G-M counter, mica window, cylindrical body, UX base, 700-900 V (VN).

1B106 "Thyrode" G-M counter, mica window, cylindrical body, coaxial base, 700-900 V (VN).

1B124 "Thyrode" gamma counter, cylindrical body ca. 12-½" long, coaxial base, 700-900 V (VN).

1B125 "Thyrode" cosmic-ray counter, cylindrical body 16" long, coaxial base, 900 V (VN).

1B126 "Thyrode" G-M counter, needle probe type, coaxial base, 900 V (VN).

6C23 [AKA X-126H]; xmtg. UHF ring-seal pulse triode, grid-on-ring, fluted anode, heater 7 VCT @ 26 A, μ 14, 600 MHz, forced-air cooled (1-10-47). Related to 8C22.

5500-SERIES SPECIALS (P. 52 ff.)

Some of the highest-numbered types (e. g., 8728-8736) are believed to tubes designed by Philips and registered by its U. S. subsidiary Amperex, but not necessarily made or sold in North America. Info on these later types by courtesy of Åke Holm.

5512 Xmtg. triode, glass ring-seal coaxial, fil. 6.3 V @ 435 A, 25 kW, 7.5 kV DC, 108 MHz (F) (9-20-49).

5566-78 Vacuum HV industrial rectifiers, all Machlett. Some are definitely dual numberings of ML-numbered items (ML-101, etc.); the rest probably are.

5566 [AKA ML-101]; fil. 22 V @ 52 A, 60 kV applied, 2.5 A.

5567-68 Fil. 10.0 V @ 11.5 A, 125 kV in, 25 mA.

5569-70 fil. 10.0 V @ 11.5 A, 140 kV in, 25 mA.

5571, 73 Fil. 13.0 V @ 12.4 A, 140 kV in , 50 mA.

5572 Fil. 10.0 V @ 17.8 A, 140 kV in, 50 mA.

5574 Fil. 13.0 V @ 12.4 A, 150 kV in, 50 mA.

5575 [AKA ML-100]; fil. 20 V @ 24 A, 150 kV PIV, 500 mA, 750 W diss., convection cooled (ML, W) (W rates at 320 mA) (3-10-48). Also see 5625.

5576 [AKA ML-200]; fil. 20 V @ 32 A, 150 kV PIV, 1.25 A, 1 kW diss., convection cooled (3-10-48).

5577-78 Fil. 13.0 V @ 12.4 A, 200 kV applied, 25 mA.

5674 Electrometer duotriode, tubular bulb w/ cap, 6-pin base, fil. 3.8 V @ 90 mA, 10 V max., for balanced bridge ckt., can measure 5 X 10^{-10} μA (GE) (11-15-47).

5882 Full-wave rectifier, octal, heater 12.6 V @ 600 mA, 350 V in, 55 mA <6S>.

5884 T3X4 submin. electrometer duotetrode, fil. 1.25 V @ 10 mA, g_m 0.015 mS (RA) (7-1-57) <5K>.

5960 [AKA S340, TD-1]; cold-cathode discharge tube (BX) (3-28-51).

5988-91 Ionization gauges (*HK*, LK) (*Electronics*, 10-50).

5988 [AKA VG-2]; plate on rod, fil. and grid wires out press; sensitivity 25 μA/micron.

5989 [AKA VG-24G]; HK-54 w/ usual base and grid-out-side; 5/16" tubulation out top; sensitivity 50 μA/micron.

5990 [AKA VG-54]; HK-54 w/ usual base and grid-out-side; 7/16" tubulation out top; sensitivity 200 μA/micron.

5991 [AKA VG-100]; all-glass type w/ all-tungsten elements (easy to outgas), horizontal cylindrical plate, 6 long base pins; sensitivity 75 μA/micron.

6102-03 Rectifiers, T-T fil. 5 V @ 7.2 A, 150 mA (W) (both 3-26-54).

6102 Cup-like glass body, 2-27/32" dia. by 2-3/16" long, large disc anode sealed into top, button fil. studs at bottom, oil-cooled, 40 kV PIV.

6103 Like 6102 but w/ 3 radiator discs added to anode for air cooling, derated to 20 kV.

6109 [AKA BS404]; apparently an adjustable corona voltage regulator, 3/8" dia. tubular metal body, glazed porcelain insulator, and electrode cap at the. Body has large setscrew to change voltage over the range 645-705 v; 55 μA max (VN).

6142 [AKA TD-9A]; gas diode voltage regulator (BX) (7-10-53).

6196 Electrometer tetrode, fil. 3.0 V @ 50 mA, 9 V (Mazda) (10-10-52). Also see 6250.

6250 Electrometer tetrode, fil. 2.5 V @ 45 mA, 9 V (Mazda) (10-10-52). Also see 6196.

6255 [AKA 56AVP]; photomultiplier.

6269 Rectifier and clipper diode, ext. anode, 2" long, flying leads for fil., heater 26.5 V @ 350 mA, 16 kV PIV, 250 mA peak (AX). Also see (UE) 588.

6356-59 Gas noise sources (BX) (all 10-9-53). All have noise ratios of 15.2 dB (*i. e.*, noise level above thermal noise), the standard value for argon fill. Also see 6684.

6356 [AKA TD10]; for 3.95-5.85 GHz in RF-49/U WG, or 5.85-6.20 GHz in RG-50/U; 70 V drop @ 250 mA.

6357 [AKA TD11]; for 8.2-12.4 GHz in RG-25/U WG, 75 V drop @ 200 mA.

6358 [AKA TD12]; for 2.60-3.95 GHz in RG-48/U WG, 80 V drop @ 250 mA.

6359 [AKA TD13]; for 18.0-26.5 GHz in RG-53/U WG, 65 V drop @ 200 mA.

6361 [AKA TT-25]; "single convectron" "misc. type" (BX) (9-4-53).

6369 ATR tube, 1B37A repackaged in half-height WG, 0.9" x 0.2" inside w/ "contact" nonpressurized flanges, for 8.5-9.0 GHz (BL, MA). Variants: 6393, 6396.

6395 Min. pentode, fil. 1.2 V @ 50 mA, gm 0.9 mS.

6436 [AKA CK1036]; T3 submin. cold-cathode rectifier, 1.5 kV PIV, 100 μA (RA) (9-12-55).

6440-41 Vacuum gauges, 10 V @ 100 mS (Adzam) (both 2-12-54).

6527 Magnetron, fixed at 9375 MHz, potted flying leads for heater, heater 7.0 V @ 600 ma, 6 kV @ 5.5 A, interchanges w/ 2J42, operable at high altitude (GE) (8-16-54).

6538-39 Trigger tubes (AX) (both 10-11-54).

6538 [AKA Z800U].

6539 [AKA Z801U].

6658 TWA, 1.7-4.0 GHz, metal shell, CO, both RF leads out one end, 10 W CW out (ITT).

6659 [AKA CK1042]; T3 submin. cold-cathode rectifier, 2.8 kV PIV, 1 mA (RA) (3-19-56).

6684 [AKA TD18]; gas noise source, 12.4-18.0 GHz in RG-91/U WG, 70 V drop @ 200 mA, 15.2 dB ("argon") noise ratio (BL) (6-13-55). Also see 6356-59.

6755 Thyratron, 4-pin base, fil. 2.5 V @ 11 A, 1.5 kV PIV, 3.2 A.

6782 [AKA TD-22]; gas noise source, 2.60-3.95 GHz, in RG-48/U WG, 45 V drop @ 250 mA, 15.2 dB ("argon") noise ratio, AC- or DC-operated (BX) (4-29-57).

6835 [AKA QK464]; storage tube, for scan conversion, etc; sotres picture or date frame, reads out 30,000 times; 600 line resolution (RA) (2-13-56).

6852 [AKA TD-24]; gas noise source, 3.3-4.9 GHz, in WR-229/U WG, 65 V drop @ 250 mA, 15.2 dB ("argon") noise ratio, AC- or DC-operated (BX).

6881 [AKA TD-21]; gas noise source, 1.12-1.70 GHz, in RG-69/U WG, 65 V drop @ 250 mA, 15.2 dB ("argon") noise ratio, DC-operated (BX) (10-28-57).

6882 [AKA TD-23]; gas noise source, 8.2-12.4 GHz in RG-52/U WG, 115 V drop @ 200 mA, DC-operated, 18.0 dB ("neon") noise ratio (BX) (7-3-61).

6912 Vidicon, 1-½" dia., for broadcast use (Resistron Labs) (9-17-56).

6932 T2X3 flat-press pentode, 6088 w/ separate suppressor lead (RA).

6964-64 Ignitron rectifiers, 16" (GE).

6964 With one grid.

6965 With two grids.

6965-66 Ignitron rectifiers, 20" (GE). 12 give 12 MW of DC @ 850 V.

6965 With one grid.

6966 With two grids.

7016 [AKA NL606]; gas-mercury rectifier, stepped tubular bulb on 4-pin base, fil. 2.5 V @ 17 A, 900 V PIV, 6.4 A (7-29-57).

7030 Rectifier/charging diode, 7" high, 3-17/32" dia., three posts out crown, fil. 13 V @ 36 A, 2.5 kV, PIV, 200 mA, 2.5 kW diss., forced-air cooled (F) (6-27-60).

7064 [AKA XP1000]; photomultiplier (DuM?) (8-5-57).

7065 [AKA 150AVP]; photomultiplier (DuM?) (8-5-57).
7100 Vacuum rectifier or clipper diode, 30 kV PIV, 3 A, 3.5 kW diss., forced-air cooled (IT) (5-21-62).
7101 [AKA TD-29]; gas noise source, 1.12-1.70 GHz, in RG-69/U WG, 65 V drop @ 250 mA, 18.0 dB ("neon") noise ratio, AC- or DC-operated (BX).
7131 Vacuum rectifier or clipper diode, 40 kV PIV, 3 A, 3 kW diss., forced-air cooled (IT).
7209 Xmtg. triode, 6442 format, high shock rating, 3 GHz, plate-pulsed at 3.5 kV @ 3.0 A (ML) (8-1-60).
7210 Xmtg. triode, 6442 format, 3 GHz, plate-pulsed at 3.5 kV @ 2.8 A, 12-sec. warmup (ML) (8-1-60).
7222 "Iatron" storage tube, 5", electrostatic focus and deflection, two writing guns (IT) (6-15-59).
7256 Magnetron, tunes 8.5-8.6 GHz, 40 kW pulse output (7-7-58).
7258 Triode-pentode, mobile-rated variant of 6AN8, heater 13.5 V @ 210 mA (GE, RCA, SY, TS) (7-14-58) <9DA>.
7322, 7620, 7621 Cermet ruggedized hydrogen thyratrons, cylindrical ring-seal bodies, anode on pin recessed into top, grid on ring, cathode on mounting flange, heater studs at base, w/ H_2 reservoir (EG&G).
7322 [AKA 1802]; 5-¼" high, 3-5/8" dia., heater 6.3 V @ 16 A, 25 kV PIV, 1.5 kA peak, 2 A, 10 min. warmup (1-19-59). H_2 reservoir takes 6.3 V @ 6 A. Functional replacement for 5948, 5949 and (often) 1257 (EG&G).
7620 [AKA HY-2]; 4-3/8" high, 2-5/16" dia., heater 5.0 V @ 7.5 A, 20 kV PIV, 500 A peak, 0.4 mA avg., 5 min. warmup. Functional replacement for 1258, 3C45, and (often) 4C35 (EG&G). H_2 reservoir takes 6.3 V @ 4 A.
7621 [AKA KU-70C, HY-2]; 2-½" high, 1.2" dia., heater 6.3 V @ 3.5 A, 8 kV, 100 A peak, 100 mA avg., 350 kW pulse output, 30-sec. warmup, functionally replaces 1258, 3C45, and 4C35 (EG&G, KL) (5-21-62). H_2 reservoir wired across heater.
7338 TWA, 2.4-3.6 GHz, metal case, both RF leads out one end, 1 kW pulse output (IT).
7339-40 TWAs, 8.0-9.6 GHz, metal case, both RF leads out one end, 1 kW CW output (IT).
7341 TWA, 8.0-9.6 GHz, 5 W CW output.
7347 TWA, 2.0-4.0 GHz, 1 kW output.
7350 TWA, 7.0-11.0 GHz, 1 kW output.
7386 Hydrogen thyratron, same ratings as 5545 (5-18-59).
7407 TWA, 6.7-11.4 GHz, 150 mW output.
7414 Magnetron, tunes 5.4-5.9 GHz, 10 kW pulse output (3-23-59).
7443 Magnetron, 5.5-5.6 GHz, 400 W output.
7444 Magnetron, tunes 5.4-5.9 GHz, 1 kW pulse output (5-15-61).
7445 Magnetron, 9.1-9.5 GHz, 100 W output.
7475 [AKA LS620]; neon voltage regulator, on Br. B4 4-pin base, 100 V @ 8 mA (MU, PH).
7484, A Magnetron, tunes 1250-1350 MHz, 2 MW output (11-23-59).
7539 [AKA 1855]; Graphechon scan-converter tube (RCA) (4-11-60).
7585 Ignitron, Size C, 5552A char., cooled by spiral water coil around arc chamber; interchanges w/ other Size C tubes (AX) (1960).
7620 See 7322.
7621 See 7322.
7632 [AKA ORP10]; photoconductor (12-28-59).
7633 [AKA ORP11]; photoconductor (12-28-59).
7665-67 Cermet hydrogen thyratrons (KL).
7665 [AKA KU-72]; 20 kV peak @ 350 A, 3.5 MW output, 300 mA avg.
7666 [AKA KU-73]; 25 kV peak @ 1 kA, 12.5 MW output, 1.5 A avg.
7667 [AKA KU-74]; fil. 6.3 V @ 22.0 A (KL), 33 kV peak @ 2 kA, 33 MW output, 4 A avg. (11-26-62).
7779 [AKA XD56]; vac. rectifier or clipper diode, 25 kV PIV, 6 A, 3 kW diss., forced-air cooled (IT) (11-28-60).
7782 [AKA KU-71]; hydrogen thyratron, 10 kV peak @ 200 A, 1 MW output, 200 mA avg. (KL) (11-26-62).
7810-12 TWAs, for 4-GHz TV radio-relay use, heaters 6.3 V (TSU) (1954). Possibly used only in Japan.
7810 Low-noise rcvg. preamp.
7811 Driver stage, 1.5 kV, 23 dB gain, 350 mW out.
7812 Output stage, 2.3 kV, 13 dB gain, 3.8 W out.
7817 [AKA XP1000]; photomultiplier (9-19-60).
7818 [AKA XP1030]; photomultiplier (10-24-60).
7820 Xmtg. triode, μ 6, 8 kV, Forced-air cooled, 4.3 kW (F) (12-12-60).
7831-32 Xmtg. cermet triodes, μ 18, 10 kV, 50 MHz (F).
7831 Water-cooled, 20 kW (2-18-63).
7832 Vapor-cooled, 30 kW (11-25-63).
7836 [AKA QE08/200H]; xmtg. triode.
7837-39 Xmtg. triodes, μ 41, 22 MHz (F) (all 9-25-61).
7037 Water-cooled, 90 kW, 17.5 kV.
7038 Forced-air-cooled, 25 kW, 25 kV.
7039 Forced-air-cooled, pulse-rated at 160 A peak, derated to 10 kW, 65 kV.
7866 [AKA KU-274]; cermet hydrogen thyratron, 50 kV peak @ 2 kA, 4 A avg., 50 MW out (KL)(12-10-62).
7869 Vacuum rectifier or clipper diode, 56 kV PIV, 4 A, 3.5 kW diss., forced-air cooled (IT) (7-3-61).
7884 [AKA YL1060]; xmtg. tetrode (AX?)
7906-07 Vacuum rectifiers or clipper diodes, 50 kV PIV, 15 A (IT) (both 2-26-62).
7906 Water-cooled, 50 kW diss.
7907 Forced-air cooled, 20 kW diss.
7971 Power pentode, ruggedized Ediswan 11E12, octal base, heater 6.3 V @ 1.6 A, gm 13 mS, 800 V, 200 mA, 20 W (2-13-61).
8045-46 Xmtg. triodes, μ 20, 16 kV, 22 MHz (F) (11-25-63).
8045 Water-cooled.
8046 Forced-air cooled.
8104, J Xmtg. triodes, μ 25, 8 kW, 6 kV, 30 MHz, water-cooled (F) (2-26-62).
8109-10 Vacuum rectifiers or clipper diodes, 60 kV PIV, 12 A (IT) (both 12-3-62).
8109 Water-cooled, 20 kW diss.
8110 Forced-air cooled, 10 kW diss.
8131-33 Xmtg. cermet triodes, μ 20, 6.5 kV, 80 MHz (F).
8131 Water-cooled.
8132 Forced-air cooled.
8133 Vapor-cooled.
8136 Min 9-pin sharp-cutoff pentode, computer-rated 6DK6 (AX, GE, RCA) (1-1-62) <7CM>. Replaced 6DK6 in vertical amps of later Tektronix 540/550-series scopes.
8146-47 Xmtg. cernet triodes, μ 17, 10 kV, 50 MHz (F) (both 4-6-64).
8146 Water-cooled, 40 kW.
8147 Air-cooled, 20 kW.
8148 Xmtg. triode, water-cooled, μ 5.5, 8 kV, 40 kW (F) (12-17-62).
8253 Hydrogen thyratron, sim. to 5C22 or 6279.
8366 Magnetron, fixed at 33.2 GHz ± 0.2 GHz, 7619 ratings - probably a retuned 7619 (SY).
8544 [AKA LCF80]; premium 6LN8.
8581 [AKA YL1270]; xmtg. tetrode.
8598 Image converter, 12 kV (TF).
8679 [AKA YL1121]; xmtg. tetrode (AX).

8680 [AKA YD1212]; xmtg. triode (AX).
8681 [AKA YJ1080]; pulse magnetron.
8728-36 Xmtg. triodes (8-8-66 for 8732-36; unknown for others).
8728 [AKA YD1150]; (PH).
8729 [AKA YD1151].
8730 [AKA YD1152].
8731 [AKA YD1160]; (AX).
8732 [AKA YD1161].
8733 [AKA YD1162].
8734 [AKA YD1173].
8735 [AKA YD1182].
8736 [AKA YD1192].
8801 [AKA YD1180]; xmtg. triode.
8812-14 Xmtg. tetrodes (AX?).
8812 [AKA YL1420].
8814 [AKA YL1440].
8813 [AKA YL1430].
8867 [AKA YD1352S]; xmtg. triode (AX).
8888 [AKA YL1470]; xmtg. tetrode.
8901 [AKA VA-98M]; reflex klystron, tunes 23.6-24.4 GHz, 2.8" long by 1" high by 1.3" deep, flying leads, 500 V, 100 mW out (VA).
8913 [AKA YD1195]; xmtg. triode.
8915 [AKA YL1520]; xmtg. tetrode.
8918 [AKA YD1342]; xmtg. triode.
8935-37 Xmtg. triodes (AX?).
8935 [AKA YD1185].
8937 [AKA YD1197].
8936 [AKA YD1187].
8952 [AKA YD1175]; xmtg. triode, fil. 5.6 V @ 130 A, 10 kW.
8958 [AKA YD1177]; xmtg. triode.
8990 Xmtg. *tetrode*.

NONREGISTERED TYPES (P. 105 ff.)

1 Mercury half-wave rectifier, ST14 bulb (RCA, by courtesy of Bob Dennison); probably a predecessor to the (vacuum) 1-V.
2B-H5 Nuvistor triode, heater 1.8 V @ 600 mA (HI).
3D-HH13 Nuvistor-style duotriode; has regular nuvistor base; envelope about 2-½ times usual height; heater 3.5 V @ 600 mA, μ 36, g_m 13 mS (TSU) <8LK>. Pictured in Hisashi Ohtsuka, The Classic Valves, 1994, p. 241. May have been US-registered as (GE, SY) 3FX7.
6D-HH13 Nuvistor-style duotriode; variant of 3D-HH13 w/ heater 6.3 V @ 300 mA. May have been US-registered as (GE, SY) 6FX7.
45 ("More Lore.") An Internet posting has referenced an ST16-bulbed tube stamped "45" and branded "Radiocoin," comprising a tetrode internally connected as a triode, and testing OK as a 45.
545, 552, 554, 588, 589 Vacuum HV diodes, rectifiers or radar clippers, ext.-anode, minimal glass bodies, flying fil. leads w/ lugs, for air or oil cooling (UE), 1956.
545 Length 2", 7/8" dia., 50 mA.
552 Length w/o leads 1-3/8", dia. 7/8" overall, anode dia. 0.55", T-T fil. 2.5 V @ 1,7 A, 20 kV PIV in air or oil, 10 mA; fits 60-A fuse clip.
554 Length w/o leads 2-¼", dia. 1.2" overall, anode dia. 0.69", heater 6.3 V @ 3.5 A, 17 kV PIV, 70 mA in air or 150 mA in oil.
588 Heater 6.3 V @ 1.65 A, 16 kv PIV, 65 mA. Also see 6269.
589 Length w/o leads 1.9, dia. 7/8" overall, anode dia. 0.6", heater 6.3 V @ 1.6 A, ratings in air 10 kV PIV @ 10 mA or 16 kV PIV @ 45 mA; ratings in oil 100 or 65 mA respectively; fits 60-A fuse clip. Possibly same as (TS) 6339.
572-3, -10, -30, -160 Audio power triodes, tubular hard-glass bulbs on ceramic UX bases, graphite anodes, T-T fils., 125 W (SV). Mu corresponds to the dash-number. These are in addition to the existing 572B.
M2057 Compactron beam power tube similar to oversized 8950, for use in RF linear amps (GE).
T-10-S Transmitting variant of 10 w/ plate cap. Also see T-30-S.
T-30-S Variant of 30 w/ cap added for plate, for use in regenerative SW receivers (Triad; described in *Radio*, June 1933, by courtesy of Bob Thomas). A plate-capped *receiving* tube!

MORE BENDIX - RED BANK TYPES (p. 120)

TD-1 [AKA 5960, q. v.]; cold-cathode discharge tube.
TD-9A [AKA 6142, q. v.]; gas diode voltage regulator.
TD-10 [AKA 6356, q. v.]; gas noise source.
TD-11 [AKA 6357, q. v.]; gas noise source.
TD-12 [AKA 6358, q. v.]; gas noise source.
TD-13 [AKA 6359, q. v.]; gas noise source.
TD-17 [AKA 5643]; q. v.]; submin. thyratron.
TD-18 [AKA 6684, q. v.]; gas noise source.
TD-19 Gas diode voltage regulator.
TD-21 [AKA 6881, q. v.]; gas noise source.
TD-22 [AKA 6782, q. v.]; gas noise source.
TD-23 [AKA 6882, q. v.]; gas noise source.
TD-24 [AKA 6852, q. v.]; gas noise source.
TD-28 Cold-cathode tetrode thyratron.
TD-29 [AKA 7101, q. v.]; gas noise source.
TD-36 [AKA 7416, q. v.]; gas diode.
TD-45 Cold-cathode discharge tube.
TD-61 [AKA 5787WA, q. v.]; voltage regulator.
TD-68 [AKA 6542, q. v.]; voltage regulator.
TE-40 [AKA 6888, q. v.]; dual-control pentode.
TE-56 [AKA 3D21B, q. v.]; pulse modulator.
TE-74 [AKA 3D21, q. v.]; pulse modulator.
TE-81 [AKA 7757, q. v.]; HV amplifier.
TE-90 [AKA 5R4WGB, q. v.]; rectifier.
TE-93 [AKA 5R4WGA, q. v.]; rectifier.
TG-25 to **TG-60** Two-electrode glass vacuum spark gaps, for protective use, tubular w/ end caps, radioactive-stabilized.
TG-25 Fires at 400 V, 3W.
TG-26 Fires at 750 V, 5 W, 1 kA surge.
TG-27 Fires at 1.0 kV, 5 W, 1 kA surge.
TG-30 Fires at 2.0 kV, 5 W, 2 kA surge.
TG-34 Fires at 5.0 kV, 5 W, 3 kA surge.
TG-37 Fires at 10.0 kV, 5 W, 4 kA surge.
TG-56 Fires at 20.0 kV, 7.5 W, 5 kA surge.
TG-60 Fires at 50 kV, 7.5 W, 6 kA surge.
TN-1 Min. noise generator, 8.5-9.6 GHz, noise source in mount, WG, transmission type.
TN-2 Min. noise generator, 8.5-9.6 GHz, noise source in mount, WG, terminated type.
TN-3 Min. noise generator, 2.7-2.9 GHz, noise source in mount, WG, terminated type.
TT-1 [AKA 6144, q. v.]; noise diode.
TT-2 [AKA 5947, q. v.]; temperature-limited diode.
TT-29 [AKA 6352, q. v.]; temperature-limited diode.
TT-30 [AKA 5845, q. v.]; temperature-limited diode.
TT-65 [AKA 5783WA, q. v.]; voltage regulator.

HEINTZ & KAUFMAN (P. 122)

VG-2 [AKA 5988, q. v.]; ionization gauge.
VG-24G [AKA 5989, q. v.]; ionization gauge.
VG-54 [AKA 5990, q. v.]; ionization gauge.
VG-100 [AKA 5991, q. v.]; ionization gauge.

KUTHE LABORATORIES TYPES (P. 123)

Later, ITT. Note that W also used numbers prefixed with "KU."

KU-14 [AKA 6777, q. v.]; hydrogen thyratron.
KU-25 Hydrogen thyratron, 5C22 (q. v.) adapted for high pulse rate.
KU-274 [AKA 7866, q. v.]; hydrogen thyratron.
KU-51 Hydrogen diode, rectifier or radar clipper diode, fil. 5.0 V @ 6.2 A, 10 kV PIV, 200 mA as rectifier.
KU-52 [AKA 8264, q. v.]; hydrogen diode.
KU-53 [AKA 7178, q. v.]; hydrogen diode.
KU-70C [AKA 7621, q. v.]; hydrogen thyratron.
KU-71 [AKA 7782, q. v.]; hydrogen thyratron.
KU-71Z [AKA 8765, q. v.]; hydrogen thyratron.
KU-72 [AKA 7665, q. v.]; hydrogen thyratron.
KU-72Z [AKA 8553, q. v.]; probably hydrogen thyratron.
KU-73 [AKA 7666, q. v.]; hydrogen thyratron.
KU-74 [AKA 7667, q. v.]; hydrogen thyratron.
KU-82 [AKA 7583, q. v.]; hydrogen thyratron.
KU-92 Cermet hydrogen diode, rectifier or radar clipper, 15 kV PIV, 500 mA as rectifier.
KU-93 [AKA 8275, q. v.]; hydrogen diode.
KU-94 Cermet hydrogen diode, rectifier or radar clipper, 20 kV PIV, 4.0 A as rectifier.
KU-99 Hydrogen thyratron, fil. 6.3 V @ 2.7 A.

SVETLANA (New section, ca. P. 152)

The SV product line, besides the 572-family (above), includes the following smaller tubes:

5U4G	811-3 and -10
6550C	833A, C
6AS7G	8560AS
6BM8	EF86/6267
811A	EL34/6CA7

There are a wide variety of larger tubes: some of "Eimac" types, some of SV's own design like the 4CX1600B and -U. There are reports of plans to introduce more "audio" types.

WESTERN ELECTRIC (P. 155 ff.)

P. 156. Frank Hagenbuch reports having a 205F stamped "FAST" (quick-heating?). This may be related to the "205FA" reported in <u>75 Years of WEstern Electric Tube Manufacturing</u>. Further info welcome.

P. 160. The 310A is normally an audio type or as a repeater tube used in telephone offices. However, it became a submarine (!) repeater tube as well (*Electronics*, Jan. 1952, p. 165). In 1952 the Western Union Telegraph Company used 16 310As in a submerged telegraph repeater, 100 miles off Long Island, in its cable from Rockaway Beach, LI, to Bay Roberts, Newfoundland. The tubes were used in two push-pull amplifiers: one working, one spare. Speed of the cable, originally 83 words per minute, was tripled.

P. 165. The 715-types were not exclusively radar pulsers. Two 715Bs were used as current regulators in the shore-end power supply for AT&T's Key West - Havana submarine cable, 1950. The transmitter in the National Bureau of Standards C2 ionospheric sounder, ca. 1951, used six 715Cs, in two driver-driver-final strings; later replaced with 4PR60s.

P. 167. The D-171225 was a copy of the Siemens C3E repeater tube. Not widely known is the fact that Sylvania also made 1000 reverse-engineered copies of this captured type, on the same "crash" basis [5].

SUBSTITUTIONS (P. 173, COL. 1)

Special hints [6] for late-vintage Zenith tube Trans-Oceanics are as follows. The "unobtainable" 1L6 can be replaced by the common 1R5 by clipping off Pin 5, which is an extra connection to the filament. The cost of this substitution is that correct alignment and dial calibration are unobtainable on the top short-wave bands even though the radio may perform well. This has also been reported [7] as working well in a Zenith 5G41 AM-only portable.

The similarly scarce 50A1 ballast in late T-Os is replaceable with a 1000-ohm 3-watt resistor. To get the voltage regulation that the 50A1 provided, a 9.1-volt zener diode (1N4739, etc.) is connected across the filament string (cathode on Pin 7 of the 3V4, anode on Pin 1 of the 1U5).

AUCTION PRICES (P. 182 ff.)

Some prices from the 1996 Rochester auction are:

BY BRAND NAME

Amrad "S", S16 style, w/ ceramic socket........25
Cunningham Audiotron, good........95
Cunningham Audiotron, good, in UX adapter 90
De Forest
 D-01A w/ can, good........80
 DV-2 w/ can, good........70
 DV-3 (2), fils. OK........40
 H, glass chips inside, unknown cond.........270
 Spherical audion, 1 fil. good........1100
Electrad diode w/ socket........35
Marathon AC-608 (5), all good........130
Marconi - De F audion, BBT, open........38
Myers RAC (2)........good w/ socket, 70; open, 43
Northern Electric
 216A, good........45
 221D (3), good........48
 DX-221, good........35
 RX-235 (pair), good........10
QRS Red-Top, NIB, w/ paper........50
RCA
 2A3 (5), NIB........260
 203As (2), aluminum base, crisp "world" boxes........30
 1608 (3), NIB........80
 1848 (?) iconoscope, 5", octal base........240
 1850A iconoscope, w/ GE box........210
 UV-201A, BBT, pretty "rainbow" getter, good emission........40
 WD-11, BB, NIB........80
 Store-display neon, cond. unk........105
Speed (Cable) 296 "Triple Twin," good........50
Westinghouse
 WL-195, good........10
 WL-706 magnetron........30
 WR-21 in box, good........190
Western Electric
 1A photocell in orig. can........110
 2A photocell, one lead off........48
 201A (*not* "industry" 201A!), good........180
 205B, 205D, and 205E, good fils........110
 205D (2), good........260
 216A (3), two OK........125
 216A, good (3 lots of 1)........60, 60, 60
 216As (3), good........140
 216As (5), good........160
 720CY magnetron, NIB........43
 2J50 magnetron, NIB........50
 VT-1, NIB, less tips........60

GENERAL

UV-207, open........45
WD-11s (2), tipped, bklt. bases, good........70
Crookes "mineral" CRT, repro, vertical on wood base........225
Crookes Maltese-Cross CRT, works........300
Disc-TV neon (De Forest 601?), dud........75

FOREIGN

CV1226 short silica valve (Br.) and French TM-30, good........170
Early Br. radar (20): CV85, CV92, CV90, CV116, CV1484, etc.;

plus some US	160
E4M Métal (French WW I), used	850
Genelex KT-66 (4), good, in orig. boxes	210
Genelex KT-88 (2), used, "excellent"	110
Marconi-Osram V24, good	100

SWEEP TUBES (P. 178 ff.)

Here is some information on the use of sweep tubes in CB linear amplifiers [8].

The Kris "Big Boomer" uses a 6KV6, driving two or (depending on model) three 6LQ6s. Three 6KV6s are in the Varmint "10 meter" amplifier, drivieng six 6LR6s.

The DNA Phantom 500 uses four 6JG6s dribving six 6LQ6s; the PD400 uses one 6JG6 driving another, driving six 6LQ6s; the Maverick uses four 6JB6s driving four 6LQ6s.

The MACO Majestic amplifier uses the (GE) M2057 Compactron.

REFERENCES

1. By courtesy of Alan Douglas and Jim Cross.
2. By courtesy of Jim Cross.
3. P. Guibord and J. Stephen, "Those Unique Rogers Tubes," Ottawa Vintage Radio Club newsletter, Vol. 4 No. 2 (Summer 1996), pp. 3-7.
4. H. F. Dart and G. C. Crom, Frequency Amplification (Scranton, PA: International Textbook Co., 1930), pp. 54-55 and 60-61 (by courtesy of Jim Cross).
5. L. A. Sibley, "'German' Tubes by Western Electric," *Radio Age*, March 1997, pp. 12-13.
6. L. A. Sibley, "Restoring the Later Trans-Oceanics," *Radio Age*, Vol. 20 No. 5 (May 1995), p. 10.
7. M. R. George, "A Tube Substitute for the Type 1L6," *Antique Radio Classified*, Vol. 14 No. 1 (Jan. 1997), pp. 12-13.
8. By courtesy of Jon Cohen.

TRENDS IN TUBE DESIGN (PP. 3-4)

Additional insight may come from the chart on the next page, which is reproduced from an RCA sales reference of July, 1935. It lists receiving tubes *not* sold by RCA, giving a glimpse at the rest of the industry. The "groups" of competitors mentioned in the table were as follows:

	GROUP																		
Maker	1	2	3	4	5	6	7	8	9	10	11	12	13	14	15	16	17	18	19
Arcturus		•				•													
Champion										•	•	•						•	
Gold Seal						•		•	•	•	•	•	•	•			•	•	
Grigsby-Grunow							•	•		•									•
Hy-Sylvania			•		•	•	•	•	•	•	•	•						•	
Hytron					•	•		•				•					•		
Johnsonburg		•			•	•				•	•	•			•	•		•	
Ken-Rad				•	•			•	•	•									
Nat'l Union	•							•				•				•	•		
Raytheon	•			•	•			•	•	•									
Republic												•	•	•	•	•	•	•	•
Sparton							•						•		•		•	•	
Triad								•											
Tung-Sol						•			•	•		•					•	•	

These are reference lists to give brief descriptions of available tubes and should not be used to establish their commercial status.

COMPETITORS RECEIVING TUBES

RECEIVING TUBES

NOT Manufactured or Merchandized By RCA Radiotron Division
(See Foot Note)

Tube Type	§ Mfgr's	Bulb Size	Base		Tube Description	Voltage
2A3H	Group 1	16	Med.	4-Pin	Power Amplifier Triode	2.5 H.
2B6	" 2	16	Med.	7-Pin	Triple Twin Power Output	2.5 H.
2F7	Ken Rad	ST12	-	7-Pin	Triode Pentode Converter	2.5 H.
6B5	Raytheon	12	Med.	(?)	Triple Twin Power Output	6.3 H.
6B6	Group 2	ST16	Med.	7-Pin	Triple Twin Power Amplifier	6.3 H.
6C7	Grig. G.	12	Small		Duplex Diode Triode (Spray Shield)	6.3 H.
6D7	Grig. G.	12	Small	7-Pin	R-F Amplifier (Spray Shield)	6.3 H.
6E6	Nat. Un.	ST14	Med.	7-Pin	Twin Triode Power Amplifier	6.3 H.
6E7	Group 19	12	Small	7-Pin	Super-Cont. R-F Ampl. (Spray Shield)	6.3 H.
6Y5S	Grig. G.	12	Small	6-Pin	Full-Wave Mercury Vapor Rectifier	6.3 H.
6Z5	Grig. G.	12	Small	6-Pin	Full-Wave Rectifier	6.3/12.6 H.
12A5	Group 5	12	Small	7-Pin	Pentode Power Amplifier	6.3/12.6 H.
**12Z5	" 6	ST12	Small	7-Pin	Full-Wave Rect. & Voltage Doubler	6.3/12.6 H.
2	Grig. G.	14	Med	5-Pin	Duplex Diode Detec. (No Spray Shield)	2.5 H.
2S	Grig. G.	14	Med.	5-Pin	Duplex Diode Detec. (Spray Shield)	2.5 H.
4	Grig. G.	12	Med	5-Pin	Duplex Diode Detec. (No Spray Shield)	2.5 H.
4S	Grig. G.	12	Med.	5-Pin	Duplex Diode Detec. (Spray Shield)	2.5 H.
10S	Triad	S17	Med.	4-Pin	Osc. or Amp.. (Plate lead-top cap)	7.5 F.
14	Group 3	S14	Med.	5-Pin	Amplifier	14.0 H.
15	" 3	S12	Small	5-Pin	Detector Oscillator (R. F. Pentode)	2.0 H.
17	" 3	S14	Medl	5-Pin	Amplifier	14.0 H.
18	" 3	S17	Med.	6-pin	Pentode Power Amplifier	14.0 H.
29	Hyg. Syl.	S12	Small	6-Pin	Detector	2.5 H.
51	Group 8	S14	Med.	5-Pin	Super-cont. R-F Amp. (= our 35)	2.5 H.
52	" 9	S14	Med.	5-Pin	Power Ampl. - Double Grid Triode	6.3 F.
64	Nat. Un.	12	Small	5-Pin	Detector and Amplifier	6.3 H.
65	Nat. Un.	12	Small	5-Pin	A. F. Amplifier	6.3 H.
67	Nat. Un.	12	Small	5-pin	Detector & Amplifier	6.3 H.
68		12	Small	5-Pin	Power Output Pentode	6.3 H.
69	Hyg. Syl.	S12	Small	6-Pin	Special Detector	6.3 H.
70	Sparton	S12	Small	6-Pin	Duplex Diode-Detector-Double Grid	6.3 H.
84	Grig. G.	12	Small	4-Pin	Half-Wave Rectifier	2.5 F.
90	Ken-Rad	S12	Small	6-Pin	Double Grid Detector (A-C)	2.5 H.
92	Ken-Rad	S12	Small	6-Pin	Double Grid Detector (D-C)	6.3 H.
96	Ken-Rad	S12	Med.	4-Pin	Half-Wave Mercury Vapor Rectifier	10.0 H.
181	Sparton	17	Med.	4-Pin	Power Amplfier	3.0 F.
182B	Group 11	S17	Med.	4-Pin	Detector Amplifier	5.0 F.
183	" 3	S17	Med.	4-Pin	Power Amplifier	5.0 F.
281M	Hytron	S19	Med.	4-Pin	Half-Wave mercury Vapor Rectifier	7.5 F.
401	Group 13	14	Med.	4-Pin	*Ampl. & Detec.-(Unimport.)	3.0 H.
403	" 14	T14	Med.	4-Pin	*Power Amplifier (")	3.0 H.
482A	Sparton	17	Med.	4-Pin	D-C Power Amplifier	5.0 F.
482B	Group 15	17	Med.	4-Pin	Power Amplifier	5.0 F.
483	" 16	17	Med.	4-Pin	Triode Power Amplifier	5.0 F.
484	" 17	14	Med.	5-pin	D-C Detector Ampl. (Unimport.)	3.0 H.
484A	Sparton	14	Med.	5-Pin	D-C Amplifier Detector	3.0 H.
485	Group 18	S14	Med.	5-Pin	Detector Amplifier	3.0 H.
486	Sparton	14	Med.	5-pin	D-C R-F Amplifier (Unimportant)	3.0 H.
950	Nat. Un.	ST14	Med.	5-Pin	Power Amplifier Pentode (.120 Amp.)	2.0 F.
951	Nat. Un.	ST12	Small	4-Pin	Screen Grid R-F Amplifier	2.0 F.
985	John'b'g	14	Med.	4-Pin	Full-Wave Merc. Vap. Rect.(Unimport.)	5.0 F.
986	John'b'g	-	Med.	4-Pin	Full-Wave Merc. Vap. Rect. (")	5.0 F.
BA	Raytheon	19	Med.	4-Pin	Full-Wave Gaseous Rect. (Unimport.)	-
BH	Raytheon	14	Med.	4-Pin	Full-Wave Gaseous Rect.	-
BR	Raytheon	12	Med.	4-Pin	Half-Wave Gaseous Rect.	-
GA	Arcturus	14	Med.	5-Pin	Power Pentode	5.0 F.
Wun.	Arcturus	12	Small	6-Pin	Detector Twin Grid	2.5 H.
Wun. B	Arcturus	12	Small	6-Pin	Detector Twin Grid	2.5 H.
Wun. A-Auto	Arcturus	12	Small	6-Pin	Detector Twin Grid	6.3 H.
FW4	Fada		-		Flashograph tube with d. a. r. c. feature	-
F257	Fada		-		Pentode (unimportant)	6.3 F.

*Top Base
**RCA License Bureau states that this tube has been cancelled.
§Two or more companies manufacturing the same type tube are listed under group numbers 1, 2, 3, etc. in the accompanying chart.

Abbreviations:
F. - Filament Type
H. - Heater Type
Wun. - Wunderlich

we hired was a salesman because, in the process of developing our service, we wanted to make sure from day one that we had customer input as to what the service should look like. This was nine months before we had anything to sell.

HOUSE: This is not conventional wisdom, right?

BECHTOLSHEIM: It depends on the product. If you're doing a product in a well-defined space where your advantage is an improvement over the *status quo*, you typically know what needs to be done. If you're doing a new type of service or it is a new generation of product that never existed before, you better find out what the match is between the product and the customer opportunity.

KRESSEL: Entrepreneurs need to understand the fundamental business-building issues in addition to product development. The product development part is usually the easy one for engineers. However, in addition they need to realistically assess the barriers to revenue generation and the steps to reaching business profitability. This involves understanding customer needs, who the buyers are, how long it takes to close an order, and how to support the customer after the sale to ensure a happy customer. Delays add to startup costs and understanding the process is the difference between starting a venture that needs $1 million to become viable and one that may need $10 million or even $50 million. An entrepreneur must set the expectations of his investors early by the maximum amount of realism.

HOUSE: So we have, on the one hand, the painkiller versus vitamin test.

KRESSEL: Right. That's the first step. Then the next test is the period of time it will take and how much money it will take you to build a cash flow break-even business.

BECHTOLSHEIM: The length of the sales cycle is a key element.

KRESSEL: And if you can't figure that out, you are asking for an investment in charity.

YANG: This all applies to the faster, better, cheaper investment, where we first go to potential customers and see if anybody cares. And, if they do, we figure out if it can be done. But in a brave new world investment, you are asking people to change the way they think about reality. And customers are not really forward-thinking. So in this type of investment, you've got to extrapolate through analogy. What other inventions is this equivalent in magnitude to? Is this the next X, Y, or Z?

If so, what are the dynamics of the industry, and how would you finance it? Because, if you're creating an industry, chances are you can't finance it or do it as a start-up; you have got to do it with some significant partners. So you must consider who might those partners be and why they would care whether this is going to happen. With some of those investments, you just hold your nose and jump.

HOUSE: So if I wanted to write a business plan that would cause you to hold your nose and jump, as opposed to hold your nose and run, what do I put in it?

They said, "How can you...bring ...that to the marketplace?" You're bound to get rebuffed.

—BERNARD V. VONDERSCHMITT

chairman, founder, and former chief executive officer, Xilinx Inc.

YANG: It would have to be really a revolutionary idea that could create an industry. And in there would be some vision of similar changes in other industries.

The Internet is a great example. If you were to say, "I want to do a search site on the Internet" when the Web was just forming, who would you go to? You couldn't test it with customers and say, "If this would happen, would you use this?" They would say, "I don't know—this is so far out." The analogy would be that, as an interactive medium, this might be the next television.

And, if so, who would care and how would you generate revenues? I think that is a reasonable analogy of something where you couldn't have really tested it when it was first coming out.

Today's opportunities

HOUSE: What are the opportunities for start-ups today? How many think that there are more opportunities today than ever? [Show of hands.] So there's no question on that one. But are they centered in just a few areas or are they all over the place? [Show of hands favoring all over the place.]

BECHTOLSHEIM: People should look at their field of expertise, because once you start a company, the clock ticks. You start spending money and you'd better get there in a hurry. You have to know what you're going to do at that point and you've got to succeed.

So people have to be familiar with the field before they jump into it. It doesn't matter whether it's communications or Internet or whatever.

HOUSE: So preparation and grounding is vital.

BECHTOLSHEIM: In fact, a great thing to do is to develop your business plan while you are in school. You will never have that much free time again. It's a great environment to think about what company makes sense. Check out the Web, learn about venture capital, start talking to people. There is a big advantage in the school setting that other people later in life do not have. That's a competitive advantage.

DALAL: I wouldn't restrict your comments to people in school because opportunities in this world are constantly being created. Those opportunities continue to occur when you are at a Microsoft or a Hewlett-Packard because the world is changing. The world has never been impacted more by technology than it is today, so there is more opportunity than ever before for solving problems or creating new industries. This is really the golden age of entrepreneurialism.

YANG: And money today is extremely plentiful. I can't think of a better time to be an entrepreneur because the opportunities are there and the money is more available than it has ever been in the history of the universe.

OR-BACH: Physics will allow us to keep doubling the density and amount of the technology every 18 months or less, and we are creating a new world every time. This creates whole new opportunities for start-ups; things that could not be done before are now feasible.

It's like a balloon, the surface is getting greater and greater as we develop it. Every day there are more and more opportunities. ◆

Although solid-state technology overwhelmingly dominates today's world of electronics, vacuum tubes are holding out in two small but vibrant areas. They do so for entirely different reasons. Microwave technology relies on tubes for their power-handling capability at high frequencies ["Tubes: still vital after all these years," Robert S. Symons, *IEEE Spectrum*, April, 1998, pp. 52–63]. The other area—the creation and reproduction of music—is a more complicated and controversial story.

The complications and controversy stem from the fact that music is played to be heard by human beings, whose nonlinear ear-brain hearing systems are far from fully understood. Since no one knows exactly how to model the human auditory system, no one knows exactly what engineering measurements are appropriate to evaluating the performance of audio equipment. A smidgen of some kinds of distortion may sound worse to the ear than larger amounts of other kinds. So ultimately, the only way to judge audio equipment is by listening to it. Hence the controversy: subjective human perception—especially when flanked by questions of artistic merit—is made to order for arguments and disputation.

Briefly stated, a commercially viable number of people find that they prefer the sound produced by tubed equipment in three areas: musical-instrument (MI) amplifiers (mainly guitar amps), some processing devices used in recording studios, and a small but growing percentage of high-fidelity equipment at the high end of the audiophile market. These areas employ vacuum tubes of the type once

The Cool Sound of Tubes

ERIC BARBOUR
Svetlana Electron Devices Inc.

known as receiving tubes, but now called simply tubes. Not only has the use of vacuum tubes in these fields defied the semiconductor tide elsewhere, but such use and demand has even surged in the course of the 1990s.

Today vacuum-thermionic devices hold sway over the US $100 million worldwide guitar amp business. One rough estimate shows a 10-percent-per-year growth in demand for tubes used in MI amplifiers and high-end audio since the late 1980s, with no apparent slackening—even during the U.S. recession of 1991–92.

Interestingly, much of the demand for audio tubes derives not from the United States, but from Asia. In Japan, Taiwan, and mainland China, tubed high-end equipment enjoys a powerful cult status, and vintage U.S. and European electric guitars and guitar amps are valued collectors' items.

Why tubes—subjective reasons

The three areas of tube audio tend to be mutually exclusive and appear not to influence each other, even though all three directly involve the production or reproduction of music. It is common to see the same tube types, such as the popular EL34 power pentode, in electric-guitar amplifiers and in high-end stereo amplifiers. Often, too, these disparate products employ similar circuit topologies.

Electric-guitar amplifiers, it is estimated, consume as many as three out of four of the world's production of audio tubes. This is hardly surprising, since the tubed guitar amp seems unshakably enthroned at the top of the rock 'n' roll world. In this case, the use of tubed amplifiers in the early rock of the 1950s and '60s caused their distinctive distortions to become the standard tonal effect for the electric guitarist. A cultural bias formed during those years among U.S. and British musicians in favor of the particular nonlinearities of those amps, which typically were quite simple and had little or no negative feedback to improve their linearity.

As documented in many books on electric-guitar technique; in magazines such as *Guitar Player, Guitar World, Vintage Guitar,* and others; and on Usenet news groups such as alt.guitar.amps, the clipping distortion and other sonic artifacts of '50s-designed tubed amplifiers supply the sonic signature required for a successful guitar amp.

Discussion of an amp's merit frequently hinges on the clipping effect, which is often described as yielding a

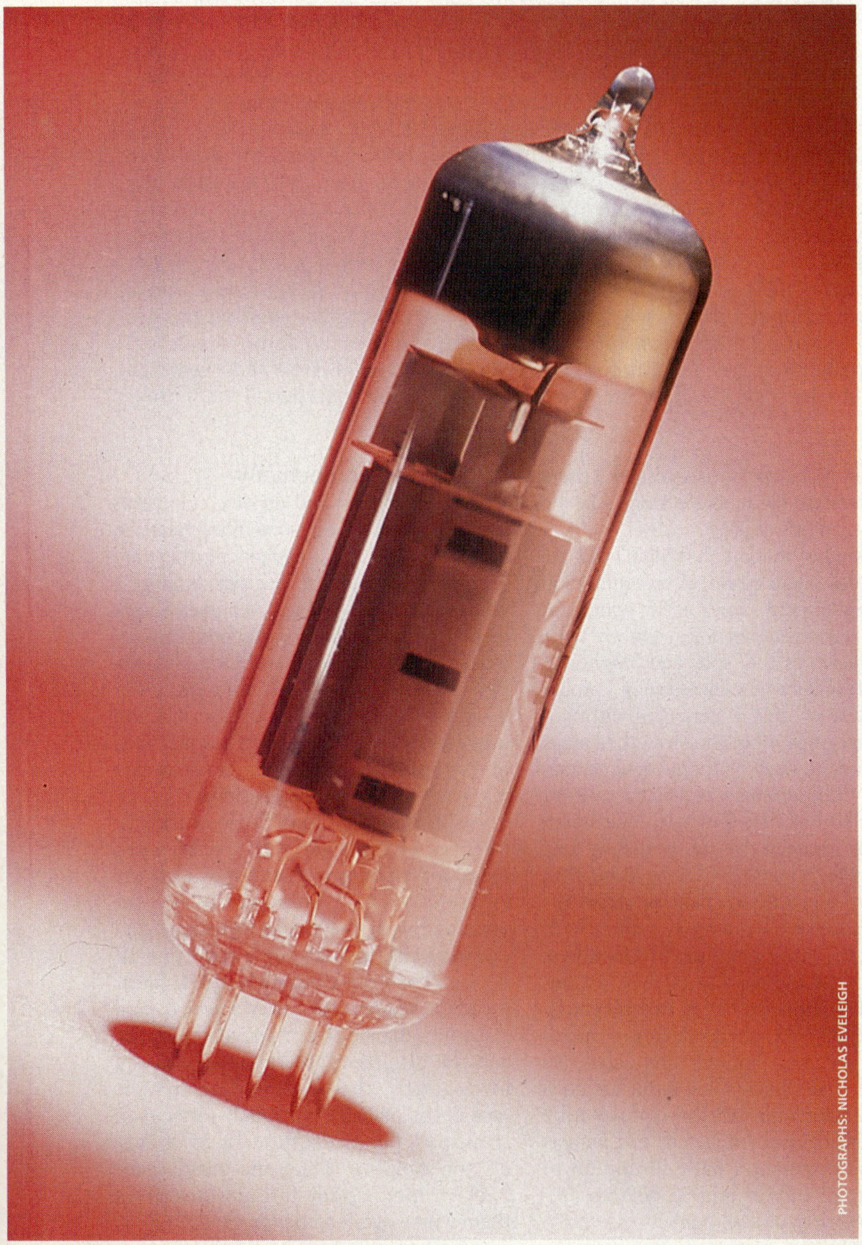

PHOTOGRAPHS: NICHOLAS EVELEIGH

ONE OF THE LAST REMAINING TUBE DOMAINS IS IN MUSIC APPLICATIONS, BUT THERE THE DEVICES FLOURISH AND EVEN INNOVATE

Two tubes in one, the Svetlana 6BM8 combines a triode and a pentode in a single glass envelope.

Defining tubes

5AR4: rectifier tube, used to convert high-voltage ac into dc plate power for amplifying tubes. One of the most common rectifier tubes, widely used in early guitar amps and hi-fi amps, it is currently manufactured in China. European designator: GZ34.

6DJ8: commonly used in modern audiophile equipment as a preamp or driver. This dual triode, gain of 35, was originally intended for use at radio frequencies in receiver front-ends and is currently manufactured in Russia and China. European designator: ECC88; special-quality designator: 6922.

6L6: first successful beam power tetrode, introduced by RCA in 1936 and made in numerous versions and variations since then. Suffixes (G, GA, GB, GC, GTY, and so on) refer to later modifications and upgradings of the basic 6L6 devices, which are currently manufactured in Russia (three versions), China (many versions), and Slovakia.

12AT7: dual triode with gain of 70, originally designed for radio-frequency applications. Because it was used by Fender in many of its popular push-pull guitar amps as the phase-splitter section, it became a standard for such use in music amplifiers. It is still manufactured in Yugoslavia. European designator: ECC81.

12AX7: most common preamplifier tube used in guitar amps, high-end audio and professional music equipment. A dual triode, gain of 100, it is still manufactured in Russia, Slovakia, and Yugoslavia. European designator: ECC83.

300B: audio power triode originated by the Western Electric division of AT&T in 1935 and used at first in movie-theater amplifiers, later in regulated power supplies. The original 300B has attracted an audiophile cult, especially in Japan and other Asian countries, so that its street price has risen as high as ¥200 000 in Tokyo, prompting the manufacture of versions in the United States, Russia, China, and Slovakia, specifically for that small market.

5881: rugged version of 6L6 beam tetrode, originally used in industrial and military equipment. Originally made by Tung-Sol Manufacturing Co. (now defunct) and other firms. Old American 5881s are sought-after collectibles. A different version is currently made at the Reflector factory in Saratov, Russia.

6550: beam-power tetrode, introduced in 1955 by Tung-Sol Manufacturing Co. (now defunct). Originally intended as a low-cost audio power tube, capable of as much as 100 W in push-pull pairs, it is popular in hi-fi amplifiers and in very large guitar and bass-guitar amplifiers. It is currently manufactured in Russia and China.

EL34: audio power pentode, introduced in early '50s by the giant Dutch firm, Philips Gloeilampenfabrieken N.V. (Eindhoven, the Netherlands). The EL34 became popular for its use in Marshall guitar amplifiers, as well as in various hi-fi amplifiers of the '50s and '60s. Currently it is being made in Russia (three versions), China (three versions), Slovakia, and Yugoslavia. U.S. designator: 6CA7. (Note: some 6CA7s are actually beam-power tetrodes, not pentodes.)

EL84: miniature power pentode, introduced by Philips in the mid-'50s. It became popular in guitar amps from its use in the Vox AC-30 and was also commonly used in smaller hi-fi amplifiers of the late '50s and early '60s. It is currently manufactured in Russia, China, Slovakia, and Yugoslavia. U.S. designator: 6BQ5.

KT66: beam-power tube manufactured by British firm M-O Valve Ltd., London, from 1937 until 1988. Similar to an uprated 6L6GC, the original KT66 is now a rare collector's item. A version is currently being made in China.

KT77: beam-power tube manufactured by M-O Valve Ltd. from 1957 to 1988. The KT77 was an exact replacement for the EL34, except for its beam-power construction. The original KT77 is now a rare collector's item. A version is currently being manufactured in China.

KT88: beam-power tube manufactured by M-O Valve Ltd. from 1957 to 1988. The KT88 was an exact replacement for the 6550, except for its higher plate ratings. The original KT88 is now a rare collector's item. Versions are currently being manufactured in China and Slovakia; a Russian version is due shortly.

sound like a brass wind instrument. The saturation distortion of the output transformer, which couples the power tubes to the speaker, also plays a key role in determining an amplifier's sound. Another amplifier parameter—its touch sensitivity—is affected by circuit nonlinearities and loose regulation of the plate-power supply.

Tubes have also been cited, albeit not without controversy, as facilitating a controlled so-called infinite sustain effect because of the way their signal compression interacts with acoustic feedback from speaker to guitar string. These effects are well-known among musicians, yet seem difficult to reproduce accurately with solid-state equipment. The many designers who have tried to build tube simulators over the past 30 years have achieved varying levels of musical and financial success. Such equipment has its supporters, but most amateur and professional guitarists remain faithful to tubed amplifiers.

As Ritchie Fliegler, vice president of marketing at Fender Musical Instruments Corp., Corona, Calif., said in a private letter last year, "This is not even a topic for discussion as far as I'm concerned...there is no substitute for tube electronics in the hearts and minds of pros."

This kind of thinking has influenced the professional-audio world of recording and mixing equipment. Since 1985, some studio engineers have been attracted by what they perceive as the "soft" and "euphonic" sound of vacuum-tube electronics—probably because of their experiences with vintage tubed amps, since modern tubed amps can be and are made without these characteristics. Tube enthusiasts usually contrast the soft sound of tubes with the harsh sound of modern digital recording and mixing, which may have more to do with the use of electrolytic coupling capacitors and inexpensive op-amp ICs than with solid-state devices in themselves. Regardless of the validity of their reasoning, studio engineers began experimenting with old tube-equipped condenser microphones, preamps, limiters, and equalizers from the 1946–70 era. The result has been twofold: street prices for vintage tube equipment have skyrocketed, and numerous small companies have sprung up to manufacture tube-equipped devices, following a variety of design practices.

Why tubes—objectively

If cries of fraud and derisive comments about "magical sound" sometimes greet the use of tubes in audio equipment, there are also highly competent electrical engineers who see definite advantages in tubed equipment. An example is John Atwood, consulting engineer and owner of One Electron Co., Santa Clara, Calif. The erstwhile designer of application-specific ICs and other solid-state logic circuits has managed to transform his hobby of tube audio design into a full-time consulting business.

In Atwood's opinion, "Some of the differences in the audio qualities between tubes and transistors have to do with the inherent physical properties of the devices and with the circuit topologies and components used with each type of device. There is no way around it: linear [triode] vacuum tubes have lower overall distortion than bipolar transistors or FETs, and the distortion products are primarily lower-order...the clipping characteristic of tubes is actually not much softer than transistors, but feedback tends to 'square-up' the clipping. Thus, the heavy feedback in most solid-state designs gives them worse overload performance.

"A low- or no-feedback design can be driven harder without audible distortion," Atwood continued. "High feedback also can lead to transient intermodulation distortion (TIM), caused by clipping or slew-rate limiting within the feedback loop." [See table on p. 29 for a comparison of the attributes of tubes and transistors in audio applications.]

Clipping distortion is not the only issue. In semiconductors, the shift of characteristics with temperature along with their relatively low maximum operating temperature has led to extensive use of Class-B amplifiers to keep power dissipation down. In many designs, the result has been audible crossover distortion, which often does not show up in published specifications. Those specs are typically based on measurements made at full power, where crossover distortion is at a minimum. Consumer tube amplifiers use Class-A or Class-AB designs, which have vanishingly low distortion even at low signal levels.

Another audio expert—Bruce Rozenblit, the owner of Transcendent Sound Co., Kansas City, Mo., and a well-known designer of tubed amplifiers—said he sees the output transformer as a major factor behind tube sound: "The warmth is created by a large component of second-order distortion, and the slow rise time of the output transformer causes a coloration that I would describe as a smoothing effect...the transformer is a nonlinear element that causes alterations of the signal in the time and frequency domains, thereby altering the sound."

Bill Whitlock, the president of Jensen Transformers Inc., Van Nuys, Calif., has a similar view. "Much of the perceived advantage of tubes is actually due to the transformers which often accompany them," he said. "Harmonic distortions tend to be rather high—1 percent or more is common—but are almost entirely low order (second and third harmonics). Transformer distortions are frequency-dependent in a way that most electronic distortions are not. Intermodulation (IM) distortions are created when a low frequency modulates the gain of a high frequency passing through the channel at the same time. The ear is much more sensitive to these IM distortions than to total harmonic distortion (THD). For electronic distortions, one normally expects IM to be three to four times the THD. In transformers, however, IM is actually about a third to a quarter of the THD."

Speaking about solid-state designs, Whitlock asserted that they "...depend on huge amounts of negative feedback to 'fix everything', including crossover distortion. Op amps commonly have open-loop THD in the 20 percent to 70 percent range. Stabilization generally requires open-loop gain to fall at 6 dB per octave. This means that, for ultrasonic input signals, the op amp has little gain margin to fix its own distortion. The ultrasonic signals, along with distortion products, are fed to the next stage for further distortion and intermodulation. This intermodulation creates audible, but non-harmonically related, artifacts which contaminate the noise floor and mask many subtle features of the music."

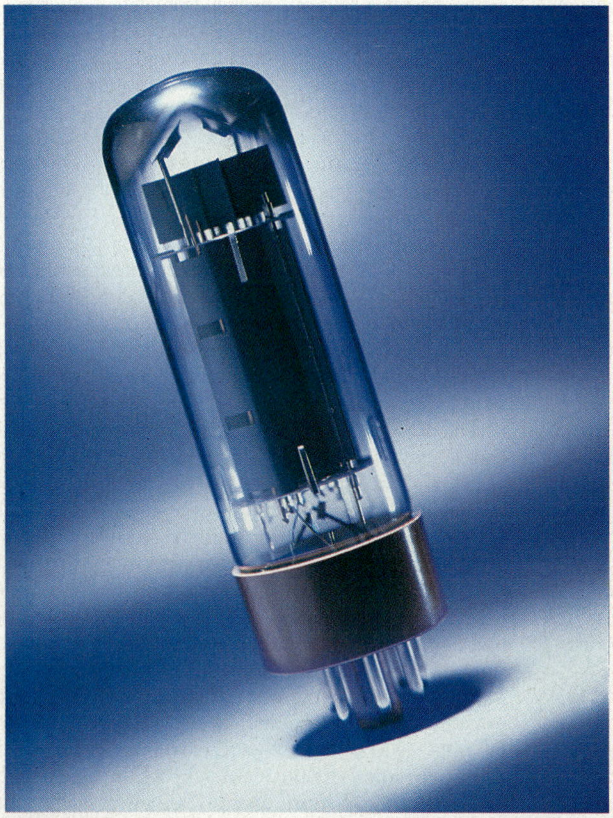
Widely used in guitar amplifiers, the octal-based Svetlana EL34 power pentode features a graphite-coated screen grid and a gold-plated control grid.

Whereas many analog-circuit designers today appear to feel that tubes are inherently noisy, Atwood disagrees. "In the audio region, 1/f noise tends to dominate over shot noise—it varies considerably from device to device," he observed. "You need to select devices for low 1/f noise. ... Microphonics, though, are a problem [with tubes]."

Whitlock said he saw things somewhat differently: "Higher operating voltages can expand one end of the dynamic-range envelope, but for audio, tubes are generally noisier (without the help of transformers) at the other end." Atwood qualified this statement with the comment that it is true only for low-impedance, low-level sources such as some (noncapacitive) microphone elements.

In addition to the active devices themselves, the passive components customarily used with them apparently also favor tube designs. According to Atwood, the low impedances of bipolar transistors necessitate the use of large (high-value) coupling capacitors, which for the most part means electrolytics. High dielectric absorption, imperfect high-frequency characteristics, and aging of electrolytics put them at a sonic disadvantage compared with the good-quality film capacitors nearly universally used for coupling the stages of tubed amps. "While it is possible to build solid-state amps with no electrolytics, it is rarely done," he said.

Whitlock said he felt that the main problem with electrolytics is actually at low frequencies: "The major problem with electrolytics is dielectric hysteresis, which produces high THD at low frequencies, where there are ever-increasing voltage swings across the capacitor." He also stated that direct coupling is the most desirable scheme, whether semiconductors or tubes are used.

Parallel views are held by guitar-amp designers. As Rick Perrotta, a longtime recording engineer and cofounder of guitar-amp manufacturer Matchless Amplifiers, Santa Fe Springs, Calif., explained, when a transformer saturates, it has a compressing effect. "When someone says that a particular guitar amplifier has a big fat sound, that is due, in part, to the output transformer saturating along with the output tubes. Solid-state amps can't perform that trick."

Guitar amp origins

It was in the '30s that the idea of attaching a transducer, amplifier, and loudspeaker to a guitar first caught on. Early guitar amps were used primarily with lap guitars, usually made of solid wood or metal with no resonant body to increase volume. Later amps were adopted by big-band guitarists, and in the '50s, amplifiers became pretty much mandatory for the amalgam of blues, country music, and jazz known as rock 'n' roll. Since transistors did not enter wide usage until about 1960, all of the originating styles of rock guitar were developed on tube amplifiers. Later,

Defining terms

Black-plate: usually refers to the 6L6GC power tube having a shiny black coating on its plate and formerly manufactured by RCA Electron Tubes Division in its Harrison, N.J., plant. Now a rare collectible, it is much sought after by guitarists for its unique "hazy" tone, due to unusual cathode and other materials. A few other RCA-made tubes, such as the 5687 and 5751 dual small-signal triodes, were also black-plate types.

DI (direct input) box: an interface device for connecting an unbalanced audio signal to the balanced input of a professional mixing console. Some DI boxes are entirely passive and provide zero signal gain, while others incorporate tubed or solid-state preamplifiers.

Head: a separate guitar amplifier chassis in its own cabinet, used with a separate speaker cabinet. A guitar amp built into a single cabinet with a speaker is known as a combo amp. The separate head and speaker give maximum flexibility and keep tubes away from vibrations produced by speakers.

Microphonics: changes in electrical characteristics of a device traceable to mechanical vibration; it causes the vibration waveform to appear in the device's output signal, hence it is a type of noise.

New old stock (NOS): a term borrowed from antique collectors to describe a tube or other component no longer being manufactured and found unused in its original factory packaging.

Smooth plate: colloquial name for Telefunken's original version of the 12AX7/ECC83 dual triode, which has large plates with no embossed features. The tube has not been manufactured since the '70s and is now much in demand for use in guitar amps and other audio equipment.

Tolex: brand name for a plastic-coated fabric covering used on U.S. guitar-amplifier cabinets in the '50s and '60s.

Touch sensitivity: a term used by electric guitarists to describe some kinds of signal-compression effects, usually caused by loose power-supply regulation or clipping distortion in a guitar amplifier. Such effects make the signal volume change greatly with the "touch" or force applied to the guitar strings by the player

Tweed: refers to guitar amplifiers made by Fender Musical Instruments through the '50s, so called because of the tweed-like, yellow-tan fabric covering their wooden cabinets.

musicians discovered that using a gain device before the guitar amp forced the amp to clip heavily; they liked the resulting sound, and it became the foundation of hard rock, later called heavy metal.

Since all the past styles of guitar playing are still regarded as musically valid, the market for guitar equipment has fragmented. Various manufacturers offer arrays of amplifiers, preamps, effects processors, and other means of electronically processing the guitar signal. Both reproductions of early equipment and innovations are available. Most remarkably, the basic designs of tube guitar amps tend to be based on a few prototypes that date from the '50s or early '60s. Extra channels or gain stages are added, tone controls are modified, sound effects (like reverberation and electronic tremolo) vary, and speaker cabinets become available in various configurations. Yet the basic circuits keep returning to the same set of paradigms.

No manufacturer of guitar amps has been as influential as Fender. Between 1946 and 1965, founder Leo Fender and his design team created most of the rock guitar sound in the form of the amplification used with their solid-body guitars, such as the famous and widely-copied Stratocaster model.

The most primitive design for a Fender amp is the Champ model. Being intended as a low-cost amplifier for students and beginners, a typical Champ uses a single 6V6GT or 6L6GC power tube. Because of its single-ended power stage—and the large amount of second-harmonic distortion thereby engendered—the Champ had a sound often described as "soft" and "lush." The amp's small, cheap output transformer saturated easily and gave very poor low-frequency response. Early Champs used 6V6GTs, were extremely primitive, and had no feedback, while later models had more complex circuits and loop feedback. In spite of their crudeness, early Champs are now valuable collector's items, and have been much imitated in recent years. The Champ sound is a standard, typical of many low-cost amps used on early rock 'n' roll records.

The most popular Fender models among serious professionals are the Bandmaster, Twin, Showman, and Bassman with push-pull 6L6GC or 5881 output tubes. The tone of these models has a ringing quality much sought after. The peculiar distortions from these amps and their matching speakers, in addition to the inherently light regulation afforded by tube rectifiers, gives distinctive inherent compression effects. This combination is what makes possible the infinite sustain effect mentioned earlier.

Even though the Bassman was intended originally for bass guitar, it was widely used for lead guitar and became possibly the most copied guitar amp in history, especially the 1959 model equipped with four 25-cm speaker drivers. The Twin Reverb model is often modified with extra gain stages for more distortion, producing a fair degree of compression and allowing the lengthy sustain of guitar notes. It served as the prototype for many modern amps with complex preamp sections.

Starting in 1962, a new sound appeared in Britain. Jim Marshall, a London music dealer, found that imported Fender amps were popular but too expensive, and so he developed his own. While his first amp was a copy of the Bassman, he later changed the output tubes to push-pull EL34s. These European tubes were true pentodes, different in electrical behavior from the beam tetrodes used in Fender amps. With the new tubes, Marshall's amps took on a tone described as very distorted and "crunchy," which is now considered the classic British blues-rock sound. Interestingly, the EL34 had reliability problems when operated in deep clipping for long periods, so in the '70s the U.S. distributor for Marshall amps switched the output tubes to 6550 beam tetrodes. As the sound of these amps was much more like very powerful Fenders, some preference arose among U.S. musicians for a "harder" sound than Marshalls give with EL34s. New distribution in the 1980s had EL34-equipped Marshalls entering the United States, as Jim Marshall preferred.

The third common guitar-amp design is that of the models AC15 and AC30 made by Vox Amplification Ltd., London. These were often used in Britain and throughout Europe, most notably by the Beatles at the peak of their popularity. The AC15 uses two push-pull EL84 output tubes, the AC30 four EL84s. Both models use self-bias of the output tubes, in Class A operation and with no negative feedback, unlike many other push-pull guitar amps. The result is a unique tone that varies greatly with string-plucking force. AC30s were made available with a Top Boost option, adding gain stages for further versatility. The Top Boost AC30 design is widely imitated by modern amp designers.

Bass guitar amplified otherwise

A bass guitar has different needs from a lead guitar. Bass is used to reinforce the song rhythmically, working with the melody at a pitch several octaves below. Nearly all bass guitars are solid-body types—simply larger versions of regular electric guitars with very thick strings. Since the bass sound is not always assisted by distortion, solid-state designs have come to hold sway over this market. Any amp with a high damping factor and capable of generating high powers at low frequencies can serve as a bass amp. Yet a ground swell of interest in tubed bass amplifiers has surfaced since 1990.

Vacuum tubes and transistors compared

Vacuum tubes: advantages

- Highly linear without negative feedback, especially some small-signal types.
- Clipping is smooth, which is widely considered more musical than transistors.
- Tolerant of overloads and voltage spikes.
- Characteristics highly independent of temperature, greatly simplifies biasing.
- Wider dynamic range than typical transistor circuits, thanks to higher operating voltages.
- Device capacitances vary only slightly with signal voltages.
- Capacitive coupling can be done with low-value, high-quality film capacitors.
- Circuit designs tend to be simpler than semiconductor equivalents.
- Operation is usually in Class A or AB, which minimizes crossover distortion.
- Output transformer in power amp protects speaker from tube failure.
- Maintenance tends to be easier because tubes can be replaced by user.

Vacuum tubes: disadvantages

- Bulky, hence less suitable for portable products.
- High operating voltages required.
- High power consumption; needs heater supply.
- Generate lots of waste heat.
- Lower power efficiency than transistors in small-signal circuits.
- Low-cost glass tubes are physically fragile.
- More prone to microphonics than semiconductors, especially in low-level stages.
- Cathode electron-emitting materials are used up in operation, resulting in short lifetimes (typically 1–5 years for power tubes).
- High-impedance devices that usually need a matching transformer for low-impedance loads, like speakers.
- Usually higher cost than equivalent transistors.

Transistors: advantages

- Usually lower cost than tubes, especially in small-signal circuits.
- Smaller than equivalent tubes.
- Can be combined in one die to make integrated circuit.
- Lower power consumption than equivalent tubes, especially in small-signal circuits.
- Less waste heat than equivalent tubes.
- Can operate on low-voltage supplies, greater safety, lower component costs, smaller clearances.
- Matching transformers not required for low-impedance loads.
- Usually more physical ruggedness than tubes (depends on chassis construction).

Transistors: disadvantages

- Tendency toward higher distortion than equivalent tubes.
- Complex circuits and considerable negative feedback required for low distortion.
- Sharp clipping, in a manner widely considered non-musical, due to considerable negative feedback commonly used.
- Device capacitances tend to vary with applied voltages.
- Large unit-to-unit variations in key parameters, such as gain and threshold voltage.
- Stored-charge effects add signal delay, which complicates high-frequency and feedback amplifier design.
- Device parameters vary considerably with temperature, complicating biasing and raising the possibility of thermal runaway.
- Cooling is less efficient than with tubes, because lower operating temperature is required for reliability.
- Power MOSFETs have high input capacitances that vary with voltage.
- Class B totem-pole circuits are common, which can result in crossover distortion.
- Less tolerant of overloads and voltage spikes than tubes.
- Nearly all transistor power amplifiers have directly-coupled outputs and can damage speakers, even with active protection.
- Capacitive coupling usually requires high-value electrolytic capacitors, which give inferior performance at audio-frequency extremes.
- Greater tendency to pick up radio-frequency interference, due to rectification by low-voltage diode junctions or slew-rate effects.
- Maintenance more difficult; devices are not easily replaced by user.
- Older transistors and ICs often unavailable after 20 years, making replacement difficult or impossible.

Early bass amps, such as Fender models, were essentially little different from guitar amps. Then the Ampeg SVT was introduced in 1969. It dwarfed previous bass amplifiers, producing 300 W from six 6146 or 6550 tubes. The SVT became a standard much imitated, especially in the last 10 years. Modern tube bass amps are usually very large, producing 200 W at least from a set of 6550 beam tetrodes—as many as 10 in some models.

Professional audio gear

For all that guitar and bass amps are targeted at professional musicians, they are often considered somewhat apart from the professional audio market. The usual definition of "pro audio" specifies sound-reinforcement and sound-recording equipment. Each kind is used both in a live musical setting and in the recording studio, yet they are somewhat differentiated by application. For stage use, ruggedness and reliability matter the most. In the studio, though, there is more concern about the specifics of the device's transfer function, which may be exploited as a sound effect.

The most general use of vacuum tubes in the recording studio is inside condenser microphones, used to record vocals. Each microphone model has its own transducer assembly or "capsule" design, each having distinctive variations in frequency response, phase shift, and signal distortion. Even the finest and most carefully designed capsules are not exactly perfect, so recording engineers have learned to exploit the imperfections artistically.

The advantage of tubes in condenser microphones is their very high input impedance, which does not load down the capsule significantly. Many recording professionals agree that the soft clipping exhibited by the tube—along with its high-voltage operation—can also be important, since some singers have very powerful voices, capable of producing peaks far beyond the dynamic range of typical solid-state electronics.

Almost as much the rage as tubed microphones are special preamps and so-called direct input boxes, which in studio signal chains serve to provide voltage gain, impedance conversion, and (in the case of direct input boxes) conversion from unbalanced to balanced connections. In spite of their lesser usage in studios, compared with tubed condenser microphones, these preamp devices are also riding high on a resurgence of popularity. Among the best-liked designs of this category are units that are solid-state but for a single 12AX7 tube operated at a very low plate voltage—as little as 12 V. This so-called starved-plate operation delivers high distortion, which some equipment designers consider the only useful characteristic of vacuum tubes in audio.

Expensive yet much favored is a category of tube preamps that are a lot like classic circuits from the '50s and '60s. This would include products from Tube-Tech (TC Electronic A/S, Risskov, Denmark), Anthony DeMaria Laboratories (ADL),

New Paltz, N.Y., and Manley Laboratories, Chino, Calif. The standard design involves push-pull circuits, frequently using 12BH7 or 12FQ7 dual triodes along with 12AX7s and 12AU7s. Input and output coupling transformers, to match impedances of 600-Ω lines to the tube circuits, are found in these devices, in spite of the difficulties of making high-quality audio transformers.

A key use of tubes in the recording chain is in signal compressors, also called limiters or leveling amplifiers. In general terms, a high-quality audio compressor consists of a preamp, a so-called sidechain with audio-signal rectifier and peak detector, and a voltage-controlled attenuator or amplifier driven by the peak-detected voltage. The control is arranged so that signal gain in the preamp rises as the signal level on the input falls, giving a narrowed dynamic range. Compressors of this type are considered mandatory in recording and production studios, for a variety of sonic effects. Rock music is often heavily compressed to give the illusion of greater loudness.

Tube compressors tend to follow the examples of vintage equipment. Among the most influential was the Teletronix LA-2A, in which a controlled attenuator, consisting of an electroluminescent panel driven directly by the tube peak detector, shone onto a light-sensitive cadmium-sulfide resistor. Such devices have compression ranges of at most 10:1—not large compared with the enormous compression ratios available with modern solid-state designs.

The limitations of the tubed compressors have apparently not affected their value. LA-2As known to sell for $4000 on the street today were worth perhaps $100 in 1980. Whereas all the makers of vintage compressors are currently out of business or no longer producing such equipment, most of the modern firms produce compressors, ranging from starved-plate types (Aphex Tubessence) to vintage-style designs.

A tube compressor of modern design is the Summit Audio DCL-200, a dual compressor of hybrid design. Its input and output circuits, as well as the peak-detecting sidechain, are solid-state, while 12AX7s are used for signal amplification.

A similar situation exists with outboard audio equalizers (EQs). Essentially complex tone controls, dedicated EQs are viewed as adjuncts to the EQ circuitry built into the channel modules of modern mixing boards. Many of the above companies also have tube-equipped EQs in their lines. Often they are based on classic designs by long-defunct firms, such as Pultec and Lang. The tube EQs often use passive RLC filters, buffered on the input and output with tubes, whereas active filters are the usual basis for modern solid-state EQs. The Summit Audio EQP-100, manufactured since 1985, uses tube gain sections and passive RLC filters combined with input-output drivers based on op amps using discrete transistors.

Other specialized areas in which tubes find application include mixers and rotating-speaker tremolo devices. The latter, pioneered in the '40s by Leslie, have remained in favor for their distinctive phasing sound. Leslie speakers are considered essential for use with the Hammond B-3 electric organ, a standard instrument for rock and jazz music for 40 years. In point of fact, Leslie speakers with tube amplifiers are still being manufactured by Hammond Suzuki Corp., Addison, Ill., while Motion Sound Inc., Salt Lake City, Utah, has introduced rotating-speaker devices that incorporate tube-based electronic circuitry.

High-end equipment

Perhaps the oddest subgenre of tube audio is the high-end audio component market. High-end equipment is aimed at the most obsessive audiophiles, famed for worrying about small details which most people ignore or cannot even hear. Yet in spite of its obscurity and its notorious marketing-driven focus, high end is the most vibrant and active area of all.

Until recently, the high-end market belonged almost entirely to solid-state equipment. The use of tubed equipment in the field started as a semi-nostalgic subculture, with the use by audiophiles of vintage hi-fi amplifiers from before 1970. Unlike users in the music performance market, audiophiles have often preferred tubes for their clean, smooth sound—in some cases, far more detailed and life-like than most early solid-state equipment.

Manufacturers of new tube equipment like Audio Research Corp., Minnetonka, Minn., and Conrad-Johnson Design Inc., Fairfax, Va., appeared during the '70s. The field was small and sedate until the late '80s, when interest (and the number of manufacturers) began to skyrocket.

Whereas the hi-fi market originated after World War II in the United States and Europe, the latest aggressive rise in high-end sales has been concentrated in the Asian countries, led by Japan. Much of the high-end equipment made in the West is exported and sold in the Far East. Indeed, Asian audiophiles see labels reading "Made in USA" and "Made in England" as badges of status and quality.

The rise of high-end sales was influenced by the statements of subjective audio reviewers, whose nontechnical and rarely rigorous listening tests at times encouraged near-hysteria among magazine readers. A positive review in a powerful magazine such as *Stereophile* can trigger hundreds or even thousands of unit sales, and turn an unknown manufacturer into an instant success. A negative review can sink a small firm just as easily (and has done so). This applies only to the small North American market. In Asia, terms such as "audio nut" and "audio mania" are often applied to the users and makers of high-end equipment. In Japan alone, the market for tube equipment could easily be five or 10 times as large as in the United States, the source of much of the equipment. The high-end market in Asia seems much more chaotic and ill-defined than elsewhere, with many audiophiles choosing to construct their own equipment from scratch, using a wide variety of tubes and/or semiconductors.

The demand for high-end tube gear in Asia has been fueled, to an even greater degree than in the United States, by an obsession with obscure sonic characteristics. Starting in 1973, Japanese "audio maniacs" were exhorted to seek out the primitive Western Electric 300B audio triode, originally used in late-'30s movie theater amplifiers. Similar edicts were issued about other antique Western Electric tubes, such as the 205D and 212E, and about '30s Western Electric amplifiers and loudspeakers. And hi-fi components made in the '50s and early '60s, by highly regarded firms such as Marantz America Inc., Roselle, Ill., and McIntosh Laboratory Inc., Binghamton, N.Y., enjoy cult status in Asia. Such obsessive desire has driven the street prices of some vintage audio to many times their original retail prices, and has caused Marantz and McIntosh to reissue some of their past models of "obsolete" tube hi-fi components.

Much of high-end is conducted in a gold-rush fashion, with companies advertising exotic connecting cables and acoustical treatment devices while making wild claims about the supernatural results achieved. The result: negative comments from the professional engineering fraternity. Items have been published in the *Journal of the Audio Engineering Society*, in electronic-industry journals such as *EE Times*, and elsewhere that attack the methods and conclusions of the audiophiles. In spite of the bad press, high-end audio was estimated in a *Wall Street Journal* article in December 1991 to sell as much as $1 billion per year worldwide, enjoying a 20 percent annual growth rate. Japan, alone, consumes $200 million per year in high-end components.

Some influence may be exerted on high-end sound by the pro-audio world. Most audiophiles appear to be between the ages of 30 and 50, and thus have been exposed to rock guitar since childhood. Perhaps more to the point, the growth of sales of tubed equipment was also facilitated by the availability of low-cost Russian, East European, and Chinese tubes. Like the guitar and pro-audio areas, high end has expanded most since 1985, about when these tubes first started to appear on the market in large quantities throughout the United States, Western Europe, and the Far East. At the same time, receiving-tube manufacture in the West has virtually ended because of the comparatively small market.

Like the guitar world, high end is hardly a monolithic market. Demand for tube amplifiers has split into three pieces, with audiophiles choosing their preferred design topologies from a list. Designers in the field usually subscribe to only one of these topologies, and battles between the camps flare up regularly. Not so long ago, nearly all high-end tube amplifiers were of conventional classic hi-fi design, using push-pull 6550s or EL34s with negative feedback of 20 dB or less. Audio Research and Conrad-Johnson are among the largest builders of these types.

In recent years, however, the field has seen the rise of the single-ended amplifier. Most of this started in Japan, with the preferred design using one or two primitive power triodes and no feedback. Distortion is substantial, though mostly second-harmonic in nature. Power output tends to be low, necessitating the use of efficient speakers. A parallel phenomenon has been the return to favor of highly efficient horn speakers, which were scorned as too "colored" or inaccurate by audiophiles—until single-ended amps became trendy. An example of a single-ended triode amplifier is the Moth Audio 304TL.

The third and least-used design is called output transformerless (OTL). Driving a low-impedance speaker directly is difficult with tubes, calling for many low-impedance triodes or tetrodes in push-pull sets. Nonetheless, some audiophiles consider the elimination of the output transformer, with its electrical nonlinearities, to be worth the effort.

A modern OTL is the Transcendent Sound amplifier, which uses eight EL509 power tetrodes to produce 80 W into an 8-Ω speaker load and is one of the few modern tube audio devices to be covered by a U.S. patent. Transcendent's Rozenblit feels that the output transformer is fine for guitar amps but not for high-fidelity: "The warmth is gone, the detail and resolution are much greater, and the rise time is much faster…[yet] I cannot duplicate the…sound with solid-state devices that is generated by my OTL. I wish I could, because I could sell 100 times as many units."

Choosing a tube

It is obvious that tube-audio designers like to use tubes from the same small group of types. In guitar amps, preamp and output tube choice are usually dictated by old Fender/Marshall/Vox designs. Preamp stages are made of 12AX7s; phase splitters and drivers are 12AX7s or else 12AT7s; and output tubes are from one of five families: 6L6GC, 6V6GT, EL34, EL84, and 6550.

For example, when Matchless amps arrived in the late '80s, their use of EF86 low-noise pentodes in their input stages befuddled musicians; the amps eventually proved themselves and came into vogue, in spite of the unexpected tube lineup. In professional audio, the standard choices are not fully settled. Vintage-style equipment uses 12AX7s, 12AU7s, 12AT7s, and 12BH7s, sometimes 12AY7s and a few others. Low-cost equipment opts for the 12AX7 because of its ready availability, since it is the most popular preamp tube in guitar amps.

Svetlana's EF86 low-noise pentode was once common in European hi-fi and guitar amps.

High-end designers in the past relied on power tubes endemic in the late '50s and '60s, primarily 6550 and EL34 types. The arrival of single-ended triode amplifiers has encouraged wide use of the 300B and other primitive triodes using directly-heated cathodes.

In recent years, tubes have developed a poor reputation for reliability. Most of it seems the result of low-quality imported tubes, primarily Chinese in origin. A side-effect has been the veneration of new old stock [see Defining Terms, p. 28] U.S. and European tubes, of types that are still being manufactured but not by the original makers. Since each manufacturer has its own proprietary formula for cathode coatings and other materials, tubes of the same type but from different manufacturers often sound slightly different. Since the standards have already been settled as being new old stock types like the RCA black-plate 6L6GC and Telefunken smooth-plate 12AX7 [see Defining Terms, p. 28], those types of tubes have increased in value along with the vintage equipment they were used in originally, even if the tubes end up in new equipment. Some boutique guitar-amp makers and high-end firms deliberately equip their new production with scarce NOS tubes, charging the customer accordingly, while professing their aural superiority and limiting the sales of their product (and, perhaps, also ensuring the product's obsolescence when tube stocks run out).

Another general tendency has been for tubed amplifiers (of both guitar and high-end types) to be designed and built by technically unqualified people. As Transcendent's Rozenblit said, "Properly designed tube equipment is extremely reliable. I have repaired many pieces of tube gear from the late '50s or early '60s where all of tubes still worked. Usually, a filter capacitor dried out and failed. The poor reputation comes from basically two factors. One is that from about 1970 to about 1990, the quality of vacuum tubes went through the floor. With the resurgence of tube audio, very high-quality tubes are now being manufactured. The second reason is that many contemporary tube designers don't know what they are doing. They are not engineers, they are salesmen. Tubes are forgiving and easy to work with, which allows the less sophisticated to easily enter the marketplace."

Coming attractions

A major stumbling block to the wide use of tubes, especially in consumer audio electronics, is the fact that most tube-equipped audio devices are made in the United States or Europe in small quantities, and tend to sell for high prices. A change in this situation may be presaged by the arrival in the high-end market of many low-cost amplifiers and preamps made in offshore factories. Firms such as Antique Sound USA Co., Seattle, Wash.; Alpha Audio Laboratory, Belmont, Calif.; and Jolida Inc., Annapolis Junction, Md., have recently begun to import low-cost amps and preamps made by Chinese or Taiwanese factories. In the tubed guitar-amplification world, imported Asian-made products have yet to make a significant impact, although most inexpensive solid-state amps are Asian-made and a few tube amps are being exported from China. New Sensor Corp., of New York City, imports a line of guitar amps built for it in the Reflector factory in Saratov, Russia—these "Sovtek" amps have proven to be popular and affordable.

Solid-state designers have long

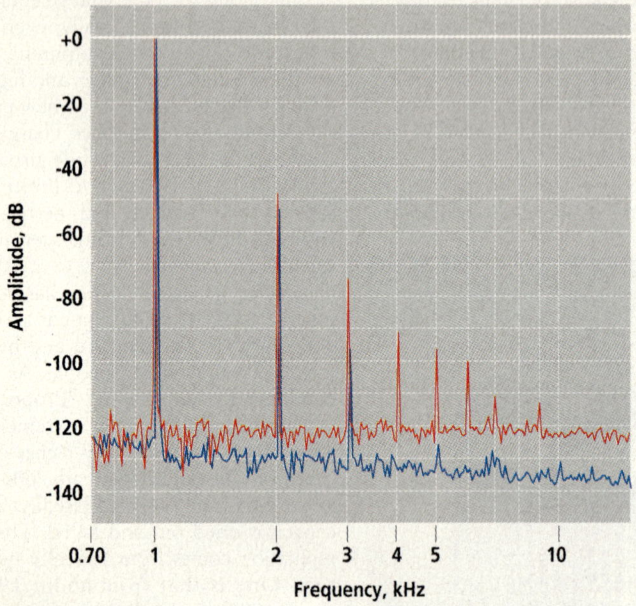

6AU6A pentode: Noise floor is about 120 dB below the fundamental; second harmonic is 48 dB down [red curve, above]. Operating conditions were taken from the resistance-coupled amplifier tables in RCA tube manual RC-21, 1961, p. 438.

6SN7GTB medium-mu triode: Noise floor is about 130 dB below the fundamental; second harmonic is 52 dB down [blue curve, above]. Operating conditions were taken from the resistance-coupled amplifier tables in RCA tube manual RC-21, 1961, p. 439. Note: the voltage gain of this device is much lower than that obtained with any of the transistors or the pentode.

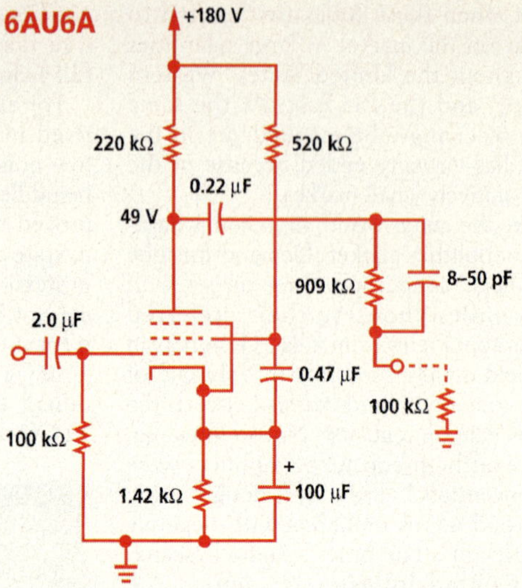

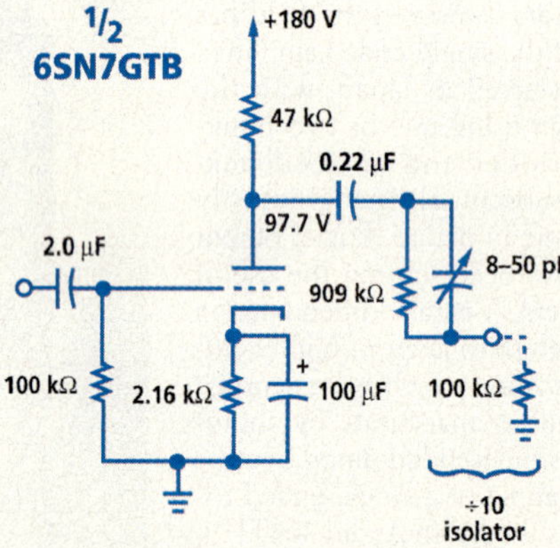

Distortion under test

Since much of the rationale for the continued use of tubes in audio equipment is based on distortion and noise, we decided to compare how several representative tubes and transistors performed in this regard. Using basic circuit designs in which these devices are typically used, we tested two tubes against four types of transistor—respectively, a medium-mu triode and a pentode against a low-voltage bipolar transistor, a low-voltage junction FET (JFET), a high-voltage bipolar transistor, and a high-voltage MOSFET. The figures show the distortion and noise spectra of each device.

All the measurements were made with an Audio Precision System 2 with its analog oscillator set to 1 kHz. The oscillator output level was adjusted to give an rms voltage of 2.00 V at the output of the test device. The output impedance of the oscillator was either 20 Ω or 600 Ω, depending on the device under test.

The test device was monitored by the System 2's fast Fourier transform (FFT) function using its 20-bit analog-to-digital converter running at 48 kilosamples per second. The FFT was synchronous (no windowing), and was averaged over 16 samples. The residual harmonics

attempted to produce tube simulators, using solid-state analog circuits like diode clippers and compressors to produce transfer functions and distortion akin to those of tubed guitar amps. Although these products have attracted only a limited following to date, the recent appearance of solid-state amplifiers based on digital signal processing (DSP) and using physical modeling algorithms, has stirred some interest among serious musicians.

To quote Art Thompson, technical editor of *Guitar Player* magazine, San Mateo, Calif.: "Transistor amps have definitely come a long way towards their ultimate goal of sounding exactly like their tube-powered counterparts, but I think most pro players and industry people have realized that this is a futile objective...." As to the success of DSP guitar amps like the Line 6 products from Fast Forward Designs Inc., Culver City, Calif., Thompson said, "We've tested some impressive tube-simulation devices, and better ones are undoubtedly on the horizon."

In the professional audio market, the final frontier for vacuum tubes looks to be in the mixing console. This device is the heart of a working studio, serving as primary signal router, EQ and effects controller, and the tool to mix down multiple recorded channels to a stereo master music program. Mixers used in most stu-

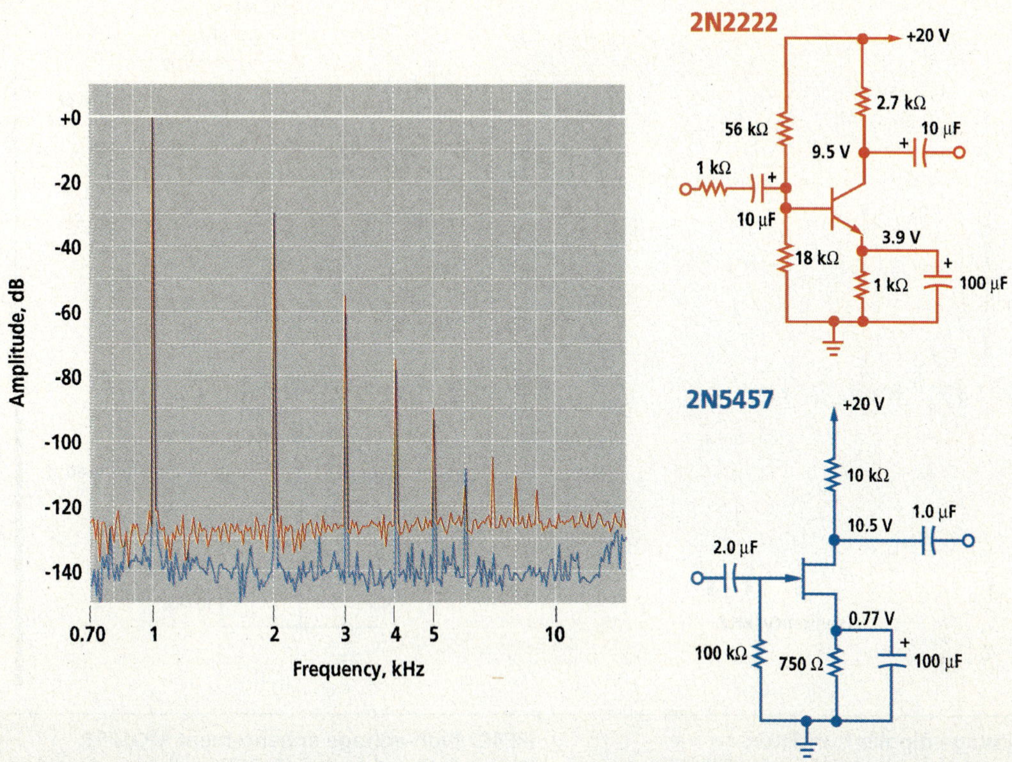

2N2222 low-voltage bipolar transistor:
The noise floor is about 125 dB below the fundamental; the second harmonic is –30 dB. The circuit was taken from the output stage of "High-Fidelity Preamplifier," p. 609 of RCA transistor manual SC-14, 1973, with a 2N2222 substituted for the (similar) 2N3242A. A 1-kΩ resistor was used to match the 20-Ω output impedance of the generator to the transistor.

2N5457 low-voltage junction FET:
Unlike the high-voltage MOSFET, the JFET has excellent noise performance (–140 dB) but poor distortion (second harmonic is only 30 dB down). Because of the lower I_{dss} of available junction FETs, the drain resistor was raised to 10 kΩ; the bias was adjusted to give about 1/2 V_{DD} at the output.

were at least 120 dB down. The input impedance of the System 2 was set to 100 kΩ. Both input and output were floating, with the test fixture separately grounded to the System 2.

In the vacuum tube test fixture, the filament voltage was dc, regulated to 6.3 V. The output of the tube was monitored through a 10:1 compensated divider so as to reduce ac and dc loading.

The summary of second-harmonic distortion levels follows:

• 6SN7GTB triode	–52 dB
• 6AU6A pentode	–48 dB
• 2N2222 low-voltage bipolar	–30 dB
• 2N5457 low-voltage JFET	–30 dB
• MJE2361 high-voltage bipolar	–46 dB
• IRF822 high-voltage MOSFET	–41 dB
• HS-11 transformer	–90 dB

Although this is not intended to be an exhaustive examination of all available semiconductors or tubes, the resulting frequency spectra lead us to some conclusions that experienced audio designers have often remarked upon in the past.

• Transistors operating on low-voltage supplies tend to have higher spectral distortion components than tubes.
• If we go to high-voltage transistors, operating on supplies comparable to those of the tubes, the distortion

Continued on p.34

dio settings are large and complex and often equipped with computer-controlled automated mix-down capability. Although expensive, large mixing desks are frequently based on low-cost op-amp ICs such as the 5534, and use electrolytic capacitors for coupling the audio signal from stage to stage.

Some recording engineers have found this scheme wanting, hence recent introductions of tubed mixers. Companies producing such mixers are Summit Audio, the UK's Manley Laboratories and TL Audio Ltd., the latter being the sole producer of full-size mixing consoles with tubes at this time. The market may expand in the future, as interest in audio-tube electronics grows.

The most interesting new development is the introduction since 1990 of new tube types specifically designed for audio applications. Russian, Slovakian, Serbian, and Chinese factories currently opt to imitate the popular audio tubes of past types, such as 6L6GC and 12AX7. Yet the market has been open to some new tube types not directly based on any U.S. or European tube or modifications thereof. These types have aroused some interest from original-equipment makers, even though they seldom fit existing equipment.

In the guitar and pro-audio markets,

Continued from p.33

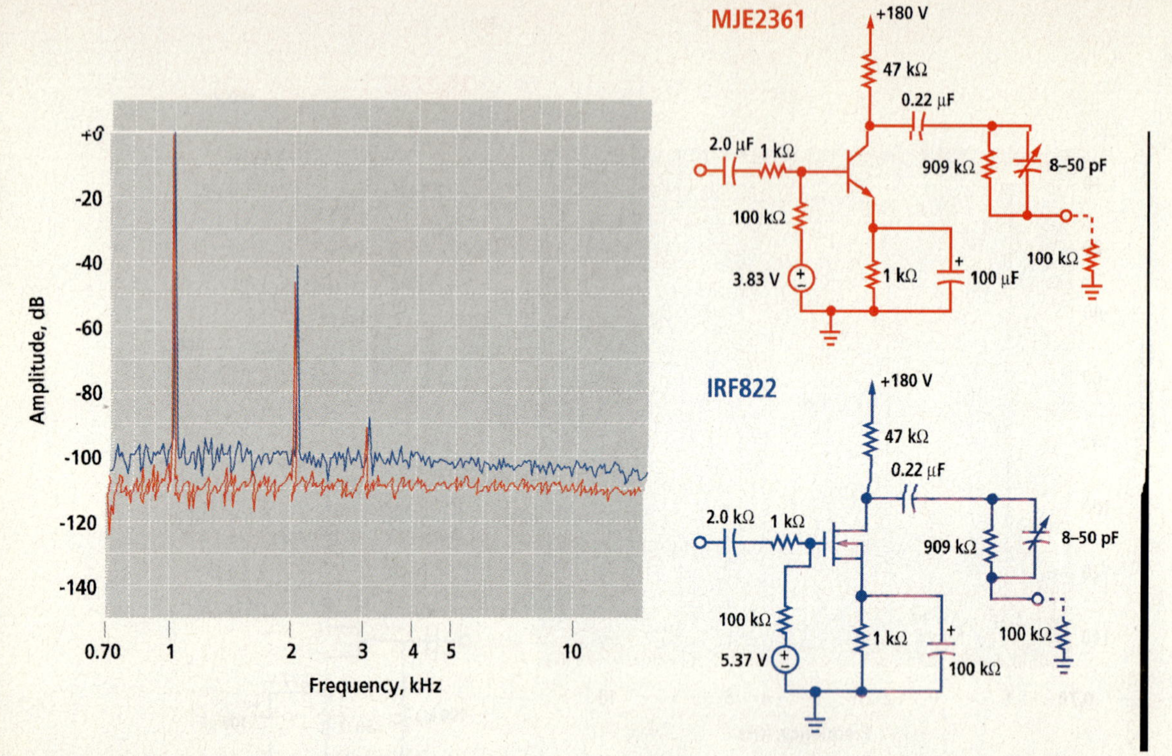

MJE2361 high-voltage bipolar transistor:
In this test, the transistor was substituted for the 6SN7GTB triode, and the bias was chosen to give the same operating point as the tube. A 1-kΩ resistor was used to match the 20-Ω output impedance of the generator to the transistor. The result: a noise floor at about –110 dB, and a second-harmonic level of –46 dB.

IRF822 high-voltage enhancement MOSFET:
When substituted for the 6SN7GTB, with bias adjusted to give the same operating point as the tube, the MOSFET exhibited excellent distortion characteristics, which were compromised by its noise floor of –100 dB—about 30 dB above the tube's. Second-harmonic distortion is 41 dB down, which is only 59 dB above the noise.

products are less objectionable. Unfortunately, the noise floor of such devices is much higher. The IRF822 was very triode-like in distortion yet suffered from a noise floor some 30 dB higher than that of the triode.

- No other active device possesses both the low distortion products and the low noise floor of the medium-mu triode—albeit at the expense of voltage gain.
- The distortion products of transformers are much lower than those of active devices, yet quite different in character. Note that the odd-order harmonic products tend to be higher in level than the even-order products—exactly the reverse of the tubes and transistors.

It should be obvious that these simple circuit designs can be improved upon, by using differential topologies with constant-current loads and negative loop feedback.

It should also be obvious that the same techniques can be applied to transistors or to tubes; and if this were done, the triode would continue to enjoy some advantages over the semiconductors—and the pentode, for that matter.

—E.B. with John Atwood

John Atwood is a consultant on tubed audio design and owns One Electron Co., Santa Clara, Calif.

new versions of classic power tetrodes and pentodes include, from the JJ/Teslovak Co., Cadca, Slovakia, the KT88S and E34LS, which are modified and uprated in power dissipation rating from their KT88 and EL34 prototypes. New high-power triodes, such as the SV811/572 series and 3CX300A1, have been introduced by Svetlana. Even more than other fields, the high-end market has seen an explosion of power triodes: the VV30B and VV52B from KR Enterprise Co., Prague, Czech Republic; Svetlana's SV811/SV572 series; the AV30B SL and AV62B SL from AVVT Technologies, Prague, Czech Republic; and many versions of the 300B from two different factories in China, two in Russia, and one in Slovakia—not to mention the revival of Western Electric and its resumed manufacture of the original 300B.

The KT90 beam tetrode, made by Ei Electronic, Nis, Serbia, has proved so popular for audiophile amps that a Chinese factory has even introduced its own version. Also winning a following in high-end audio are some Russian tubes unknown in the West until recently, such as the 6C33C-B from Electronpribor Ulyanov, Ulyanovsk, Russia. More new types are expected to be introduced, and some out-of-production types will be reintroduced, over the next several years.

Vacuum tubes enjoy an unshakable position in the design of guitar amplification. Their use in the recording studio and in home audio often seems like a fad, but it has very old roots and is therefore unlikely to disappear overnight, as fads do. Technical reasons for the use of tube electronics in some applications look solid and are well-defended both by professional designers and by the users, who define and drive the marketplace. So long as musical tastes demand them, tubed audio equipment and replacement tubes will be produced well into the next century, and probably thereafter. ◆

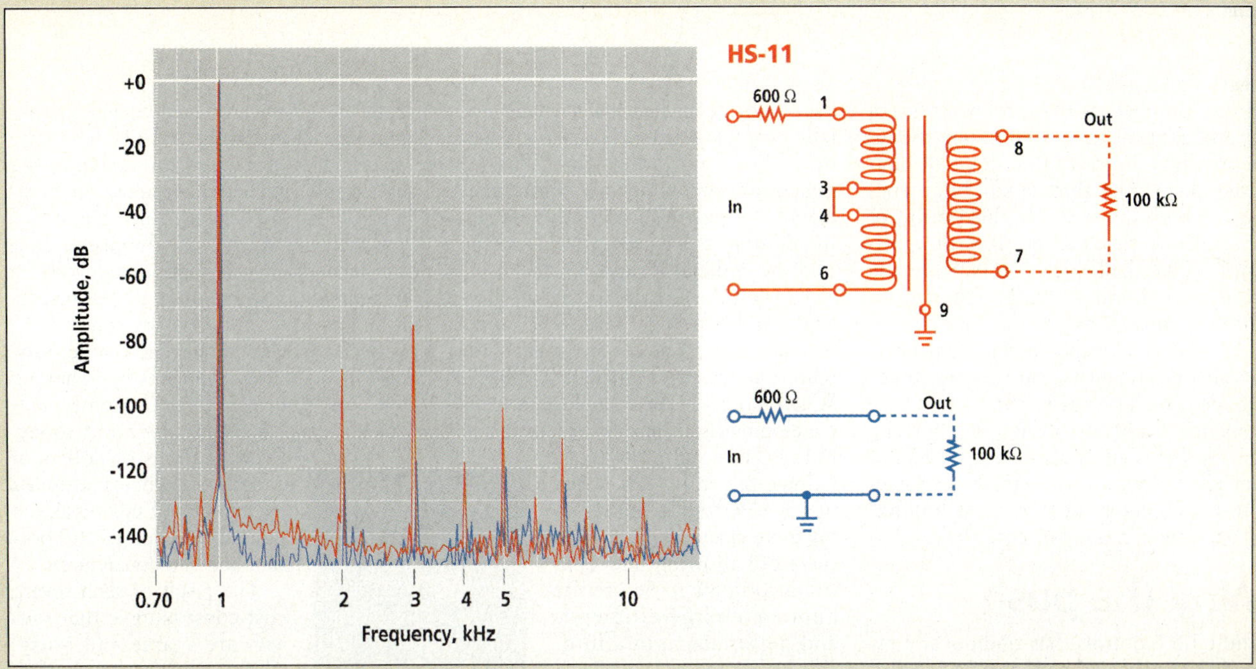

Triad HS-11 input transformer:

Source impedance was set to 600 Ω. The transformer was loaded by the 100-kΩ input impedance of an Audio Precision analyzer. This test was included because some professional audio experts commented on the unusual behavior of audio coupling transformers in tube circuits—specifically, that the odd harmonics tend to be stronger than the even ones. The HS-11 is typical of small input transformers used to couple a 600-Ω balanced line to a tube grid.

Residual noise plus distortion of the analyzer, with only a 600-Ω resistor in series, is at least 118 dB down.

To probe further

Well-written articles on the subjective and objective characteristics of tube audio electronics include: "Why tubes sound so good," by Doug Fearn, *Pro Audio Review*, January/February 1996, Vol. 2, no. 1, p. 15; "The grounded ear" Peter Sutheim's column in *The Audio Amateur*, Issue 3, 1980, p. 34; and "Tubes versus transistors—Is there an audible difference?" by Russell O. Hamm in *Journal Of the Audio Engineering Society*, May 1973, Vol. 21, no. 4, p. 267 (reprinted in *Glass Audio*, Issue 4, 1992, p. 16 on).

An unusual approach to testing for distortion products in power amplifiers is taken in "Spectral contamination measurement," by Deane Jensen and Gary Sokolich, presented at the 85th AES Convention, 3–6 November 1988 (Audio Engineering Society, New York City, 1988). Spectral distortion products of transistor amplifiers are compared, along with those from a McIntosh MC-30 tube amplifier, whose spectral plot is quite different from those of solid-state amplifiers.

In professional recording, tube equipment is a small but growing field. Recent articles about it include: "Tube processors—The outboard renaissance," by Loren Alldrin, *MIX* magazine, May 1997, Vol. 21, no. 5, p. 108; "A reader's guide to vintage gear," by Fletcher, *MIX*, November 1996, p. 84; "Retro-Active," by Michael Molenda, *Electronic Musician*, June 1995, p. 36; "The retro movement," by Sue Sillitoe, *Audio Media*, November 1994, p. 48; and "The vacuum tube rides again," by Walter Sear, *MIX*, May 1994, p. 24.

Textbooks on professional audio equipment design using tubes are a quite recent phenomenon. The most useful include *Principles of Power* and *The Ultimate Tone*, by Kevin O'Connor (Power Press Publishing Co., London, Ont., Canada, 1996); *Vacuum Tube Guitar And Bass Amplifier Theory*, by Tino Zottola (self-published, 1997); and Dan Torres' *Inside Tube Amps* (Sparpco Inc., San Mateo, Calif., 1996). An unusual reference book, with hundreds of schematic diagrams for tube-based music equipment is Aspen Pittman's *The Tube Amp Book*, edition 5 (Groove Tubes Inc., Sylmar, Calif., 1995).

Suitable for novices is *Beginner's Guide To Tube Audio Design*, by Bruce Rozenblit (Audio Amateur Press, Peterborough, N.H., 1997). All the books are available from Antique Electronic Supply, 6221 S. Maple Ave., Tempe, AZ 85283.

A taste of the fascinating and obscure world of "boutique" guitar amplifiers is afforded by Art Thompson's "20 mule duel: we wrangle a herd of new boutique amps," in *Guitar Player*, February 1997, Vol. 31, no. 2, p.118.

Plenty of publications intensively cover vacuum-tube audio electronic design. Their emphasis is usually on hi-fi audio, although material about guitar amps and professional studio electronics appears regularly. English-language publications include *Glass Audio* (Box 176, Peterborough, NH 03458) and *Vacuum Tube Valley* (1095 E. Duane Ave., Suite 106, Sunnyvale CA 94086).

Magazines that concentrate on high-end audio design include *Sound Practices* (Box 180562, Austin, TX 78718), *Ultra-High Fidelity* (Box 158, Cheshunt, Herts., EN7 6UH, UK), and *Valve* (Box 2786, Poulsbo, WA 98370).

Non-English tube-audio publications include:

- The Italian *Audion* (Piazza Madonna Aldobrandini 7, 50123 Firenze) and *Costruire Hi-Fi* (Via Toti 9, 20010 Bareggio, Italy).
- The Japanese *MJ Audio Technology* (Seibundo Shinkosha Publishing Co., 13-7, Yayoicho 1-chome, Nakano-Ku, Tokyo 164, Japan).
- The French *Musique et Technique* (Bureau de Depot 1050, Brussels 5, Belgium).
- The German *Hi-Fi Scene* (Dufourstrasse 165, CH-8008 Zurich, Switzerland).

About the author

Eric Barbour has been an applications engineer with Svetlana Electron Devices Inc., Portola Valley, Calif., since July 1996. His work involves testing and characterizing new vacuum-tube types and constructing and testing amplifier circuitry for Svetlana audio and RF tubes. Earlier, he was a senior technician and then an engineer with the U.S. Department of Energy, Intelligent Electronics Co., and Dionex Corp.

A staff editor of *Vacuum Tube Valley magazine*, Sunnyvale, Calif., since its founding in 1995, he also contributes to *Glass Audio* magazine, Peterborough, N.H.

Spectrum editor: Michael J. Riezenman

Home appliances get smart

CLEVERER SENSING AND CONTROL CAPABILITIES ARE INCREASING THE INTELLIGENCE OF HOME APPLIANCES

consumer

Vivek V. Badami & Nicolas W. Chbat
General Electric Co.
Corporate R&D Center

For some years now, refined sensing and controls have been raising the IQ of household appliances. All that stops them from shining still brighter is cost, which limits the sophistication of the sensors that can be used in appliances. The challenge is to improve the control technology while adhering to the cost constraints.

A "smart" appliance is one that incorporates novel sensing and control algorithms [Fig. 1] so as either to automate some manual function (such as estimating the clothes load in a washer), to economize on energy use, or to introduce something completely new (such as boiling-water detection on an electric cooktop).

Why the push

The term "controls" has traditionally referred to knobs and buttons on a panel of an appliance. In what follows, controls also refers to the integration of sensors and algorithms in a feedback arrangement to improve the performance of the appliance.

The incentives to smarten up domestic appliances are several.

Energy efficiency is one goal. Progressive improvements in this aspect of appliances are mandated in an ongoing series of regulations from the U.S. Department of Energy and are encouraged by electrical utilities [Fig. 2]. These pressures have had a significant impact. While improved sensors and controls are not responsible for all the gains, they have played a large role in them and will surely play a bigger one in the future.

The environment is another factor. Here, regulation and public sentiment have both pushed to lower environmentally harmful emissions, such as chlorofluorocarbons and detergents. The consumer, too, should not be forgotten, ever hungry for new features at the old price. In addition, the marketplace is seeing a trend toward upscale appliances.

Last but not least is the availability of enabling technology. Electronic controllers have more features and greater reliability than electromechanical controllers. What's more, sensor technology, driven by developments in micromachining technology, and actuator technology, driven by less pricey power electronics, are melding more closely with ICs and therefore allowing new control algorithms to be used to improve energy efficiency and performance in affordable hardware.

Together, these factors have bred a distinct evolution in the control technology of household appliances, which have moved from open-loop electromechanical controls, to electronic controls plus simple sensing, and to closed-loop systems in which sensors are integrated with controls.

Problems with the interface

Along with these advances, however, has arisen concern over the user-interface. Most appliances present the user with a range of dials or buttons on which to enter set points for various process parameters; in a washer, these would be water level, load size, and fabric type. In fact, most users rarely ever change the settings. It would be preferable to allow the user to enter higher-level, or "performance," goals, and leave the lower-level details to the machine. In a washer, for example, the user would set energy consumption, minimum wear, or washability, instead of water level and load size. Appliance control technology will continue to evolve in this direction, where multiple sensor data is integrated with supervisory control.

Because of consumer and marketplace pressures, domestic appliances are among the most price-sensitive electromechanical systems mass-manufactured today. Even fractions of dollars of added cost when multiplied by millions of units makes a huge difference to the bottom line of manufacturers.

The result of such severe cost-constraints is that sensors are sparse and noisy, while actuators may be few and restricted in their effects (in a refrigerator, for example, the compressor cannot cool the compartments infinitely quickly); and limited processor capabilities limit the sophistication of the control logic. (The simple processors used in appliance control usually have 4- to 8-bit words and less than 32KB of ROM and 256B RAM—on the small side for sophisticated control algorithms.)

In addition, models of appliance control are not available. A control model is a mathematical relationship from sensors to actuators. Because it has to run in real time, it needs to be somewhat simple. Unfortunately, even though models do exist for other aspects of appliances, such as detergent action, mechanical subsystem